THE PHYSICS OF POLARIZED TARGETS

Magnetic resonance is a field that has expanded to a wide range of disciplines, both in basic research and in its applications, and polarized targets have played an important role in this growth. This volume covers the range of disciplines required for understanding polarized targets, focusing in particular on the theoretical and technical developments made in dynamic nuclear polarization (DNP), nuclear magnetic resonance (NMR) polarization measurement, high-power refrigeration and magnet technology. Beyond particle and nuclear physics experiments, dynamically polarized nuclei have been used for experiments involving structural studies of biomolecules by neutron scattering and by NMR spectroscopy. Emerging applications in magnetic resonance imaging (MRI) are also benefiting from the sensitivity and contrast enhancements made possible by DNP or other hyperpolarization techniques. Topics are introduced theoretically using language and terminology suitable for scientists and advanced students from a range of disciplines, making this an accessible resource to this interdisciplinary field.

TAPIO O. NIINIKOSKI is the former leader of the Polarized Targets Group at CERN in Geneva and is the author or co-author of more than 380 scientific papers. He was member and project leader of the European Muon Collaboration, Spin Muon Collaboration, CERN μSR Collaboration and ATLAS Collaboration. He was Spokesman of the RD39 Collaboration and Leader of CERN Central Cryogenic Laboratory until 2009.

THE PHYSICS OF POLARIZED TARGETS

TAPIO O. NIINIKOSKI

CAMBRIDGE
UNIVERSITY PRESS

University Printing House, Cambridge CB2 8BS, United Kingdom

One Liberty Plaza, 20th Floor, New York, NY 10006, USA

477 Williamstown Road, Port Melbourne, VIC 3207, Australia

314–321, 3rd Floor, Plot 3, Splendor Forum, Jasola District Centre, New Delhi – 110025, India

79 Anson Road, #06–04/06, Singapore 079906

Cambridge University Press is part of the University of Cambridge.

It furthers the University's mission by disseminating knowledge in the pursuit of education, learning, and research at the highest international levels of excellence.

www.cambridge.org
Information on this title: www.cambridge.org/9781108475075
DOI: 10.1017/9781108567435

© Tapio Niinikoski 2020

This publication is in copyright. Subject to statutory exception and to the provisions of relevant collective licensing agreements, no reproduction of any part may take place without the written permission of Cambridge University Press.

First published 2020

Printed in the United Kingdom by TJ International Ltd. Padstow Cornwall

A catalogue record for this publication is available from the British Library.

ISBN 978-1-108-47507-5 Hardback

Cambridge University Press has no responsibility for the persistence or accuracy of URLs for external or third-party internet websites referred to in this publication and does not guarantee that any content on such websites is, or will remain, accurate or appropriate.

To Ulla

Contents

Preface ix
Acknowledgements xi

1 Introduction to Spin, Magnetic Resonance and Polarization   1
2 Resonance and Relaxation of Interacting Spin Systems   43
3 Electron Paramagnetic Resonance and Relaxation   98
4 Dynamic Nuclear Polarization   155
5 Nuclear Magnetic Resonance and Relaxation   199
6 NMR Polarization Measurement   245
7 Polarized Target Materials   283
8 Refrigeration   339
9 Microwave and Magnet Techniques   393
10 Other Methods of Nuclear Spin Polarization   421
11 Design and Optimization of Polarized Target Experiments   436

Appendices 471
Index 509

Preface

Spin is the Rosetta stone of physics. It has proved many theoretical models and disproved others; it is inconceivable to study and explain matter, radiation and interactions without including the spin of the constituent particles and structures. For example, the Pauli exclusion principle is needed to understand the structure of atoms, and this is based on the knowledge that electrons are spin ½ fermions of which only one can occupy a given quantum state. Atomic electrons pair with opposite spins so as to build magnetically neutral matter, the most common state of solids and fluids. Also the conduction electrons in metals may form Cooper pairs with opposite spins, which explains superconductivity. These pairs behave as bosons with zero or integer spin. Quantum statistics of these half-integer or integer spin particles is the basis of low-temperature properties of quantum fluids and solids.

The Pauli exclusion principle is also seen in the electron degeneracy pressure of white dwarfs and in the neutron degeneracy pressure of the neutron stars that stabilize these astronomical objects. The degeneracy pressure generates the repulsive magnetic force due to the exchange interaction. It is only the gravity force that makes the neutron star collapse into a black hole.

Going deeper into matter, the understanding of nuclear structures requires the introduction of the spin for their constituent protons and neutrons, the nucleons. These, in turn, are built of quarks and gluons which are spin ½ fermions for the first and spin 1 bosons for the latter; this knowledge is vital for the understanding of the structure and interactions of nucleons in nuclear matter.

In the Foreword of his book *Spin in Particle Physics*, Elliott Leader states that 'spin is an essential and fascinating complication in the physics of elementary particles', a reversal of the slogan of the 1960s that 'spin is an inessential complication of particle physics'. The slogan, an anecdote, can be understood from the historic fact that in the 1960s new particles were discovered almost on a monthly basis by analyzing bubble chamber pictures from which the particle masses could be determined. This was clearly more rewarding than wrestling with the complicated equipment and enormous statistics required for the study of the interactions of these particles in fine detail.

Spin polarized targets, the subject of this book, are an important tool and part of such complicated equipment. Other tools include polarized sources and polarized secondary beams, and the acceleration, storage and polarimetry of such beams. Our focus here is on the polarized solid targets that are particularly suitable for scattering experiments in secondary beams, but also important for the experiments in intense primary beams of protons and electrons; their intensity has developed steadily to reach now levels close to that of unpolarized beams.

In the course of the development of the polarized target techniques in the 1960s and 1970s, many disciplines were brought together: solid state physics and chemistry, electron spin spectroscopy, NMR spectroscopy, cryogenics, microwave, magnet and radiofrequency techniques. All of these needed to be pushed beyond their known limits in order to satisfy the growing needs of the particle physics experiments. Truly multidisciplinary skills were required.

The techniques developed for polarized targets helped also these scientists, in turn, in their own specialized disciplines. Among their achievements are the studies of nuclear magnetism and pseudomagnetism started in the 1970s, of large biomolecules by spin contrast variation started in the 1980s and of high-resolution NMR spectroscopy using DNP of rare spins more recently. Also, dissolution DNP was invented for the signal enhancement in MRI. These are good examples of the cross-fertilization and spin-offs resulting from the multidisciplinary collaborations.

The book is organized in the following way. The first two chapters describe the generic spin and its resonance in a high magnetic field, first without other interactions and then including interactions among themselves and with the constituents of the solid lattice. These chapters can be skipped by those who are already experts in magnetic resonance. Chapter 3 then focuses on the behavior of paramagnetic (unpaired) electron spins as is needed for the understanding of dynamic nuclear polarization in solids, which is the topic of Chapter 4.

Nuclear spins and their resonance in the solid lattice are discussed in Chapter 5, before describing the measurement of nuclear spin polarization in Chapter 6. Chapter 7 deals with the preparation and handling of the solid polarized target materials, and Chapter 8 focuses on the refrigeration of such materials during dynamic polarization using microwave irradiation. The required magnet and microwave techniques are the topics of Chapter 9.

Chapter 10 lists briefly other methods of generating high nuclear spin polarization, also used in particle physics experiments but, in particular, as used in new chemical and biomedical applications. Finally, the design and optimization of experiments with polarized targets are discussed in Chapter 11, focusing specifically on high-energy particle scattering experiments.

Acknowledgements

I am greatly indebted to Olli V. Lounasmaa, Nicholas Kurti and Michel Borghini, who encouraged and guided me during my first steps in the polarized target science.

In my work to develop, design, construct and operate polarized targets I have benefited from fruitful discussions and helpful advice of a large number of talented physicists and accelerator experts whose experiments at CERN and elsewhere required polarized targets. Among them I wish to name in particular Kors Bos, Franco Bradamante, Gerry Bunce, Don Crabb, Louis Dick, Giuseppe Fidecaro, Maria Fidecaro, Bernard Frois, Erwin Gabathuler, Ray Gamet, Lau Gatignon, Dietrich Von Harrach, Peter Hayman, Vernon Hughes, Alan Krisch, Franz Lehar, Alain Magnon, Gerhard Mallot, Aldo Penzo, Hans Postma, Ludwig van Rossum, Terry Sloan, Heinrich Stuhrmann, Fred Udo, Rüdiger Voss, Stephen Watts and Dave Websdale.

I am thankful for the enthusiastic participation and skilful expertise of the following scientists in the various polarized target projects that were the source of this book: Peter Berglund, Willem de Boer, Geoff Court, Steve Cox, Hans Glättli, Patrick Hautle, Dan Hill, Naoaki Horikawa, Yoshikazu Ishikawa, Shigeru Ishimoto, Yuri Kisselev, Mike Krumpolc, Jukka Kyynäräinen, Salvatore Mango, Akira Masaike, Werner Meyer, Kimio Morimoto, Pierre Roubeau, Stephen Trentalange, Tom Wenckebach, Paul Weymuth and John Wheatley. Their contributions have been vitally important in these projects.

Among the many engineers and technicians who have supported the projects at CERN, I would like to thank in particular Jacques André, Jean-Louis Escourrou, George Gattone, Charles Policella, Jean-Michel Rieubland, Adriaan Rijllart, Michel Uldry and Jean Zambelli.

During the production of the manuscript the expert help and advice of the professional staff of Cambridge University Press are gratefully acknowledged.

1
Introduction to Spin, Magnetic Resonance and Polarization

In this chapter, we shall review the mathematical formalism required for the understanding of the spin physics of polarized targets. Particular focus is given to the problems treating the situations that are favorable for obtaining high polarizations: high magnetic field and low lattice temperature.

In the following sections we shall first discuss the concept of the spin and magnetic moment and work out in detail some standard quantum mechanical problems involving these variables. The quantum statistics of a system of spins is then overviewed, before briefly introducing the thermodynamics of spin systems. Most of these can be found in well-known textbooks of quantum mechanics, such as those of Dicke and Wittke [1] and of Landau and Lifshitz [2], and of magnetic resonance, such as Abragam [3], Goldman [4], Abragam and Goldman [5] and Slichter [6]. The main justification for presenting textbook material is that we need to make frequent reference to this basic formalism. Three further reasons are:

(1) to introduce a consistent notation and vocabulary;
(2) to refer uninitiated readers to the basic source literature for further reading; and
(3) to introduce the SI units.

There are differences in the way how some basic entities are defined in the textbooks, and therefore a consistent notation and vocabulary are useful in developing the theory of spin dynamics.

Magnetic resonance is one of the last fields of physics where the old Gaussian units are still commonly used, or they are mixed with the MKSA units, which is a subset of SI units. Because the SI units have been almost exclusively used for more than 25 years in most other fields of physics, we have made an effort to extend this to magnetic resonance. We shall also refer to Appendix A.1 where the SI unit system is compared with CGS Gaussian system (Tables A1.1 and A1.2). In the same appendix the fundamental physical quantities and variables, relevant for magnetic resonance, are defined in Table A.1.3, and the physical constants are listed in Table A.1.4, both in the SI system of units.

The basic results and terminology of this chapter will be used in Chapter 2 to describe various interactions of spin systems in general, and those of electron spin systems more specifically in Chapter 3. The basic groundwork is equally important for dynamic nuclear polarization (DNP) which is the subject of Chapter 4 and for nuclear magnetic resonance (NMR) that is discussed in Chapter 5.

1.1 Quantum Mechanics of Free Spin

1.1.1 Spin

The angular momentum vector **J** has the same units as the Planck constant $\hbar$ and can therefore be expressed for a rigid body as

$$\mathbf{J} = \hbar \mathbf{I}, \tag{1.1}$$

where the vector **I** is called *spin*. The components of **I** are unitless numbers in classical mechanics, whereas in quantum mechanics they are unitless operators performing rotations about the three coordinate axes. Macroscopic rigid bodies can have a large spin, the components of which can be incremented or decremented in steps of 1 that is very small in comparison with the length of **I**, whereas elementary particles have a definite maximum projection I of **I** on any coordinate axis. This maximum projection I is called the *intrinsic spin*, or briefly the spin.

The concept of the intrinsic spin of an elementary particle was controversial until Dirac's relativistic theory of electron became accepted after the experimental discoveries of the positron and of the creation and annihilation of electron-positron pairs. Since then spin has played a fundamental role in particle physics, proving and disproving many theories. The most famous proofs are probably those of the quantum electrodynamics (QED) based on the Lamb shift of atomic hydrogen and on the anomalous magnetic moment of the electron, and the tests of unified electroweak theories based on the accurate measurements of parity violation parameters in atomic, nuclear and high-energy interactions.

It has been suggested that the intrinsic spin may have a still deeper meaning in physics through general relativity [7, 8], possibly explaining the existence of the three types of charged leptons, the electron e, the muon μ and the tau lepton τ.

For the main purpose of this book, we do not need to specify whether the spin of a particle is due to intrinsic angular momentum, or due to the motion of a complex composite system (such as quarks and gluons or partons in a hadron, or nucleons in a nucleus). Landau and Lifshitz [2] discuss this in the context of nonrelativistic quantum mechanics. They note that the law of conservation of angular momentum is a consequence of the isotropy of space in both classical and quantum mechanics. They remark, however, that in quantum mechanics the classical *definition* of the angular momentum $\mathbf{r} \times \mathbf{p}$ of a particle has no direct significance owing to the fact that the position **r** and momentum **p** cannot be simultaneously measured. In other words, neither **r** nor **p** of the constituents has significance for observations of a complex system of particles, with a probe which does not break the structure.

Thus, a stable composite particle, in a definite internal state with given internal energy, has also an angular momentum of definite magnitude J, due to the motion of the constituent particles. This angular momentum can have $2I + 1$ orientations in space. With this understanding of the angular momentum, its origin becomes unimportant, and Landau and Lifshitz [2] thus arrive at the concept of an 'intrinsic' angular momentum which must be ascribed to a particle regardless of whether it is 'composite' or 'elementary'.

When discussing the dynamics of a system made of these composite (in the sense described above) particles, such as nuclei or electrons in a solid lattice built of ions, atoms or molecules, the origin of the angular momentum of the stable composite becomes unimportant; however, a reserve must be made on electronic spins with regard to spin-lattice relaxation, for example.

This 'intrinsic' angular momentum which is not connected with the dynamics of the solid material is called the *spin* to distinguish it from the *orbital angular momentum*. Paramagnetic electrons in a solid are said to possess an *effective spin* when only the lowest magnetic states of the ground-state multiplet are populated; in this case the term 'spin' must be understood as a shorthand notation.

The complete wave function of a particle with a spin depends on the three coordinates of the particle and on the spin variable. The spin variable is a discrete one, and it gives the projection of the intrinsic angular momentum on a selected direction in space. The selection of this direction is often the key problem to be solved. Only in a steady high magnetic field, this direction is parallel or close to the field vector.

1.1.2 Spin and Magnetic Dipole Moment

In classical electromagnetic theory, the magnetic moment[1] $\bar{\mu}$ of a volume containing currents with density $\mathbf{j}_m$ is (see, for example, Ref. [9] p. 130):

$$\bar{\mu} = \frac{1}{2} \int_\tau (\mathbf{r} \times \mathbf{j}_m) d\tau, \qquad (1.2)$$

where $\mathbf{r}$ is the vector pointing to the volume element $d\tau$. If the currents can be considered as charge densities ρ_e moving with a velocity $\mathbf{u}$, the magnetic moment becomes

$$\bar{\mu} = \frac{1}{2} \int_\tau \rho_e (\mathbf{r} \times \mathbf{u}) d\tau. \qquad (1.3)$$

This resembles the mechanical angular momentum

$$\mathbf{J} = \int_\tau \rho_m (\mathbf{r} \times \mathbf{u}) d\tau \qquad (1.4)$$

of mass densities ρ_m moving at a velocity $\mathbf{u}$. If the system is composed of identical particles with mass m and charge e, the gyromagnetic ratio, γ, defined as

---

[1] As discussed in Appendix A.1, in SI units the unit of magnetic moment is $[\mu] = Am^2$; the magnetic energy of the dipole is then $E = \boldsymbol{\mu} \cdot \mathbf{B}$, magnetic field being expressed in $[B] = Vs/m^2 = $ Tesla.

$$\gamma = \frac{\mu}{J}, \qquad (1.5)$$

becomes

$$\gamma = \frac{e}{2m} \qquad (1.6)$$

for the case where μ and J are parallel.

In classical mechanics there is no general reason why these vectors should be parallel, but in quantum mechanics this is the case for closed systems. However, when adding the quantum mechanical angular momentum vectors of the system, the resultant magnetic momentum vector does not generally align with the angular momentum vector. A way of understanding this is that because the magnetic moment component perpendicular to J cannot be determined simultaneously with that along J (it can be thought to undergo rapid rotation around the axis), the only observable is the projection of μ along J. This gives rise to the structural g-factor, which is particularly important in electron spin resonance.

In the case of a complex structure, the gyromagnetic ratio is written in the terms of the g-factor as

$$\gamma = \frac{ge}{2m}, \qquad (1.7)$$

where the factor g contains the entire description of the magnetic structure. For electrons we also often write

$$\gamma = \frac{g\mu_B}{\hbar}, \qquad (1.8)$$

where we have introduced the fundamental constant Bohr magneton

$$\mu_B = \frac{\hbar e}{2m_e}. \qquad (1.9)$$

We note here that for a negatively charged pointlike particle the gyromagnetic factor and the magnetic moment are always negative. In the literature the symbol β is often used for the Bohr magneton, but we reserve here this symbol for the inverse spin temperature.

Free pointlike charged particles (such as electron or muon) have the g-factor close to the Dirac value $g = 2$. The deviations are often given using an anomalous magnetic moment a defined by

$$g = 2(1+a). \qquad (1.10)$$

The deviation can be measured to a high accuracy using a Penning trap for electrons or a storage ring for muons; comparisons with theoretical calculations have given important proofs of QED and QCD and restrained the limits of any substructure of the leptons [10].

As was discussed in the previous subsection, the spin angular momentum variable is a discrete one. Therefore, the magnetic moment projection on the axis of quantization also has only discrete values. This quantum mechanical fact, which will be discussed below in this chapter, is seen in a striking way in magnetic resonance measurements which are the basis of a large industry today.

The gyromagnetic ratio and the magnetic moment can be positive or negative, depending on not only the sign of the charge of the pointlike particle but also the structure of the complex system made of constituents. In this book we shall always assume, however, that the magnetic moment of the nucleus or electron is parallel or antiparallel to its spin angular momentum, depending on the sign of the gyromagnetic ratio:

$$\widehat{\vec{\mu}} = \gamma \hbar \hat{\mathbf{I}}, \tag{1.11}$$

where the vectors $\widehat{\vec{\mu}}$ and $\hat{\mathbf{I}}$ now are taken as quantum mechanical operators. The three components of these vectors can be mathematically represented by the so-called spin matrices, which will be discussed below.

1.1.3 Spin Operator Algebra

For simplicity, we shall eliminate here the vector notations but maintain the operator symbols with circumflex for a while in order to make the operators clearly distinct from constants. The spin operator $\hat{I}$ thus has the projections $\hat{I}_j (j = x, y, z)$ along the three coordinate axes in the same way as the angular momentum operator $\hat{J} = \hbar \hat{I}$. The algebra with operators requires the knowledge of their commutation relations[2] which, for the rotation operators, are obtained by considering infinitesimally small (elementary) rotations about the coordinate axes. For example, by performing a small rotation first around the x-axis and then around the y-axis, and then rotations about the same axes in reverse order and direction, the net result is a small positive rotation about the z-axis. The same can be achieved by comparing small rotations around the x- and y-axes, with rotations made around y- and x-axes. The difference of these two is a small rotation about the z-axis. This can be mathematically represented as a commutation relation

$$\hat{I}_x \hat{I}_y - \hat{I}_y \hat{I}_x \equiv \left[\hat{I}_x, \hat{I}_y\right] = i\hat{I}_z.$$

Cyclic permutation of the subscripts gives the following commutation relations for the spin operator components:

$$\begin{aligned} \left[\hat{I}_x, \hat{I}_y\right] &= i\hat{I}_z, \\ \left[\hat{I}_y, \hat{I}_z\right] &= i\hat{I}_x, \\ \left[\hat{I}_z, \hat{I}_x\right] &= i\hat{I}_y. \end{aligned} \tag{1.12}$$

---

[2] See, for example, Landau and Lifshitz [2] Chapters IV and VIII, and Dicke and Wittke [1] Chapters 9 and 12.

It is remarkable that these equations, based on the sole assumption that the space is isotropic, result in all the physics of spin. Notably, because the spin components do not commute, only one of them can be measured at a time, and the remaining two being not simultaneously measurable. Some other immediate results are briefly reviewed below.

The square of the 'magnitude' of $\hat{I}$

$$\hat{I}^2 = \hat{I}_x^2 + \hat{I}_y^2 + \hat{I}_z^2 \tag{1.13}$$

commutes with all three components of I as a consequence of Eq. 1.12:

$$\left[\hat{I}^2, \hat{I}\right] = 0, \tag{1.14}$$

i.e. $\hat{I}^2$ and any one of the components of $\hat{I}$ are simultaneously measurable.

Instead of $\hat{I}_x$ and $\hat{I}_y$ it is often more convenient to use the complex combinations

$$\hat{I}_\pm = \hat{I}_x \pm i\hat{I}_y, \tag{1.15}$$

which satisfy, based on Eq. 1.12 directly, the relations

$$\left[\hat{I}_+, \hat{I}_-\right] = 0, \tag{1.16}$$

$$\left[\hat{I}_z, \hat{I}_\pm\right] = \pm \hat{I}_\pm, \tag{1.17}$$

and

$$\hat{I}_\pm \hat{I}_\mp = \hat{I}^2 - \hat{I}_z^2 \pm \hat{I}_z. \tag{1.18}$$

Let us assume that m is the eigenvalue of $\hat{I}_z$:

$$\hat{I}_z \psi = m \psi. \tag{1.19}$$

The operators $\hat{I}_\pm$ are now seen to be ladder operators with respect to the eigenvalue m of $\hat{I}_z$, because

$$\hat{I}_z(\hat{I}_+\psi) = (m+1)(\hat{I}_+\psi), \tag{1.20}$$

which can be obtained using Eq. 1.17 or Eq. 1.12 directly.

Because of relation 1.14, the eigenfunction ψ can be chosen so that it simultaneously satisfies Eq. 1.19 and

$$\hat{I}^2 \psi = a\psi, \tag{1.21}$$

where a is the square of the magnitude (i.e. length squared) of the spin vector, which we shall evaluate now. Firstly, because both expectation values and their sum

$$\langle \hat{I}_x^2 \rangle + \langle \hat{I}_y^2 \rangle$$

must be positive, from Eq. 1.13 it is clear that the expectation values of $\hat{I}^2$ and $\hat{I}_z$ satisfy

$$\langle \hat{I}^2 \rangle \geq \langle \hat{I}_z^2 \rangle,$$

which gives
$$a \geq m^2. \tag{1.22}$$

As a consequence of Eq. 1.20, the difference $2I$ of the greatest $(+I)$ and least $(-I)$ possible eigenvalue m of $\hat{I}_z$ must be an integer; I may take then any half-integer[3] value 0, 1/2, 1, 3/2, etc., and

$$m = -I, \; -I+1, \ldots I-1, I. \tag{1.23}$$

If m has its maximum value I, then $\hat{I}_+\psi = 0$, and, using Eq. 1.18,

$$\hat{I}_-\hat{I}_+\psi = 0 = \left(\hat{I}^2 - \hat{I}_z^2 - \hat{I}_z\right)\psi; \tag{1.24}$$

from this and Eq. 1.19 with $m = I$, we get

$$a = \left\langle \hat{I}^2 \right\rangle = I(I+1). \tag{1.25}$$

The eigenvalue of the operator $\hat{I}^2$ is therefore $I(I+1)$, where I is called the spin quantum number and gives the maximum projection of the spin vector along an axis. Speaking of spin I therefore means speaking of a vector with magnitude $\sqrt{I(I+1)}$ and maximum projection on any axis of I. For simplicity we shall use in the rest of the book, unless ambiguities or clarity require otherwise, the same symbol I to denote the spin vector operator $\hat{I}$ and the spin quantum number I (or spin in short); possible confusions between these should become clarified by the context.

1.1.4 Matrix Representation of the Spin Operator

A quantum mechanical operator can be represented by a matrix acting upon a state vector which represents the wave function. The elements or components of these have a direct physical significance and can be related to the expectation values of the observables.

The wave function of a particle with spin I has $2I + 1$ components; the squares of the magnitudes of these components give the probability of the magnetic states m. The spin operator in matrix representation has $(2I + 1)\cdot(2I + 1)$ elements

$$\begin{aligned}
\left(I_x\right)_{m,m-1} &= \left(I_x\right)_{m-1,m} = \frac{1}{2}\sqrt{(I+m)(I-m+1)}, \\
\left(I_y\right)_{m,m-1} &= -\left(I_y\right)_{m-1,m} = \frac{i}{2}\sqrt{(I+m)(I-m+1)}, \\
\left(I_z\right)_{m,m} &= m,
\end{aligned} \tag{1.26}$$

where m is the magnetic quantum number; the rest of the elements are zero.

---

[3] The orbital angular momentum operator L can take only integer values of L_z, which is the consequence of restricting the form of the wave function to represent simple orbital motion. This restriction is by no means valid for complex wave functions such as that of the nucleon, whose constituents undergo relativistic motion.

The most important case is spin $I = 1/2$, for which the matrix components of the spin vector operator of Eq. 1.26 are:

$$I_x = \frac{1}{2}\begin{bmatrix} 0 & 1 \\ 1 & 0 \end{bmatrix};$$

$$I_y = \frac{1}{2}\begin{bmatrix} 0 & -i \\ i & 0 \end{bmatrix}; \quad (1.27)$$

$$I_z = \frac{1}{2}\begin{bmatrix} 1 & 0 \\ 0 & -1 \end{bmatrix}.$$

These 2×2 matrices are called Pauli spin operators, often denoted by $\sigma \equiv 2I$; their direct multiplication gives

$$\sigma^2 = 3 \cdot \mathbf{1}, \quad (1.28)$$

where **1** is the unit matrix

$$\mathbf{1} \equiv \begin{bmatrix} 1 & 0 \\ 0 & 1 \end{bmatrix}.$$

Equation 1.28 is clearly compatible with the value given by Eq. 1.25 for I. Moreover, similar direct multiplication yields

$$(\sigma \cdot \mathbf{a})(\sigma \cdot \mathbf{b}) = \mathbf{a} \cdot \mathbf{b} + i\sigma \cdot \mathbf{a} \times \mathbf{b}, \quad (1.29)$$

where **a** and **b** are any constant vectors. Furthermore, by replacing **a** and **b** by any unit vector **e** we get immediately

$$(\sigma \cdot \mathbf{e})^{2p} = 1 \quad (1.30)$$

and

$$(\sigma \cdot \mathbf{e})^{2p+1} = \sigma \cdot \mathbf{e}. \quad (1.31)$$

According to relations 1.29–1.31, any scalar polynomial of the components of σ can be reduced to terms independent of σ and to a term linear in σ; furthermore, any scalar function of σ reduces to a linear function, if it can be expanded as a Taylor series. These relations will be used in calculating traces involving the density matrix, without resorting to the so-called high-temperature approximation. This is a very important property of the Pauli spin operator for the theory of DNP at low temperatures, where high polarizations can be obtained.

In the case of $I = 1/2$, both Eqs. 1.25 and 1.28 yield $\langle \hat{I}^2 \rangle = \frac{3}{4}$. If the spin is in a state where one of its components (say, in the z-direction) has its maximum value of $\langle \hat{I}_z \rangle = \frac{1}{2}$, then I_x and I_y have zero expectation values, but $\langle \hat{I}_x^2 \rangle = \langle \hat{I}_y^2 \rangle = \frac{1}{4}$. As will be seen in Section 1.1.6, this can be understood by the precession of the spin vector which makes the components perpendicular to the axis of quantization oscillate sinusoidally.

For spin $I = 1$ we get the matrix representations of the components:

$$I_x = \frac{1}{\sqrt{2}} \begin{bmatrix} 0 & 1 & 0 \\ 1 & 0 & 1 \\ 0 & 1 & 0 \end{bmatrix};$$

$$I_y = \frac{i}{\sqrt{2}} \begin{bmatrix} 0 & 1 & 0 \\ -1 & 0 & 1 \\ 0 & -1 & 0 \end{bmatrix}; \qquad (1.32)$$

$$I_z = \begin{bmatrix} 1 & 0 & 0 \\ 0 & 0 & 0 \\ 0 & 0 & -1 \end{bmatrix}.$$

These will be used explicitly in an example in Chapter 5.

1.1.5 Magnetic Energy Levels

Let us now consider a particle with spin I in a magnetic field B_0, with the field vector lying along the z-axis so that $\mathbf{B}_0 = \mathbf{k} B_0$. The spin is associated with the dipole moment $\hat{\mu} = \hbar \gamma \hat{I}$, where the gyromagnetic factor γ is defined by Eq. 1.6. The magnetic energy of the particle is then (in operator form)

$$\hat{\mathcal{H}} = -\hat{\vec{\mu}} \cdot \mathbf{B}_0 = -\hbar \gamma B_0 \hat{I}_z, \qquad (1.33)$$

and the Schrödinger equation

$$\hat{\mathcal{H}} \psi = E \psi \qquad (1.34)$$

has $2I + 1$ eigenvalues

$$E_m = -m \hbar \gamma B_0, \qquad (1.35)$$

because the eigenvalues of $\hat{I}_z$ go from $m = -I$ to $m = +I$, as was shown in Eq. 1.23. The magnetic energy level splitting is often visualized as shown by Figure 1.1a.

1.1.6 Larmor Precession

The time-dependent Schrödinger equation

$$\hat{\mathcal{H}} \psi = i\hbar \frac{\partial \psi}{\partial t} \qquad (1.36)$$

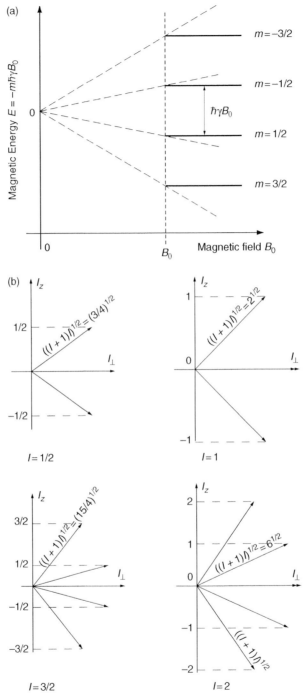

Figure 1.1 (a) The magnetic energy levels for a free particle with spin $I = 3/2$ and gyromagnetic factor γ in a steady magnetic field B_0. (b) The possible projections of the spin of a free particle in a steady field along the z-axis of the field and perpendicular to it, for spins $I = 1/2, 1, 3/2$ and 2. The perpendicular component rotates in its plane with undetermined phase angle

can be directly integrated by inserting Eq. 1.33 and using the I_z of Eq. 1.26. Taking spin $I = 1/2$ and choosing initial conditions so that the spin at $t = 0$ points in the positive x-direction (has the maximum component I_x), we get

$$\psi(0) = \frac{1}{\sqrt{2}} \begin{bmatrix} 1 \\ 1 \end{bmatrix}, \quad (1.37)$$

which satisfies

$$I_x \psi(0) = \frac{1}{2} \psi(0).$$

The solution of Eq. 1.36 is found straightaway:

$$\psi(t) = \frac{1}{\sqrt{2}} \begin{bmatrix} e^{-i\omega_0 t/2} \\ e^{+i\omega_0 t/2} \end{bmatrix}, \quad (1.38)$$

where the angular frequency ω_0 has the value

$$\omega_0 = -\gamma B_0. \quad (1.39)$$

The expectation values

$$\langle I_j \rangle = \psi * I_j \psi, \quad j = x, y \quad (1.40)$$

of the Pauli spin matrix components x and y of Eqs. 1.27 can be now evaluated by directly inserting them to Eq. 1.40; the resulting

$$\langle I_x \rangle = \frac{1}{2} \cos \omega_0 t,$$
$$\langle I_y \rangle = \frac{1}{2} \sin \omega_0 t \quad (1.41)$$

show that the projection of the spin vector in the x–y plane freely rotates at an angular frequency $\omega_0 = -\gamma B_0$ which is the Larmor precession frequency. A coordinate transformation to a frame rotating at this frequency removes the time dependence of the rotating components; in this frame the spin has an eigenvalue $I^\dagger \psi = \frac{1}{2}\psi$, i.e. $\langle I^\dagger \rangle$ is time independent. The transformation to the rotating frame is a key technique in the understanding of saturation in magnetic resonance; it will be discussed in more detail in the next section.

In the above example, the initial state of the spin was $I_x = 1/2$. The preparation of the spin into the wanted initial state can be made using several techniques. The theoretically simplest one requires rotating the field by 90° much faster than the Larmor precession. Technically this is possible only at low field values. In high fields a transverse field oscillating at the angular frequency ω_0 can be applied for such a duration that the spin pointing originally in the z-direction is tilted by 90° (this is called a 90° pulse), or by sweeping the frequency of a small transverse field slowly to ω_0 and then reducing its amplitude to zero (adiabatic demagnetization to zero field in rotating frame). These techniques will be discussed in Chapters 2 and 11.

The classical treatment of the above example gives exactly same result for ω_0, and is best done by using the rule of transforming the time derivative $d\mathbf{A}(t)/dt$ of vector $\mathbf{A}$ to a frame rotating at speed ω:

$$\frac{d\mathbf{A}}{dt} = \frac{\partial \mathbf{A}}{\partial t} + \vec{\omega} \times \mathbf{A}(t). \tag{1.42}$$

The classical equation of motion of magnetic moment in a field $\mathbf{B}_0$ is obtained by equating the time change of the angular momentum $d\mathbf{J}/dt$ to the magnetic torque $\vec{\mu} \times \mathbf{B}_0$. By using the gyromagnetic ratio of Eq. 1.5 we get

$$\gamma \frac{\partial \mathbf{J}}{\partial t} = \mathbf{J} \times \left(\mathbf{B}_0 + \frac{\vec{\omega}}{\gamma} \right). \tag{1.43}$$

By selecting $\vec{\omega} = -\gamma \mathbf{B}_0$, the explicit time dependence of $\mathbf{J}$ disappears. In this frame $\mathbf{J}$ is a stationary vector, which means that in the original frame its component perpendicular to $\mathbf{B}_0$ rotates in the plane normal to $\mathbf{B}_0$ at the angular frequency $\omega_0 = -\gamma B_0$ of Eq. 1.39, the Larmor frequency.

1.1.7 Spin Resonance and Rotating Frame

In Chapters 1 to 6 we shall often use exponential operators of the form e^A. Let us assume operators A and B and their commutator $C = [A,B] = AB - BA$, which is assumed to commute with both operators so that $[C,A] = [C,B] = 0$. We shall need the following general relations concerning such operators A, B and C:

$$e^{A+B} = e^A e^B e^{-C/2} \tag{1.44}$$

and

$$e^{A+B} = e^{C/2} e^B e^A. \tag{1.45}$$

In the special case of commuting operators $C = 0$ and their exponential functions are seen to commute so that

$$e^{i(A+B)} = e^{iA} e^{iB}. \tag{1.45'}$$

For commuting A and B we also have

$$Ae^B = e^B A. \tag{1.46}$$

The commutation rules of the exponential operators simplify the solution of the Schrödinger equation if the Hamiltonian has no explicit time dependence. Introducing the exponential operator $U = \exp(-i\omega t I_z)$ one can derive the following transformation rules using the spin matrices and their commutation relations:

1.1 Quantum Mechanics of Free Spin

$$U^{-1}I_z U = I_z;$$
$$U^{-1}I_\pm U = I_\pm e^{\pm i\omega t}. \tag{1.47}$$

The first of these follows immediately from Eq. 1.46, because I_z commutes with itself and therefore with its exponential. The second one can be proven by noting first that the commutation rule 1.17 directly gives

$$I_z I_\pm = \pm I_\pm + I_\pm I_z = I_\pm(\pm\mathbf{1} + I_z),$$

where $\mathbf{1}$ is a unit matrix of same dimension as $\hat{I}_z$. Multiplying $n-1$ times from the left we then obtain

$$I_z^n I_\pm = I_\pm(\pm\mathbf{1} + I_z)^n.$$

Expanding now U^{-1} into Taylor series and using the above we find

$$U^{-1}I_\pm = \left[1 + \frac{i\omega t}{1!}I_z + \ldots + \frac{(i\omega t)^n}{n!}I_z^n + \ldots\right]I_\pm$$

$$= I_\pm e^{i\omega t(\pm 1 + I_z)} = I_\pm e^{\pm i\omega t}U^{-1}$$

which, after multiplying from the right by U, yields the second one of Eqs. 1.47. Expressing now I_x and I_y in the terms of $I_\pm$ from definition 1.15, it is straightforward to show that

$$U^{-1}I_x U = I_x \cos\omega t + I_y \sin\omega t;$$
$$U^{-1}I_y U = -I_x \sin\omega t + I_y \cos\omega t. \tag{1.47'}$$

These are seen to represent rotation of the x–y plane about the z-axis by an angle of ωt. The use of the transformation to such a rotating frame is illustrated with the following example, which gives the exact solution of a basic problem in magnetic resonance.

Let us consider a particle with spin I in a high steady field $\mathbf{B}_0$ along z-axis, superposed by a small field $\mathbf{B}_1(t) = B_1\exp(i\omega t)$ perpendicular to it and rotating with angular frequency ω. The spin Hamiltonian is then

$$\mathcal{H} = \mathcal{H}_0 + \mathcal{H}_1 = \hbar\gamma\left[B_0 I_z + B_1\left(I_x \cos\omega t + I_y \sin\omega t\right)\right]$$

and the Schrödinger equation

$$i\frac{\partial\psi}{\partial t} = \mathcal{H}\psi$$

$$= -\gamma\left[B_0 I_z + B_1\left(I_x \cos\omega t + I_y \sin\omega t\right)\right]\psi \tag{1.48}$$

$$= \left(\omega_0 I_z + \frac{\omega_1}{2}\left[I_+ e^{-i\omega t} + I_- e^{+i\omega t}\right]\right)\psi$$

will now have to be solved (here we have written $\omega_1 = -\gamma B_1$). To find the exact solution, transformation to a rotating frame will be done by the substitution

$$\psi = U\psi_e = e^{-i\omega I_z t}\psi_e.$$

The new equation

$$i\hbar \frac{\partial \psi_e}{\partial t} = \{U^{-1}\mathcal{H}U - iU^{-1}\dot{U}\}\psi_e \tag{1.49}$$

can be simplified by the rules 1.44–1.46 and the commutation relations 1.12; the resulting equation

$$i\frac{\partial \psi_e}{\partial t} = \{(\omega_0 - \omega)I_z + \omega_1 I_x\}\psi_e \tag{1.50}$$

has the solution

$$\psi_e = e^{-i[(\omega_0-\omega)I_z + \omega_1 I_x]t}\psi_e(0), \tag{1.51}$$

where $\psi_e(0) = \psi(0)$. Using Eq. 1.48 this can be written in the form

$$\psi = e^{-i(\omega_0-\omega)I_z t} e^{-ia(\mathbf{e}\cdot\mathbf{I})t}\psi(0), \tag{1.52}$$

where

$$a = -\text{sign}(\gamma)\sqrt{(\omega_0 - \omega)^2 + \omega_1^2} \tag{1.53}$$

and **e** is a unit vector with components

$$e_x = \frac{\omega_1}{a},\ e_y = 0,\ e_z = \frac{\omega_0 - \omega}{a}. \tag{1.54}$$

We can now determine the probability amplitude $A_{m'm}$ of finding the spin in the state m' at time t, after initially in state m at $t = 0$:

$$A_{m'm} = \langle m'|e^{-i(\omega_0-\omega)I_z t}e^{-ia(\mathbf{e}\cdot\mathbf{I})t}\psi(0)|m\rangle,$$

which gives the probability $P_{m'm} = |A_{m'm}|^2$ for transition $m \to m'$:

$$P_{m'm} = |\langle m'|e^{-ia(\mathbf{e}\cdot\mathbf{I})t}\psi(0)|m\rangle|^2. \tag{1.55}$$

For spin 1/2 this becomes, using the Pauli spin matrices 1.27 and their multiplication rules 1.28–1.31,

$$P_{-\frac{1}{2},\frac{1}{2}} = \sin^2\theta \sin^2\frac{at}{2}, \tag{1.56}$$

1.1 Quantum Mechanics of Free Spin

where $\sin\theta = \omega_1/a$. By examining this result, one may show that in the frame rotating at angular frequency ω, the frequency of the field B_1 in the laboratory frame, the spin axis (along which I has maximum projection) precesses at an angular frequency 1.53 around a constant effective field

$$\mathbf{B}_e = \left(B_0 - \frac{\omega}{\gamma}\right)\mathbf{k} + B_1\mathbf{i} \tag{1.57}$$

of magnitude

$$|\mathbf{B}_e| = \frac{a}{|\gamma|}$$

and angle

$$\theta = \arctan\frac{\omega_1}{\omega_0 - \omega}$$

with respect to the z-axis. The exact resonance occurs at frequency $-\omega_0$; the spin then rotates around the effective field $\mathbf{B}_e = B_1\mathbf{i}$ at a frequency ω_1, inverting the direction of its z-component with intervals

$$\frac{T}{2} = \frac{\pi}{\omega_1} = \frac{\pi}{\gamma B_1}.$$

We note that we now know the exact time evolution of one of the components of the spin; the components perpendicular to that remain unknown because of their unknown phase angle at the time when the rotating field was turned on.

In solving the above problem, we made a transformation to a rotating coordinate system, a technique that allows to treat many problems in an elegant way. We shall return back to this technique in several sections of this book.

The generalization of the above treatment to arbitrary spin is tedious and we shall omit it here.

The motion of the spin axis is illustrated in Figure 1.2 without rotating field (in the stationary frame) and in the rotating frame with $\omega = \omega_0$. The classical motion of an angular momentum associated with a parallel magnetic moment is exactly same; the only difference is that once the component of the spin in the direction of the quantum mechanical spin axis is determined, the components perpendicular to it exist but their phase angles cannot be measured, whereas the classical angular momentum can be fully described in such a way that it does not have undetermined perpendicular components. A way of proving this mathematically involves the transformation from the quantum description to the classical one, by going to the limit $I \to \infty$, at which the perpendicular component with undetermined phase angle will become zero.

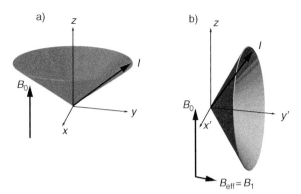

Figure 1.2 (a) Larmor precession of spin I in steady field B_0 oriented along the z-axis in the stationary coordinate frame xyz; (b) Larmor precession of spin I in effective field B_{eff} in the coordinate frame $x'y'z$ rotating at the Larmor frequency ω_0

The important results 1.56 and 1.57 are exact and are valid in any time scale; the time-dependent perturbation theory gives a different (and unphysical) behavior near the resonance of a single (or non-interacting) spin. However, the transition probability

$$W_{m,m-1} = \frac{2\pi}{\hbar^2}\left|\langle m-1|\mathcal{H}_1|m\rangle\right|^2 f(\omega) = \frac{\pi}{2}\gamma^2 B_1^2 \left|\langle m-1|I^+|m\rangle\right|^2 f(\omega)$$
$$= \frac{\pi\omega_1^2}{2}(I+m)(I-m+1)f(\omega),$$
(1.58)

calculated using time-dependent perturbation theory, correctly describes the behavior of an assembly of spins which has a resonance shape function $f(\omega)$ due to interactions among the spins; this treatment obviously requires that $Wt \ll 1$. It can be also shown[4] that as long as $Wt \ll 1$, the exact solution convoluted with the lineshape function is equivalent to Eq. 1.58.

For $Wt \gg 1$ and for intermediate times, the evolution of the magnetization of a general spin system under saturation requires the introduction of relaxation mechanisms and the use of a quantum statistical treatment. Equation 1.58 is then valid only in some special cases such as in liquids where frequent collisions of the molecules carrying spins destroy the phase coherence of the spins and result in rapid spin-lattice relaxation. In solids this does not happen, and it was only after the pioneering work by Redfield [11] and Provotorov [12] that the saturation of magnetic resonance in solids was well understood. An introductory discussion of quantum statistics will be given below, and the quantum statistical treatment of saturation in solids will be discussed in Chapter 2.

---

[4] See the excellent and instructive discussion of Ref. [3], pp. 27–32.

1.2 Quantum Statistics of a System of Spins

1.2.1 Polarization and Spin Temperature

Up till now we have discussed the behavior of one isolated spin in a magnetic field which may have a small time-dependent transverse component. We have thus avoided all questions related with the interactions among the spins and with their environment, and we have also simplified the previous treatment by not introducing any macroscopic effects. We shall now introduce the mathematical basis for the quantum statistical treatment of problems related with spin systems, which allows to take both into account rigorously.

Let us consider a large array of spins I in a solid sample at high magnetic field $\mathbf{B}_0$ (unless mentioned otherwise, we shall always assume high external field along the z-axis). The spin Hamiltonian $\mathcal{H}$ is the energy operator of the particle and it is often written in the general form

$$\begin{aligned} \mathcal{H} &= -\hbar \gamma B_0 I_z + \mathcal{H}_1 \\ &= \hbar \omega_0 I_z + \mathcal{H}_1, \end{aligned} \quad (1.59)$$

where we have introduced the definition of the Larmor frequency ω_0 from Eq. 1.39. Here the first part on the right-hand side describes the magnetic dipole energy in the steady field discussed in Section 1.1.5, and the second part $\mathcal{H}_1$ includes all other interactions, such as those with any other particles with spin and magnetic dipole moment, with time-dependent external fields, or interactions of a possible nuclear electric quadrupole moment with an electric field gradient tensor. There are many other possible small contributions to $\mathcal{H}_1$; these will be discussed in some detail in Chapter 2. We emphasize that the definition of a high field is that the magnetic dipole energy operator, the Zeeman Hamiltonian, dominates the second part $\mathcal{H}_1$.

Let us first ignore the small $\mathcal{H}_1$ and hence assume that the spin system is composed of many non-interacting individual spins with the Zeeman Hamiltonian $\mathcal{H}_z = -\hbar \gamma B_0 I_z$, giving rise to the magnetic levels $E_m = -m\hbar\gamma B_0$ of Eq. 1.35. The generally uneven distribution of populations among the magnetic levels m gives rise to the need of parameters defining the average orientation in a spin system. Because spin I can be in one of the $2I+1$ states, it is clear that $2I$ parameters are needed to fully describe the relative (arbitrary) populations of these levels. The orientation parameters are most generally defined as expectation values of irreducible spin tensors [13], of which the vector polarization (or simply polarization)

$$P_I = \frac{\langle I_z \rangle}{I} \quad (1.60)$$

and tensor polarization (or alignment)

$$A_I = \frac{\langle 3I_z^2 - I^2 \rangle}{I^2} = \frac{3\langle I_z^2 \rangle - I(I+1)}{I^2} \quad (1.61)$$

are the simplest ones. In the above expressions the expectation value $\langle Q \rangle$ of operator Q is the average over the sample volume. For spin 1/2 the polarization P is the only non-zero one; for spin 1 the parameters P and A completely define the distribution of the states in thermal equilibrium. The P may vary from -1 to $+1$ for all spins, while A (for spin 1) may get values from -2 to 1. For spin 3/2 and above, the higher parameters remain small (in thermal equilibrium distribution) unless P is near unity.

In high field the Zeeman energy $\mathcal{H}_Z = -\hbar\gamma B_0 I_z$ is much larger than the remaining part $\mathcal{H}_1$ of the spin Hamiltonian, according to our definition. This suggests the use of perturbation theory for the calculation of the behavior of the spins interacting with their environment. Moreover, as the material is composed of order $N_A \approx 10^{23}$ particles with spin, it is entirely impractical to sum up explicitly the Hamiltonians 1.59 for all individual spins. This forces one to introduce statistical methods in the treatment of the behavior of the spin system.

The introduction of quantum statistics and perturbation theory can be made in a purely formal way, and one can beautifully demonstrate the validity and power of these in the treatment of a very wide variety of problems. Rather than resorting to such formalism at this point, we shall first introduce two simple and plausible concepts:

(i) Spin relaxation processes do exist and will reasonably fast bring the system to the thermal equilibrium corresponding to a definite distribution of populations of the magnetic energy levels $E_m = -m\hbar\gamma B_0$ of Eq. 1.35;

(ii) In the steady state, the probability p_m that a magnetic state m of energy E_m of a spin will be occupied is proportional to the Boltzmann factor:

$$p_m \propto e^{-\frac{E_m}{k_B T}},$$

where T is the temperature of the spin system, and k_B is the Boltzmann constant.

The spin relaxation processes of concept (i) will be discussed in Section 2.4; they are very important from the point of view of DNP as well. The concept (ii) will be formally derived in Section 1.2.2.

The Boltzmann distribution leads trivially to the polarization

$$P_I = \frac{\langle I_z \rangle}{I} = \frac{1}{I} \frac{\sum_{m=-I}^{+I} m p_m}{\sum_{m=-I}^{+I} p_m} = \frac{\sum_{m=-I}^{+I} m e^{-\frac{E_m}{k_B T}}}{I \sum_{m=-I}^{+I} e^{-\frac{E_m}{k_B T}}}; \qquad (1.62)$$

here the sums are over the possible spin states m. This can be also expressed using hyperbolic functions by

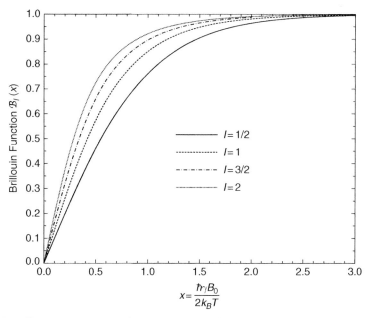

Figure 1.3 Brillouin function for spins $I = \frac{1}{2}, 1, 3/2$ and 2

$$P_I = \mathcal{B}_I(x) = \frac{2I+1}{2I}\coth(2I+1)x - \frac{1}{2I}\coth x;$$
$$x = \frac{\hbar\gamma B_0}{2k_B T},$$
(1.63)

which is the well-known Brillouin function. Figure 1.3 shows the function for some values of I. For spin $I = 1$ the Brillouin function can be also written

$$P_1 = \mathcal{B}_1(x) = \frac{4\tanh x}{3+\tanh^2 x}.$$
(1.63')

When the magnetic level populations are in thermal equilibrium corresponding to the Boltzmann distribution, the alignment A_I is uniquely related with the polarization P_I. Provided that quadrupole effects do not broaden the levels substantially, this high-field relation for spin $I = 1$ is simply

$$A_1 = 2 - \sqrt{4 - 3P_1^2}.$$
(1.63'')

Negative values of alignment therefore cannot be obtained if the spin system is in internal thermal equilibrium. Relationship 1.63'' is shown in Figure 1.4.

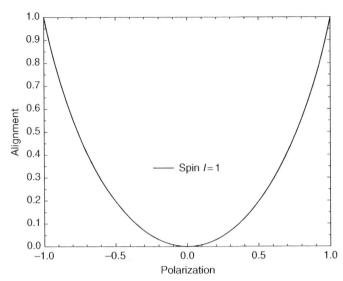

Figure 1.4 Relationship for spin $I = 1$ between the vector polarization and alignment (also called tensor polarization), when the magnetic level populations are in internal thermal equilibrium. We note that alignment higher than 1 or negative can only be obtained when the magnetic level populations are distributed non-thermally

For higher spin in thermal equilibrium, the alignment and other higher orientation parameters are also uniquely related with the polarization, but the relationships are transcendental functions of complexity increasing with the spin and with the order of the parameter.

The spin polarization is associated with the static magnetization or magnetic moment density:

$$M_I = \frac{1}{V}\sum_i \mu_z^i = \hbar\gamma_I n_I \langle I_z \rangle = \hbar\gamma_I n_I I P_I, \qquad (1.64)$$

where the sum is over the spins I in the volume V, and $n_I = N_I/V$ is the number density of the particles with the spin I in the sample. By defining[5] the static susceptibility

$$\chi_0 = \frac{\mu_0 M}{B_0}, \qquad (1.65)$$

where $\mu_0 = 4\pi \cdot 10^{-7}$ Vs/Am is the permeability of free space, we see that

$$\chi_0 = \frac{\mu_0 \hbar \gamma_I n_I I}{B_0} P_I \qquad (1.66)$$

---

[5] In SI system, the usual definition is $\mathbf{B} = \mu_0(\mathbf{H} + \mathbf{M}) = (1 + \chi_0)\mu_0\mathbf{H}$, which deviates slightly from the definition of Eq. 1.65. In high external field the difference is insignificant in dilute paramagnetic systems and totally negligible in nuclear magnetism, even at high polarizations.

is linear in P_I. This result is important for the measurement of polarization.

For spin $I = 1/2$ the Brillouin function reduces to $\mathcal{B}_{1/2}(x) = \tanh x$, giving

$$P_{\frac{1}{2}} = \tanh\frac{\hbar\gamma_I B_0}{2k_B T} = \tanh\left(\frac{1}{2}\alpha\omega_0\right), \tag{1.67}$$

where we have defined the inverse spin temperature α

$$\alpha = \frac{\hbar}{k_B T}, \tag{1.68}$$

and expressed the magnetic field in the terms of the Larmor frequency of Eq. 1.39, which is defined to be always positive when calculating level populations or polarization.[6]

At high temperature[7] (i.e. when $P \ll 1$), the expression of P_I simplifies to

$$P_I = \frac{\alpha\gamma_I B_0(I+1)}{3} = \frac{C}{T};$$
$$C = \frac{\hbar\omega_0(I+1)}{3k_B}, \tag{1.69}$$

which is the well-known Curie law of the static paramagnetic susceptibility proportional to T^{-1}:

$$\chi_0 = \frac{\mu_0 n_I \hbar^2 \gamma_I^2 I(I+1)}{3k_B T}. \tag{1.70}$$

The static longitudinal susceptibility is thus always positive (at any positive spin temperature), which means that the thermal equilibrium magnetization is always parallel to the magnetic field, according to Eq. 1.65. The spin polarization, as defined in Eq. 1.62, has then the same sign as the spin temperature when the gyromagnetic ratio γ is positive, and opposite sign when γ is negative; this is the consequence of defining a temperature to describe the energy level population ratios for the magnetic moment, the orientation of which can be parallel or antiparallel to the spin angular momentum vector.

In the case of high spin densities with large magnetic moments, the magnetization and therefore the static susceptibility deviate slightly from the Curie law. The magnetization of the spins I (we assume that there are no other spin species in the sample) is given by the exact form of Eq. 1.64, where the P_I at high temperature is defined exactly by Eq. 1.69. The Larmor precession frequency depends on the internal static field $B_{int} = \mu_0 H_{int}$, where[8]

---

[6] The negative sign in Eq. 1.39 follows from the need to have the lowest magnetic level of the electron (which has negative γ) occur when $m = -1/2$. This is entirely arbitrary and has historic origin.

[7] This assumption is not done unless stated explicitly.

[8] This ignores the fact that for like spins the molecular field is higher by a factor of 3/2, resulting from the correct treatment of the dipolar fields and interactions. Furthermore, the discussion presented here disregards the facts that crystal structure influences the molecular field, and that the sample shape also contributes to the internal field. These will be discussed in Sections 1.3.4 and 1.3.6, and in Chapters 2 and 5. In addition, the influence of the interactions of nuclear moments with the electrons is ignored.

$$H_{int} = H + M, \quad (1.71)$$

which gives

$$\omega_0 = -\gamma B_{int} = -\gamma \mu_0 H_{int}.$$

Using these we can rewrite the magnetization of Eq. 1.64:

$$M = \frac{T_C}{T} H_{int},$$

where we have defined the Curie temperature T_C

$$T_C = \frac{\mu_0 n_I \hbar^2 \gamma_I^2 I(I+1)}{3k_B}. \quad (1.72)$$

The static susceptibility can now be solved

$$\chi_0 = \frac{M}{H} = \frac{M}{H_{int} - M} = \frac{T_C}{T - T_C}. \quad (1.73)$$

We remind that it is required that $T \gg T_C$ and that $P \ll 1$ in order that Eq. 1.73 be valid. These conditions are not always met in polarized targets, as we shall discuss below. Equation 1.73 resembles closely the Curie–Weiss law, with the difference that no molecular field coefficient was introduced in the relation between the static magnetization and the internal field in Eq. 1.71.

We note that if we write in Eq. 1.69 $\omega_0 = \gamma B_0$, the equation is exact for all situations where $P \ll 1$, because the Larmor frequency measures the internal field rather than the applied field. It offers a sound basis for the measurement of spin polarizations via calibration in thermal equilibrium at high temperature. This will be discussed in Chapter 6.

It is instructive to examine the size of the correction of Eq. 1.73 to the Curie law 1.70 in polarized target materials. Table 1.1 gives T_C in a typical material 1-butanol doped with 5% of water and 10^{20} spins/cm^3 of a paramagnetic complex EHBA-Cr(V):

Table 1.1 Parameters related with the laws of Curie and of Curie–Weiss at 2.5 T magnetic field, for 1-butanol glass doped with 10^{20} spins/cm^3 of a paramagnetic complex EHBA-Cr(V). For the nuclear spins, the last column also gives the thermal equilibrium polarization at 1 K temperature.

Particle	Spin	$\omega_0/2\pi$ (Hz)	n_I (spins/cm^3)	T_C (K)	C (Eq. 1.69) (K)
Cr(V)	1/2	69.3×10^9	1.0×10^{20}	7.676×10^{-4}	1.6629
proton	1/2	106.5×10^6	0.79×10^{23}	1.427×10^{-6}	2.5508×10^{-3}
deuteron	1	16.35×10^6	0.79×10^{23}	9.002×10^{-8}	5.2311×10^{-4}

We see that the Curie temperature causes insignificant correction for the nuclear susceptibilities at 1 K temperature. The same is not exactly true for the electronic spins, but because their polarization is nearly complete at or below 1 K temperature, in a 2.5 T field, the Curie–Weiss law becomes insignificant. None of the spin systems has a risk of undergoing spontaneous magnetization, because the range of the temperatures in the polarized targets is well above T_C. Note, however, that if some phase separation and crystallization of the paramagnetic compound would occur, there could be a magnetic phase transition in the microcrystals, if the molecular field effect would be strong in them.

The last column in Table 1.1 lists the coefficient C of Eq. 1.69 which gives the nuclear spin polarization $P = C/T$ at high temperature in 2.5 T field. If the Curie–Weiss law is replaced by the Curie law, the error in the proton polarization in thermal equilibrium at 1 K is 1.4 ppm, which is insignificant in view of the accuracy with which the lattice temperature can be determined.

1.2.2 Density Matrix

The above equations are practical for handling static situations in high field and at any temperature (apart from the cases where high temperature was specifically required). For spin dynamics, involving time-dependent fields and interactions among the spins, the density matrix techniques[9] offer suitable tools and allow to also define the concept of spin temperature in the low effective field $\mathbf{B}_e$ (equivalent of Eq. 1.57 for single spin) of the rotating frame. It also enables the rigorously correct handling of the calculation of saturation, lineshapes and relaxation times in solids.

The density matrix ρ of a system which is an assembly of a large number N of constituents is defined most generally by its elements

$$\rho_{ij} = \int \psi_i \psi_j \, d\tau, \tag{1.74}$$

where integration is over the space in order to average over the ensemble. The expectation value of an operator Q can be shown to be

$$\langle Q \rangle = \mathrm{Tr}\{\rho Q\}, \tag{1.75}$$

and the time variation of the density matrix obeys the von Neumann–Liouville equation

$$i\hbar \frac{d\rho(t)}{dt} = [\mathcal{H}, \rho(t)], \tag{1.76}$$

which can be obtained directly from the Schrödinger equation. In case of a time-independent Hamiltonian, writing the density matrix $\rho(0)$ at $t = 0$, $\rho(t)$ is then given by

---

[9] For more complete introductory discussion, see Ref. [4].

$$\rho(t) = e^{-i\mathcal{H}t/\hbar} \rho(0) e^{+i\mathcal{H}t/\hbar}, \qquad (1.77)$$

which leaves the eigenvalues and diagonal elements of $\rho(t)$ constant in time. The off-diagonal elements of $\rho(t)$ exhibit undamped oscillatory behavior.

Averaging over the lattice variables can be assumed in many cases, so that $\rho(t) = \text{Tr}_{\text{lattice}} \{\rho_{\text{spin+lattice}}(t)\}$ involves only the spin variables of the system. In evaluating the dipolar interaction energy, the sum over the lattice variables is, however, mostly left explicitly in the expressions.

We may use often the following relationships for calculating the steady-state equilibrium of the observable Q:

$$\frac{d\langle Q \rangle}{dt} = \left\langle \frac{i}{\hbar}[\mathcal{H}, Q] + \frac{\partial Q}{\partial t} \right\rangle, \qquad (1.78)$$

which is easily derived from Eqs. 1.75 and 1.76.

If the Hamiltonian does not have explicit time dependence, i.e.

$$\frac{\partial \langle \mathcal{H} \rangle}{\partial t} = 0,$$

which states that the forces of the system are conservative, then it follows directly from Eq. 1.78 that

$$\frac{d \langle \mathcal{H} \rangle}{dt} = 0 \qquad (1.79)$$

which states that the total energy is conserved.[10]

The Boltzmann (or canonical) distribution is obtained by maximizing the entropy

$$S = -k_B \, \text{Tr} \, \{\rho \ln \rho\} \qquad (1.80)$$

at constant energy $E = \text{Tr} \{\rho \mathcal{H}\}$. The resulting density matrix is

$$\rho = Z^{-1} e^{-\frac{\mathcal{H}}{kT}}, \qquad (1.81)$$

where Z is called the partition function which performs the normalization $\text{Tr}\{\rho\} = 1$ and is given by

$$Z = \text{Tr}\left\{ e^{-\frac{\mathcal{H}}{kT}} \right\}. \qquad (1.81a)$$

The density matrix has the diagonal elements

$$\rho_i = A e^{-\frac{\mathcal{H}_i}{kT}}. \qquad (1.82)$$

---

[10] See Ref. [1] p. 125 or Ref. [14] p. 34.

1.2 Quantum Statistics of a System of Spins

where we have written

$$\mathcal{H}_i = \langle i | \mathcal{H} | i \rangle \tag{1.83}$$

using the eigenfunctions $|i\rangle$ of $\mathcal{H}$.

The matrix ρ is also sometimes called the spin polarization matrix; the resemblance of its diagonal elements in Eq. 1.82, with the classical expression of magnetic level population and polarization of Eq. 1.62, is obvious.

Throughout this book we are interested in high polarizations, which require high field and low temperature. Most of the results of spin dynamics in the literature are obtained by using the so-called high-temperature approximation of the density matrix

$$\rho = Z^{-1}\left[1 - \frac{\mathcal{H}}{k_B T}\right], \tag{1.84}$$

which requires that the first term in the Taylor expansion be small; this is obviously not valid at low temperatures. However, this linearized density matrix allows to calculate explicit rate equations for the transient behavior of spin temperatures. Experiments in the low-temperature regime have shown that in many favorable cases the conclusions based on the high-temperature approximation 1.84 are valid also at low temperatures, at least qualitatively.

The strength of the density matrix formulation of the statistical state of an array of spins lies in the fact that it allows the definition of the spin temperature in the rotating frame in which the transverse RF field is static. The spin temperature may be very much different from the lattice temperature of the solid material. This concept has been verified many times experimentally and it is the only way to theoretically explain many subtle phenomena in NMR and EPR. We shall frequently return to this in the Chapters 2–5.

We note that the spin temperature can be either positive or negative because the energy spectrum of the eigenvalues of the spin Hamiltonian is limited upwards. The diagonal elements of the density matrix are always positive and sum up to 1, whereas the off-diagonal elements are smaller by several orders of magnitude and oscillate in time at a frequency close to the Larmor frequency.

To illustrate the use of the density matrix in explicit form, we shall calculate the polarization $P = \langle I_z \rangle / I$ of Eq. 1.60 for an assembly of N identical spins $I_i = I$:

$$P = \frac{\langle I_z \rangle}{NI} = \frac{\left\langle \sum_{i=1}^{N} I_z^i \right\rangle}{NI}.$$

These nuclear spins are bound in a solid medium with which they are in thermal equilibrium, but they are not interacting strongly with each other (the spin density can be thought to be very small). Adding up the individual spin Hamiltonians of Eq. 1.33, we get then the simple total Hamiltonian

$$\mathcal{H}_{tot} = \hbar\omega_0 \sum_i I_z^i,$$

where the lattice interaction part and the spin-spin interaction part are assumed to be negligibly small (all relaxation times are long but not infinite). Before evaluating the density matrix of Eq. 1.81 we must average the total Hamiltonian over the ensemble of the N spins (which is equivalent to averaging over the lattice variables), because the state of any spin i can be regarded as independent of the states of all other spins:[11]

$$\mathcal{H}_0 = \frac{1}{N}\mathcal{H}_{tot} = \frac{1}{N}\sum_{i=1}^{N}\hbar\omega_0 I_z^i = \hbar\omega_0 I_z,$$

where the last step is not only a shorthand notation but means that the z-component of averaged spin matrix can be used to represent spin in the Hamiltonian. The density matrix is now

$$\rho = \frac{e^{-\mathcal{H}_0/kT}}{\text{Tr}\{e^{-\mathcal{H}_0/kT}\}} = \frac{e^{-\alpha\omega_0 I_z}}{\text{Tr}\{e^{-\alpha\omega_0 I_z}\}}. \tag{1.85}$$

We may now calculate the expectation value of I_z using Eq. 1.75:

$$\left\langle \sum_{i=1}^{N} I_z^i \right\rangle = \text{Tr}\left\{\rho \sum_{i=1}^{N} I_z^i\right\} = \frac{\text{Tr}\left\{e^{-\alpha\omega_0 I_z}\sum_{i=1}^{N} I_z^i\right\}}{\text{Tr}\{e^{-\alpha\omega_0 I_z}\}}. \tag{1.86}$$

We note that $\left[I_z^i, I_z^j\right] = 0$ because the spins do not interact and use Eqs. 1.44–1.46 for rearranging the terms under the trace so that we may evaluate the diagonal components.

To obtain the diagonal elements of Eq. 1.82 we shall consider the eigenvalue m of I_z in eigenstate ψ_m for one of the independent spins:

$$I_z\psi_m = m\psi_m. \tag{1.87}$$

By applying the operator I_z $n-1$ times from the left, we get

$$I_z^n\psi_m = m^n\psi_m; \tag{1.88}$$

for all functions $f(aI_z)$, which can be expanded into a Taylor polynomial, we shall then find

$$f(aI_z)\psi_m = f(am)\psi_m \tag{1.89}$$

---

[11] This follows from the assumption that the spins are not strongly interacting. If we do not make such an assumption, the interaction piece must be also included in the Hamiltonian. This situation may be valid at very low spin temperatures, corresponding to nearly complete spin polarization. The argument has been developed using the notions of basic quantum mechanics in Ref. [6], pp. 157–160.

and

$$f(aI_z)I_z\psi_m = f(am)m\psi_m. \qquad (1.90)$$

By inserting these under the traces of Eq. 1.67, with $a = -\alpha\omega_0$ and

$$f(aI_z) = e^{-\alpha\omega_0 I_z}$$

we get

$$P = \frac{N\mathrm{Tr}\{I_z e^{-\alpha\omega_0 I_z}\}}{NI\mathrm{Tr}\{e^{-\alpha\omega_0 I_z}\}} = \frac{\sum_{m=-I}^{+I} m e^{-\alpha\omega_0 m}}{I\sum_{m=-I}^{+I} e^{-\alpha\omega_0 m}}, \qquad (1.91)$$

which is identical to Eq. 1.62.

Although this result can be derived in many simpler ways, the above one is based on the formal quantum statistics, and it demonstrates several key features of the density matrix techniques which will be needed later on in this book. One of these is that there was no need to resort to the 'high temperature' approximation to obtain the result; the reason why we found no complexity arising from the exponential function was that the operator in question (I_z) commutes with the entire Hamiltonian, and therefore with the density matrix.

In high external field, the above total Hamiltonian, the so-called Zeeman energy Hamiltonian, is usually much larger than the terms describing the spin-spin interactions and the lattice interactions. These can then be considered as perturbations to the main problem, which can be solved first using the simplified density matrix. Further examples on the use of the density matrix will be given in the following chapters.

1.3 Thermodynamics of Spin Systems

1.3.1 Thermodynamic Functions and Measurement of Spin Temperature

Let us consider a system of N_I spins I which is sufficiently isolated from its surroundings so that it can reach an internal thermodynamic equilibrium at an inverse spin temperature

$$\beta = \frac{\hbar}{k_B T}. \qquad (1.92)$$

The equilibrium entropy per spin, corresponding to the maximum of Eq. 1.80, is

$$\begin{aligned}\frac{S}{N_I k_B} &= -\mathrm{Tr}\{\rho\ln\rho\} = -\mathrm{Tr}\left\{\rho\ln\frac{\exp(-\mathcal{H}/k_B T)}{Z}\right\}, \\ &= \frac{1}{k_B T}\mathrm{Tr}\{\rho\mathcal{H}\} + \ln Z\,\mathrm{Tr}\{\rho\} = \frac{\beta}{\hbar}\langle\mathcal{H}\rangle + \ln Z\end{aligned} \qquad (1.93)$$

where the partition function Z was given by Eq. 1.81 and the expectation value of the energy, obtained from Eq. 1.75, is

$$\langle \mathcal{H} \rangle = \mathrm{Tr}\{\rho \mathcal{H}\}. \tag{1.94}$$

The fact that entropy must always be positive, regardless of the spin temperature for example, follows directly from its quantum statistical definition. This is extremely general and independent of the nature of the subsystem, because the entropy is proportional to the logarithm of the number of quantum states which must be always greater than unity.[12] Similarly, the quantum statistics directly gives the result that entropy is an additive quantity, i.e. that the entropy of a composite system is the sum of the entropies of its parts. Moreover, the maximum entropy of a spin system corresponds to totally disordered spins which have all $(2I+1)^{N_I}$ diagonal elements of the total density matrix equal to 1, yielding therefore the maximum spin entropy $S/k_B = \ln(2I+1)^{N_I} = N_I \ln(2I+1)$.

We have shown in detail the steps leading to the final result of Eq. 1.93 because in the literature this formula is sometimes given with different signs. We have chosen the present formula as it makes evident that the first term is always negative, whereas the second term is always positive because Z is always greater than 1.

The partition function Z of Eq. 1.81 alone can be used for deriving all thermodynamic variables of a quantum statistical system. The Gibbs free energy, for example, is

$$G = -k_B T \ln Z. \tag{1.95}$$

The change of the free energy is the work done on the body in a reversible isothermal process. The free energy is related to the total energy and entropy by

$$G = E - TS, \tag{1.96}$$

which can be obtained directly from Eqs. 1.93 and 1.95, and the entropy is

$$\frac{S}{k_B} = \frac{\partial}{\partial T}(T \ln Z) = \ln Z + \frac{T}{Z}\frac{\partial Z}{\partial T}, \tag{1.95'}$$

which is easily verified to give the same result as Eq. 1.93.

The absolute temperature, or simply the temperature, of a body is defined by the second law of thermodynamics

$$\frac{1}{T} = \frac{dS}{dE}. \tag{1.97}$$

The temperature as well as entropy are purely statistical quantities and are meaningful only for macroscopic bodies. Definition 1.97 can be used for the experimental determination of the spin temperature by supplying a known quantity ΔQ of heat to an isolated spin system and measuring the corresponding change ΔS in entropy, yielding

$$T = \frac{\Delta Q}{\Delta S}. \tag{1.98}$$

---

[12] For a more complete account, see the excellent discussion on entropy by Landau and Lifshitz [14], pp. 22–32.

The known quantity of heat can be supplied by NMR techniques which will be discussed in subsequent chapters, whereas the entropy can be measured from the nuclear polarization at such high field that the Zeeman energy of Eq. 1.59 dominates the interaction energy. The transition to the high field and back must be done adiabatically, i.e. sufficiently fast so that the system does not exchange too much energy with the surroundings, but also slowly enough so that the internal thermal equilibrium is conserved in the spin system during the field ramp.

The specific heat C_B at constant magnetic field is obtained from the entropy

$$C_B = T \frac{\partial S}{\partial T}\bigg|_B. \tag{1.98'}$$

This parameter is important for magnetic cooling applications.

The above treatment of the thermodynamic functions and relations is extremely general, although not very transparent. The relations derived up till now can be used for handling all interactions, even in the case of phase transitions. We shall now move to dealing with the more specific and simpler high-field situation where more explicit results can be obtained, valid also at low spin temperatures.

1.3.2 Entropy in High Magnetic Field

When the Zeeman energy is large in comparison with the interaction energy so that the latter can be ignored, the density matrix is given to a good approximation by

$$\rho = Z^{-1} \exp(-\beta \omega_0 I_z) \tag{1.99}$$

and the partition function is

$$Z = \text{Tr}\{\exp(-\beta \omega_0 I_z)\}, \tag{1.100}$$

which satisfies the requirement $\text{Tr}\{\rho\} = 1$. Using these and the definition of entropy 1.80 we find

$$\begin{aligned}\frac{S}{k_B} &= -\text{Tr}\{\rho(-\beta \omega_0 I_z - \ln Z)\} = \beta \omega_0 \langle I_z \rangle + \ln Z \\ &= \beta \omega_0 IP + \ln\left(\sum_{m=-I}^{I} \exp(-\beta \omega_0 m)\right).\end{aligned} \tag{1.101}$$

By recalling from Eq. 1.39 that $\omega_0 = -\gamma B_0$, and by using the same definition of $x = -\beta \omega_0 / 2$ as in the Brillouin function of Eq. 1.63, the expression for the high-field entropy per spin becomes

$$\frac{S}{k_B} = -2IxP + \ln\left(\sum_{m=-I}^{I} \exp(2xm)\right). \tag{1.102}$$

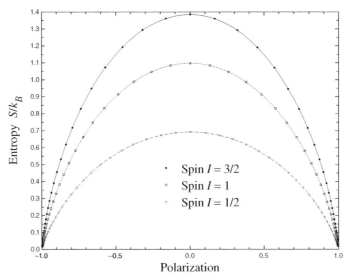

Figure 1.5 Entropy of non-interacting spin systems in internal thermal equilibrium, in a high field and in the absence of possible magnetic phase transitions that can be expected in low effective field at polarizations approaching 1 or −1

We note that if γ changes sign, then both x and P change sign and therefore the first term in Eq. 1.102 stays always negative. The same is true when the spin temperature changes sign.

It is straightforward but tedious to show that the sum under the logarithm can be expressed in the terms of the hyperbolic functions:

$$\sum_{m=-I}^{I} \exp(2xm) = \frac{\sinh(2I+1)x}{\sinh x}. \qquad (1.103)$$

Inserting this and using in the Brillouin function we may finally write the high-field entropy explicitly

$$\frac{S}{k_B} = x\left(\coth x - (2I+1)\coth(2I+1)x\right) + \ln\frac{\sinh(2I+1)x}{\sinh x}. \qquad (1.104)$$

The relation between the entropy and polarization must then be solved by eliminating x from Eqs. 1.63 and 1.104 numerically, if the entropy must be determined from the measured high-field polarization. As was said above, this is the basis for the measurement of the absolute spin temperature when an accurate secondary thermometer is not available.

At high temperatures, the first term tends to zero as $-x^2$, whereas the second term converges to $\ln(2I+1)$. This latter result can also be inferred directly from the definition of the entropy of Eq. 1.80, because the diagonal elements of the density matrix tend to 1 when the temperature goes to infinity, and its off-diagonal elements are 0 in equilibrium. The upper bound of the entropy is therefore $\ln(2I + 1)$, i.e. the logarithm of the number of possible quantum states that the spin I spin can occupy.

For spin 1/2 the entropy can be expressed in the terms of polarization directly:

$$\frac{S}{k_B} = \ln 2 - \frac{1}{2}\left(\ln(1-P^2) + P\ln\frac{(1+P)}{(1-P)}\right). \tag{1.105}$$

Figure 1.5 shows the high-field entropy per spin for the cases of $I = 1/2$, 1 and 3/2. The entropy per spin reaches its maximum of $\ln(2I + 1)$ at $P = 0$. We see that the entropy becomes zero for both positive and negative maximum polarization as can be expected, because both cases correspond to maximal order in the spin system.

1.3.3 Interacting Spin System in the High-Temperature Approximation

The high-temperature approximation can be used in situations where the exponential in the density matrix of Eq. 1.81 is so small that it can be accurately described with its Taylor series truncated after the first two terms:

$$\rho \cong \frac{1}{Z}(1 - \beta\mathcal{H}), \tag{1.106}$$

where **1** is a unity matrix of the same dimension as the Hamiltonian. This situation arises when β is small enough, i.e. the temperature must be high. In this case the energy of the spin system is

$$\langle\mathcal{H}\rangle = \text{Tr}\{\rho\mathcal{H}\} = \frac{1}{Z}\text{Tr}\left\{\mathcal{H} - \frac{\beta}{\hbar}\mathcal{H}^2\right\} = -\frac{\beta}{\hbar(2I+1)}\text{Tr}\{\mathcal{H}^2\}, \tag{1.107}$$

where we have used the knowledge that $\text{Tr}\{\mathcal{H}\} = 0$ and that $Z = 2I + 1$ at high temperatures. The entropy is obtained by inserting this energy to Eq. 1.93:

$$\frac{S}{N_I k_B} = -\text{Tr}\{\rho\ln\rho\} = \ln Z + \frac{\langle\mathcal{H}\rangle}{k_B T} = \ln(2I+1) - \frac{\beta^2}{\hbar^2 Z}\text{Tr}\{\mathcal{H}^2\}. \tag{1.108}$$

In this expression we have again emphasized the fact that the entropy per spin is evaluated, by showing N_I explicitly on the left; this is done because the spin density enters in the interaction piece which is a part of the general spin Hamiltonian.

We shall now introduce spin interactions, although their detailed discussion will be left to Chapter 2. Let us take the important case where the dominant spin interaction is due to the magnetic dipolar force between the nuclei. The Hamiltonian is then

$$\mathcal{H} = \mathcal{H}_z + \mathcal{H}_1 = -\hbar\gamma B I_z + \mathcal{H}_D, \tag{1.109}$$

where $\mathcal{H}_D$ is the so-called dipolar Hamiltonian which describes the interaction of the dipole moment with the dipolar field. The dipolar field itself is the sum of magnetic fields arising from the neighboring magnetic dipoles and has both static and fluctuating components. With this the trace of the squared Hamiltonian becomes

$$\text{Tr}\{\mathcal{H}^2\} = \hbar^2\gamma^2 B^2 \text{Tr}\{\hat{I}_z^2\} + \text{Tr}\{\mathcal{H}_D^2\} = \hbar^2\gamma^2 \text{Tr}\{\hat{I}_z^2\}\left(B^2 + B_{loc}^2\right), \qquad (1.110)$$

where we have defined the local field B_{loc}

$$\frac{B_{loc}^2}{B^2} = \frac{\text{Tr}\{\mathcal{H}_D^2\}}{\text{Tr}\{\mathcal{H}_Z^2\}} = \frac{\text{Tr}\{\mathcal{H}_D^2\}}{\hbar^2\gamma^2 B^2 \text{Tr}\{\hat{I}_z^2\}}, \qquad (1.111)$$

which measures the mean strength of the dipolar field. This field and its evaluation will be discussed in more detail in the next chapter, and we only need to know here that B_{loc} has a value around 10^{-4} T for most nuclear spins with a high density typical of solid target materials.

The trace in the denominator can be evaluated immediately, based on the commutation rules and algebra of the spin operator. This is done by first evaluating the trace of the square of the spin operator $\text{Tr}\{\hat{I}^2\}$ (we now use explicitly the operator symbol to avoid confusion). From Eq. 1.25 we learned that for all magnetic states m the eigenvalue of $\hat{I}^2$ is

$$\hat{I}^2\psi = I(I+1)\psi, \qquad (1.112)$$

which means that the matrix representation of the squared spin vector has diagonal elements all of which have the magnitude $I(I+1)$. Because there are altogether $2I+1$ diagonal elements, the trace sums up to

$$\text{Tr}\{\hat{I}^2\} = I(I+1)(2I+1). \qquad (1.113)$$

On the other hand, we know that the trace of an operator is independent of the representation so that the traces of the squares of all spin components must be equal, from where it follows that

$$\text{Tr}\{\hat{I}_x^2\} = \text{Tr}\{\hat{I}_y^2\} = \text{Tr}\{\hat{I}_z^2\} = \frac{1}{3}I(I+1)(2I+1). \qquad (1.114)$$

The entropy can now be written, using the fact that at high temperatures $Z = 2I+1$,

$$\frac{S}{N_I k_B} = \ln(2I+1) - \frac{\beta^2\gamma^2 I(I+1)}{3}\left(B^2 + B_{loc}^2\right). \qquad (1.115)$$

To evaluate more easily the magnitude of the deviation of the entropy from its maximum value, we may finally insert the Curie temperature of Eq. 1.72:

$$\frac{S}{N_I k_B} = \ln(2I+1) - \frac{T_C}{T}\frac{1}{n_I k_B T}\frac{B^2 + B_{loc}^2}{\mu_0}, \qquad (1.116)$$

where n_I is the number density of the spins I. The entropy reduction from the maximum is seen to be the ratio of the magnetic and thermal energy densities, multiplied by the ratio T_C/T. The Curie temperatures, listed in Table 1.1, are around 1 µK, and it is therefore clear that extremely low spin temperatures are required for a substantial entropy reduction in external magnetic fields which are of the order of the local field B_{loc}. At high field and high polarization, however, the entropy may be substantially reduced from the maximum, although Eq. 1.116 is not exactly valid then.

The specific heat is now obtained from Eq. 1.116 using Eq. 1.98':

$$\frac{C_B}{N_I k_B} = \frac{T_C}{2n_I k_B T^2} \frac{B^2 + B_{loc}^2}{\mu_0}. \quad (1.116')$$

In the above it was implicitly assumed that the spin interactions feel the same temperature as the Zeeman energy reservoir. This requires that the Zeeman levels are broadened so that a sufficient overlap occurs, and is true only at rather low fields, of the order of B_{loc}. The internal thermal equilibrium of the two reservoirs is established in a time scale of T_2, the decay time of the free-precession NMR signal, whereas the Zeeman and dipolar reservoirs relax towards the lattice temperature at very much slower rates T_{1Z} and $T_{1D} \approx T_{1Z}/3$, respectively. When the magnetic field is changed, the Zeeman reservoir is cooled or heated if the field change happens at a constant entropy, whereas the dipolar reservoir is not influenced by the field change. Similarly, if an adiabatic demagnetization or magnetization is made in the rotating frame at high enough effective field, the dipolar reservoir is unaffected, whereas the Zeeman reservoir is cooled or heated. The specific heat which is experimentally measurable (by using a resonant method) is then

$$\frac{C_B}{N_I k_B} = \frac{T_C}{2n_I k_B T^2} \frac{B^2}{\mu_0}. \quad (1.116'')$$

Goldman [4] and Abragam and Goldman [5] discuss the quasi-equilibrium of the spin systems at high fields and make the reasonable assumption, well verified by several types of experiments, that the systems can be described by the density matrix with two temperatures

$$\rho = \frac{\exp(-\beta_Z \omega_0 I_z - \beta \mathcal{H}'_D)}{\text{Tr}\{\exp(-\beta_Z \omega_0 I_z - \beta \mathcal{H}'_D)\}}. \quad (1.117)$$

In the high field $\mathcal{H}_D$ is replaced by the so-called secular part of the dipolar Hamiltonian $\mathcal{H}'_D$, which has the important property that it commutes with the Zeeman Hamiltonian. The reasons for this will be discussed in Chapter 2.

At high temperatures and fields, the Hamiltonian and density matrix become

$$\mathcal{H} = \hbar \omega_0 I_z + \mathcal{H}'_D;$$
$$\rho = \frac{1 - \beta_Z \omega_0 I_z - \beta \mathcal{H}'_D}{Z} \quad (1.118)$$

and in the above low-field equations the squared local field must be replaced by

$$B'^2_{loc} = \frac{\text{Tr}\{\mathcal{H}'^2_D\}}{\hbar^2 \gamma^2 \text{Tr}\{I_z^2\}}. \quad (1.119)$$

Let us now consider the non-adiabaticity arising from the dipolar interactions when the field is lowered to such a value that thermal contact is established between the reservoirs. This field B_m is called the 'mixing field' and was discussed in [15]. Demagnetization is started from the initial field and temperature B_i and β_i, and the two reservoirs are initially at the same temperature. During the field ramp the Zeeman temperature is evolving as $\beta_z = \beta_i B_i / B$ until the ramp is stopped at the mixing field value and the two inverse temperatures are allowed to equalize again to a common final value β_f, which is obtained by requiring that during the thermal mixing the total energy of the spin system must stay constant:

$$\text{Tr}\{\mathcal{H}\} = \text{Tr}\left\{\left(1 + \beta_i \frac{B_i}{B_m}\omega_m I_z - \beta_i \mathcal{H}'_D\right)\mathcal{H}\right\} = \text{Tr}\left\{\left(1 + \beta_f \omega_m I_z - \beta_f \mathcal{H}'_D\right)\mathcal{H}\right\}. \quad (1.120)$$

Ignoring the small term $\beta_i \mathcal{H}'_D$ it is straightforward to calculate the final inverse temperature [15]

$$\beta_f = \frac{\beta_i B_i B_m}{B_m^2 + B'^2_{\text{loc}}} = \beta_i \frac{B_i}{\sqrt{B_m^2 + B'^2_{\text{loc}}}} \times \frac{B_m}{\sqrt{B_m^2 + B'^2_{\text{loc}}}}, \quad (1.121)$$

where the last factor describes the non-adiabaticity which in materials such as silver can be about 0.99 at $B_m = 0.25\,\text{mT}$ [15]. The losses in polarization and temperature in this case are 1%, and the loss of entropy is 2% because of its T^{-2} dependence. Such non-adiabaticities can lead to important corrections in the measurement of the absolute temperature which requires the cycling between high and low fields.

If thermal mixing is performed after a first demagnetization and the Zeeman reservoir is then again cooled at high field before a second demagnetization, the dipolar reservoir stays cold during this field cycle. A second thermal mixing then results in a much lower loss in polarization or temperature because the dipolar reservoir already was much colder.

Similar experimental findings have been made in the rotating frame. The non-adiabaticities here have a practical importance for the polarization losses occurring during the adiabatic reversal of polarization.

1.3.4 Thermodynamics at High Polarization

At low spin temperature, which is synonym of high polarization and strong reduction of entropy, the handling of entropy becomes more complicated, both conceptually and in the terms of the mathematical tools. Abragam and Goldman [5] introduce the density matrix 1.117 and the partition function

$$Z(\beta_z, \beta) = \text{Tr}\left\{\exp\left(-\beta_z \omega_0 I_z - \beta \mathcal{H}'_D / \hbar\right)\right\} \quad (1.122)$$

with two inverse temperatures. These are the Lagrange multipliers of the maximization of the entropy while keeping energy constant. The partition function yields the Zeeman and dipolar energies

$$\hbar\omega_0\langle I_z\rangle = \hbar\omega_0\mathrm{Tr}\{\rho I_z\} = -\hbar\frac{\partial}{\partial\beta_Z}\ln Z(\beta_Z,\beta);\qquad(1.123)$$

$$\langle\mathcal{H}'_D\rangle = \mathrm{Tr}\{\rho\mathcal{H}'_D\} = -\hbar\frac{\partial}{\partial\beta}\ln Z(\beta_Z,\beta).\qquad(1.124)$$

The problem is that both energies depend on both temperatures, although in the high-field region the two temperatures are separately constants of motion. This results from the fact that the two Hamiltonians depend on the same spin variables, and therefore their expectation values are not independent of each other, except in the high-temperature limit where the dipolar energy stays practically constant. We are therefore not dealing with the simpler situation where there are two independent heat reservoirs.

In addition to the energies, the partition function 1.122 yields again all thermodynamic quantities such as the entropy

$$\frac{S}{k_B} = -\mathrm{Tr}\{\rho\ln\rho\} = \ln Z(\beta_Z,\beta) + \beta_Z\omega_0\langle I_z\rangle + \beta\langle\mathcal{H}'_D\rangle/\hbar.\qquad(1.125)$$

A diagrammatic method allows to calculate the expansion of these in the terms of β [5] for a given lattice symmetry and sample shape. These calculations are beyond our present scope, and we recommend reading the impressive results on the non-linear effects on the dipolar energy, spin polarization, free energy, entropy and transverse susceptibility obtained by the team of Abragam and Goldman [5].

1.3.5 Cooling by Adiabatic Demagnetization of a Spin System

The adiabatic demagnetization of a paramagnetic electron spin system was the primary method for reaching the millikelvin temperature region until the end of 1960s when dilution refrigeration became a practical method of continuous cooling. At the same time superconducting magnet technology had rapid progress and fields over 5 T in large volumes became soon available. These enabled the practical application of the nuclear magnetic cooling which requires precooling at temperatures below 20 mK and fields in excess of 5 T.

Interestingly, the theoretical understanding and practical tools for the quantum statistical treatment of spin systems under strong saturation also developed at the end of the 1960s [16]. These led to practical equipment for the DNP yielding spin temperatures of the order of few millikelvin in 2 T to 5 T fields, and to the understanding of the behavior of the spin temperature in the rotating frame under strong saturation. The latter gave the basis for the adiabatic demagnetization in the rotating frame (ADRF).

This wealth of methods offered interesting possibilities for the exploration of the untouched land of nuclear magnetism which could now be approached by two different schemes: (1) Demagnetization of a metallic sample precooled itself in a high field to a temperature below 1 mK by the demagnetization of a large nuclear refrigerator;

and (2) demagnetization in the rotating frame of the nuclear spins in a dielectric sample, pre-cooled by dynamic nuclear cooling to a spin temperature of a few millikelvin at high field.

The reversal of polarization after demagnetization was of interest to the research teams undertaking work in nuclear magnetism. In static field this can be accomplished by a diabatic field reversal, and in the rotating frame it can be obtained by a fast passage through the resonance line. It has also a practical potential, although yet largely unexplored, for polarized targets.

Adiabatic Demagnetization in the Laboratory Frame and Cooling of Conduction Electrons

Equation 1.104 shows that when the magnetic field is changed adiabatically (at constant entropy) in the high-field region, the ratio B/T must stay constant and the temperature must decrease proportional to the field. The strongly reduced nuclear spin entropy therefore can be used for cooling the electrons in a metal, if the thermal contact between the conduction electrons and the nuclear spin system is good. This is the case in many metals such as copper. Moreover, the thermal conductivity of the electronic system is quite good so that copper and other similar metallic materials can be used for cooling samples attached to it.

The thermal contact between the nuclear spins and the conduction electrons is good because of the relatively fast spin relaxation time obeying the relation

$$\tau_1 = \frac{\kappa}{T_e}, \quad (1.126)$$

where κ is the Korringa constant and T_e is the temperature of the conduction electrons. The Korringa constant depends on the magnetic field at low field values by

$$\kappa(B) = \kappa_\infty \frac{B^2 + B_{loc}^2}{B^2 + \alpha B_{loc}^2}, \quad (1.127)$$

and also on temperature at very high polarization and low field [17].

When the demagnetization is carried out from the initial field and temperature B_i and T_i to the final field B_f, the final spin temperature is

$$T_f = \frac{B_f}{B_i} T_i, \quad (1.128)$$

assuming that the final field is so high that the thermal mixing between the Zeeman and dipolar reservoirs is slow. The nuclear spin system will then start to warm up due to the inevitable heat leak. The evolution of the spin temperature is obtained using the specific heat of the high-temperature approximation of Eq. 1.116″ by integration

$$t = \int_0^t dt = \int_{T_f}^T \frac{C_B}{\dot{Q}} dT = \frac{B_f^2 T_C}{2\mu_0 \dot{Q}/V} \left(\frac{1}{T_f} - \frac{1}{T} \right), \quad (1.129)$$

which yields a linear increase of the inverse spin temperature if the heat leak is constant. It is clear that the demagnetization should be stopped at the highest possible field which can yield the required temperature, because the experimental time depends on the square of the field value. Similarly, the control of the heat leak is a determining factor for the ultimate temperature which can be reached.

During this evolution the electron temperature will settle in a quasi-equilibrium with the spin temperature; assuming exponential relaxation these temperatures are related by [15]

$$\frac{T_e}{T} - 1 = \frac{2\kappa\mu_0 \dot{Q}/V}{B_f^2 T_C}. \tag{1.130}$$

The lowest T_e which can be reached is then equal to $2T$, and it is obtained by demagnetizing to

$$B_f^{opt} = \sqrt{\frac{2\kappa\mu_0 \dot{Q}}{T_C V}}. \tag{1.131}$$

When demagnetizing to a field where mixing with the dipolar (or interaction) reservoir takes place, the cooling of this reservoir causes some increase of entropy. In the high temperature approximation, the adiabatic efficiency of Eq. 1.121 was

$$\frac{\Delta S}{S} = 1 - \frac{B_m}{\sqrt{B_m^2 + B_{loc}'^2}}, \tag{1.132}$$

when thermal mixing between the Zeeman and dipolar energy reservoirs was made at the field B_m. An adiabatic sweep below this field is possible when the sweep speed is slow in comparison with the mixing time constant which is a rapidly decreasing function of decreasing B. A slow reversal of the field, however, would result in an adiabatic increase of temperature after zero crossing, without any possibility of obtaining negative temperatures.

Diabatic Reversal of Polarization in the Laboratory Frame

When a rapid field flip is made from a value B to $-B$ (opposite direction), in the high-temperature model the final temperature is related to the initial one by

$$T_f = -T_i \frac{B^2 + B_{loc}^2}{B^2 - B_{loc}^2}, \tag{1.133}$$

because only the Zeeman temperature is reversed and not the dipolar one; the equation is obtained by recalling that energy is conserved after the field flip. The flip is therefore followed by an irreversible increase of entropy when the Zeeman reservoir warms the interaction reservoir to a negative temperature. Experiments with Ag show [15] that for $P < 0.6$ the loss of polarization is 4% as predicted by Eq. 1.133 for a 250 μT flip field, but when the polarization is higher, the flipping efficiency is strongly reduced so that the final polarization cannot exceed -0.65 even for the initial polarization of 0.85. The decrease can be partly understood by the fact that the high-temperature approximation overestimates the heat capacity of the Zeeman reservoir at high P, while the heat capacity of the interaction reservoir may be underestimated. The irreproducibility of the results may also suggest that if the entropy is below the critical value, transitions between the magnetic phases can cause large irreversibility during the field flipping.

Adiabatic Demagnetization in the Rotating Frame (ADRF)

It should be first noted that in metals the ADRF is impractical because the high-frequency field does not penetrate the sample but rather heats up the electronic system. On the other hand, in dielectric samples the flipping or ramping of the static field, ramped first to a low value, is impractical because the sample must contain paramagnetic electronic spins which cause rapid nuclear spin relaxation at low magnetic fields; these paramagnetic spins are required for obtaining a high DNP.

The motion of an isolated spin in a high static field B_0 superposed with a small transverse rotating field B_1 could be understood as precession about an effective field B_e defined by Eq. 1.57. The isolated, i.e. non-interacting, spin can be manipulated reversibly by changing the effective field slowly in comparison with the effective Larmor precession frequency in the rotating frame. In the more practical case of interacting spins, the relaxation rates between the lattice, the Zeeman reservoir and the spin interaction reservoir must be taken into account in evaluating in the outcome of ADRF.

If the transverse field is small compared with the effective field, we have $B_e = (\Delta\omega/\omega_0)B_0$ as the effective field; if this field is large in comparison with the linewidth, no thermal mixing occurs between the Zeeman and dipolar reservoirs, and the two reservoirs can be at different temperatures in the time scale $t \approx T_1$, which can be hours or even days.

When the effective field approaches the local field, the situation changes. If in addition the strength of the transverse field is larger than that required for the linear approximation, but still small in comparison with the local field, one must use the Provotorov equations in order to determine the time evolution and equilibrium values of the temperatures of the various energy reservoirs. We shall therefore leave the detailed discussion of ADRF to the Chapter 2, where the dynamic behavior of interacting spins is better understood. Here we shall focus on the static results and on the similarities between the ADRF and adiabatic demagnetization in the laboratory frame.

Let us now perform ADRF from such a large effective field that no thermal mixing occurs, down to the mixing field

$$b_m = B_0 + \frac{\omega_m}{\gamma} = \frac{\omega_m - \omega_0}{\gamma}, \tag{1.134}$$

where the relaxation between the Zeeman and dipolar interaction energy reservoirs is reasonably fast in comparison with the spin-lattice relaxation. The relaxation takes place at a constant total energy, and the final inverse temperature, in the high-temperature approximation, is analogous with that of Eq. 1.121:

1.3 Thermodynamics of Spin Systems

$$\beta_f = \beta_i \frac{B_i}{\sqrt{b_m^2 + B_1^2 + B_{loc}'^2}} \times \frac{\sqrt{b_m^2 + B_1^2}}{\sqrt{b_m^2 + B_1^2 + B_{loc}'^2}}, \tag{1.135}$$

where the second term on the right describes again the non-adiabaticity due to the mixing. The losses in entropy can again be of the order of 1% in favorable cases. If the effective field is further reduced so slowly that the dipolar and Zeeman temperatures are always in good equilibrium with each other, the common inverse temperature undergoes reversible changes with the effective field:

$$\beta_f = \beta_i \frac{B_i}{\sqrt{b^2 + B_1^2 + B_{loc}'^2}} \times \frac{b_m}{\sqrt{b_m^2 + B_{loc}'^2}}, \tag{1.136}$$

where it was taken into account that the rotating field B_1 must be very small in comparison with both the local field and the mixing field.

The ADRF can be stopped at any effective field and the transverse field can be adiabatically reduced to zero. If this is done at the frequency ω_0, all available Zeeman order is transformed to dipolar order, and if the initial polarization is sufficiently high, magnetic ordering may take place in the nuclear spin system.

The above expressions are further slightly modified due to the finite relaxation rates between the various energy reservoirs, to be discussed in Chapter 2.

Polarization Reversal by Adiabatic Passage in Rotating Frame

If the frequency or field sweep is continued through zero effective field, the Zeeman temperature will be adiabatically reversed. In this process there is no adiabatic loss equivalent to that of the field flip given by Eq. 1.133, because in the rotating frame one can proceed adiabatically through zero effective longitudinal field without losses due to relaxation, provided that the spin-lattice relaxation times are much longer than the time spent in the passage.

In practice, the strength of the transverse field and the sweep rate of the steady field are optimized so that the losses due to relaxation are minimized. The polarization loss in the reversal by polarization reversal by adiabatic passage in rotating frame (APRF) is then reduced to that due to the loss of entropy when performing the thermal mixing, and to the loss due to other spins whose temperature remains untouched during the passage. At very high polarization the above results based on the high-temperature approximation are qualitatively valid, but nuclear magnetic phase transition phenomena may reduce the efficiency of the polarization reversal, as will be noted below.

The reversal can be performed starting from positive or negative polarization and spin temperature, and the sweep of the frequency or field can be started from above or below, with the same results. To reduce losses due to thermal mixing, however, it is best to perform thermal mixing at positive Zeeman temperature when the initial polarization is positive, and at negative Zeeman temperature when the initial polarization is negative.

A more complete discussion of the polarization reversal by APRF, involving the relaxation phenomena, will be given in Chapters 2 and 11.

1.3.6 Nuclear Magnetic Phase Transitions

The susceptibility of Eq. 1.73 diverges at temperature T_C which suggests that a nuclear spin system could undergo a magnetic phase transition in the vicinity of T_C. When a more precise estimation is made including the first non-linear term beyond the high-temperature approximation, the susceptibility reads [15]

$$\chi_L = \chi_0 \frac{1}{1-(R+L-D)\chi_0} = \frac{T_C}{T-(R+L-D)T_C}, \quad (1.137)$$

where $\chi_0 = T_C / T$, $L = 1/3$ is the Lorenz constant, and R is the strength of other possible interactions, notably the Ruderman–Kittel interaction and exchange interaction in metals. D is the demagnetizing factor in the direction of the external field, which depends on the shape of the sample. L and D depend on the dipolar interactions. The divergence can be considerably shifted from the simplest prediction, and phase transitions can happen at both positive and negative temperatures.

More refined calculations can be made taking into account the high polarization of the nuclei. These are based on the mean-field models, on the spherical model, on high-temperature expansion and other quantum spin theories and on Monte Carlo simulations, and yield realistic and even quantitative estimates for the antiferromagnetic transition temperature at zero field [15].

Lounasmaa and coworkers in Otaniemi have pioneered the study of nuclear magnetism in simple metals such as Cu, Ag and Rh, reviewed in [15]. They have determined the experimental zero-field transition temperatures $T_N = 58 \pm 10$ nK for natural Cu and 0.56 ± 0.06 nK for natural Ag. The experimental critical fields are $270 \pm 10\,\mu$T and $100 \pm 10\,\mu$T, respectively, extrapolated to $T = 0$ K. The lower and upper critical entropies for the transition which has a metastable coexistence of paramagnetic and antiferromagnetic phases are $S_{c1} = (0.48 \pm 0.04)\,k_B \ln 4$ and $S_{c2} = (0.61 \pm 0.03)\,k_B \ln 4$ for Cu, both isotopes of which have $I = 3/2$.

In Ag at negative temperatures, ferromagnetic ordering is expected, and this has been observed below the experimental critical temperature of $T_c = -1.9 \pm 0.4$ nK. Monte Carlo simulations give a theoretical value close to this. The critical entropy for obtaining the ferromagnetic transition is $S_c = 0.82\,k_B \ln 2$ (both Ag isotopes with spin have $I = 1/2$).

Lounasmaa and Oja [15] also reviewed the work done by other groups on nuclear magnetism in Tl, Sc and AuIn$_2$ where the exchange interaction is strong. This should lead to higher transition temperatures and makes the spin-lattice relaxation so fast that the conduction electrons are in thermal equilibrium with the nuclear spins. While the interpretation of the results for Tl and Sc are difficult, AuIn$_2$ shows a first-order ferromagnetic phase transition at a surprisingly high Curie point of $T_C = 35\,\mu$K [18].

The nuclear magnetic ordering in the rotating frame was first observed in CaF_2 by the Saclay group in 1969 [19, 20]. A transition to ordered state was observed in the NMR dispersion signal when ADRF was performed to zero effective field, with initial polarizations $-P \gg 0.3$. It was not possible to determine the transition temperature in these experiments, but it was found that when the steady field was along the |100| crystal axis, antiferromagnetic order resulted at $T < 0$, whereas a ferromagnetic order with sandwich domain structure was found for |111| direction. For positive temperatures a transition to helical order was found with field along the |111| axis. Neutron scattering experiments at Saclay with LiH similarly revealed an antiferromagnetic transition, with $T_N = -1.1\,\mu K$ [21–23]. The Leiden group [24] studied $Ca(OH)_2$ and found a transition to ferromagnetic state with $T_c = -0.9 \pm 0.2\,\mu K$ [25]. This was also obtained at initial polarizations $-P > 0.3$ before ADRF.

From the foregoing it appears clear that nuclear spin ordering cannot take place in a high magnetic field, where the DNP is performed. In the dilute electron spin system, however, ordering is quite possible, if the spin concentration is high enough, or if the spins are not uniformly distributed. In such a case the exchange interaction will strongly shorten the electron and nuclear spin-lattice relaxation times, which entail fairly low DNP.

The nuclear magnetic ordering is likely to cause substantial loss of polarization when performing the reversal of polarization by adiabatic passage. This will be discussed in Chapter 11 where polarization reversal techniques in polarized targets are described.

References

[1] R. H. Dicke, J. P. Wittke, *Introduction to Quantum Mechanics*, Addison-Wesley, Reading, MA, 1966.
[2] L. D. Landau, E. M. Lifshitz, *Quantum Mechanics*, 3rd ed., Pergamon Press, Oxford, 1977.
[3] A. Abragam, *The Principles of Nuclear Magnetism*, Clarendon Press, Oxford, 1961.
[4] M. Goldman, *Spin Temperature and Nuclear Magnetic Resonance in Solids*, Clarendon Press, Oxford, 1970.
[5] A. Abragam, M. Goldman, *Nuclear Magnetism: Order and Disorder*, Clarendon Press, Oxford, 1982.
[6] C. P. Slichter, *Principles of Magnetic Resonance*, 3rd ed., Springer-Verlag, Berlin, 1990.
[7] C. N. Yang, Integral formalism for gauge fields, *Phys. Rev. Lett.* **43** (1974) 445–447.
[8] C. N. Yang, The spin, in: G. M. Bunce (ed.) *Proc. 5th Int. Symp. on High Energy Spin Physics*, American Institute of Physics, New York, 1983, pp. 1–3.
[9] W. K. H. Panofsky, M. Phillips, *Classical Electricity and Magnetism*, Addison-Wesley, Reading, MA, 1962.
[10] S. J. Brodsky, S. D. Drell, Anomalous magnetic moment and limits on fermion substructure, *Phys. Rev. D* **22** (1980) 2236–2243.
[11] A. G. Redfield, Nuclear magnetic resonance saturation and rotary saturation in solids, *Phys. Rev.* **98** (1955) 1787–1809.
[12] B. N. Provotorov, Magnetic resonance saturation in crystals, *Soviet Phys. – JETP* **14** (1962) 1126–1131.

[13] R. J. Blin-Stoyle, M. A. Grace, Oriented nuclei, in: *Handbuch der Physik*, Springer, Berlin, Heidelberg, 1957.
[14] L. D. Landau, E. M. Lifshitz, *Statistical Physics*, 2nd revised and enlarged ed., Pergamon Press, Oxford, 1968.
[15] A. S. Oja, O. V. Lounasmaa, Nuclear magnetic ordering in simple metals at positive and negative nanokelvin temperatures, *Rev. Mod. Phys.* **69** (1997) 1–136.
[16] M. Borghini, Spin-temperature model of nuclear dynamic polarization using free radicals, *Phys. Rev. Lett.* **20** (1968) 419–421.
[17] P. Jauho, P. V. Pirilä, Spin-lattice relaxation due to electrons at very low temperatures, *Phys. Rev. B* **1** (1970) 21–24.
[18] T. Herrmannsdörfer, F. Pobell, Spontaneous nuclear ferromagnetic ordering of In nuclei in $AuIn_2$: Part I: Nuclear specific heat and nuclear susceptibility, *J. Low. Temp. Phys.* **100** (1995) 253–279.
[19] M. Chapellier, M. Goldman, V. H. Chau, A. Abragam, Production et observation d'un état antiferromagnétique nucléaire, *C.R. Acad. Sci.* **268** (1969) 1530–1533.
[20] M. Chapellier, M. Goldman, V. H. Chau, A. Abragam, Production and observation of a nuclear antiferromagnetic state, *J. Appl. Phys.* **41** (1970) 849–853.
[21] Y. Roinel, V. Bouffard, C. Fermon, et al., Phase diagrams in ordered nuclear spins in LiH: A new phase at positive temperature?, *J. Physique* **48** (1987) 837–845.
[22] Y. Roinel, G. L. Bachella, O. Avenel, et al., Neutron diffraction study of nuclear magnetic ordered phases and domains in lithium hydride, *J. Physique Lett.* **41** (1980) 123–125.
[23] Y. Roinel, V. Bouffard, G. L. Bachella, et al., First study of nuclear antiferromagnetism by neutron diffraction, *Phys. Rev. Lett.* **41** (1978) 1572–1574.
[24] J. Marks, W. T. Wenckebach, N. J. Poulis, Magnetic ordering of proton spins in $Ca(OH)_2$, *Physica B* **96** (1979) 337–340.
[25] C. M. Van der Zon, G. D. Van Velzen, W. T. Wenckebach, Nuclear magnetic ordering in $Ca(OH)_2$. III. Experimental determination of the critical temperature, *J. Physique* **51** (1990) 1479–1488.

2

Resonance and Relaxation of Interacting Spin Systems

In this chapter we shall first outline the types of interaction of spins, which are most important for solid polarized targets: the magnetic dipole interaction, the quadrupole interaction, the spin-orbit interaction, the hyperfine interaction and some other direct and indirect spin interactions. These, and in particular the dipolar interaction, are then used in the discussion of the magnetic resonance phenomena, such as the resonance lineshape and saturation. The relaxation of spins, which is phenomenologically introduced already in the saturation, is then overviewed in greater depth, before closing with a section on sudden and adiabatic changes of spin systems in the rotating frame.

2.1 Interactions of Spin Systems

2.1.1 Magnetic Dipole Interactions among Spins in Solids

Let us first calculate the magnetic field due to a classical magnetic dipole moment $\vec{\mu}$ placed at the origin. This is obtained most conveniently from its magnetic scalar potential at the point **r**: [1]

$$V_m(\mathbf{r}) = \frac{\mu_0}{4\pi r^3} \vec{\mu} \cdot \mathbf{r}, \tag{2.1}$$

the gradient of which gives the magnetic induction:

$$\mathbf{B} = -\nabla V_m(\mathbf{r}) = \frac{\mu_0}{4\pi} \nabla \left(\frac{\vec{\mu} \cdot \mathbf{r}}{r^3} \right) = -\frac{\mu_0}{4\pi} \left(\frac{\vec{\mu}}{r^3} - 3\vec{\mu} \cdot \mathbf{r} \frac{\mathbf{r}}{r^5} \right). \tag{2.2}$$

Choosing the $\theta = 0$ axis of the spherical polar coordinate system along the direction of the dipole (see Figure 2.1a), we obtain

---

[1] The magnetic scalar potential is defined as the energy needed to move a unit magnetic pole from infinity to point r, θ. The magnetic scalar potential is a single-valued function only if there are no currents present. However, upon taking the gradient for obtaining the magnetic induction **B**, the terms involving currents disappear.

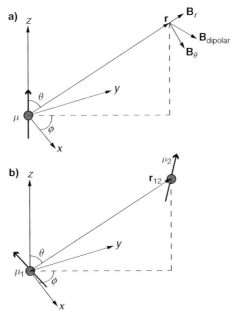

Figure 2.1 (a) Magnetic dipole in the polar coordinate system; (b) two magnetic dipoles in the polar coordinate system

$$B_r = -\frac{\partial V_m}{\partial r} = \frac{\mu_0 \mu}{2\pi r^3} \cos\theta,$$
$$B_\theta = -\frac{\partial V_m}{r\partial \theta} = \frac{\mu_0 \mu}{4\pi r^3} \sin\theta. \qquad (2.3)$$

The local field due to a dipole, at constant radius r, has thus a magnitude

$$\frac{\mu_0}{4\pi r^3}\mu \leq |B_{loc}| \leq \frac{\mu_0}{2\pi r^3}\mu \qquad (2.4)$$

and a direction which varies from antiparallel in the horizontal plane to parallel on the $\theta = 0$ axis. The lower limit applies in the plane perpendicular to the dipole moment and the upper one on the axis $\theta = 0$.

It is interesting to note the strength of this field at atomic distances, in the case of usual hydrocarbon polarized target materials. For example, glassy 1-butanol has the density $\rho = 0.97\,\mathrm{g/cm^3}$, and the atomic density of hydrogen is

$$n_H = \frac{f_H N_A \rho}{M_H} = 0.79 \cdot 10^{23}\ \mathrm{cm^{-3}}, \qquad (2.5)$$

where $f_H = 0.135$ is the free hydrogen fraction of all nucleons in the material, and M_H is the molar weight of hydrogen. The average distance between the protons is therefore

2.1 Interactions of Spin Systems

$$\overline{r} = n_H^{-1/3} = 0.233 \text{ nm}, \tag{2.6}$$

and we may immediately calculate the strength of the local field Eq. 2.3, due to the nearest neighbor proton, by introducing the magnetic moment of Eq. 1.7 and reminding ourselves of Eq. 1.25:

$$\frac{\mu_0}{4\pi}\sqrt{\langle\mu_p^2\rangle}n_H \leq |B_{loc}| \leq \frac{\mu_0}{2\pi}\sqrt{\langle\mu_p^2\rangle}n_H = 2\mu_0 n_H \hbar\gamma_p\sqrt{I_p(I_p+1)} = 0.3846 \text{ mT}. \tag{2.7}$$

In the Gaussian units the local field is therefore in the range 2 gauss to 4 gauss.

If two classical magnetic dipoles $\vec{\mu}_1$ and $\vec{\mu}_2$ are separated by the vector $\mathbf{r}_{12}$, we may now calculate the magnetic energy of the dipole $\vec{\mu}_2$ in the field of the dipole $\vec{\mu}_1$ using Eq. 2.2:

$$E_{21} = \vec{\mu}_1 \cdot \mathbf{B}_{21} = \frac{\mu_0}{4\pi}\left[\frac{\vec{\mu}_1 \cdot \vec{\mu}_2}{r_{12}^3} - \frac{3(\vec{\mu}_1 \cdot \mathbf{r}_{12})(\vec{\mu}_2 \cdot \mathbf{r}_{12})}{r_{12}^5}\right]. \tag{2.8}$$

We note that the dipolar energy is invariant under the inversion of $\mathbf{r}$ and under the inversion of the indices 1 and 2. The dipole energy E_{12} of the moment $\vec{\mu}_1$ in the field of $\vec{\mu}_2$ is therefore equal to E_{21}.

In discussing the dipolar spin interactions, two important cases must be treated somewhat differently: the case of low external fields $B_0 \approx B_{loc}$ and the case of high external fields $B_0 \gg B_{loc}$. We are here interested mainly in high external fields, which have potential for high dynamic polarization and where perturbation theory can be easily applied. This consists of solving the Zeeman Hamiltonian first (Eqs. (1.19) and (1.23)) for each spin and then treating the dipolar term as a small perturbation. Before the application of the perturbation, each of the spins in the pair therefore is found in a state with definite magnetic quantum number m_1 and m_2.

In high field the energy of Eq. 2.8 has various components, with different time variations due to the Larmor precession of the magnetic dipole moments. The components along the static field give rise to a static part, whereas the transverse components give parts that oscillate at the sum and difference frequencies of the Larmor precessions of the spins in the pair.

The quantum mechanical expression for the dipolar energy is obtained by replacing the classical vectors $\vec{\mu}_{1,2}$ by their operator counterparts of Eq. 1.11. Doing this and reorganizing slightly the expression, we shall obtain the well-known dipolar Hamiltonian

$$\mathcal{H}_{21} = \frac{\mu_0 \hbar^2 \gamma_1 \gamma_2}{4\pi r_{12}^3}\left[\hat{I}_1 \cdot \hat{I}_2 - 3(\hat{I}_1 \cdot \mathbf{r}_{12}^0)(\hat{I}_2 \cdot \mathbf{r}_{12}^0)\right], \tag{2.9}$$

where we have deliberately kept the spins and the gyromagnetic factors distinct. By rewriting the spin operators in component form, and by expressing the I_x and I_y in terms of the raising and lowering operators of Eq. 1.15, we may rewrite Eq. 2.9

$$\mathcal{H}_{21} = \frac{\mu_0 \hbar^2 \gamma_1 \gamma_2}{4\pi r_{12}^3}\left[A + B + C + D + E + F\right] \tag{2.10}$$

with

$$A = I_{1z}I_{2z}\left(1 - 3\cos^2\theta\right), \tag{2.11a}$$

$$B = -\frac{1}{4}\left(I_1^+ I_2^- + I_1^- I_2^+\right)\left(1 - 3\cos^2\theta\right), \tag{2.11b}$$

$$C = -\frac{3}{2}\left(I_1^+ I_{2z} + I_{1z} I_2^+\right)\sin\theta \cos\theta\, e^{-i\phi}, \tag{2.11c}$$

$$D = -\frac{3}{2}\left(I_1^- I_{2z} + I_{1z} I_2^-\right)\sin\theta \cos\theta\, e^{+i\phi} = C^*, \tag{2.11d}$$

$$E = -\frac{3}{4} I_1^+ I_2^+ \sin^2\theta\, e^{-2i\phi}, \tag{2.11e}$$

$$F = -\frac{3}{4} I_1^- I_2^- \sin^2\theta\, e^{+2i\phi} = E^*, \tag{2.11f}$$

where we have expressed the unit vectors $\mathbf{r}°$ in terms of the polar coordinates θ and ϕ (see Figure 2.1b) and omitted the circumflex above the spin operators for simplicity (as usual). This form of presenting the dipolar Hamiltonian is particularly convenient when calculating the magnetic resonance absorption lineshape, because it enables us to separate the relevant terms for each possible resonance line from those which give insignificant effects.

For a large array of spins, the dipolar Hamiltonian is obtained by performing the double sum over the spin pair indices in Eqs. 2.11a–f. Using quantum statistics, we may then calculate lineshapes, transition rates and relaxation times in various spin systems. We shall return to these in the subsequent sections.

In summing over all spins, we shall meet two different kinds of spin pairs: the like spin pairs (with $I_j = I_k$ and $\gamma_j = \gamma_k$) and the unlike spin pairs (with different gyromagnetic factors and, possibly, spins). The high-field energy levels reveal the significant difference between these two cases. The frequencies corresponding to the transitions representing the various terms in Eqs. 2.11a–f are substantially different, as one may expect from the classical picture. Let us first discuss the various terms in the simplified case of two spins 1/2. We note that A, which is completely diagonal in the matrix representation, describes the static part of the interaction, due to the static component of the local field. The term B has no diagonal elements in the matrix representation and causes the simultaneous reversal of the two spins, which must have opposite orientations. These two terms make no change in the total magnetization, and according to first-order perturbation theory, they contribute in first order to the splitting of the energy levels of the system. The terms A and B commute with the unperturbed Hamiltonian in the case of like spins, whereas only A commutes with $\mathcal{H}_0$ in the case of unlike spins. Therefore, in the latter case B is discarded from the secular part of the Hamiltonian, which enters in the calculation of the lineshapes and transition probabilities.

The term B is the 'flip-flop' term which causes fast rearrangement of dipolar energy for like spins and entails cross-relaxation for spins with slightly different Larmor frequencies. This term is very important for dynamic nuclear polarization when the electron resonance line has inhomogeneous broadening.

The terms C and D mix states with total magnetic quantum number differing by one unit. Their effect is seen to produce the correct eigenstates of the total Hamiltonian (comprising of the Zeeman and the dipolar terms) as small admixtures with states differing by one unit of angular momentum $|\Sigma\, m_j\rangle + \alpha |\Sigma\, m_j - 1\rangle$, where $\alpha \ll 1$. A perpendicular resonant field causes transitions due to these terms in first-order perturbation theory proportional to this α.

The terms E and F join states with both spins down to those with both spins up; they cause transitions in which $\Delta m = \pm 2$. The terms C, D, E and F therefore have only off-diagonal elements, and they produce admixtures of the zero-order (unperturbed) states into the exact states. The admixtures can be calculated using the second-order perturbation theory; these are proportional to α^2. These terms are important in the 'solid effect' mechanism for dynamic nuclear polarization.

Table 2.1 summarizes the states linked by the dipolar interaction terms. The terms A and B shift energy levels and therefore broaden the resonance lines. In the case of like spins, the term B allows cross-relaxation transitions to happen between the neighboring spins, leading to spin diffusion. In the case of unlike spins, the term B does not contribute to the broadening of the two main resonance lines, because it does not commute with the unperturbed Hamiltonian but it allows second-order cross-transitions to occur near the angular frequency $|\omega_1 - \omega_2|$.

The spin Hamiltonian which is relevant in the spin dynamics under saturation (weak or strong) near the main resonance at the Larmor frequency in high field is now

$$\mathcal{H} = \mathcal{H}_Z + \mathcal{H}'_D, \tag{2.12}$$

where the part representing the Zeeman energy is

$$\mathcal{H}_Z = \hbar B_0 \sum_j \gamma_j I_j, \tag{2.13}$$

and the secular part of the dipolar spin Hamiltonian is

$$\mathcal{H}'_D = \frac{\mu_0 \hbar^2}{8\pi} \sum_{j \neq k} \frac{\gamma_j \gamma_k}{r_{jk}^3} \left(1 - 3\cos^2\theta\right)\left(3 I_{jz} I_{kz} - I_j \cdot I_k\right). \tag{2.14}$$

This can be seen to commute with the unperturbed Zeeman Hamiltonian $\mathcal{H}_Z = -\hbar \omega_0 I_z$. When j and k label unlike spins, we have to remove $I_{jz} I_{kz} - I_j \cdot I_k$ from the term in the last

Table 2.1 States linked with the terms of the dipolar interaction of Eqs. 2.11a–f.

Dipolar interaction term	$\Delta(m_1 + m_2)$	Δm_1	Δm_2
A	0	0	0
B	0	± 1	$-(\pm 1)$
C	1	0, 1	1, 0
D	-1	$-1, 0$	$0, -1$
E	2	1	1
F	-2	-1	-1

parentheses, because the term B for unlike spins influences magnetic resonance only at frequencies $|\omega_j - \omega_k|$, assumed to be far from the Larmor frequency of the species under consideration.

In a high magnetic field the dipolar energy reservoir represented by the Hamiltonian of Eq. 2.14 is isolated from the Zeeman energy, and these two reservoirs can therefore stay at different temperatures for much longer periods than time T_2 required by the establishment of the thermal distribution itself in each of the reservoirs. A transverse magnetic field, rotating at a frequency close to the Larmor frequency, may 'mix' rapidly these reservoirs, and therefore the dipolar interaction leads to several peculiar resonance phenomena, which will be discussed in Section 2.3. The establishment of the thermal equilibrium by relaxation will be discussed in Section 2.3, and phenomena related with strong transverse fields will be treated in Section 2.4.

2.1.2 Quadrupole Interaction

A nucleus in a state of definite angular momentum has a charge distribution with cylindrical symmetry. Classically this is understood by the effect of averaging of the charge distribution due to rotation. Upon reorientation of the angular momentum, the quadrupole energy thus depends only on the difference between the charge distributions parallel and transverse to the angular momentum vector. This difference is measured with the constant eQ, where Q is the quadrupole moment of the nucleus and one of the nine components of the general quadrupole operator needed in the full description of the interaction of the charge distribution of the nucleus with the generalized electric field gradient tensor. The electric field gradient tensor can be diagonalized by transforming to a suitable coordinate system $0xyz$; this simplifies the quadrupole Hamiltonian to

$$\mathcal{H}_Q = \frac{eQ}{4I(2I-1)}\left[V_{zz}(3I_z^2 - I^2) + (V_{xx} - V_{yy})(I_x^2 - I_y^2)\right], \qquad (2.15)$$

where

$$V_{zz} = \frac{\partial^2 V}{\partial z^2},$$

$$V_{xx} = \frac{\partial^2 V}{\partial x^2},$$

$$V_{yy} = \frac{\partial^2 V}{\partial y^2}.$$

Because Laplace's equation $V_{xx} + V_{yy} + V_{zz} = 0$ is valid for the electric field due to the atomic electrons in the nucleus, the gradient tensor can be characterized by two parameters

$$eq = V_{zz} \equiv \frac{\partial^2 V}{\partial z^2}, \qquad (2.16)$$

called the electric field gradient, and

$$\eta = \frac{V_{xx} - V_{yy}}{V_{zz}} \equiv \frac{\frac{\partial^2 V}{\partial x^2} - \frac{\partial^2 V}{\partial y^2}}{\frac{\partial^2 V}{\partial z^2}}, \quad (2.17)$$

which is called the asymmetry parameter. Here V is the potential giving rise to the electric field. The detailed derivation of the Hamiltonian (Eq. 2.15) requires the use of Glebsch–Gordan coefficients, irreducible tensor operators and the Wigner–Eckart theorem, and is beyond the scope of this book. We refer the reader interested in the derivation to the monographs of Abragam [1] and Slichter [2].

The electric field gradient is often axially symmetric, in which case the z-axis can be conveniently chosen along the symmetry axis. This gives $\eta = 0$, which simplifies the calculation of the quadrupolar lineshape. The resulting lineshapes in high field will be discussed in Chapter 5.

Nuclear spins in states $m = \pm 1/2$ have opposite values of the projections of the spin vector on the z-axis. The quadrupole energy is same in these two states, if the charge distribution is axially symmetric. No quadrupole shift nor broadening can therefore be observed in transitions between the states $m = \pm 1/2$. Furthermore, if the nucleus is in a definite state of spin 1/2 and parity, its charge distribution is spherically symmetric and has therefore no quadrupole moment. Consequently, the spin 1/2 nuclei cannot have quadrupole effects in their magnetic resonance.

In order to determine the energy levels, it is more convenient to write the Hamiltonian with Zeeman and quadrupole interactions in the coordinate system where the z'-axis is parallel to the magnetic field. We may choose to rotate the coordinates about the y-axis so that

$$I_{z'} = I_z \cos\theta + I_x \sin\theta,$$

where θ is the polar angle between the field and the principal axis of the electric field gradient tensor. Assuming that $\eta = 0$, we get

$$\mathcal{H} = -\hbar\gamma B I_{z'} + \mathcal{H}_Q, \quad (2.18)$$

where

$$\mathcal{H}_Q = \frac{e^2 qQ}{8I(2I-1)} \left[(3c^2 - 1)(3I_{z'}^2 - I(I+1)) + 3c\sqrt{1-c^2}\{I_{z'},(I_+ + I_-)\} + \frac{\sqrt{1-c^2}}{2}(I_+^2 + I_-^2) \right]. \quad (2.19)$$

Here $c = \cos\theta$ and $I_\pm = I_{x'} \pm iI_{y'}$. Because the Zeeman term dominates in high field, the quadrupole term can be treated as a perturbation, yielding the magnetic energy levels $E_m = E_m^{(0)} + E_m^{(1)} + E_m^{(2)} + \ldots$ with shifts from first- and second-order perturbation calculations given by [3]

$$E_m^{(0)} = -\hbar\gamma Bm, \qquad (2.20)$$

$$E_m^{(1)} = \hbar\omega_Q(3c^2-1)\left[3m^2-a\right], \qquad (2.21)$$

$$E_m^{(2)} = -\frac{9\hbar\omega_Q^2}{2\omega_L}m\left[4c^2(1-c^2)(8m^2-4a+1)+(1-c^2)^2(-2m^2+2a-1)\right] \qquad (2.22)$$

in the case that the gradient tensor is axially symmetric so that $\eta = 0$. Here $\omega_L = \gamma B$ is the Larmor frequency, θ is the angle between the magnetic field vector and the principal axis of the electric field gradient tensor, and

$$\omega_Q = \frac{e^2qQ}{8\hbar I(2I-1)} \; ; \; a = I(I+1) \; ; \; c = \cos\theta. \qquad (2.23)$$

We note that the first-order energy level shifts are symmetric with respect to m, which splits the resonance line to $2I$ components. In contrast to this, the second-order energy level shifts are antisymmetric with respect to m, which leaves the first-order frequency differences for symmetric transitions $m-1, m$ and $-(m-1), -m$ unchanged. The second-order term, however, changes the powder lineshape substantially, if ω_Q/ω_L is not much smaller than 1, because of the more complex dependence on $\cos\theta$.

The quadrupolar lineshapes are discussed in Chapter 5, where also the exact energy levels are calculated, valid at all field values and without perturbation theory, for the case of axially symmetric field gradient tensor. The implications of the quadrupole lineshape to the measurement of polarization are discussed in Chapter 6.

Because the nuclear quadrupole moment couples to the electric field gradient which depends on the molecular and lattice structure, thermal excitations such as phonons can couple with the Hamiltonian (Eq. 2.15) causing spin-lattice relaxation in solids. Theoretical estimates, however, are in disagreement with the observed much shorter relaxation times in cubic crystalline solids (see Ref. [1], p. 414). This will be briefly discussed in Section 2.3.4.

2.1.3 Spin-Orbit Interaction

The magnetic moment of an unpaired electron moving in an electric field **E** is coupled to the orbital motion with the Hamiltonian

$$\mathcal{H}_{SO} = \frac{e\hbar}{2m_e^2c^2}\mathbf{S}\cdot(\mathbf{E}\times\mathbf{p}), \qquad (2.24)$$

which is called in short the spin-orbit Hamiltonian. If the electric field is a function of radius only so that

$$\mathbf{E}(\mathbf{r}) = \frac{\mathbf{r}}{r}E(r) \qquad (2.25)$$

then we have

$$\mathbf{E} \times \mathbf{p} = \frac{E(r)}{r} \mathbf{r} \times \mathbf{p} = \frac{E(r)}{r} \hbar \mathbf{L}, \tag{2.26}$$

where **L** is the orbital angular moment operator. The spin-orbit Hamiltonian (2.24) can now be written in terms of the spin-orbit coupling constant λ as

$$\mathcal{H}_{SO} = \lambda \mathbf{L} \cdot \mathbf{S}. \tag{2.27}$$

The Hamiltonian describing an unpaired electron belonging to an atom can then be written in the general form

$$\mathcal{H} = \frac{p^2}{2m} + V_0 + \mu_B \mu_0 \mathbf{H} \cdot \mathbf{L} + g_e \mu_0 \mathbf{H} \cdot \mathbf{S} + \lambda \mathbf{L} \cdot \mathbf{S}, \tag{2.28}$$

where p is the momentum and V_0 the atomic potential of the electron. Note that we have used here the magnetic field strength **H** rather than the magnetic induction **B** in order to be able to separate the magnetic field due to the orbiting electron under the spin-orbit coupling term. In a free hydrogen-like atom with spherically symmetric electric field, this interaction leads to the splitting of the states of given L and S to degenerate levels, and to the well-known Landé g-factor

$$g_J = 1 + \frac{J(J+1) - L(L+1) + S(S+1)}{2J(J+1)}$$

which varies from 2 to 1 with increasing L.

For solid materials the symmetry of the electric field is lifted by the potential which binds the paramagnetic atom, so that the Hamiltonian must be written as

$$\mathcal{H} = \frac{p^2}{2m} + V_0 + V_1 + \mu_B \mu_0 \mathbf{H} \cdot \mathbf{L} + g_e \mu_0 \mathbf{H} \cdot \mathbf{S} + \lambda \mathbf{L} \cdot \mathbf{S}, \tag{2.29}$$

where V_1 is the crystal potential of an unpaired electron belonging to an atom, ion or molecule. The treatment of the problem now depends on whether the crystal potential term is smaller or larger than the spin-orbit coupling. This has been addressed comprehensively by Slichter ([2], pp. 503–516), and we shall limit ourselves here to quote the main points and results.

When the spin-orbit term dominates, the treatment is somewhat similar to that of the free atom, in which case $\mathbf{J} = \mathbf{L} + \mathbf{S}$ is a good quantum number, with values between $L + S$ and $|L - S|$ and eigenvalues of the spin-orbit Hamiltonian

$$E_{SO} = \frac{\lambda}{2} \left[J(J+1) - L(L+1) - S(S+1) \right]. \tag{2.30}$$

The effect of V_1 is then taken into account by assuming that it arises from fixed charges replacing the lattice around the atom, so that the potential can be expanded into spherical harmonics. Taking only the lowest terms, two cases illustrate the symmetries of main importance:

$$V_1 = A(x^2 - y^2) \tag{2.31}$$

or
$$V'_1 = B(3z^2 - r^2), \tag{2.32}$$

where A and B are constants. The first one corresponds to a field due to charges symmetrically in the x-y plane and the second to an axially symmetric field. Using the Wigner–Eckart theorem one can show that the potential can be replaced now by an operator

$$\mathcal{H}_1 = C_J(J_x^2 - J_y^2) \tag{2.33}$$

or
$$\mathcal{H}_1 = C'_J(3J_z^2 - J^2). \tag{2.34}$$

The magnetic field terms can now be treated analogously to the free atom case which allows their replacement by the Zeeman operator

$$\mathcal{H}_z = g_J \mu_B \mu_0 \mathbf{H} \cdot \mathbf{J} \tag{2.35}$$

which can be combined with the above Eqs. 2.33 and 2.34 to yield an effective Hamiltonian

$$\mathcal{H}_{\text{eff}} = C_J(J_x^2 - J_y^2) + g_J \mu_B \mu_0 \mathbf{H} \cdot \mathbf{J} \tag{2.36}$$

or
$$\mathcal{H}'_{\text{eff}} = C'_J(3J_z^2 - J^2) + g_J \mu_B \mu_0 \mathbf{H} \cdot \mathbf{J} \tag{2.37}$$

which are seen to closely resemble the Hamiltonian (Eq. 2.18) of a nucleus with electric quadrupole moment. Such a Hamiltonian yields, using first-order perturbation theory, magnetic energy levels and Larmor frequencies which depend on the angle between the magnetic field and the axis of symmetry of the field due to the lattice. In the case of spin-orbit coupling this anisotropy can be very large when the coefficient C_J is large. This situation is often seen with rare-earth ions.

Slichter ([2], p. 514) estimates the coefficient for the case of axial field and $J = 3/2$ as

$$C'_{3/2} = -\frac{2}{15} B \langle r^2 \rangle, \tag{2.38}$$

where $\langle r^2 \rangle$ is the average value of r^2 for the p-states.

In the case that the crystal potential is much larger than the spin-orbit coupling, strong quenching of the orbital momentum happens due to the strength of the crystal field where the state of the electron cannot be described by a wave function with a definite value of L, i.e. L is not a 'good' quantum number. In terms of quantum mechanics, the requirement for the quenching of the angular momentum is that the eigenfunction describing the state of the electron must be possible to choose so that it is purely real in zero magnetic field, which happens when the state is non-degenerate. Physically this means that the electric field of the crystal or molecule makes the electron orbit precess so that the electric current is reversed periodically, which averages the resulting field to zero. When a magnetic field is applied, however, some directions of the orbital moment are preferred and the electric current gives a non-zero field which couples to the electron spin. Such a quenched orbital moment gives rise to a g-factor which does not deviate markedly from g_e. This situation is often seen in the free radicals in dielectric materials.

Slichter [2] explains this situation by treating first the truncated spin-orbit Hamiltonian

$$\mathcal{H} = \frac{p^2}{2m} + V_0 + V_1 + g_e \mu_0 \mathbf{H} \cdot \mathbf{S} \tag{2.39}$$

in zero magnetic field, where the effect of V_1 is to lift the orbital degeneracy. The resulting three energy levels of the *p*-state electron have a twofold degeneracy due to the electron spin. The two remaining terms of Eq. 2.29

$$\mu_B \mu_0 \mathbf{H} \cdot \mathbf{L} + \lambda \mathbf{L} \cdot \mathbf{S} \tag{2.40}$$

are then treated perturbatively to second order, because their first-order effects vanish. The interference of these terms can be shown to lead to an effective Hamiltonian

$$\mathcal{H}_{\text{eff}} = g_e \mu_B \mu_0 \left(\frac{\lambda}{E_x - E_z} S_y H_y + \frac{\lambda}{E_x - E_y} S_z H_z \right), \tag{2.41}$$

where the energy differences in the denominators refer to a specific case where the crystal field was created by placing two negative charges symmetrically on the *y*-axis and two positive charges symmetrically on the *x*-axis which makes the crystal field vanish on the *z*-axis.

Combining Eq. 2.41 with the electron Zeeman Hamiltonian leads to

$$\mathcal{H} = \mu_B \mu_0 \left(H_x g_{xx} S_x + H_y g_{yy} S_y + H_z g_{zz} S_z \right) \tag{2.42}$$

with

$$g_{xx} = g_e, \, g_{yy} = g_e \left(1 - \frac{\lambda}{E_z - E_x} \right), \, g_{zz} = g_e \left(1 - \frac{\lambda}{E_y - E_x} \right). \tag{2.43}$$

This can be understood as an anisotropic *g*-tensor which makes the resonant frequency dependent on the relative orientations of the applied field and the coordinate system of the crystal potential. It correctly describes the fact that for positive λ the shift in *g* is down and for negative λ it is up. The treatment does not, however, give any values for λ itself, which requires the knowledge of the atomic or molecular structure and the wave functions of the electron in its ground state and excited states.

The spin-orbit interaction for nuclear spins is briefly discussed by Abragam ([1], p. 174). In solids the nuclear orbital momentum is strongly quenched and vanishes in first order as for the paramagnetic electrons. The second-order interference term remains negligibly small because of the smallness of the spin-orbit coupling, caused by the large nuclear mass. The resulting frequency shift is overwhelmed by diamagnetic and chemical shifts and has not been experimentally observed. When the nuclear orbital motion is not quenched, however, the orbital and nuclear magnetic moments can be of the same magnitude. An extreme case is free molecular hydrogen in a beam where these are comparable.

2.1.4. Hyperfine Interaction

The hyperfine interaction arises from the scalar magnetic coupling $\gamma_e \gamma_n \mathbf{S} \cdot \mathbf{I}$ between the magnetic moments of an unpaired electron and a nearby nucleus. It results from the overlap of the wave functions and can be represented in the general form[2]

$$\mathcal{H}_{HF} = \int d\tau \, |\psi(r)|^2 \gamma_e \gamma_n \hbar^2 \left\{ \frac{8\pi}{3} \mathbf{I} \cdot \mathbf{S} \delta(\mathbf{r}) + \frac{3(\mathbf{I} \cdot \mathbf{r})(\mathbf{S} \cdot \mathbf{r})}{r^5} - \frac{\mathbf{I} \cdot \mathbf{S}}{r^3} \right\}, \tag{2.44}$$

where the first term of the integrand is derived from the Dirac equation and is non-zero for s-electrons (or when the electron wave function has s-component). The second and third terms, which arise from the dipolar coupling at distance r, are non-zero for electrons which are not in a pure s-state. The integration is over the electron coordinates.

The first term is also called the Fermi contact interaction. It gives rise to the isotropic hyperfine tensor which manifests itself as a splitting of the magnetic levels without broadening. The other terms depend on the relative orientations of $\mathbf{r}$ and $\mathbf{B}$ and lead to an anisotropic hyperfine tensor which splits and broadens the resonance lines.

The expression (Eq. 2.44) is linear in I_x, I_y and I_z, and S_x, S_y and S_z, and it can be shown that it can be reduced to the general form

$$\mathcal{H}_{hf} = \mathbf{S} \cdot \tilde{\mathbf{A}} \cdot \mathbf{I} \equiv A_x I_x S_x + A_y I_y S_y + A_z I_z S_z. \tag{2.45}$$

If there are several hyperfine nuclei in the molecule, very complicated EPR line structures may arise. In the case of only isotropic couplings the magnetic levels are

$$E = \left(g\mu_B B + \sum_i a_i m_i \right) m_S. \tag{2.46}$$

In such cases the isotropic hyperfine constants a_i may be resolved in dilute liquid samples. The EPR spectra with hyperfine interactions will be discussed in Chapter 3.

The hyperfine interaction will also be seen in the resonance of the hyperfine nuclei, which are split in the same way. The resonance frequencies are in the isotropic case and in high field

$$\omega = \gamma_n B_0 + \frac{a m_S}{\hbar}, \tag{2.47}$$

---

[2] For derivation, see, for example ([2], p. 517).

where m_s is the projection of the electron spin on the direction of the magnetic field. The NMR of hyperfine nuclei will be treated in more detail in Chapter 5.

2.1.5 Other Spin Interactions

Although their effects are small and hardly visible in polarized targets, we shall briefly review some second-order effects involving the interaction of nuclear spins with the non-magnetic electronic systems in solids. These effects are due to the interaction of the electronic system with the applied field and with the nuclear spins, which may induce an internal field at the same or another nucleus. The magnetic interaction of the nuclear spin is therefore modified, leading to frequency shifts and smearing of the resonance frequencies. Among the resulting effects are the chemical shift in diamagnetic materials, and the Knight shift in metallic substances. We shall also briefly discuss the indirect coupling between nuclear spins in molecules due to their interaction with the electrons, and the exchange interaction between nearby paramagnetic electrons.

Chemical Shift of Nuclear Magnetic Resonance Frequency

High-resolution NMR shows that the resonance frequency of each nucleus depends on its chemical environment, with shifts in the range of 10 ppm for protons. The relative shift is constant and is expressed as

$$\omega = (1-\sigma)\omega_L, \tag{2.48}$$

where ω_L is the Larmor frequency of the bare nucleus. However, as the field cannot be measured with high absolute accuracy, usually the shift is defined with respect to a known resonance line whose shift is thus arbitrarily defined to be zero.

The range of the chemical shift increases with the atomic weight; for fluorine the shift ranges 600 ppm. The shift is a powerful tool for the chemical analysis of diamagnetic materials.

The theoretical model for the chemical shift derived by Ramsey [4, 5] has been reviewed thoroughly by Abragam ([1], p. 175) and Slichter ([2], p. 92), and we shall only briefly discuss the basic principles before stating the results.

Because the influence of the nuclear dipole field on the electronic structure of the atom or molecule is extremely weak, one may treat the chemical shift in two parts, firstly determining the perturbation caused by the external field on the electronic orbits and then dealing with the interaction of the nuclear spin with the sum of the external field and the fields due to the electronic orbits. The energy perturbation due to this interaction is

$$E_{chem} = -\mu_0 \hbar \gamma \, \mathbf{I} \cdot \int \frac{\mathbf{r} \times \mathbf{j}_0(\mathbf{r})}{r^3} d\tau, \tag{2.49}$$

which can be seen to correspond to the interaction of the nuclear dipole moment with a field arising from the current density $\mathbf{j}_0$, by comparing the integral with Eq. 1.2.

The formal calculation of the current density $\mathbf{j}_0$ belongs to the domain of theoretical molecular chemistry and we shall just note here that the two leading contributions in diamagnetic substances

$$\sigma = \sigma_d + \sigma_p \qquad (2.50)$$

are due to the diamagnetic shielding current and the paramagnetic moment, respectively. The paramagnetic moment is theoretically understood as a contribution from virtual excited states of the electrons. These terms have always opposite signs and they are of the same order of magnitude, which forces high accuracy in the theoretical modelling of the chemical shifts.

The chemical shifts are anisotropic so that they can be best measured in liquid state. Progress in high-resolution pulsed NMR has enabled recently their measurement also in solid samples, and it is interesting that such measurements in dynamically polarized materials have been made possible so that rare nuclei have become accessible for the experiment. This will be discussed in Chapter 11.

Knight Shift in Metals

The coupling of the nuclear spin with conduction electrons in metals leads to a frequency shift which ranges from 2.5×10^{-4} for ^{7}Li to 2.5×10^{-2} for ^{199}Hg, and which obeys the following experimental observations:

1. The NMR frequency shift $\Delta\omega$ is upwards with few exceptions.
2. The shift scales with the field.
3. The shift is almost independent of temperature.
4. The relative shift increases with increasing nuclear charge Z.
5. The shift is isotropic in cubic metals.

The Knight shift can be theoretically understood by describing the conduction band of the electron system as a weakly interacting (or even non-interacting) degenerate Fermi gas and calculating the field induced at the nucleus due to the influence of the external field. The electron wave functions are described by the Bloch functions

$$\psi_{\mathbf{k}s} = u_{\mathbf{k}}(\mathbf{r}) e^{i\mathbf{k}\cdot\mathbf{r}} \psi_s, \qquad (2.51)$$

where Ψ_s is the electron spin wave function and the plane wave part $\exp(i\mathbf{k}\cdot\mathbf{r})$ is modulated by the function $u(\mathbf{r})$ which has the periodicity of the lattice and which peaks at the nuclei positioned at $\mathbf{r} = 0$.

The resulting frequency shift is

$$K = \frac{\Delta\omega}{\omega} = \frac{8\pi}{3} \left\langle \left| u_{\mathbf{k}}(0) \right|^2 \right\rangle_{E_F} \chi_e^s, \qquad (2.52)$$

where the total spin susceptibility of the electrons is defined by

$$\langle \mu_z \rangle = \chi_e^s H_0, \qquad (2.53)$$

and the index E_F refers to the Fermi surface, in the vicinity of which the electrons contribute to the susceptibility.

We note immediately that Eq. 2.53 qualitatively explains all of the listed features. The result is a positive constant, and because of the Fermi distribution, the susceptibility of the conduction electrons also is independent of temperature. With heavier nuclei the function $u(\mathbf{r})$ peaks more heavily at $\mathbf{r} = 0$, which explains their larger effect. Also, the quantitative calculations are in excellent agreement with the theory and its refinements.

In non-cubic metals, the Knight shift depends on the relative orientations of the crystalline axes and the applied field, and a tensor part must be added to the constant K. In axially symmetric crystals, such as those with tetragonal structure, the tensor causes broadening similar to the g-shift in electron resonance. The broadening is smaller than the shift itself.

The interaction between the conduction electrons and the nuclear spins also leads to a fast spin-lattice relaxation in metals, which was mentioned in Section 1.3.5 and which will be discussed further in Section 5.3.4.

Indirect Coupling between Nuclear Spins: Molecular Spin Effects

Nuclear spins have an effect on the electronic system on the molecular level, which can be rather strong, with the well-known extreme cases of H_2 and D_2. In heavier molecules with dissimilar nuclei, the effects on nuclear spins are called pseudo-dipolar and pseudo-exchange couplings. An example of a molecule where the symmetry effects lead to molecular isomerism is solid CH_4 with its molecular ortho, meta and para states with total nuclear spin of 2, 1 and 0. In solid NH_3 the indirect coupling leads to a typical pseudo-exchange term mediated by the hyperfine coupling of the electron spins with the nuclear spins. The hyperfine coupling appears because the degeneracy of the s-part of the electron wave function is lifted by the nuclear spin coupling. A good introductory discussion to the indirect coupling between nuclear spins has been given by Slichter ([2], pp. 133–143).

The pseudo-dipolar coupling has the same bilinear spin dependence as the usual dipolar coupling (Eq. 2.9):

$$\mathcal{H}_{12} = \left[\hat{I}_1 \cdot \hat{I}_2 - 3\left(\hat{I}_1 \cdot \mathbf{r}_{12}^0\right)\left(\hat{I}_2 \cdot \mathbf{r}_{12}^0\right) \right] B_{12}, \qquad (2.54)$$

where the function B_{12} falls off as $(1/r_{12})^3$ and vanishes if the electronic wave function at the nucleus has no non-s character. As for the usual dipolar coupling, the pseudo-dipolar coupling only influences the broadening of the lineshape but does not split or shift the lines.

The pseudo-exchange coupling

$$J_{pseudo} = A_{12} I_1 \cdot I_2, \qquad (2.55)$$

where A_{12} is a constant, results from the s-character of the electron wave function at the nuclei 1 and 2. Similar to the Fermi contact term of the hyperfine interaction Hamiltonian, it leads to the isotropic splitting of NMR lines, which will be discussed in Chapter 5.

The scalar coupling (Eq. 2.55) in solids narrows the nuclear spin resonance line for like spin pairs and broadens it for unlike spin pairs. In both cases the lineshape tends towards Lorentzian. A typical example is the proton NMR line of solid NH_3, split in three overlapping lines that become partially resolved at very high polarization.

Exchange Interaction for Electron Pairs

In molecules and solids electron spins pair off so that, apart from conduction electrons in metals and inner unpaired electrons in transition metal ions, unpaired paramagnetic electrons tend to be a rare exception. The pairing of the electron spins due to the Pauli principle is thus an extremely strong phenomenon and is due to the fact that the electrons are identical. Another case is the nuclear spin pairing in the ground state of the H_2 molecule.

Pairing of identical particles occurs also for paramagnetic electrons in solids, if they are not completely isolated from each other. As the paramagnetic electron wave functions in many molecules cover the whole molecule and extend slightly outside, molecule or ion pairs at close distance in solids may have a small overlap of the wave functions so that pairing and exchange effects can appear.

The scalar coupling

$$\mathcal{H}_{exch} = J S_1 \cdot S_2 \qquad (2.56)$$

between electron spins in solids is due to electric rather than the magnetic forces which gave rise to indirect scalar coupling between nuclei. The exchange interaction leads to magnetic ordering in substances where the unpaired electrons are organized in a regular lattice and to transformation into spin glasses in less ordered structures. The magnetic ordering may result in a ferromagnetic or antiferromagnetic structure depending on the lattice symmetry and several other parameters.

In paramagnetic substances the electron exchange Hamiltonian leads to a fast spin-lattice relaxation which may be related with the rapid interchange of the z-components of oppositely oriented spin pairs. This may be associated with substantial line narrowing or broadening effects as well. In polarized targets the spin exchange is likely to set a limit to the paramagnetic concentration at about 10^{21} cm^{-3}.

2.2 Magnetic Resonance of Interacting Spins in Solids

2.2.1 Power Absorbed from a Linearly Polarized Transverse Field

Let us consider a system of interacting spins in a high external field B_0 and subjected to a linearly polarized transverse oscillating field $B_x = 2B_1 \cos\omega t = B_1(e^{i\omega t} + e^{-i\omega t})$, which can be described as a superposition of rotating and counterrotating components, of such low amplitude that the response of the spin system remains linear. The conditions under which linearity is preserved will be discussed in Section 2.2.2. We shall calculate here first the resulting power absorption using the transverse magnetization on the one hand and the magnetic level transition probability (1.58) on the other, and find an expression for the absorption part of the transverse susceptibility.

2.2 Magnetic Resonance of Interacting Spins in Solids

The transverse complex susceptibility function is defined by the relation between the rotating transverse field and the rotating transverse magnetization:

$$\mathbf{M}_\perp = M_x + iM_y = B_1 \mu_0^{-1} e^{i\omega t} \left[\chi'(\omega) - i\chi''(\omega) \right], \quad (2.57)$$

where χ' and χ'' are the real the imaginary parts of the function; these parts are called dispersion and absorption, respectively. Using this the average power absorbed in unit volume is

$$\frac{\dot{Q}}{V} = \frac{1}{2}\overline{\mathrm{Re}\left\{ \mathbf{B}^*(t) \frac{\partial \mathbf{M}(t)}{\partial t} \right\}} = \frac{1}{2} \omega \mu_0^{-1} B_1^2 \chi''(\omega), \quad (2.58)$$

which explains the commonly used term 'absorption' part of the spin susceptibility. The term 'dispersion' follows from the fact that the velocity of propagation of an electromagnetic wave which depends on this term becomes disperse because of the frequency dependence of the term. Consequently, a sharp electromagnetic pulse, containing a wide spectrum of frequencies, turns gradually into one which is more smeared in the time domain, during propagation in the disperse medium.

The power absorption counted from the number of spin flips per unit time and volume is readily obtained from the transition probability of Eq. 1.58 and the magnetic level populations p_m (which do not need to satisfy the Boltzmann distribution required in Eq. 1.62):

$$\frac{\dot{Q}}{V} = -n\hbar\omega \sum_{-I+1}^{+I} W_{m,m-1}(p_{m-1} - p_m)$$

$$= -\frac{n\hbar\omega\omega_1^2 \pi}{2} f(\omega) \sum_{-I+1}^{+I} (I+m)(I-m+1)(p_{m-1} - p_m), \quad (2.59)$$

where $n = N/V$ is the number of resonant spins I per unit volume. After evaluation of the sum which is rather tedious but straightforward, and applying Eq. 1.91 we get

$$\frac{\dot{Q}}{V} = -\frac{\pi n\hbar\omega\omega_1^2}{2} f(\omega) \sum_{-I}^{+I} -2mp_m = \pi n\hbar\gamma^2 B_1^2 \omega f(\omega) IP(I). \quad (2.60)$$

By equating the power absorbed by the spin system to the power lost (Eq. 2.58) from the field due to the absorption part of the complex susceptibility, we shall obtain the relationship

$$\chi''(\omega) = \frac{2}{\pi} \mu_0 n\hbar\gamma^2 f(\omega) IP(I) \quad (2.61)$$

between the absorption part of the transverse susceptibility and the normalized lineshape function. If the spin polarization is expressed in terms of the static susceptibility of Eq. 1.66, we shall find a simpler equation

$$\chi''(\omega) = \frac{\pi}{2} \omega_0 f(\omega) \chi_0. \quad (2.62)$$

If the width of the lineshape function in a high field is due to a distribution of the Larmor frequencies ω_0 and is not small in comparison with the dipolar width, then the above expression should be replaced by

$$\chi''(\omega) = \frac{\pi}{2}\omega f(\omega)\chi_0, \qquad (2.63)$$

which is valid even if the spin temperature is not constant in the spectrum. The tendency towards internal thermal equilibrium, however, is a strong phenomenon in equilibrium with the lattice, and usually becomes even stronger during dynamic polarization, as will be shown in Chapter 4. We also find that

$$\int_0^\infty \frac{\chi''(\omega)}{\omega}d\omega = \frac{\pi}{2}\chi_0 \int_0^\infty f(\omega)d\omega = \frac{\pi}{2}\chi_0, \qquad (2.64)$$

because the lineshape function was normalized.

Integrating Eq. 2.61 yields

$$P(I)\int_0^\infty f(\omega)d\omega \equiv P(I) = \frac{2}{\pi\mu_0 n\hbar\gamma^2 I}\int_0^\infty \chi''(\omega)d\omega, \qquad (2.65)$$

which holds for narrow absorption lines and is used for the measurement of polarization.

We note that no assumption needs to be made on the temperature or even on the thermal equilibrium in the spin system, because the definition of the polarization was invoked in direct relation to the level populations without a distribution function.

The above results are surprisingly accurate, as will be seen later in this chapter. This, however, is somewhat coincidental although it is difficult to point out the small inaccuracies which tend to cancel out at the end. The main disadvantage of the above remains that it does not permit to introduce spin interactions in a natural way. The lineshape cannot therefore be obtained from the first principles. Furthermore, it was implicitly assumed without justification that all magnetic energy levels were broadened by the same distribution function. A better theoretical introduction will therefore be given for the absorption lineshape in the next section.

2.2.2 General Properties of the Complex Susceptibility

The general linear response theory (see Ref. [6], pp. 302–306) allows one to relate the transient and steady-state responses of a system on the one hand and the real and imaginary parts of a function which relates the response of a system to excitation on the other hand. The theorem applies to all resonant systems satisfying certain criteria and is commonly used in magnetic and dielectric (optical) resonance, in circuit theory and in particle scattering.

The response is linear as long as the susceptibility at frequency ω is measured in time that is short in comparison with the time constant for the change of the Zeeman and dipolar temperatures due to the RF field. This will be discussed in detail in Section 2.2.4 based on Provotorov's equations.

2.2 Magnetic Resonance of Interacting Spins in Solids

A lengthy but instructive presentation has been given by Slichter (see Ref. [2] p. 59) on the linear response and transverse susceptibility. The treatment is based on the high-temperature approximation.

In this section we shall first outline the main points of the derivation given by Abragam and Goldman for the free precession signal and its relationship with the transverse susceptibility in the case of dipolar interactions [6]. This treatment deviates from the usual one in the way that the dipolar temperature is allowed to deviate from the Zeeman temperature, which is possible when the susceptibility is measured in a time which is short in comparison with the time required for the equalization of the two temperatures. This requirement is easy to satisfy when the exciting transverse field is very small, and the requirement can be quantified using the treatment of Provotorov, which is the subject of the next section.

We shall then calculate the absorption part of the transverse susceptibility from the formula based on the transition probability between magnetic levels broadened by unspecified spin interactions satisfying certain criteria and relate this to the formulas obtained using the dipolar interactions.

Response to a Pulse of Transverse Field

In high field and at high temperature our system of spins has the dipolar spin Hamiltonian (Eq. 2.14) and density matrix

$$\rho = 1 - \frac{\hbar \omega_0}{kT_Z} I_z - \frac{1}{kT_D} \mathcal{H}'_D, \tag{2.66}$$

which is invariant under rotation about the z-axis so that it has the same expression in the rotating frame, in which a pulse of field B_1 along the x-axis is produced during a short time τ. This induces a rotation of the spin vectors around the x-axis by an angle of

$$\theta = -\gamma B_1 \tau,$$

which transforms the initial Hamiltonian so that the initial density matrix becomes

$$\rho_0 = 1 - \frac{\hbar \omega_0}{kT_Z} \left(\cos\theta \, I_z - \sin\theta \, I_y \right) - \frac{1}{kT_D} \left[\frac{1}{2} \left(\cos^2\theta - 1 \right) \mathcal{H}'_D - i\cos\theta \, \sin\theta \left[I_x, \mathcal{H}'_D \right] + \mathcal{H}_2 \right] \tag{2.67}$$

whose evolution after the pulse is given by

$$\rho(t) = e^{-i\mathcal{H}'_D t/\hbar} \rho_0 e^{i\mathcal{H}'_D t/\hbar}, \tag{2.68}$$

and where the part $\mathcal{H}_2$ is a combination of terms $I^1_\pm I^2_\pm$ whose products with $I_{x,y}$ have a trace equal to zero.

The precessing magnetization, which results from the transverse spin components, follows their time evolution, given in the rotating frame by

$$\langle I_y(t) \rangle = \text{Tr}\{\rho(t) I_y\} = \frac{\hbar \omega_0}{kT_Z} \sin\theta \, \text{Tr}\{I_y^2\} G(t) \tag{2.69}$$

and

$$\langle I_x(t)\rangle = \text{Tr}\{\rho(t)I_x\} = \frac{\hbar\omega_0}{kT_D}\cos\theta\sin\theta\,\text{Tr}\{I_x^2\}\frac{d}{dt}G(t), \qquad (2.70)$$

where the function $G(t)$ is given by

$$G(t) = \frac{\text{Tr}\{I_y(t)I_y\}}{\text{Tr}\{I_y^2\}} = \frac{\text{Tr}\{e^{-i\mathcal{H}_D't/\hbar}I_y e^{i\mathcal{H}_D't/\hbar}I_y\}}{\text{Tr}\{I_y^2\}} \qquad (2.71)$$

and is called the free induction decay (FID) function. We note that the FID shape is given entirely by the dipolar Hamiltonian, whereas the size of the magnetization is linearly proportional to the rotation angle of the excitation pulse, provided that it is very small. In the expressions in Eqs. 2.69 and 2.70 the fact has been used that the product $\rho(t)I_y$ has only one term with non-zero trace. It should be noted also that there is no analytical method to calculate the function $G(t)$, but one has to rely on numeric methods in estimating it from the sum of the dipolar interaction terms over a reasonable fraction of the lattice around a given spin.

Response to Continuous Excitation by a Transverse Rotating Field

In the general theory of linear response formulated by Kubo and Tomita [7], the above FID signal of rotating magnetization is the response $\mathcal{R}_0(t)$ to the excitation pulse

$$\mathcal{E}_0(t) = \theta\delta(t),$$

and this response, in the complex notation and when θ is very small, is

$$\mathcal{R}_0(t) = \langle I_x(t)\rangle + i\langle I_y(t)\rangle = \theta\text{Tr}\{I_x^2\}\left\{\beta\frac{dG(t)}{dt} + i\beta_Z\omega_0 G(t)\right\}, \qquad (2.72)$$

where we have simplified the notation by using the inverse spin temperatures β and β_Z.

The continuous-wave excitation in the rotating frame is written in the complex notation as

$$\mathcal{E}(t) = \gamma B_1(t) = \omega_1 e^{-i\Delta t},$$

where $\Delta = \omega_0 - \omega$. The linear response to it in the rotating frame is proportional to the complex susceptibility:

$$\mathcal{R}(t) \equiv \omega_1\chi(-\Delta)e^{-i\Delta t}, \qquad (2.72')$$

which can be written in terms of $\mathcal{R}_0(t)$

$$\frac{1}{\omega_1}\mathcal{R}(t) = e^{-iDt}\int_{-\infty}^{+\infty}\frac{1}{\theta}\mathcal{R}_0(t')e^{+iDt'}dt'$$

$$= \frac{1}{\theta}e^{-iDt}\int_{-\infty}^{+\infty}\left[\langle I_x(t')\rangle + i\langle I_y(t')\rangle\right]e^{+iDt'}dt'$$

$$= \text{Tr}\{I_x^2\}e^{-iDt}\int_{-\infty}^{+\infty}\left\{\beta\frac{dG(t')}{dt} + i\beta_Z\omega_0 G(t')\right\}e^{+iDt'}dt$$

2.2 Magnetic Resonance of Interacting Spins in Solids

using the result (Eq. 2.72). The real and imaginary parts of the integral in the last form can be obtained by integrating the derivative of $G(t')$ in parts:

$$\int_{-\infty}^{+\infty} \frac{dG(t')}{dt'} e^{+i\Delta t'} dt' = \left[G(\infty) - G(-\infty) \right] e^{+i\Delta t'} - i\Delta \int_{-\infty}^{+\infty} G(t') \, e^{+i\Delta t'} dt'$$

$$= -i\Delta \int_{-\infty}^{+\infty} G(t') \, e^{+i\Delta t'} dt',$$

where the first term is zero because the FID response is damped to zero at infinity, and there is no response before the pulse at $t' = 0$. The response to the continuous excitation can now be written

$$\mathcal{R}(t) = \omega_1 \mathrm{Tr}\{I_x^2\} \left[\beta_Z \omega_0 - \beta \Delta \right] e^{-i\Delta t} \int_{-\infty}^{+\infty} iG(t') e^{+i\Delta t'} dt',$$

where it has been taken into account that the integration below zero yields zero, because there is no FID response before the excitation pulse. The integral therefore needs to be carried out from zero to infinity. By comparing this with Eq. 2.72' we can identify the spin components in the rotating frame as:

$$\langle I_x(\Delta) \rangle = -\mathrm{Tr}\{I_x^2\} \pi \omega_1 (\beta_Z \omega_0 - \beta \Delta) g'(\Delta) \tag{2.73}$$

and

$$\langle I_y(\Delta) \rangle = \mathrm{Tr}\{I_x^2\} \pi \omega_1 (\beta_Z \omega_0 - \beta \Delta) g(\Delta) \tag{2.74}$$

where

$$g(\Delta) = \frac{1}{\pi} \int_0^\infty G(t) \cos \Delta t \, dt \tag{2.75}$$

and

$$g'(\Delta) = \frac{1}{\pi} \int_0^\infty G(t) \sin \Delta t \, dt. \tag{2.76}$$

It is notable that in the above the susceptibility does not obey the usual Fourier transform relation with the pulse response, but the function $g(\Delta)$ does. This follows from the form of the density matrix where the Zeeman and dipolar temperatures were allowed to be different but constant during the irradiation, because the RF field was assumed to be vanishingly small so that no mixing between the two parts of the Hamiltonian happens (it can be thought to be infinitely slow). The conditions under which this is valid can be obtained from the Provotorov equations which will be discussed in the next section.

Properties of the Complex Susceptibility at High Temperatures

The function $g(\Delta)$ is symmetric about $\Delta = 0$, i.e. $g(-\Delta) = g(\Delta)$, and its integral is $\int_{-\infty}^\infty g(\Delta) d\Delta = 1$. In the limit that the dipolar and Zeeman temperatures are equal and constant, one obtains the usual relationship between the absorption and dispersion parts of the complex susceptibility and the lineshape function $g(\Delta)$:

$$\chi'(-\Delta) = \frac{1}{\omega_1}\langle I_x(\Delta)\rangle = -\mathrm{Tr}\{I_x^2\}\pi\frac{\hbar(\omega_0-\Delta)}{kT}g'(\Delta), \qquad (2.77)$$

$$\chi''(-\Delta) = \frac{1}{\omega_1}\langle I_y(\Delta)\rangle = \mathrm{Tr}\{I_x^2\}\pi\frac{\hbar(\omega_0-\Delta)}{kT}g(\Delta). \qquad (2.77')$$

In this case the well-known Kramers–Krönig relations between the real and imaginary parts of the susceptibility follow directly from the above. Their proof is given in Refs. [1], [2] and we shall not repeat it here; the relations are

$$\chi'(\omega) - \chi'(\infty) = \frac{1}{\pi}\mathcal{P}\int_{-\infty}^{+\infty}\frac{\chi''(\omega')}{\omega'-\omega}d\omega', \qquad (2.78)$$

$$\chi''(\omega) = -\frac{1}{\pi}\mathcal{P}\int_{-\infty}^{+\infty}\frac{\chi'(\omega')-\chi'(\infty)}{\omega'-\omega}d\omega', \qquad (2.79)$$

where $\mathcal{P}$ denotes the principal value of the integral which avoids divergence due to the poles of the integrands, and which follows from the derivation of the equations based on the use of the integral theorem of Cauchy. The principal value of the integrals can be evaluated by taking the limit

$$\mathcal{P}\int_{-\infty}^{+\infty}\frac{f(\omega')}{\omega'-\omega}d\omega' = \lim_{\varepsilon\to 0}\left[\int_{-\infty}^{\omega-\varepsilon}\frac{f(\omega')}{\omega'-\omega}d\omega' + \int_{\omega+\varepsilon}^{+\infty}\frac{f(\omega')}{\omega'-\omega}d\omega'\right]. \qquad (2.80)$$

The Kramers-Krönig relations allow one to compute one of the parts of the complex susceptibility if the other is known completely. Often, however, complete measurement of neither one is possible. We then need to know some general properties of the transverse susceptibility of the spin system to be able to restrict the region of measurement. Firstly, we know that absorption is zero at zero frequency; this is also directly obtained from Eq. 2.79. Furthermore, the same equation shows that absorption is antisymmetric about zero frequency, and Eq. 2.78 indicates that dispersion is symmetric about zero frequency. Secondly, direct integration of Eq. 2.79 gives the static susceptibility which is the dispersion at zero frequency:

$$\begin{aligned}\chi'(0) - \chi'(\infty) \equiv \chi_0 - \chi'(\infty) &= \mathcal{P}\frac{1}{\pi}\int_{-\infty}^{+\infty}\frac{\chi''(\omega')}{\omega'}d\omega' \\ &= \frac{2}{\pi}\int_0^{+\infty}\frac{\chi''(\omega')}{\omega'}d\omega'.\end{aligned} \qquad (2.81)$$

This confirms that the static susceptibility is proportional to the integral of the lineshape function, Eq. 2.65, and gives the basis for the determination of polarization from the integrated NMR signal.

Assuming that $\chi'(\infty) = 0$ and a very narrow absorption line which can be approximated by antisymmetric delta function

$$\chi''(\omega) = \frac{\pi}{2}\omega_0\chi_0\left[\delta(\omega-\omega_0)-\delta(-\omega-\omega_0)\right] \quad (2.82)$$

satisfying the Eq. 2.74, we may integrate Eq. 2.71 to obtain the dispersion

$$\chi'(\omega)-\chi'(\infty) = \frac{1}{\pi}\mathcal{P}\int_{-\infty}^{+\infty}\frac{\pi}{2}\omega_0\chi_0\frac{\delta(\omega'-\omega_0)-\delta(-\omega'-\omega_0)}{\omega'-\omega}d\omega'$$
$$= \frac{1}{2}\omega_0\chi_0\left(\frac{1}{\omega_0-\omega}+\frac{1}{\omega_0+\omega}\right) = \chi_0\frac{\omega_0^2}{\omega_0^2-\omega^2}. \quad (2.83)$$

We note that this satisfies $\chi'(0) = \chi_0$ only if $\chi'(\infty) = 0$, which can be expected because it was assumed for obtaining the absorption function. The dispersion function is plotted in Figure 2.2, where we may see that it is sizable at frequencies far from the resonance, and that it is a symmetric function about $\omega = 0$. It becomes evident from the form of the integral that for narrow absorption lines the dispersion must behave much like Eq. 2.83 at frequencies that are more than several linewidths away from the Larmor frequency.

If $\chi'(\infty) = 0$, then Eq. 2.81 is identical to Eq. 2.64. This is not coincidental although the linear response theory was not introduced explicitly when calculating the response of the spin system to the oscillating transverse field.

The symmetry properties of the absorption and dispersion functions can be used for rewriting the Kramers and Krönig relationships in a way which is practical for analytical or numeric evaluations of one when the other is known:

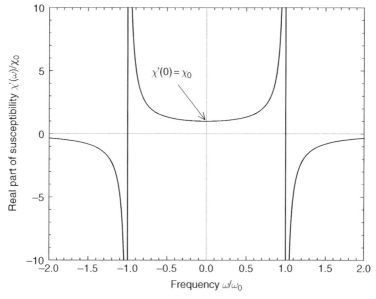

Figure 2.2 Dispersion lineshape for the case of narrow absorption line at $\omega = \omega_0$. Note that the symmetric dispersion function extends far from the wings of the absorption and has the value of static susceptibility at zero frequency

$$\chi'(\omega) - \chi'(\infty) = \frac{2}{\pi} P \int_0^{+\infty} \frac{\chi''(\omega')}{\omega'^2 - \omega^2} \omega' d\omega', \qquad (2.84)$$

$$\chi''(\omega) = -\frac{2\omega}{\pi} P \int_0^{+\infty} \frac{\chi'(\omega') - \chi'(\infty)}{\omega'^2 - \omega^2} d\omega'. \qquad (2.85)$$

The Kramers and Krönig relationship between χ' and χ'' is independent of the form of the Hamiltonian describing the interactions and it remains valid at low temperatures and at high polarization because no other assumptions than thermal equilibrium need to be made on the Hamiltonian and the density matrix describing the spin system.

Relation between the Transition Probabilities and Complex Susceptibility

Let us now assume the Boltzmann distribution of the magnetic level populations at high temperature and rewrite the susceptibility from Eqs. 2.58, 2.59 and 1.58 in the form

$$\chi''(\omega) = \frac{\hbar \omega \pi}{kTZ} \sum_{a,b} e^{-E_a/kT} |\langle a|\mu_x|b\rangle|^2 \delta(E_a - E_b - \hbar\omega), \qquad (2.86)$$

where Z is the partition function (the sum of all Boltzmann factors, needed for normalizing the level populations so that their sum is 1). By using the integral form of the delta function

$$\delta(x) = \frac{1}{2\pi\hbar} \int_{-\infty}^{+\infty} e^{-ixt/\hbar} dt, \qquad (2.87)$$

we get

$$\chi''(\omega) = \frac{\omega}{2kTZ} \int_{-\infty}^{+\infty} \sum_{a,b} e^{-E_a/kT} \langle a|\mu_x|b\rangle \langle b|\mu_x|a\rangle e^{i(E_a-E_b)t/\hbar} e^{-i\omega t} dt. \qquad (2.88)$$

Knowing that the states $|a\rangle$ and $|b\rangle$ are eigenfunctions of the Hamiltonian $\mathcal{H}$ we find a more formal way of writing the above as

$$\chi''(\omega) = \frac{\omega}{2kTZ} \int_{-\infty}^{+\infty} \sum_{a,b} \langle a|e^{-\mathcal{H}/kT} e^{i\mathcal{H}t/\hbar} \mu_x e^{-i\mathcal{H}t/\hbar}|b\rangle \langle b|\mu_x|a\rangle e^{-i\omega t} dt. \qquad (2.89)$$

The summation over a and b is in the matrix representation a trace

$$\sum_{a,b} \langle a|e^{-\mathcal{H}/kT} e^{i\mathcal{H}t/\hbar} \mu_x e^{-i\mathcal{H}t/\hbar}|b\rangle \langle b|\mu_x|a\rangle = \mathrm{Tr}\{e^{-\mathcal{H}/kT} e^{i\mathcal{H}t/\hbar} \mu_x e^{-i\mathcal{H}t/\hbar} \mu_x\}, \qquad (2.90)$$

where, in the high temperature approximation, $e^{-\mathcal{H}/kT}$ is replaced by unity. Furthermore, by defining the operator

$$\mu_x(t) = e^{i\mathcal{H}t/\hbar} \mu_x e^{-i\mathcal{H}t/\hbar}, \qquad (2.91)$$

we get a very compact expression

$$\chi''(\omega) = \frac{\omega}{2kTZ} \int_{-\infty}^{+\infty} \mathrm{Tr}\{\mu_x(t)\mu_x\} e^{-i\omega t} dt. \quad (2.92)$$

Comparing now with Eq. 2.70 shows that $\mathrm{Tr}\{\mu_x(t)\mu_x\}$ is the correlation function which gives the FID envelope.

Defining the shape function

$$g(\omega) = \frac{\chi''(\omega)}{\omega} = \frac{1}{2kTZ} \int_{-\infty}^{+\infty} \mathrm{Tr}\{\mu_x(t)\mu_x\} e^{-i\omega t} dt \quad (2.93)$$

and taking its Fourier transform we get

$$\frac{1}{2kTZ} \mathrm{Tr}\{\mu_x(t)\mu_x\} = \frac{1}{2\pi} \int_{-\infty}^{+\infty} g(\omega) e^{i\omega t} d\omega. \quad (2.94)$$

Setting $t = 0$ this becomes

$$\frac{1}{2kTZ} \mathrm{Tr}\{\mu_x(0)\mu_x\} = \frac{1}{2\pi} \int_{-\infty}^{+\infty} g(\omega) d\omega, \quad (2.95)$$

which is the static susceptibility. The higher moments of the lineshape function can now be evaluated by taking the nth derivative of Eq. 2.94 at $t = 0$ which yields

$$\frac{1}{2kTZ} \frac{d^n}{dt^n} \mathrm{Tr}\{\mu_x(t)\mu_x\}\bigg|_{t=0} = \frac{i^n}{2\pi} \int_{-\infty}^{+\infty} \omega^n g(\omega) d\omega. \quad (2.96)$$

The normalized moments of the shape function become

$$\langle \omega^n \rangle = \frac{\int_{-\infty}^{+\infty} \omega^n g(\omega) d\omega}{\int_{-\infty}^{+\infty} g(\omega) d\omega} = \frac{\dfrac{d^n}{dt^n} \mathrm{Tr}\{\mu_x(t)\mu_x\}\bigg|_{t=0}}{\mathrm{Tr}\{\mu_x(0)\mu_x\}} \quad (2.97)$$

which, by using the more explicit form (Eq. 2.91) and the von Neumann–Liouville equation (1.76), can be put to the form

$$\langle \omega^n \rangle = \left(\frac{i}{\hbar}\right)^n \frac{\mathrm{Tr}\{\overbrace{[\mathcal{H},[\mathcal{H},[\ldots[\mathcal{H},\mu_x]\ldots]\}}^{n \text{ times}}}{\mathrm{Tr}\{\mu_x^2\}}. \quad (2.98)$$

The commutators under the trace can now be evaluated for a given Hamiltonian. This form is frequently used for obtaining the moments of the dipolar Hamiltonian perturbing the Zeeman Hamiltonian, in which case the commutation rules of the spin operators enable to systematically obtain explicit formulas which can be numerically evaluated, although rather tediously for the fourth and higher moments. We shall return to this in the next section.

We emphasize that the relations for the absorption lineshape in this section were obtained in the high-temperature approximation. Abragam and Goldman [6] have extended the formal theory of lineshapes to low temperatures which is of interest for polarized targets where high nuclear polarization must be measured. We shall outline their derivation and discuss the results in Chapter 5.

2.2.3 Van Vleck's Method of Moments for Dipolar Lineshape

We shall now proceed to discuss the method which Van Vleck [8] originally developed to evaluate the moments of the absorption lineshape in the special case of broadening by the dipolar interactions only in crystalline solids. The Hamiltonian is then the sum of the Zeeman and dipolar parts of Eq. 2.12

$$\mathcal{H} = \mathcal{H}_Z + \mathcal{H}'_D, \tag{2.99}$$

where only the secular terms A and B of Eq. 2.14 were retained of the total dipolar Hamiltonian (Eq. 2.9), written here for like spins so that $\gamma_j = \gamma_k = \gamma$:

$$\mathcal{H}'_D = \frac{\mu_0 \hbar^2 \gamma^2}{8\pi} \sum_{j \ne k} \frac{1}{r_{jk}^3} \left(1 - 3\cos^2\theta\right)\left(3 I_{jz} I_{kz} - I_j \cdot I_k\right). \tag{2.100}$$

This was justified because the remaining terms lead to absorption at frequencies 0 and $2\omega_0$ which are far from the main absorption line centered around ω_0, the moments of which one wishes to measure and calculate.

The moments of the lineshape function $g(\omega)$ were defined by Eq. 2.97 as

$$\langle \omega^n \rangle = \frac{\int_{-\infty}^{+\infty} \omega^n g(\omega) d\omega}{\int_{-\infty}^{+\infty} g(\omega) d\omega},$$

but it is more practical to define also

$$\langle \Delta^n \rangle = \frac{\int_{-\infty}^{+\infty} (\omega - \langle \omega \rangle)^n g(\omega) d\omega}{\int_{-\infty}^{+\infty} g(\omega) d\omega} \tag{2.101}$$

which can be immediately seen from Eqs. 2.97 and 2.101 to satisfy

$$\langle \Delta \rangle = 0,$$
$$\langle \Delta^2 \rangle = \langle \omega^2 \rangle - \langle \omega \rangle^2. \tag{2.102}$$

2.2 Magnetic Resonance of Interacting Spins in Solids

In the case of dipolar interactions and high temperature, it can be shown in fact that all odd moments $\langle \Delta^n \rangle$ vanish. The second moment is also often denoted by $D^2 = \langle \Delta^2 \rangle$, where D is called the dipolar frequency.

We shall outline here briefly how their second and fourth moments are calculated. For a more complete presentation, we refer to [1] and [2].

We shall first note that the integral of the lineshape function and first moment $\langle \omega \rangle$ can be evaluated quite generally at high temperature [2] based on the fact that the dipolar Hamiltonian is Hermitian. The results are

$$\int_{-\infty}^{+\infty} g(\omega)\,d\omega = \frac{1}{2\hbar}\mathrm{Tr}\{\mu_x^2\} = \frac{1}{2\hbar}\gamma^2\hbar^2 \frac{I(I+1)}{3} N(2I+1)^N, \tag{2.103}$$

$$\langle \omega \rangle = \frac{\int_{-\infty}^{+\infty} \omega g(\omega)\,d\omega}{\int_{-\infty}^{+\infty} g(\omega)\,d\omega} = \frac{\frac{\omega_0}{2\hbar}\mathrm{Tr}\{\mu_x^2\}}{\frac{1}{2\hbar}\mathrm{Tr}\{\mu_x^2\}} = \omega_0, \tag{2.104}$$

where N is the number of spins counted for the evaluation. This shows that the Larmor frequency is the centroid of the lineshape function, which could be already expected because the moment $\langle \Delta \rangle$ and all odd moments vanish, meaning that the absorption line is symmetric about its centroid.

The second and fourth moments are now obtained from Eq. 2.98:

$$\langle \Delta^2 \rangle = \left(\frac{i}{\hbar}\right)^2 \frac{\mathrm{Tr}\{[\mathcal{H}'_D,[\mathcal{H}'_D,\mu_x]]\}}{\mathrm{Tr}\{\mu_x^2\}} = -\frac{1}{\hbar^2} \frac{\mathrm{Tr}\{[\mathcal{H}'_D,I_x]^2\}}{\mathrm{Tr}\{I_x^2\}}, \tag{2.105}$$

$$\langle \Delta^4 \rangle = \frac{1}{\hbar^4} \frac{\mathrm{Tr}\{[\mathcal{H}'_D,[\mathcal{H}'_D,I_x]]^2\}}{\mathrm{Tr}\{I_x^2\}}. \tag{2.106}$$

The commutator of the dipolar Hamiltonian and I_x is

$$[\mathcal{H}'_D, I_x] = \frac{\mu_0 \hbar^2 \gamma^2}{8\pi} \sum_{j\neq k} \frac{1}{r_{jk}^3} \left[(A_{jk}+B_{jk}), \sum_i I_{ix}\right]$$

$$= \frac{\mu_0 \hbar^2 \gamma^2}{8\pi} \sum_{j\neq k} \frac{1}{r_{jk}^3}(1-3\cos^2\theta_{jk})\frac{3}{2}[I_{jz}I_{ky}+I_{kz}I_{jy}]. \tag{2.107}$$

In the evaluation of the commutator use is made of the fact that $I_{jx}+I_{kx}$ commutes with $I_j \cdot I_k$. Using the basic commutation rules (Eq. 1.12) we get now

$$[\mathcal{H}'_D, I_x] = i\frac{3\mu_0 \hbar^2 \gamma^2}{16\pi}\sum_{j\neq k} \frac{1-3\cos^2\theta_{jk}}{r_{jk}^3}(I_{jz}I_{ky}+I_{kz}I_{jy})$$

and

$$Tr\left\{[\mathcal{H}'_D, I_x]^2\right\} = -\frac{\mu_0^2 \hbar^2 \gamma^4}{2(4\pi)^2} I^2(I+1)^2 (2I+1)^N \sum_{j \neq k} \left(\frac{1-3\cos^2\theta_{jk}}{r_{jk}^3}\right)^2 \quad (2.108)$$

The double summation converges rapidly and can be put to the form involving a single sum only

$$\sum_{j \neq k} \left(\frac{1-3\cos^2\theta_{jk}}{r_{jk}^3}\right)^2 = N \sum_k \left(\frac{1-3\cos^2\theta_{jk}}{r_{jk}^3}\right)^2, \quad (2.109)$$

and the denominator of the second moment was already given in Eq. 2.103 above:

$$Tr\{I_x^2\} = \sum_j Tr\{I_{jx}\}^2 = \frac{I(I+1)}{3} N (2I+1)^N. \quad (2.110)$$

The second moment for a single spin species is now finally

$$\langle \Delta^2 \rangle = \frac{3}{4} \left(\frac{\mu_0}{4\pi}\right)^2 \hbar^2 \gamma^4 I(I+1) \sum_k \left(\frac{1-3\cos^2\theta_{jk}}{r_{jk}^3}\right)^2. \quad (2.111)$$

For a powder made of small crystals randomly oriented or for a glassy material the term $(1-3\cos^2\theta_{jk})^2$ can be averaged before taking the sum; in this case the second moment becomes

$$D^2 = \langle \Delta^2 \rangle = \frac{3}{5} \left(\frac{\mu_0}{4\pi}\right)^2 \hbar^2 \gamma^4 I(I+1) \sum_k \frac{1}{r_{jk}^6} \quad (2.112)$$

and if the distances between the nearest nuclei with spin I are uniform and equal to d, the sum can be evaluated for a simple cubic lattice

$$\sum_k \frac{1}{r_{jk}^6} = \frac{8.5}{d^6} \quad (2.113)$$

which gives

$$\langle \Delta^2 \rangle = 5.1 \left(\frac{\mu_0}{4\pi}\right)^2 \hbar^2 \gamma^4 I(I+1) \frac{1}{d^6}. \quad (2.114)$$

This is a result which will be used frequently in this and the following chapters.

The fourth moment is very tedious to evaluate and we give the result [8] only

$$\langle \Delta^4 \rangle = \frac{1}{9} \left(\frac{\mu_0}{4\pi}\right)^4 \hbar^4 \gamma^8 I^2(I+1)^2 \left\{ 3 \left(\sum_k b_{jk}^2\right)^2 - \frac{1}{5}\left(8 + \frac{3}{2I(I+1)}\right) \sum_k b_{jk}^4 - \frac{1}{3N} \sum_{j \neq k \neq l} b_{jk}^2 (b_{jl} - b_{kl})^2 \right\}$$

where the indices of the triple summation must be unequal, and

$$b_{jk} = \frac{3}{2} \frac{1-3\cos^2\theta_{jk}}{r_{jk}^3}. \quad (2.115)$$

The two last sums converge more rapidly than the first which is the same as in the second moment; a fairly accurate result is obtained by including only the nearest neighbors in them. If they are omitted, we have the ratio

$$\frac{\langle \omega^4 \rangle}{\langle \omega^2 \rangle^2} = 3, \tag{2.116}$$

which is the same as that evaluated for Gaussian lineshape. It can be therefore expected that the tails of the dipolar lineshape go to zero at least as rapidly as the Gaussian of the same half-width.

As an example, the normalized Gaussian function

$$f(\Delta) = \frac{1}{\sigma\sqrt{2\pi}} e^{-\Delta^2/2\sigma^2}$$

gives the following results for the even moments

$$M_2 = \sigma^2; M_4 = 3\sigma^4; M_{2n} = 1 \cdot 3 \cdots (2n-1)\sigma^{2n}. \tag{2.117}$$

The ratio

$$\frac{M_4}{(M_2)^2} = 3 \tag{2.118}$$

gives a rather sensitive test for the assumption that an experimental lineshape is close to a Gaussian. The odd moments are all zero by definition that the line is symmetric and centered about ω_0. The half-width at half-maximum (HWHM) for a Gaussian is $\Delta_{\frac{1}{2}} = \sigma\sqrt{2\ln 2} = 1.1774\,\sigma$ which deviates only slightly from the square root of the second moment.

The second and higher even moments of the Lorentzian function

$$f(\Delta) = \frac{\delta}{\pi} \frac{1}{\delta^2 + \Delta^2} \tag{2.119}$$

diverge because the Lorenzian function goes to zero too slowly. Such a function is unphysical for a spin system in a solid, because it would require that the spins could approach each other to infinitesimally small distance, rather than the atomic distances which are in the order of few Å. If this distance of nearest approach is assumed to result in a cut-off α in the frequency spectrum of absorption around the resonance, the following results are obtained for the moments

$$M_2 = \frac{2\alpha\delta}{\pi}; M_4 = \frac{2\alpha^3\delta}{3\pi}; \frac{M_4}{(M_2)^2} = \frac{\pi\alpha}{6\delta}. \tag{2.120}$$

The modified Lorentzian function

$$f(\Delta) = \frac{\delta^{1.5}}{C\pi} \frac{1}{\delta^{2.5} + |\Delta|^{2.5}} \tag{2.120'}$$

is often used to numerically simulate experimental NMR absorption spectra in glassy alcohol and diol target materials because the function fits the features of the experimental signals better that the Gaussian or Lorentzian functions. There is, however, no theoretical justification for this shape function, and its second and higher moments diverge. Similar to the Lorentzian shape, cut-off frequencies must be introduced if the moments are required for the analysis of experimental absorption lines.

The symmetric functions

$$f(\Delta) = \frac{\delta^{2c-1}}{C\pi} \frac{1}{\left(\delta^2 + \Delta^2\right)^c} \qquad (2.120'')$$

have a finite second moment for $c \geq 1$ and a finite fourth moment for $c \geq 3$. They are special cases of the functions

$$f(\Delta) = \frac{\delta^{bc-1}}{C\pi} \frac{1}{\left(m + x^b\right)^c} \qquad (2.121)$$

which has finite moments of order lower or equal to $bc - 1$. The lineshape parameter of these functions approaches the value 3 when c goes to infinity. Their properties will be discussed in Appendix A6.

2.2.4 Provotorov Equations and Saturation

Provotorov [9] derived in 1961 equations which describe the evolution of the Zeeman and dipolar temperatures when the RF field B_1 is no longer small enough to justify the simplifying approximations of the linear response theory, where the Zeeman and dipolar Hamiltonians are unaffected by the presence of the perturbation. It is, however, assumed that the temperature is high, i.e. the Curie law (1.69) is valid, and that the RF field is smaller than the local field so that

$$\omega_1 = |\gamma B_1| \ll D \equiv |\gamma B'_L|. \qquad (2.122)$$

The local field and the dipolar frequency D are given by

$$\left(\gamma B'_L\right)^2 = D^2 = \frac{\text{Tr}\left\{\mathcal{H}'^2_D\right\}}{\text{Tr}\left\{I^2_z\right\}} \cong \left\langle \Delta^2 \right\rangle. \qquad (2.123)$$

In the frame rotating at the frequency $\omega = \omega_0 - \Delta$ of the RF field the effective Hamiltonian is

$$\mathcal{H}^* = \hbar \Delta I_z + \mathcal{H}'_D + \hbar \omega_1 I_x. \qquad (2.124)$$

Before the application of the RF field the spin system is in equilibrium with the lattice and the (high-temperature) density matrix is

2.2 Magnetic Resonance of Interacting Spins in Solids

$$\rho = 1 - \beta_Z \omega_0 I_z - \beta \mathcal{H}'_D, \tag{2.125}$$

where we have defined the inverse temperatures for the Zeeman and the dipolar interactions

$$\beta_Z = \frac{\hbar}{kT_Z};$$
$$\beta = \frac{\hbar}{kT_D} \tag{2.126}$$

in the stationary frame. Because of the invariance of the Hamiltonian under rotation about the z-axis, the density matrix has the same form in the rotating frame:

$$\rho^* = 1 - \alpha \Delta I_z - \beta \mathcal{H}'_D. \tag{2.127}$$

Here we have defined the inverse Zeeman temperature in the rotating frame as

$$\alpha = \frac{\omega_0}{\Delta} \beta_Z. \tag{2.128}$$

The energy without RF field is, using Eq. 1.75,

$$\langle \mathcal{H}^* \rangle = Tr\{\rho^* \mathcal{H}^*\} = -(\alpha \Delta^2 + \beta D^2) Tr\{I_z^2\}. \tag{2.129}$$

When turning the RF field on, the Hamiltonian and the density matrix are slightly modified, but as long as the inequality (Eq. 2.122) is satisfied, the energy is conserved because the effective Hamiltonian in the rotating frame has no explicit time dependence; this was shown by Eq. 1.79. The behavior of the spin temperatures after turning the RF field on is now obtained by taking the time derivative of Eq. 2.129:

$$\frac{d}{dt}\langle \mathcal{H}^* \rangle = -\left(\Delta^2 \frac{d}{dt}\alpha(t) + D^2 \frac{d}{dt}\beta(t)\right) Tr\{I_z^2\} = 0. \tag{2.130}$$

The time evolution of the polarization in the rotating frame can be expressed in terms of the change of the Zeeman temperature:

$$\frac{d}{dt}\langle I_z \rangle = -Tr\{I_z^2\} \Delta \frac{d}{dt}\alpha(t). \tag{2.131}$$

On the other hand, this can also be derived using Eq. 1.78

$$\frac{d}{dt}\langle I_z \rangle = \left\langle \frac{i}{\hbar}[\mathcal{H}^*, I_z] + \frac{\partial I_z}{\partial t} \right\rangle = \frac{i}{\hbar}\langle [\mathcal{H}^*, I_z] \rangle. \tag{2.132}$$

where the partial time derivative of the operator I_z is zero because it has no explicit time dependence. Its expectation value can be evaluated using Eq. 2.124 which yields

$$\frac{d}{dt}\langle I_z \rangle = \omega_1 \langle I_y \rangle. \tag{2.133}$$

This, in turn, can be related to the inverse temperatures using $\langle I_y \rangle$ from Eq. 2.74

$$\frac{d}{dt}\langle I_z \rangle = \text{Tr}\{I_x^2\} \pi \omega_1^2 \Delta \big[\alpha(t) - \beta(t)\big] g(\Delta). \tag{2.134}$$

Combining Eqs. 2.131 and 2.134 we find the time evolution of α

$$\frac{d}{dt}\alpha(t) = -\pi \omega_1^2 g(\Delta) \big[\alpha(t) - \beta(t)\big] \tag{2.135}$$

and that of β by inserting this to Eq. 2.130

$$\frac{d}{dt}\beta(t) = -\pi \omega_1^2 \frac{\Delta^2}{D^2} g(\Delta) \big[\beta(t) - \alpha(t)\big]. \tag{2.136}$$

These are the well-known Provotorov equations. We see immediately that the time constant in Eq. 2.135 is given by the usual inverse transition probability, going up inversely proportional to the absorption lineshape function as the RF frequency is shifted away from the resonance. The time constant for the inverse dipolar temperature of Eq. 2.136 goes up more slowly. The solution of the coupled differential equations is simple and yields the common rate or inverse time constant

$$\tau^{-1} = \pi \omega_1^2 g(\Delta) \left[1 + \frac{\Delta^2}{D^2}\right], \tag{2.137}$$

by which the two inverse temperatures exponentially converge towards the common equilibrium value

$$\beta_{eq} = \beta_L \frac{1 + \dfrac{\omega_0 \Delta}{D^2}}{1 + \dfrac{\Delta^2}{D^2}}, \tag{2.138}$$

if the temperatures immediately after turning the RF field on were $\beta(0) = \beta_L$ and $\alpha(0) = \beta_L \omega_0 / \Delta$. Figure 2.3 shows the steady-state inverse temperature and the rate or inverse time constant to reach this value.

The linear response approximation in Section 2.2.2 required that the susceptibility be measured in a time t_{exp} which is short in comparison with both time constants appearing in the above equations, i.e.

$$t_{exp} \ll \left(\pi \omega_1^2 g(\Delta)\right)^{-1}; \; t_{exp} \ll \left(\pi \omega_1^2 \frac{\Delta^2}{D^2} g(\Delta)\right)^{-1}. \tag{2.139}$$

These are more strict at or close to the resonance, and because $1/T_2 \approx D \approx g(0)^{-1}$, we get the requirement for protons with $t_{exp} \approx 10\,\text{ms}$ and $T_2 \approx 30\,\mu\text{s}$

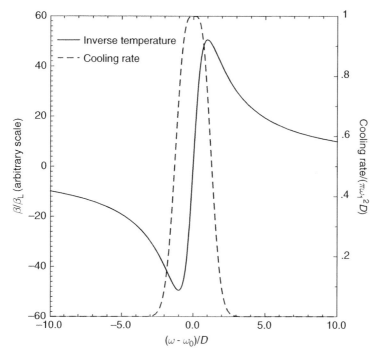

Figure 2.3 Inverse steady-state spin temperature β and the inverse time constant for reaching equilibrium, as a function of frequency deviation from the center of an ESR line with Gaussian shape. We note that the extrema of β are reached close to $\omega - \omega_0 = \pm D$ and that the cooling rates at these frequencies are high, while the rate falls close to zero beyond $\pm 3D$

$$\omega_1 \ll \sqrt{\frac{1}{\pi T_2 t_{\exp}}} \approx 10^3 \text{s}^{-1} \quad \text{or} \quad B_1 \ll 4\,\mu\text{T}. \tag{2.140}$$

The field close to the wire of the NMR coil is sometimes of the same order as the above limit,[3] and therefore the lineshape distortions resulting from a transverse field not satisfying Eq. 2.139 will be discussed below and in Chapter 5.

The conditions of validity of the Provotorov equations themselves can be obtained from the implicit assumption which was made above, namely that the dipolar and Zeeman systems are always in internal thermal equilibrium. The time constant for these is T_2, which should be short in comparison with the time constants in the Provotorov equations. This results in the requirement $|\omega_1| \ll D$ which is equivalent to $B_1 \ll 1$ mT for protons in polarized target materials at 2.5 T field. For electrons the transverse field limit is higher in proportion with the ratio of the gyromagnetic factors and is beyond the values which can be applied in the laboratory in samples at low temperatures.

Although the Provotorov equations were derived using the high-temperature approximation, the conclusions which were made from them are semi-quantitatively valid at low

---

[3] For example, if the NMR coil wire carries a current of 0.3 mA, the transverse field strength at 1 mm radius is 0.06 µT.

temperatures, which can be seen in the various steps of their derivation. One might attempt to modify them phenomenologically by introducing the specific heats of the Zeeman and dipolar reservoirs in Eq. 2.130, which states that energy is conserved, and by modifying Eq. 2.74 to correspond to low temperatures. The resulting equations would be non-linear in inverse temperature but might allow one to gain insight to the rate of approach of equilibrium in the limit of very high polarizations where the populations of the highest or lowest magnetic levels are close to 1 and the others have populations close to zero. This will be discussed qualitatively in Chapter 4.

The lattice interactions were included in the density matrix of the original work by Provotorov [9], who showed how the effect of spin-lattice relaxation in the rotating frame can be introduced in Eqs. 2.134 and 2.135. Under the assumption that the spin-lattice relaxation rates are slow in comparison with all other rates, the equations are then modified to

$$\frac{d}{dt}\alpha(t) = -\pi\omega_1^2 g(\Delta)\left[\alpha(t) - \beta(t)\right] - \frac{1}{T_{1Z}}\left[\alpha(t) - \alpha_0\right] \qquad (2.141)$$

and

$$\frac{d}{dt}\beta(t) = \pi\omega_1^2 \frac{\Delta^2}{D^2} g(\Delta)\left[\alpha(t) - \beta(t)\right] - \frac{1}{T_{1D}}\left[\beta(t) - \beta_L\right], \qquad (2.142)$$

which can be also obtained intuitively by considering the balance of energy. Here the inverse lattice temperature β_L is small in comparison with β and is usually omitted in the second equation; it cannot be omitted in the first one when α is high and the Zeeman relaxation time T_{1Z} is long in comparison with the other rates.

In his original paper [9] Provotorov calculated the NMR lineshape resulting from saturation, which does no longer correspond to the linear response approximation. The absorption and dispersion parts of the transverse susceptibility become then

$$\chi''(\Delta) = \frac{\pi}{2}\mu_0\hbar\gamma^2 n I P(I) \frac{g(\Delta)}{1 + \varepsilon(\Delta)T_{1Z}\left[1 + \frac{\Delta^2 T_{1D}}{D^2 T_{1Z}}\right]} \qquad (2.143)$$

$$\chi'(\Delta) = -\frac{\pi}{2}\mu_0\hbar\gamma^2 n I P(I) \frac{g'(\Delta) - \varepsilon(\Delta)T_{1D}\Delta/D^2}{1 + \varepsilon(\Delta)T_{1Z}\left[1 + \frac{\Delta^2 T_{1D}}{D^2 T_{1Z}}\right]} \qquad (2.144)$$

where

$$\varepsilon(\Delta) = \omega_1^2 \pi g(\Delta) \qquad (2.145)$$

and $g(\Delta)$, $g'(\Delta)$ are the normalized lineshape functions of Eqs. 2.75 and 2.76, which also obey the Kramers–Krönig relations

$$g'(\Delta) = -\frac{1}{\pi}\mathcal{P}\int_{-\infty}^{+\infty}\frac{g(\Delta')}{\Delta' - \Delta}d\Delta'. \qquad (2.146)$$

We note that the dispersion signal gets an admixture of absorption at strong saturation, and that both signals are weakened. The measurement of both signals allows one to determine the dipolar and Zeeman relaxation times T_{1D} and T_{1Z} which are difficult to predict theoretically. Unfortunately, such measurements have not yet been made in polarized target materials at low temperatures, and one has to resort to estimates yielding $T_{1D}/T_{1Z} = 1/3$, which will be discussed in the next section.

If the measurement of transverse susceptibility is not made fast in comparison with the spin-lattice relaxation time, undistorted lineshapes can only be expected if the saturation $\varepsilon(\Delta)T_{1Z}$ is very small.

Eq. 2.138 is somewhat unphysical because the spin system was assumed totally isolated from the lattice. The solution of the original Provotorov Eqs. 2.141 and 2.142 will be discussed in Chapter 4, but we shall write already here the result for the inverse dipolar and Zeeman temperatures:

$$\beta = \beta_L \frac{\omega}{\Delta} \frac{\varepsilon(\Delta)T_{1D}\frac{\Delta^2}{D^2}}{1+\varepsilon(\Delta)T_{1Z}\left(1+\frac{T_{1D}}{T_{1Z}}\frac{\Delta^2}{D^2}\right)} + \beta_L. \tag{2.140'}$$

$$\alpha = \beta_L \frac{\omega}{\Delta} \frac{1+\varepsilon(\Delta)T_{1D}\frac{\Delta^2}{D^2}}{1+\varepsilon(\Delta)T_{1Z}\left(1+\frac{T_{1D}}{T_{1Z}}\frac{\Delta^2}{D^2}\right)} + \beta_L. \tag{2.141'}$$

Here it was assumed that the relaxation of the dipolar and Zeeman temperatures in the rotating frame take place so that they feel the same lattice temperature, i.e. $\alpha_L = \beta_L$, and $\alpha_0 = \beta_L \cdot \omega_0/\Delta$. At high saturation the two inverse temperatures again approach a common value

$$\alpha = \beta = \beta_L \left(1 + \frac{\frac{\omega\Delta}{D^2}}{\frac{T_{1Z}}{T_{1D}} + \frac{\Delta^2}{D^2}}\right), \tag{2.141''}$$

which is only slightly different from Eq. 2.138 if the ratio T_{1Z}/T_{1D} does not deviate much from the theoretical value of 3.

The inclusion of the spin-lattice relaxation allows one to obtain the correct dependence of the spin temperature on the magnitude of the applied transverse field. With a spin system perfectly isolated from the lattice, any field strength produces the same end result, only the time required for reaching the equilibrium changes. This is clearly unphysical and calls for the inclusion of the spin-lattice effects.

The Provotorov equations have been used for determining the relaxation times from the continuous-wave NMR signals under small saturation. They are equally important in the evaluation of errors in polarization measurement based on integrated absorption signals (Chapters 5 and 6). These equations are also the basis for the qualitative and semi-quantitative understanding of dynamic nuclear polarization using the cooling of the dipolar interactions (Chapter 4).

2.3 Relaxation

In polarized target materials the electron spin-lattice relaxation is vitally important because it provides the mechanism for the electron Zeeman temperature to reach a steady-state equilibrium both in the absence and in the presence of a saturating rotating field, as was discussed above. The nuclear spin-lattice relaxation in polarized targets happens mainly through their contact with the electronic spins. We shall discuss the relaxation of the electronic and nuclear spins separately in Chapters 3 and 5 and limit ourselves in this section to the general aspects and mechanisms of spin relaxation phenomena.

2.3.1 *Spin-Lattice Relaxation of the Zeeman Energy*

We shall first discuss the spin transition rates in the absence of external transverse RF field, between the populations of the magnetic energy levels E_m of a spin system at high temperature where the Curie law (1.69) is valid. Basing his model on transition rates [10], Gorter showed that the inverse spin temperature β obeys

$$\frac{d\beta}{dt} = \frac{\beta_L - \beta}{T_1} \qquad (2.147)$$

with

$$\frac{1}{T_1} = \frac{\frac{1}{2}\sum_{m,n} W_{mn}\left(E_m - E_n\right)^2}{\sum_n E_n^2}, \qquad (2.148)$$

which leads to an exponential approach towards the inverse lattice temperature with the time constant T_1. Additional requirements for the above are that the transition rates W_{mn} are independent of the populations of the energy levels, and that at all times the spin system is in internal thermal equilibrium so that the level populations obey the Maxwell–Boltzmann statistics. Because the Curie law states that at high temperatures the polarization P is proportional to the inverse temperature of the spins, it also approaches exponentially its equilibrium value P_0 with the same time constant.

The calculation of the transition rates in the Eq. 2.148 can be made in a formal way based on generalized Provotorov equations, as has been shown by Abragam and Goldman [6]. Using the Liouville formalism and memory functions, they derive dynamic equations for a system with Hamiltonian

$$\mathcal{H} = \mathcal{H}_0 + V$$

where $\mathcal{H}_0 = \mathcal{H}_1 + \mathcal{H}_2$ and V is a small perturbation not commuting with any of the parts of the Hamiltonian, but

$$\text{Tr}\{\mathcal{H}_1\} = \text{Tr}\{\mathcal{H}_2\} = \text{Tr}\{\mathcal{H}_1\mathcal{H}_2\} = 0; \quad [\mathcal{H}_1, \mathcal{H}_2] = 0.$$

The time evolution of the inverse temperatures of the two parts of the Hamiltonian can then be written as

$$\frac{d\beta_1}{dt} = -W(\beta_1 - \beta_2);$$
$$\frac{d\beta_2}{dt} = -\varepsilon W(\beta_2 - \beta_1) \qquad (2.149)$$

with

$$\varepsilon = \frac{\text{Tr}\{\mathcal{H}_1^2\}}{\text{Tr}\{\mathcal{H}_2^2\}};$$

$$W = \frac{\int_0^\infty \text{Tr}\{[\mathcal{H}_1, V] e^{-it\mathcal{H}_0/\hbar} [V, \mathcal{H}_1] e^{it\mathcal{H}_0/\hbar}\} dt}{\text{Tr}\{\mathcal{H}_1^2\}}. \qquad (2.150)$$

Here W can be identified as a transition rate which can be evaluated for different processes such as saturation, cross relaxation, spin-lattice relaxation and so on. It is evident that the internal thermal equilibrium within the two systems must be rapid in comparison with the other time constants in the above equations. Provotorov derived similar equations for the specific case of cross relaxation [11] using the relevant Hamiltonians and density matrices directly.

In spin-lattice relaxation the lattice is assumed to be at a constant temperature, by virtue of its high heat capacity or, at low temperatures, by continuous cooling through solid-helium boundary to a bath of liquid helium, which itself is part of a continuous cooling cycle. In this case only the first of Eq. 2.148 is non-zero; it can be rewritten for a spin operator $\mathcal{H}_1 = Q$ as

$$\frac{d}{dt}\langle Q \rangle = -W(\langle Q(t) \rangle - \langle Q_L \rangle) \qquad (2.151)$$

where $\langle Q_L \rangle$ is the value of the operator in thermal equilibrium with the lattice. Writing the perturbation as a product $V = FA$ where F is a lattice operator of the lattice Hamiltonian $\mathcal{F}$ and A a spin operator, Abragam and Goldman [6] obtain the rate

$$W = -\frac{\text{tr}\{\rho(\mathcal{F})F^2\}}{\text{Tr}\{Q^2\}} \int_0^\infty \sum_{m,n} |\langle m|[Q,A]|n\rangle|^2 G(t) e^{-i(E_m - E_n)t/\hbar} dt \qquad (2.152)$$

where

$$G(t) = \frac{\text{tr}\{FF(t)\}}{\text{tr}\{FF\}}; \quad F(t) = e^{-i\mathcal{F}t/\hbar} F e^{+i\mathcal{F}t/\hbar} \qquad (2.153)$$

is the lattice correlation function of the variable F and $\rho(\mathcal{F})$ is the lattice density matrix, which cannot be simplified in the high temperature approximation like the spin density matrix. The lattice correlation decays from 1 at $t = 0$ to zero in a characteristic time τ_c

$$\tau_c = \frac{1}{\text{tr}\{\rho(\mathcal{F})FF\}} \int_0^\infty \text{tr}\{\rho(\mathcal{F})FF(t)\} dt. \qquad (2.154)$$

Although the decay is not necessarily exponential, approximating it so yields

$$W = -\frac{\text{tr}\{\rho(\mathcal{F})F^2\}}{\text{Tr}\{Q^2\}} \sum_{m,n} |\langle m|[Q,A]|n\rangle|^2 \frac{\tau_c}{1 + (E_m - E_n)^2 \tau_c^2/\hbar^2} \qquad (2.155)$$

In the above expressions 'tr' means trace over the lattice variables, and 'Tr' trace over the spin variables; the high-temperature approximation was used for the spin density matrix. The density matrix for the lattice cannot be expanded into Taylor series and truncated, because the energy spectrum of the phonons extends practically to zero and the phonons obey the Bose–Einstein statistics.

Phonons are the dominant lattice excitations in dielectric crystalline solids at low temperatures. The dominance is witnessed by the low-temperature specific heat which often obeys very well the prediction of the Debye model

$$C = 1944 \left(\frac{T}{\Theta_D}\right)^3 \frac{\text{J}}{\text{mol K}} \qquad (2.156)$$

where Θ_D is the Debye temperature, listed in Table A4.1 for several solids. The model is based on the approximation that the spectrum of phonons is cut off at the phonon energy

$$\hbar\Omega_D = k\Theta_D, \qquad (2.157)$$

that there are three modes of oscillation, one longitudinal and two transverse, and their dispersion is linear with a velocity of propagation, which is the same for all modes and for all orientations of the wave vector with respect to the lattice directions. The reasons why the model holds well at low temperatures (defined by $T \ll \Theta_D/4$) are that these conditions are rather well satisfied for low momentum phonons which dominate due to their Bose–Einstein statistics, and that their velocity is the acoustic velocity v_a in this limit. However, it is well known that the velocities of the transverse and longitudinal phonons are not equal and depend also on orientation in most crystals. This is taken into account by defining v_a as an average acoustic velocity. This point has been well described, for instance, by Landau and Lifshitz [12], whose presentation we shall briefly outline below.

The spectral density of phonon states in a crystal of volume V is

2.3 Relaxation

$$\sigma(\omega) = \frac{3V}{2\pi^2} \frac{\omega^2}{v_a^3} \tag{2.158}$$

and the Debye cut-off frequency is

$$\Omega_D = \sqrt[3]{6\pi^2} \frac{v_a}{a}; \quad a = \sqrt[3]{\frac{V}{N}}, \tag{2.159}$$

where a is the average lattice spacing of the monoatomic (or monomolecular) solid. Using these the spectral density becomes

$$\sigma(\omega) = \frac{9N\omega^2}{\Omega_D^3} \tag{2.160}$$

For diatomic or more complex ionic solids the low-temperature specific heat tells that the relevant spacing is obtained from the molar volume and Avogadro number

$$a = \sqrt[3]{\frac{V_m}{N_A}}. \tag{2.161}$$

The phonons are propagating lattice vibrations which can be represented classically by acoustic plane waves displacing atoms from their equilibrium positions by

$$u_j^s = u_j e^{i(\mathbf{k} \cdot \mathbf{r}_s - \omega t)}. \tag{2.162}$$

characterized by a wave vector $\mathbf{k}$ and angular frequency ω which is the frequency of oscillation of the atoms in the unit cell labelled by the vector $\mathbf{s}$. The atom and the direction of its displacement are labelled by j which takes three r values, where r is the number of atoms in the unit cell. The position vectors can be expressed as

$$\mathbf{r}_s = s_1 \mathbf{a}_1 + s_2 \mathbf{a}_2 + s_3 \mathbf{a}_3 \tag{2.163}$$

where s_i are integers and $\mathbf{a}_i$ are the basic vectors defining the unit cell. It follows from the form of Eq. 2.162 that one may add to the exponential any integer multiple of 2π without changing the displacement; this phase factor can be expressed as $2\pi \mathbf{b} \cdot \mathbf{r}_s$ with

$$\mathbf{b} = p_1 \mathbf{b}_1 + p_2 \mathbf{b}_2 + p_3 \mathbf{b}_3 \tag{2.164}$$

where p_i are any positive or negative integers or zero, and

$$\mathbf{b}_1 = \mathbf{a}_2 \times \mathbf{a}_3 / v; \quad \mathbf{b}_2 = \mathbf{a}_3 \times \mathbf{a}_1 / v; \quad \mathbf{b}_3 = \mathbf{a}_1 \times \mathbf{a}_2 / v; \quad v = \mathbf{a}_1 \cdot \mathbf{a}_2 \times \mathbf{a}_3 \tag{2.165}$$

define the basis vectors of what is called the reciprocal lattice.

The quantum treatment of the lattice involves the replacement of the definite displacement of the lattice points by quantized sound waves or phonons propagating in the lattice with definite energy and direction. These are represented by displacement operators similar to Eq. 2.162

$$u_j^s = u_j e^{i\mathbf{k}\cdot\mathbf{r_s}} + u_j^\dagger e^{i\mathbf{k}\cdot\mathbf{r_s}}. \tag{2.166}$$

where u_n, $u_n^\dagger$ are Hermitian conjugate operators with the only non-zero matrix elements

$$\langle n|u_j|+1\rangle = \langle n+1|u_j^\dagger|n\rangle^* = \left(\frac{\hbar\omega}{2M}\right)^{\frac{1}{2}} e^{-i\omega t}\sqrt{n+1}, \tag{2.167}$$

where n is the number of phonons in that state. The phonon Hamiltonian is obtained by summing Eq. 2.166 over the indices n and $\mathbf{s}$.

The phonon energy is

$$E_{ph} = \hbar\omega \tag{2.168}$$

and the wave vector determines the quasi-momentum

$$\mathbf{p} = \hbar\mathbf{k} \tag{2.169}$$

which is defined only within an arbitrary additive constant vector $2\pi\hbar\omega\mathbf{b}$. The velocity of a phonon is given by the group velocity

$$\mathbf{v} = \frac{\partial\omega}{\partial\mathbf{k}} = \frac{\partial E_{ph}(\mathbf{p})}{\partial\mathbf{p}}, \tag{2.170}$$

which is exactly analogous to the relation between the energy, momentum and velocity of a real particle and is independent of energy and momentum for the low-temperature phonons.

The freely propagating phonons interact with each other and with other possible excitations of the lattice because the oscillators are not perfectly harmonic at larger displacement. This enables the phonons to collide with each other elastically or inelastically, and the system to reach thermal equilibrium corresponding to Bose–Einstein statistics which follows from the possibility that any number of phonons may occupy a given state. The mean number of phonons in a quantum state defined by E_{ph} and $\mathbf{p}$ is then

$$n_\mathbf{p} = \frac{1}{e^{E_{ph}(\mathbf{p})/kT} - 1}. \tag{2.171}$$

The Hamiltonian describing the coupling of the spin and phonon systems can be expressed as

$$\mathcal{H}_{ph} = \hbar \sum_q F^q A^q, \tag{2.172}$$

where F^q and A^q are the lattice and spin operators, respectively. The lattice operators can be expanded in the terms of the stresses due to the phonons causing displacements of the unit cells of the lattice from their equilibrium positions

$$F = F_0 + F_1 W + F_2 W^2 + F_3 W^3 + \ldots \tag{2.173}$$

where the terms represent sums over the indices *i, k* of the product of tensors F_{nik} which are n^{th} derivatives of the energy with respect to the stress tensor

$$W_{ik} = \frac{1}{2}\left(\frac{\partial u_i}{\partial x_k} + \frac{\partial u_k}{\partial x_i}\right), \qquad (2.174)$$

and the powers of the stress tensor itself. The matrix element of W for the emission of a phonon of frequency ω is

$$\langle n|W|n+1\rangle = \langle n|\frac{\partial}{\partial x}u|n+1\rangle \cong \langle n|u|n+1\rangle|\mathbf{k}| = \left(\frac{(n+1)\hbar}{2\pi M\omega}\right)^{\frac{1}{2}}\frac{\omega}{v_a}, \qquad (2.175)$$

where we assumed low temperature so that

$$v_a = |\mathbf{v}| = \left|\frac{\partial E_{ph}(\mathbf{p})}{\partial \mathbf{p}}\right| = \frac{E_{ph}}{p} = \frac{\omega}{k}.$$

In the language of the second quantization, the terms $F_n W^n$ describe processes with n phonons created or destroyed. For $n = 1$, one phonon is either absorbed or emitted. The next term describes the simultaneous absorption or emission of two phonons, or absorption of one and emission of another phonon which is called the Raman process. In each process energy is conserved and the momentum conservation is within the additive constant $2\pi\hbar\mathbf{b}$; in an inelastic collision with two phonons in the initial and final states, for example, we must have

$$\mathbf{p}_1 + \mathbf{p}_2 = \mathbf{p}'_1 + \mathbf{p}'_2 + 2\pi\hbar\mathbf{b}. \qquad (2.176)$$

The mean wavelength of the phonon at 1 K temperature is

$$\lambda = \frac{v_a}{\omega/2\pi} \cong \frac{hv_a}{kT} = 240 \text{ nm} \qquad (2.177)$$

for a typical acoustic velocity of 5000 m/s in a solid. This is about three orders of magnitude more than the size of the unit cell even for fairly large molecules. At these temperatures, therefore, all the atoms of the molecule move in the same direction, and the molecules are not subjected to a large time-varying stress. This stress induces time variation on the spin-orbit coupling and therefore spin-lattice relaxation.

Let us now calculate the transition probability between two magnetic levels m and m' of a nucleus or electron with spin and magnetic moment coupled to the lattice described by the phonon Hamiltonian (Eq. 2.175). We shall calculate the order of magnitude of the matrix element Y coupling the two states $|m,n\rangle$ and $|m',n+1\rangle$ of the spin-lattice system described by the coupling Hamiltonian $AF_1 W$:

$$Y \approx \langle m|A|m'\rangle\langle n|F_1 W|n+1\rangle \cong \langle n|u|n+1\rangle F_1|\mathbf{k}| = F_1\left(\frac{(n+1)\hbar}{2\pi M\omega}\right)^{\frac{1}{2}}\frac{\omega}{v_a}, \qquad (2.178)$$

where it was assumed that $A \approx 1$. The transition probability resulting from this is

$$P_1 = \frac{1}{T_1} \approx \frac{2\pi}{\hbar}|Y|^2 \rho(E_{ph}) \tag{2.179}$$

which can be evaluated using Eqs. 2.159 and 2.178 yielding the rate of the direct process

$$\frac{1}{T_1} \approx 2F_1^2 \frac{(n+1)\hbar}{2M\omega}\left(\frac{\omega_0}{v_a}\right)^2 \frac{9N\omega_0^2}{\Omega_D^3}. \tag{2.180}$$

The number of phonons in the final state is given by Eq. 2.171 which is not necessarily large in the case of electrons in a high field and very low temperature. We shall get finally

$$\frac{1}{T_1} \approx 9\Omega_D \left(\frac{\omega_0}{\Omega_D}\right)^3 \left(\frac{F_1}{\Omega_D}\right)^2 \frac{\hbar\Omega_D}{mv_a^2}\left(\frac{1}{e^{\hbar\omega_0/kT}-1}+1\right), \tag{2.181}$$

where $m = M/N$ is the total mass of the atoms in the unit cell.

For nuclear Larmor frequencies, the last term can be quite large at temperatures where DNP is performed, but the smallness of the other terms makes the direct relaxation for nuclear spins much slower than those taking place through the electron-spin interactions, at all but the lowest temperatures attainable in dilution refrigerators. This will be discussed in Chapter 5 together with several other spin-lattice relaxation mechanisms that are specific to nuclear spins only.

For electrons around 1 K, the direct process dominates often, and we shall calculate numerical values and compare with experimental results in Chapter 3.

In the Raman process a phonon with frequency ω is absorbed and another with frequency $\omega' = \omega - \omega_0$ is emitted. The probability for this to happen is a sum over all possible frequencies of the product of probabilities of the transition and of the first photon to exist:

$$P_2 = \int_{\omega_0}^{\Omega_D} P_2^{(1)}\sigma(\omega)d\omega \tag{2.182}$$

where

$$P_2^{(1)} = \frac{2\pi}{\hbar}|Y_2|^2 \frac{\sigma(\omega')}{\hbar} \tag{2.183}$$

is the transition probability with emission of a phonon with frequency $\omega' = \omega - \omega_0$ after the first phonon was absorbed.

The matrix element for this transition is roughly

$$Y_2 \approx F_2 \frac{\omega\omega'}{v_a^2}\left[\frac{\hbar(n+1)}{2\pi M\omega}\frac{\hbar n'}{2\pi M\omega'}\right]^{\frac{1}{2}}, \tag{2.184}$$

which yields

$$P_2 = \frac{1}{T_1^{Raman}} = \frac{2}{\pi}\left(\frac{F_2}{v_a^2}\right)^2 \int_{\omega_0}^{\Omega_D} (\omega\omega')^2 \frac{\hbar(n+1)}{2M\omega} \frac{\hbar n'}{2M\omega'} d\omega \qquad (2.185)$$

using the spectral densities of Eq. 2.160. Using the Bose–Einstein populations of Eq. 2.171, changing the variables so that the integration is from 0 to $\Omega_D - \omega_0$ gives

$$\frac{1}{T_1^{Raman}} = P_2 = \frac{9^2}{\pi}\left(\frac{\hbar F_2}{mv_a^2}\right)^2 \int_0^{\Omega_D-\omega_0} \frac{\omega'^3(\omega'+\omega_0)^3}{\Omega_D^6} \frac{e^{\hbar(\omega'+\omega_0)/kT}}{\left(e^{\hbar\omega'/kT}-1\right)\left(e^{\hbar(\omega'+\omega_0)/kT}-1\right)} d\omega'. \qquad (2.186)$$

The relevant factors can be taken out of the integral by changing again the variables to

$$x = \frac{\hbar\omega'}{kT}; \; x_0 = \frac{\hbar\omega_0}{kT}; \; X_D = \frac{\hbar\Omega_D}{kT} = \frac{\Theta_D}{T}. \qquad (2.187)$$

which give

$$\frac{1}{T_1^{Raman}} = \frac{9^2}{\pi}\Omega_D\left(\frac{\hbar F_2}{mv_a^2}\right)^2\left(\frac{T}{\Theta_D}\right)^7 \int_0^{X_D-x_0} \frac{x^3(x+x_0)^3 e^{x+x_0}}{(e^x-1)(e^{x+x_0}-1)} dx. \qquad (2.188)$$

In the case of nuclear spins at low temperatures, x_0 is very small and the integral converges towards a definite value

$$I_6 \equiv \int_0^\infty \frac{x^6 e^x}{(e^x-1)^2} dx = 732.4870 \qquad (2.189)$$

which leads to a rapidly vanishing relaxation rate as the temperature is lowered well below 1 K. This case will be discussed in Chapter 5. The Zeeman relaxation time due to the Raman process becomes independent of magnetic field in the low-temperature limit.

In Figure 2.4 we see how the function under the integral (Eq. 2.189) peaks at about $x = 6$ and then rapidly approaches zero. The inset shows how the integral converges rapidly to zero as a function of its lower limit, allowing to conclude that, for all practical purposes, the approximation of extending the integration to infinity is good for $\Theta_D/T > 30$ which covers all polarized target applications.

For electron spins at low temperatures, the simplification of ignoring x_0 is doubtful because we may have $x_0 > 1$ at low temperatures; at 0.5 K and 2.5 T, for example, $x_0 = 7$. The functions under the integrals of Eqs. 2.188 and 2.189 are plotted in Figure 2.5 to show the large difference and the convergence for this value. The extension of the integration to infinity is accurate for $\Theta_D/T > 40$ and thus valid for all polarized target applications.

Performing the integration numerically for $0 \leq x_0 \leq 10$ and then fitting the results yields the following dependence on x_0 for the integral (Eq. 2.188):

$$\int_0^\infty \frac{x^3(x+x_0)^3 e^{x+x_0}}{(e^x-1)(e^{x+x_0}-1)} dx = P_6(x_0)\int_0^\infty \frac{x^6 e^x}{(e^x-1)^2} dx = P_6(x_0)I_6. \qquad (2.190)$$

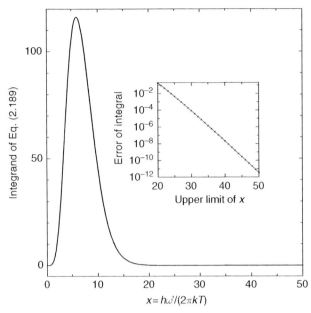

Figure 2.4 Convergence of the integral of Eq. 2.189 for such low temperatures that $\Theta_D/T > 30$. This concerns nuclear spin relaxation

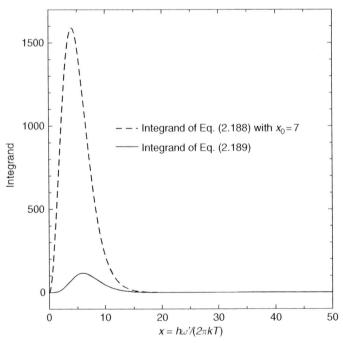

Figure 2.5 Difference between the integrands for $x_0 = 7$ in the case of electron spins at low temperatures. This shows that x_0 in the integrand cannot be ignored in Eq. 2.188. The influence of x_0, however, can be numerically evaluated (see below)

2.3 Relaxation

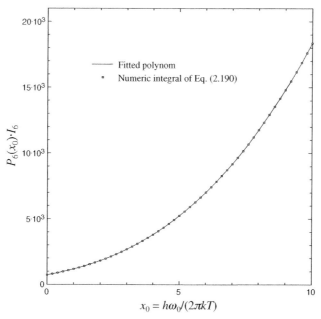

Figure 2.6 Numeric fit of the dependence of the definite integral of Eq. 2.190 on the parameter x_0

with an accuracy of about 1%. Here P_6 is a polynomial of third degree

$$P_6(x_0) = 1 + 0.4991 x_0 + 0.1016 x_0^2 + 8.8964 \cdot 10^{-3} x_0^3. \quad (2.191)$$

The integral and the fit are shown in Figure 2.6. This gives a field dependence and additional temperature dependence for the electron spin-lattice relaxation.

The relative importance of the Raman process for electron spin relaxation will be discussed in Chapter 3, together with processes which are specific to electronic spins only.

2.3.2 Relaxation of Transverse Magnetization and Dipolar Relaxation

The phenomenological Bloch equations are valid only in spin systems where the dipolar fields are averaged out, in particular by the rapid motion of the molecules in gases and liquids with low viscosity. They are especially useful in the description of spin systems under pulsed NMR measurements, a topic which is not dealt with in this book. Although these equations are clearly not obeyed in solid polarized targets, it is instructive to start from them when discussing the transverse relaxation time. The Bloch equations are

$$\frac{dM_z}{dt} = \frac{M_0 - M_z}{T_1} + \gamma (\mathbf{M} \times \mathbf{B})_z ;$$
$$\frac{dM_{x,y}}{dt} = \gamma (\mathbf{M} \times \mathbf{B})_{x,y} - \frac{M_{x,y}}{T_2}, \quad (2.192)$$

and they define the decay time constants T_1 and T_2 for the longitudinal and transverse components of the magnetization due to spins. Here T_2 is the time constant of the free precession signal decay (FID), which is exponential when the above equations describe the system correctly. In such systems T_2 depends on the relative velocities of the molecular constituents of the gaseous or liquid sample.

In solids, the assumptions behind the Bloch equations are not valid, and the time evolutions of the components of the magnetization are determined by the Provotorov equations, which were discussed in Section 2.2.4. The decay of the free precession signal is not exponential; in Section 2.2.2 it was shown that the time evolutions of the transverse components, in the domain of linear response, are the real and imaginary parts of the Fourier transform of the lineshape function. The time scale of the FID, however, is given by

$$T_2 \cong \frac{1}{\gamma B_{loc}} = \frac{1}{\mu_0 n \hbar \gamma^2 \sqrt{I(I+1)}} \tag{2.193}$$

using B_{loc} from Eq. 2.7. Here the subscript of T_2 is used only to refer to the transverse magnetization decay, and it has nothing to do with the random process of dephasing, which is involved in the case of gases and liquids where molecules undergo random motion.

One may view the FID of transverse magnetization in solids by recalling that spins permanently feel the oscillating transverse magnetic field of the same array of neighboring spins, which has the effect of keeping the transverse components in phase. The spatial variation and fluctuation of the longitudinal dipolar field, however, has a dephasing effect, which ultimately causes the disappearance of the transverse magnetization. It is quite common that in a crystalline solid the transverse magnetization components in the rotating frame may change sign, even several times, during their FID towards zero.

The spin-lattice effects in the rotating frame were discussed in an instructive way by Slichter ([2] p. 242), along the lines of the original work by Redfield [13]. Three basic relaxation equations can be defined in the rotating frame, one for each of the longitudinal and transverse magnetizations, and one for the dipolar energy:

$$\frac{\partial M_z}{\partial t} = \frac{M_0 - M_z}{T_{1z}};$$

$$\frac{\partial M_x}{\partial t} = \frac{-M_x}{T_{1x}}; \tag{2.194}$$

$$\frac{\partial \langle \mathcal{H}'_D \rangle}{\partial t} = \frac{\langle \mathcal{H}'_D \rangle_l - \langle \mathcal{H}'_D \rangle}{T_{1D}}.$$

Here M_y is not considered, as it does not exchange energy with the lattice, and the partial derivatives indicate the explicit changes due to the lattice interactions only. The equilibrium value of the dipolar energy $\langle \mathcal{H}'_D \rangle_l$ is that when the spins are at the temperature of the lattice. Clearly, T_{1z} can be identified to be equal to the usual T_1, but T_{1x} is not T_2 or similar to it, because exchange of energy with the lattice is involved. T_{1x} must therefore also be of similar order of magnitude as T_{1z}, because it determines how quickly the magnetization in

2.3 Relaxation

the rotating frame will find its longitudinal equilibrium value along the effective field B_{eff}. The dipolar relaxation time T_{1D} must also depend on the value of T_1, because there is no direct contact between the dipolar system and the lattice, but the relevant energy transfer must depend on the energy exchange between the Zeeman reservoir and the lattice.

The total energy in the rotating frame is

$$\langle E \rangle = -M_z b_0 - M_x B_1 + \langle \mathcal{H}'_D \rangle, \tag{2.195}$$

and its derivative with respect to the lattice couplings is

$$\frac{\partial \langle E \rangle}{\partial t} = -b_0 \frac{\partial M_z}{\partial t} - B_1 \frac{\partial M_x}{\partial t} + \frac{\partial \langle \mathcal{H}'_D \rangle}{\partial t}; \tag{2.196}$$

the assumption that the magnitude of the magnetization M always lies along the effective field then leads, using Eq. 2.192, to the equations for the magnetization and common spin temperature

$$\frac{\partial M}{\partial t} = \frac{1}{T_1}\left(M_{eq} - M\right);$$

$$\frac{d\beta}{dt} = \frac{1}{T_1}\left(\beta_{eq} - \beta\right), \tag{2.197}$$

where, when neglecting $\langle \mathcal{H}'_D \rangle_l$, the common time constant is

$$\frac{1}{T_1} = \frac{1}{b_0^2 + B_1^2 + B'_{loc}}\left(\frac{b_0^2}{T_{1z}} + \frac{B_1^2}{T_{1x}} + \frac{B'_{loc}}{T_{1D}}\right), \tag{2.198}$$

and the equilibrium values are

$$M_{eq} = M_0 \frac{B_{eff} b_0 / T_{1z}}{\dfrac{b_0^2}{T_{1z}} + \dfrac{B_1^2}{T_{1x}} + \dfrac{B'^2_{loc}}{T_{1D}}};$$

$$\beta_{eq} \propto \frac{1}{\sqrt{b_0^2 + B_1^2 + B'^2_{loc}}}. \tag{2.199}$$

In the above reasoning it was assumed that the thermal equilibrium of the subsystems, and between them, is established much more rapidly than T_1, which requires a transverse field strength much higher than that of the linear response domain. No upper limit, however, was possible to define in a natural way, because the density matrix formulation was not used for the problem.

The treatment here does not lead to theoretical methods for determining the various spin-lattice relaxation times, but rather to the understanding on how the various field strengths and the frequency offset may influence the time required for the evolution of the system towards dynamic equilibrium. Another important result is that of the equilibrium magnetization, which can be seen to follow the effective field so that its projection on the steady field will change sign when b_0 changes sign.

If the rotating field will be suddenly removed, the transverse magnetization will decay in the time scale of T_2 defined roughly by Eq. 2.193, rather than T_{1x}.

2.3.3 Cross-relaxation and Thermal Mixing

These two phenomena are important in the description of systems of spins with two or more species with different Larmor frequencies, and in the presence of a transverse RF field.

Cross-relaxation involves transitions between one or several pairs of spins undergoing a simultaneous flip so that angular momentum is conserved and the possible energy unbalance is compensated by a change in the dipolar energy. The probability of such transitions can be large in a high-density spin system, because of the conservation of both energy and angular momentum.

We have already discussed the two-spin transitions within like-spin systems which lead to line broadening in a solid and into the establishment of a dipolar temperature, both in the absence of and during RF saturation. We shall now treat the case when there are two spin species with spins I and S having nearly equal gyromagnetic ratios γ_1 and γ_2; the resonance lines do not overlap. Their numbers in the sample are N_1 and N_2 and they have spin-lattice relaxation times $T_1^{(1)}$ and $T_1^{(2)}$. The spin Hamiltonian in high field is

$$\mathcal{H} = \mathcal{H}_Z + \mathcal{H}'_D + \mathcal{H}_{RF} + \mathcal{H}_L \tag{2.200}$$

which is composed of the Zeeman part

$$\mathcal{H}_Z = \hbar\omega_1 I_z + \hbar\omega_2 S_z, \tag{2.201}$$

of the RF part

$$\mathcal{H}_{RF} = \hbar\omega_{RF}\left(I^+ e^{i\omega t} + I^- e^{-i\omega t}\right), \tag{2.202}$$

where $\omega_{RF} = \gamma_1 B_1$ and B_1 is the RF field strength, of the lattice interactions

$$\mathcal{H}_L = \hbar\sum_q F^{q,1} A^{q,1} + \hbar\sum_q F^{q,2} A^{q,2}, \tag{2.203}$$

and of the secular part of the dipolar interaction Hamiltonian, which will be subdivided in the following way:

$$\mathcal{H}'_D = \mathcal{H}'_{1D} + \mathcal{H}'_{2D} + \mathcal{H}'_{IS} + \mathcal{H}'_{CR}. \tag{2.204}$$

Here the first and second terms on the right are the secular dipolar Hamiltonians if only the species 1 or 2 had a magnetic moment, respectively. These terms commute with I_z and S_z simultaneously. The third term

$$\mathcal{H}'_{IS} = \sum_{i<k} a_{ik} I_{iz} S_{kz} \tag{2.205}$$

2.3 Relaxation

also commutes with I_z and S_z simultaneously and contributes to the linewidth of both species. The fourth term

$$\mathcal{H}'_{CR} = -\frac{1}{4}\sum_{i<k} a_{ik}\left(I_i^+ S_k^- + I_i^- S_k^+\right) = \mathcal{H}^+_{CR} + \mathcal{H}^-_{CR} \quad (2.206)$$

leads to the cross-relaxation between the two species but does not contribute to linewidths.

Provotorov [11] wrote the density matrix for this system in the high-temperature approximation, transformed it to the interaction representation where each component of the Hamiltonian becomes

$$\mathcal{O}(t) = e^{i\mathcal{H}t/\hbar}\mathcal{O}e^{-i\mathcal{H}t/\hbar} \quad (2.207)$$

and calculated its time evolution using Eq. 1.76:

$$\frac{d\rho}{dt} = -\frac{\pi}{2}\hbar\omega_{RF}^2\left[\left(I^+(t)\right)_{-\Delta_1}\left[\left(I^-(t)\right)_{\Delta_1},\rho\right]\right] - \frac{2\pi}{\hbar^2}\left[\left(\mathcal{H}_D^+(t)\right)_{-\Delta_{12}}\left[\left(\mathcal{H}_D^-(t)\right)_{\Delta_{12}},\rho\right]\right]$$

$$-\frac{2\pi}{\hbar^2}\sum_{j=1}^{2}\left[\left(\mathcal{H}_{jD}^+(t)\right)_{-\omega_j}\left[\left(\mathcal{H}_{jD}^-(t)\right)_{\omega_j},\rho\right]\right] \quad (2.208)$$

$$-\frac{2\pi}{\hbar^2}\sum_{j=1}^{2}\int_{-\infty}^{\infty}\left[\left(F^{1j}(t)\right)_{-\omega}\left(A^{1j}(t)\right)_{\Delta_j}\left[\left(F^{1j}(t)\right)_{\omega}\left(A^{1j}(t)\right)_{-\Delta_j},\rho\right]\right],$$

where the operators $\left(\mathcal{O}(t)\right)_\omega$ are defined by the transforms

$$\left(\mathcal{O}(t)\right)_\omega = \frac{1}{2\pi}\int_{-\infty}^{\infty}dt\, e^{i(\mathcal{H}_D+F)t/\hbar}\mathcal{O}(t)e^{-i(\mathcal{H}_D+F)t/\hbar}e^{i\omega t} \quad (2.209)$$

Provotorov then showed that the Hamiltonian can be diagonalized under the following conditions:

$$\frac{H_1}{H_{loc}} \ll 1;\quad \frac{\gamma_1 H_{loc}}{\Delta_{12}} \ll 1;\quad \frac{\gamma_1 H_{loc}}{\omega_{1,2}} \ll 1;\quad \frac{1}{\gamma_1 H_{loc} T_1^{(1,2)}} \ll 1. \quad (2.210)$$

As a consequence of this and the fact that some of the parts of the Hamiltonian commute, the density matrix can be put to the form

$$\rho(t) = Ce^{\alpha(t)I_z + \beta(t)S_z + \gamma(t)\mathcal{H}'_D + \mathcal{H}'_L/kT_0} \quad (2.211)$$

where

$$C^{-1} = (2I+1)^{N_1}(2S+1)^{N_2}\,\mathrm{Tr}\left\{e^{-\mathcal{H}'_L/kT_0}\right\}. \quad (2.212)$$

Using the relations

$$\frac{dI_z(t)}{dt} = \mathrm{Tr}\left\{I_z\frac{d\rho}{dt}\right\};\quad \frac{dS_z(t)}{dt} = \mathrm{Tr}\left\{S_z\frac{d\rho}{dt}\right\};\quad \frac{d\mathcal{H}'_D(t)}{dt} = \mathrm{Tr}\left\{\mathcal{H}'_D\frac{d\rho}{dt}\right\} \quad (2.213)$$

he now obtained the following set of differential equations for the time evolution of the inverse temperatures α, β and γ:

$$\frac{d\alpha}{dt} = -W_{RF}(\alpha - \gamma\hbar\Delta_1) - \frac{1}{T_{12}}(\alpha - \beta - \gamma\hbar\Delta_{12}) - \frac{1}{T_{s1}}(\alpha - \gamma\hbar\omega_1) + \frac{1}{T_1^{(1)}}(\alpha_0 - \alpha)$$

$$\frac{d\beta}{dt} = \frac{\delta}{T_{12}}(\alpha - \beta - \gamma\hbar\Delta_{12}) - \frac{\delta}{T_{s2}}(\beta - \gamma\hbar\omega_2) + \frac{\delta}{T_1^{(2)}}(\beta_0 - \beta); \qquad (2.214)$$

$$\frac{d\gamma}{dt} = \hbar\Delta_1\varepsilon W_{RF}(\alpha - \gamma\hbar\Delta_1) + \frac{\hbar\Delta_{12}\varepsilon}{T_{12}}(\alpha - \beta - \gamma\hbar\Delta_{12}) +$$

$$+ \frac{\hbar\omega_1\varepsilon}{T_{s1}}(\alpha - \gamma\hbar\omega_1) + \frac{\hbar\omega_2\varepsilon}{\delta T_{s1}}(\beta - \gamma\hbar\omega_2) + \frac{1}{T_1''}(\gamma_0 - \gamma)$$

where

$$W_{RF} = \frac{\pi}{2}\hbar\omega_{RF}^2 \frac{Tr\{I_{-\Delta_1}^+ I_{\Delta_1}^-\}}{Tr\{I_z^2\}} = \frac{\pi}{2}\hbar\omega_{RF}^2 g(\Delta_1), \qquad (2.215)$$

$$\frac{1}{T_{12}} = \frac{2\pi}{\hbar^2} \frac{Tr\{(\mathcal{H}_D^+)_{-\Delta_{12}}(\mathcal{H}_D^-)_{\Delta_{12}}\}}{Tr\{I_z^2\}}, \qquad (2.216)$$

$$\frac{1}{T_{sj}} = \frac{2\pi}{\hbar^2} Tr\{(\mathcal{H}_{jD}^+)_{-\omega_j}(\mathcal{H}_{jD}^-)_{\omega_j}\}, \qquad (2.217)$$

$$\delta = \frac{Tr\{I_z^2\}}{Tr\{S_z^2\}}; \quad \varepsilon = \frac{Tr\{I_z^2\}}{Tr\{\mathcal{H}_D'^2\}}, \qquad (2.218)$$

and the spin-lattice relaxation times have definitions analogous to the previous section.

General solutions for these equations have not been given, but with the usual relaxation times one may see that there will be a short relaxation time for the dipolar temperature β to approach equilibrium with the Zeeman temperatures α and γ, and a longer relaxation time for the system to reach a steady-state dynamic equilibrium. Provotorov notices that in the absence of RF field the short relaxation time will be

$$\tau_D = \frac{T_{12}}{1 + \delta + \varepsilon\Delta_{12}^2\hbar^2} \qquad (2.219)$$

when the small terms proportional to $(T_{js})^{-1}$ are omitted in the equations.

Rodak [14] has calculated, using the above equations, the steady-state temperature of the second spin system during RF irradiation as

$$\frac{\beta - \beta_0}{\beta_0} = -\frac{\omega_1 + \Delta_1}{\omega_1 + \Delta_{12}} \frac{\Delta_1 \Delta_{12} + H_{loc}^2 \left(\gamma_1^2 + \gamma_2^2 \frac{N_2}{N_1} \right) \frac{T_1^{(1)}}{T_1''}}{\left[\Delta_1^2 + H_{loc}^2 \left(\gamma_1^2 + \gamma_2^2 \frac{N_2}{N_1} \right) \frac{T_1^{(1)}}{T_1''} \right] \left(1 + \frac{N_2 T_{12}}{N_1 T_1^{(2)}} \right) + \left(\Delta_1 - \Delta_{12} \right) \frac{N_2 T_1^{(1)}}{N_1 T_1^{(2)}}}.$$

(2.220)

He notices that when both terms on the right side are positive, the saturation of line 1 results in the heating of the spins 2. The nominator of the second term, however, may become negative when

$$\Delta_1 \Delta_{12} < -H_{loc}^2 \left(\gamma_1^2 + \gamma_2^2 \frac{N_2}{N_1} \right) \frac{T_1^{(1)}}{T_1''}. \tag{2.221}$$

This leads to the cooling of the second spin system. The example which was considered was two electron lines (hyperfine lines, for example); in this case, however, the cooling amounted to 10% at best.

Thermal Mixing

When the transverse field rotates at a speed close to the Larmor frequency so that $\Delta \approx D$, the temperatures of the Zeeman and dipolar systems converge rapidly towards a common value, as was shown by the Provotorov equations. This occurs because the frequency differences in the cross-transitions of the like spins involve a spectrum which includes Δ. The same can be achieved by lowering the static field to a value in the vicinity of B_{loc}, rather than lowering the effective field to B_{eff} by the application of the rotating transverse field. The only difference in the two approaches is that the dipolar Hamiltonian must be truncated at high static field, whereas it cannot be truncated at low static field.

If there are two spin species initially at high field and at different temperatures, lowering the static field has similar effects when the dipolar widths are such that the two resonance frequency spectra overlap. This has been studied in LiF and was reviewed by Abragam [1].

A similar phenomenon can be observed at a high static field when applying a rotating transverse field at a frequency which is equal to the sum or the difference of the two Larmor frequencies. The two Zeeman temperatures in their rotating frames will then approach a common value at a rate which depends on the magnitude of the transverse field, and on the lineshapes of the two spin species. The process is irreversible and takes place at constant Zeeman energy, and the final temperatures can be obtained using the heat capacities of the two spin systems [15]. If the line widths are so broad that also the dipolar temperatures will get equalized, this must be taken into account in analysing the outcome.

2.3.4 Nuclear Spin-Lattice Relaxation in Solids

In Section 2.3.1 it was pointed out that there is no direct contact between the nuclear spins and the phonons of the lattice at low temperatures, even at the highest values of magnetic fields obtainable in the laboratory. The nuclear spin-lattice relaxation must therefore proceed via interactions with the electronic system, a fact which is most important for the DNP which is based on the off-resonance saturation of the same electronic system.

The other possible mechanisms of nuclear spin-lattice relaxation have adverse effects in DNP, because they produce a 'leakage' mechanism for the nuclear polarization. The leading causes for leakage relaxation are unwanted residual paramagnetic impurities. Other causes, such as the direct contact with phonons, or via nuclear spin-orbit coupling or nuclear quadrupole interaction coupled to lattice phonons, have been shown to produce negligible relaxation rates in comparison with the residual paramagnetic impurities [1].

The spin-lattice relaxation has been formulated using the generalized Provotorov equations ([6] pp. 49–52), which provide an elegant method for arriving at the right expressions for the relaxation of each of the rates towards lattice.

The experimental nuclear spin-lattice relaxation will be discussed in Chapter 5.

2.4 Interacting Spins in a Strong Transverse Field

We shall discuss here spin systems at high static field and in the presence of such high transverse field that the Provotorov equations are not adequate. This topic is of particular interest for pulsed NMR, which is not used for nuclear spin polarization measurement for obvious reasons. Here the topic is of interest for the adiabatic reversal of nuclear spin polarization, which is relevant for polarized targets.

When the strength of the transverse field, rotating at a frequency close to the Larmor precession, is of the same order of magnitude as the local field, $B_1 \approx B'_{loc}$, the thermal equilibrium between the Zeeman and dipolar reservoirs is obtained in a time of the order of $1/D$. The Hamiltonian of the system is

$$\mathcal{H} = \hbar\omega_0 I_z + \mathcal{H}'_D + \hbar\omega_1\left(I_x \cos\omega t + I_y \sin\omega t\right). \tag{2.222}$$

When this is transformed to the rotating frame, one obtains

$$\mathcal{H}_{eff} = e^{i\omega t}\mathcal{H}e^{-i\omega t} - \hbar\omega I_z = \hbar\Delta I_z + \mathcal{H}'_D + \hbar\omega_1 I_x. \tag{2.223}$$

This can now be used for calculating the energy and entropy in the rotating frame, in order to evaluate the resulting spin temperature after a sudden change in the Hamiltonian and during a slow adiabatic change of the Hamiltonian. In the former case the energy is conserved, while the second takes place at constant entropy.

The energy in the rotating frame is

$$\left\langle \mathcal{H}_{eff}\right\rangle = \text{Tr}\left\{\rho_{initial}\mathcal{H}_{eff}\right\} = \text{Tr}\left\{\rho_{final}\mathcal{H}_{eff}\right\} = -\beta\hbar^{-1}\text{Tr}\left\{\mathcal{H}^2_{eff}\right\}, \tag{2.224}$$

2.4 Interacting Spins in a Strong Transverse Field

which yields for the final spin temperature

$$\beta = -\frac{\text{Tr}\{\rho_{initial}\mathcal{H}_{eff}\}}{\text{Tr}\{\mathcal{H}_{eff}^2\}} \qquad (2.225)$$

where the initial density matrix in the rotating frame features different initial temperatures for the Zeeman and dipolar reservoirs, and the final density matrix in the rotating frame is described by a common inverse spin temperature β:

$$\rho = \frac{e^{-\beta\mathcal{H}_{eff}}}{\text{Tr}\{e^{-\beta\mathcal{H}_{eff}}\}} \cong 1 - \beta\mathcal{H}_{eff}. \qquad (2.226)$$

The trace in the denominator of Eq. 2.225 is

$$\text{Tr}\{\mathcal{H}_{eff}^2\} = \text{Tr}\{I_z^2\}\hbar^2(\Delta^2 + \omega_1^2 + D^2) \qquad (2.227)$$

where $D = \gamma B'_L$.

Adiabatic Demagnetization in the Rotating Frame

It should be first noted that in metals the adiabatic demagnetization in the rotating frame (ADRF) is impractical because the high-frequency field does not penetrate the sample but rather heats up the electronic system. On the other hand, in dielectric samples the flipping or ramping of the static field, ramped first to a low value, is impractical because the sample must contain paramagnetic electronic spins which cause rapid nuclear spin relaxation at low magnetic fields; these paramagnetic spins are required for obtaining a high dynamic nuclear polarization (DNP).

The motion of an isolated spin in a high static field B_0 superposed with a small transverse rotating field B_1 could be understood as precession about an effective field B_{eff} defined by Eq. 1.57. The isolated, i.e. non-interacting, spin can be manipulated reversibly by changing the effective field slowly in comparison with the effective Larmor precession frequency in the rotating frame. In the more practical case of interacting spins, the relaxation rates between the lattice, the Zeeman reservoir and the spin interaction reservoir must be taken into account in evaluating in the outcome of ADRF.

If the transverse field is small compared with the effective field, we have $B_{eff} = (\Delta\omega/\omega_0)B_0$ as the effective field; if this field is large in comparison with the linewidth, no thermal mixing occurs between the Zeeman and dipolar reservoirs, and the two reservoirs can be at different temperatures in the time scale $t \approx T_1$ which can be hours or even days.

When the effective field approaches the local field, the situation changes. If in addition the strength of the transverse field is larger than that required for the linear approximation, but still small in comparison with the local field, one must use the Provotorov equations in order to determine the time evolution and equilibrium values of the temperatures of the various energy reservoirs. We shall therefore leave the detailed discussion of ADRF to

Chapter 11, after the dynamic behavior of interacting spins is better understood. Here we shall focus on the static results and on the similarities between ADRF and adiabatic demagnetization in the laboratory frame.

Let us now perform ADRF from such a large effective field that no thermal mixing occurs, down to the mixing field

$$b_m = B_0 + \frac{\omega_m}{\gamma} = \frac{\omega_m - \omega_0}{\gamma}, \qquad (2.228)$$

where the relaxation between the Zeeman and dipolar interaction energy reservoirs is reasonably fast in comparison with the spin-lattice relaxation. The relaxation takes place at constant total energy, and the final inverse temperature, in the high-temperature approximation, is analogous with that of Eq. 1.121:

$$\beta_f = \beta_i \frac{B_i}{\sqrt{b_m^2 + B_1^2 + B'^2_{loc}}} \times \frac{\sqrt{b_m^2 + B_1^2}}{\sqrt{b_m^2 + B_1^2 + B'^2_{loc}}}, \qquad (2.229)$$

where the second term on the right describes again the non-adiabaticity due to the mixing. The losses in entropy can again be of the order of 1% in favorable cases. If the effective field is further reduced so slowly that the dipolar and Zeeman temperatures are always in good equilibrium with each other, the common inverse temperature undergoes reversible changes with the effective field:

$$\beta_f = \beta_i \frac{B_i}{\sqrt{b^2 + B_1^2 + B'^2_{loc}}} \times \frac{b_m}{\sqrt{b_m^2 + B'^2_{loc}}}, \qquad (2.230)$$

where it was taken into account that the rotating field B_1 must be very small in comparison with both the local field and the mixing field.

The ADRF can be stopped at any effective field and the transverse field can be adiabatically reduced to zero. If this is done at the frequency ω_0, all available Zeeman order is transformed to dipolar order, and if the initial polarization is sufficiently high, magnetic ordering may take place in the nuclear spin system.

The above expressions are further slightly modified due to the finite relaxation rates between the various energy reservoirs, to be discussed in Chapter 11.

Polarization Reversal by Adiabatic Passage in Rotating Frame

If the frequency or field sweep is continued through zero effective field, the Zeeman temperature will be adiabatically reversed. This is called adiabatic passage in rotating frame (APRF). In this process there is no adiabatic loss equivalent to that of the field flip given by Eq. (1.133), because in the rotating frame one can proceed adiabatically through zero effective longitudinal field without losses due to relaxation, provided that the spin-lattice relaxation times are much longer than the time spent in the passage.

In practice the strength of the transverse field and the sweep rate of the steady field are optimized so that the losses due to relaxation are minimized. The polarization loss in the reversal by APRF is then reduced to that due to the loss of entropy when performing the thermal mixing, and to the loss due to other spins whose temperature remains untouched during the passage. At very high polarization the above results based on the high-temperature approximation are qualitatively valid, but nuclear magnetic phase transition phenomena may reduce the efficiency of the polarization reversal, as will be also noted below.

The reversal can be performed starting from positive or negative polarization and spin temperature, and the sweep of the frequency or field can be started from above or below, with the same results. To reduce losses due to thermal mixing, however, it is best to perform thermal mixing at positive Zeeman temperature when the initial polarization is positive, and at negative Zeeman temperature when the initial polarization is negative.

Polarization reversal by adiabatic passage in the rotating frame will be discussed further in Chapter 11.

References

[1] A. Abragam, *The Principles of Nuclear Magnetism*, Clarendon Press, Oxford, 1961.
[2] C. P. Slichter, *Principles of Magnetic Resonance*, 3rd ed., Springer-Verlag, Berlin, 1990.
[3] G. M. Volkoff, Second order nuclear quadrupole effects in single crystals, *Can. J. Phys.* **31** (1953) 820–836.
[4] N. F. Ramsey, Magnetic shielding of nuclei in molecules, *Phys. Rev.* **78** (1950) 699–703.
[5] N. F. Ramsey, Chemical effects in nuclear magnetic resonance and in diamagnetic susceptibility, *Phys. Rev.* **86** (1952) 243–246.
[6] A. Abragam, M. Goldman, *Nuclear Magnetism: Order and Disorder*, Clarendon Press, Oxford, 1982.
[7] R. Kubo, K. Tomita, General theory of magnetic resonance, *J. Phys. Soc. Japan* **9** (1954) 888.
[8] J. H. Van Vleck, The dipolar broadening of magnetic resonance in solids, *Phys. Rev.* **74** (1948) 1168–1183.
[9] B. N. Provotorov, Magnetic resonance saturation in crystals, *Soviet Phys. – JETP* **14** (1962) 1126–1131.
[10] C. J. Gorter, *Parametric Relaxation*, Elsevier, New York, 1947.
[11] B. N. Provotorov, A quantum statistical theory of cross relaxation, *Soviet Phys. – JETP* **15** (1962) 611–614.
[12] L. D. Landau, E.M. Lifshitz, *Statistical Physics*, 2nd revised and enlarged ed., Pergamon Press, Oxford, 1968.
[13] A. G. Redfield, Nuclear magnetic resonance saturation and rotary saturation in solids, *Phys. Rev.* **98** (1955) 1787–1809.
[14] M. I. Rodak, Effect of magnetic resonance saturation on cross relaxation, *Soviet Phys. – JETP* **18** (1964) 500–502.
[15] M. Goldman, S. F. J. Cox, V. Bouffard, Coupling between nuclear Zeeman and electronic spin-spin interactions in dielectric solids, *J. Phys. C: Solid State Phys* **7** (1974) 2940–2952.

3
Electron Paramagnetic Resonance and Relaxation

3.1 Electron Spin and Magnetic Moment

The atomic electron is presently regarded as a pointlike particle. Direct tests of quantum electrodynamics (QED) in electron-positron collisions confirm the absence of structure down to the distance scale of 2×10^{-18} m. The electron g-factor is therefore close to the Dirac value of 2; the best experimental value [1] is obtained using a single-electron Penning trap with a cavity cooled to 100 mK temperature:

$$\frac{g_e}{2} = 1.00115965218085(76). \qquad (3.1)$$

This limits the substructure to the scale of 1×10^{-18} m, and it is presently the most accurately known fundamental constant.

The relative deviation of g_e from 2 is called a, the anomalous magnetic moment. For the free electron it is, from the above result,

$$a \equiv \frac{g_e}{2} - 1 = 1.15965218085(76) \cdot 10^{-3},$$

which deviates by less than 4×10^{-9} from the theoretical value [2, 3] involving eighth-order QED and small weak and hadronic corrections. This agreement gives still more strict bounds to any substructure [4, 5] given that there is no presently known reason for a large cancellation which could cause the substructure to give a g-factor of a pointlike particle.

The electrons in closed shells of atoms, and usually the electrons which participate in valence-bonding between atoms, form pairs with antiparallel spins. These pairs have no total magnetic moment. The closed-shell electrons have no orbital magnetic moment either. Unpaired electrons tend to be rare because pairing is energetically favored, and therefore paramagnetism of materials and compounds is rather an exception than a rule.

Transition metal ions have incomplete inner shells which may exhibit a magnetic moment. These ions may substitute similar non-magnetic ions in diamagnetic salts, thus forming a diluted paramagnetic substance. The transition metals include the iron group

(Ti, V, Cr, Mn, Fe, Co, Ni, Cu, Zn), the palladium group, the rare earth group (Ce, Pr, Nd, Pm, Sm, Eu, ..., Yb), the platinum group and the actinides.

As an example, the Ce^{3+} ion has one $4f$ electron outside the closed shells which have the electronic structure of a Xe atom. In a free ion the spin-orbit coupling lifts partially the $(2S+1)(2L+1) = 2 \times 7 = 14$-fold degeneracy of this $4f$ electron into multiplets with $J = L+S = 7/2$ and $J = L - S = 5/2$. An ion bound in a crystal experiences the crystal field which has a certain symmetry; this field lifts partially the degeneracy of these multiplets depending on the symmetry. In the double nitrate of cerium and magnesium (CMN), for example, the crystal field has trigonal symmetry at the site of Ce^{3+}; this splits the ground-state multiplet with $J = 5/2$ into three doublets. Such doublets occur in ions with odd number of electrons and they have the special property that the remaining two-fold degeneracy cannot be lifted by an electric field but only by a magnetic field. These are called Kramers doublets and the two states into which the field splits the doublet are called Kramers conjugate states. In CMN the spacing between the three doublets is more than 30 K so that at 1 K temperature only the lowest doublet is populated.

The static paramagnetism under these conditions can be described as that of a fictitious 'effective spin' $S = \frac{1}{2}$ (see also Section 1.1.1), but the magnetic moment deviates from that of a free electron. The g-factor is strongly anisotropic in CMN with a maximum value $g_\perp = 1.83$ when the field is perpendicular to the crystal axis, and a minimum value $g_\parallel = 0.0236$ when the field is parallel to it.

CMN is well known to low-temperature physicists, because it has a fairly low density of electron spins which entails a magnetic phase transition temperature only below 2 mK temperature. It can therefore be used for cooling by adiabatic demagnetization to the millikelvin range of temperatures [6, 7]. It has also been used for thermometry based on the Curie law, down to the same temperature range.

The effective paramagnetic electron spins of transition metal ions have usually values of S between 1/2 and 5/2. The effective spin is the result of the interaction of the electron spin with the orbital moment, which may take values between 0 and 3. In polarized targets the interesting materials are only those which have $S = 1/2$, because the higher spin materials have complicated magnetic level structures which lead to broad resonance lines, with a rich structure. It is also important that the ion can be diluted by embedding it in a non-magnetic crystal, because otherwise the interactions between the spins are too strong for dynamic nuclear polarization (DNP) and can even lead to a magnetic phase transition at the low temperatures required.

Other materials with paramagnetic spins of interest for polarized targets include the following categories:

(1) stable free radicals such as triphenylmethyl, TEMPO, porphyrexide and DPPH, soluble in a glass-forming matrix (Section 3.6.1);
(2) radiolytic free radicals created in situ by radiation damage, such as $\cdot NH_2$ (Section 3.6.3);

(3) metallo-organic compounds such as propanediol-Cr(V) and other pentavalent Cr complexes; Ti and V also form similar compounds of potential use in polarized targets (Section 3.6.2);
(4) paramagnetic transition metal ions substituting non-magnetic ions in ionic crystals, such as Nd^{3+} ion impurity in lanthanum magnesium double nitrate (LMN);
(5) vacancies such as a negative ion vacancy filled by an electron (F-centers), a positive ion vacancy with an electron missing from an adjacent ion (V_1-centers) or more complicated defect states such as divacancies;
(6) interstitial or substitutional point defects in diamagnetic crystals such as N atoms in diamond, Al in SiO_2, H in CaF_2 or Ag in KCl;
(7) localized donor and acceptor states in semiconductors, for example, P in Si;
(8) optically induced triplet states in molecules which are diamagnetic in their ground state;
(9) conduction electrons in metals (although these are not paramagnetic in the strict sense of the term).

In analogy with the transition metal ions, the interactions of the effective spin of the above paramagnetic electrons with the lattice[1] lead to the following phenomenological changes of its paramagnetic resonance, comparing with free electrons:

(i) frequency shift (g-shift due to spin-orbit interaction);
(ii) splitting into several lines due to interaction with nuclear spins (hyperfine splitting);
(iii) broadening due to dipolar interactions.

The g-shifts observed in solids, gases and liquids are often much larger than the anomalous magnetic moment but smaller than those of the free transition metal ions. As was discussed above and in Section 2.1.3, this shift is caused by atomic, molecular and crystal field interactions, and its theoretical understanding requires second-order perturbation theory for evaluating the interference term of the orbital momentum interaction with the external field and with the electron spin. The g-shift as well as the hyperfine splitting often exhibit substantial anisotropy and, in solids other than single crystals, these lead to the inhomogeneous line broadening, which may exceed the dipolar (homogeneous) broadening by a large factor.

In the following discussion of electron paramagnetic resonance (EPR), we shall not go in such a mathematical detail as in Chapter 2. The molecular structure and lattice interactions complicate the EPR often so much that the rigorous analysis is very difficult, if not impossible (in closed form). However, because EPR is the key to DNP, we shall try to enlighten the phenomenology with some simplified theoretical models and examples. For those who wish to gain a deep insight into the EPR of transition ions, we recommend reading the book of Abragam and Bleaney [8]. The EPR of radiolytic free radicals is treated by Pshetzhetskii et al. [9] and by Roginskii and Tupikov [10], and the basic EPR techniques are treated by Wertz and Bolton [11] and by Eaton et al. [12].

---

[1] The matrix of the material is called here lattice, even for glassy, liquid or gaseous disordered states.

3.2 EPR Absorption Spectrum

3.2.1 *The g-Shift*

The origin of the *g*-shift of the resonance frequency of an unpaired electron bound in an atom, a molecule or a crystal is the coupling of its magnetic moment to the motional magnetic field in the electric field of the atom, molecule or crystal. This gives rise to the spin-orbit coupling that was discussed in Section 2.1.3:

$$\mathcal{H} = \frac{e\hbar}{2mc^2}\mathbf{S}\cdot(\mathbf{E}\times\mathbf{p}), \quad (3.2)$$

which in the radial electric field of a free atom becomes $\mathcal{H} = \lambda\,\mathbf{L}\cdot\mathbf{S}$. In a free hydrogen-like atom, this interaction leads to the Landé *g*-factor

$$g = 1 + \frac{J(J+1) - L(L+1) + S(S+1)}{2J(J+1)},$$

which varies from 2 to 1 with increasing *L*.

In atoms and ions bound to molecules or crystals, the spherical symmetry of the electric field experienced by the unpaired electron is destroyed. As a consequence of this, the state of the electron cannot be described by a wave function with a definite value of *L*, i.e. *L* is not a 'good' quantum number. Consequently, the orbital moment is quenched[2] and the *g*-factor does not deviate markedly from g_e. In free radicals and complexes used in polarized targets this deviation rarely exceeds 0.5%; in molecular and ionic crystals consisting of small molecules, however, much larger deviations can be found, as was discussed in Section 3.1.

For some free paramagnetic atoms the *g*-factor can be as large as 6; this results from the interaction of an unpaired inner-shell electron with the rest of the electronic system, which becomes polarized. In crystalline materials, transition metal ion substitutional impurities such as Ce^{3+}, discussed above, usually have *g*-factors very far from 2, with a large anisotropy. A case well studied in the field of polarized targets is the LMN in which dilute Nd impurities have a large anisotropic *g*-factor; this will be discussed in more detail in Section 3.6.

Because the *g*-factor of the electron reflects the magnetic coupling with the lattice, it is important for the spin-lattice relaxation and for the DNP.

Due to the anisotropy of the molecular or crystal field, the *g*-factor is anisotropic and therefore dependent on the orientation of the magnetic field vector. As was discussed in Section 2.1.3, the motion of the electron in the superposition of the centrally symmetric Coulomb field and the anisotropic crystal field causes a motional magnetic field, the interaction of which with the magnetic moment of the electron can be expressed with the spin Hamiltonian

---

[2] For a detailed discussion on the quenching of the orbital moment, see, for example, Refs. ([9] pp. 20–22 and [13] pp. 89–92).

$$\mathcal{H} = \mu_B \left(B_x g_{xx} S_x + B_y g_{yy} S_y + B_z g_{zz} S_z \right), \tag{3.3}$$

where the g-tensor is diagonalized, and $\mu_B = e\hbar/2m_e$ is the Bohr magneton. The dyadic notation

$$\tilde{\mathbf{g}} = \mathbf{i} g_{xx} \mathbf{i} + \mathbf{j} g_{yy} \mathbf{j} + \mathbf{k} g_{zz} \mathbf{k} \tag{3.4}$$

is practical for writing the spin Hamiltonian in the form

$$\mathcal{H} = \mu_B \mathbf{B} \cdot \tilde{\mathbf{g}} \cdot \mathbf{S}. \tag{3.5}$$

In an oriented single crystal, the resonance frequency

$$\omega = \frac{g(\theta, \varphi) \mu_B B}{\hbar} \tag{3.6}$$

is now dependent on the Euler angles θ, φ in the following way:

$$g(\theta, \varphi) = \sqrt{g_{xx}^2 \sin^2\theta \sin^2\varphi + g_{yy}^2 \sin^2\theta \cos^2\varphi + g_{zz}^2 \cos^2\theta}. \tag{3.7}$$

The g-tensor is often axially symmetric, due to the axial symmetry of the molecule or crystal. We may then write

$$g_{xx} = g_{yy} = g_\perp;$$
$$g_{zz} = g_\parallel$$

where $g_\parallel$ and $g_\perp$ are the values of the g-factor for a magnetic field oriented parallel and perpendicular to the symmetry axis of the molecule or crystal. We then have

$$g(\theta) = \sqrt{g_\perp^2 - (g_\perp^2 - g_\parallel^2)\cos^2\theta}. \tag{3.7'}$$

If the orientation of the molecular axis is isotropic and random such as in glassy, polycrystalline or powder materials, the number of spins with axis in the direction of the solid angle element $d\Omega$ is constant. The lineshape function then obeys

$$f(\omega)d\omega = \frac{d\Omega}{4\pi} = \frac{1}{2} d\cos\theta. \tag{3.8}$$

Solving $\cos\theta$ from Eq. 3.7 and inserting $g(\theta)$ from Eq. 3.7' yield for axially symmetric g-tensor

$$f(\omega) = \frac{1}{2}\frac{d}{d\omega}\cos\theta = \frac{1}{2}\frac{d}{d\omega}\frac{1}{\sqrt{g_\perp^2 - g_\parallel^2}}\sqrt{g_\perp^2 - \left(\frac{\hbar\omega}{\mu_B B}\right)^2}. \tag{3.9}$$

The lineshape function after normalization becomes

$$f(\omega) = \frac{\omega}{\sqrt{(\omega_\perp^2 - \omega_\parallel^2)(\omega_\perp^2 - \omega^2)}} \quad \text{for } \omega \text{ between } \omega_\parallel \text{ and } \omega_\perp; \tag{3.10}$$

$$f(\omega) = 0 \qquad \text{elsewhere.}$$

Here $\omega_\perp = g_\perp \mu_B B/\hbar$ and $\omega_\parallel = g_\parallel \mu_B B/\hbar$, and the lineshape, valid for electronic spin $S = 1/2$ in isotropic polycrystalline or glassy materials, is shown in Figure 3.1.

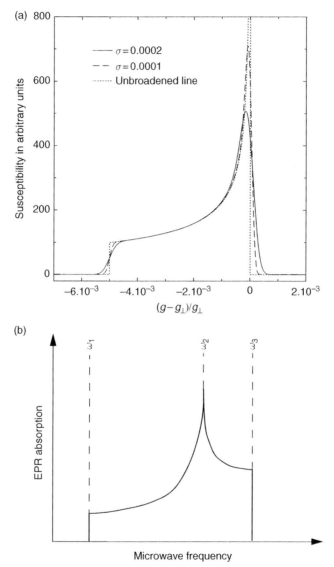

Figure 3.1 (a) EPR lineshapes in a glassy solid due to broadening by axially symmetric g-tensor, Eq. 3.10, with different contributions from dipolar broadening. The hyperfine interactions with nuclei in the paramagnetic molecule are assumed very small. The anisotropy of the g-tensor is 0.5%, typical for Cr(V) complexes with diols. The Gaussian broadening corresponds roughly to the normal and deuterated forms of the complex and the solvent matrix. (b) Unbroadened EPR lineshape in a glassy solid when the g-tensor is not axially symmetric (see Eq. 3.17)

The EPR spectra are experimentally determined usually by scanning the field at fixed microwave frequency. The lineshape function under the same assumptions then becomes

$$f(B) = \frac{B_\| B_\perp^2}{B^2 \sqrt{\left(B_\|^2 - B_\perp^2\right)\left(B^2 - B_\perp^2\right)}} \quad \text{for } B \text{ between } B_\perp \text{ and } B_\|; \tag{3.11}$$

$$f(B) = 0 \quad \text{elsewhere.}$$

For narrow lines the shapes 3.10 and 3.11 are almost identical (but inversed about the centroid, of course).

In Eq. 3.10 it was assumed that the resonance at each orientation or 'spin packet' is infinitely sharp; for real lines these spin packets are better represented by symmetric shape functions $h(\omega - \omega')$ representing dipolar and other possible spin-spin interactions. The resulting lineshape is then obtained by the convolution

$$\bar{f}(\omega) = \int_0^\infty f(\omega') h(\omega - \omega') d\omega'; \tag{3.12}$$

such a shape is superimposed in Figure 3.1. The convoluted shape functions cannot be calculated in closed form in general, and numerical methods must be used if such shapes are needed for the purpose of fitting experimental signal shapes, for example. An approximate analytical function, however, can be obtained for small g-anisotropy and Lorentzian broadening, by writing

$$f(\omega) = \frac{\omega}{\sqrt{\left(\omega_\perp^2 - \omega_\|^2\right)\left(\omega_\perp^2 - \omega^2\right)}} \cong \frac{\omega}{\sqrt{\left(\omega_\perp^2 - \omega_\|^2\right) 2\bar{\omega}\left(\omega_\perp - \omega\right)}},$$

where $\bar{\omega} = (\omega_\perp + \omega_\|)/2$ is an average frequency between the poles of the lineshape. Direct integration with

$$h(\omega - \omega') = \frac{T_2}{\pi} \frac{1}{1 + (\omega - \omega')^2 T_2^2}$$

then gives the convolution

$$\bar{f}(\omega) = \frac{T_2}{\pi \sqrt{\left(\omega_\perp^2 - \omega_\|^2\right) 2\bar{\omega}}} \int_{\omega_\|}^{\omega_\perp} \frac{\omega' d\omega'}{\left[1 + (\omega - \omega')^2 T_2^2\right] \sqrt{\omega_\perp - \omega'}}.$$

If the paramagnetic molecule has no axial symmetry and we denote $g_{xx} = g_1$, $g_{yy} = g_2$, $g_{zz} = g_3$, with $g_1 < g_2 < g_3$, finding the lineshape is more complicated. The result for the normalized shape is

$$f(B) = \frac{2}{\pi} \frac{B_1 B_2 B_3}{B^2 \sqrt{(B_2^2 - B_3^2)(B_1^2 - B^2)}} K(k) \quad \text{for } B \text{ between } B_3 \text{ and } B_2;$$

$$f(B) = \frac{2}{\pi} \frac{B_1 B_2 B_3}{B^2 \sqrt{(B^2 - B_3^2)(B_1^2 - B_2^2)}} K\left(\frac{1}{k}\right) \quad \text{for } B \text{ between } B_2 \text{ and } B_1; \quad (3.13)$$

$$f(B) = 0 \quad \text{elsewhere}$$

where

$$k^2 = \frac{(B_1^2 - B_2^2)(B^2 - B_3^2)}{(B_1^2 - B^2)(B_2^2 - B_3^2)} \quad (3.14)$$

and K is the elliptic integral

$$K(k) = \int_0^{\pi/2} \frac{d\alpha}{\sqrt{1 - k^2 \sin^2 \alpha}}. \quad (3.15)$$

The function in Eq. 3.13 has a sharp peak at B_2, where $k = 1$ and $K(1) = 1$, which give

$$f(B_2) = \frac{2}{\pi} \frac{B_1 B_3}{B_2 \sqrt{(B_2^2 - B_3^2)(B_1^2 - B_2^2)}}. \quad (3.16)$$

If $B_1 = B_2$ we have $k = 0$ and $K(0) = \pi/2$, which yields the same result as Eq. 3.11.
The corresponding normalized lineshape as a function of frequency is

$$f(\omega) = \frac{2}{\pi} \frac{\omega}{\sqrt{(\omega_3^2 - \omega_2^2)(\omega^2 - \omega_1^2)}} K(k) \quad \text{for } \omega_1 \leq \omega \leq \omega_2;$$

$$f(\omega) = \frac{2}{\pi} \frac{\omega}{\sqrt{(\omega_3^2 - \omega^2)(\omega_2^2 - \omega_1^2)}} K\left(\frac{1}{k}\right) \quad \text{for } \omega_2 \leq \omega \leq \omega_3; \quad (3.17)$$

$$f(\omega) = 0 \quad \text{elsewhere}$$

with

$$k^2 = \frac{(\omega_2^2 - \omega_1^2)(\omega_3^2 - \omega^2)}{(\omega^2 - \omega_1^2)(\omega_3^2 - \omega_2^2)} \quad (3.18)$$

and K is the same elliptic integral as in Eq. 3.15. The lineshape resulting from Eq. 3.17 is shown in Figure 3.1b; this is characterized by a sharp peak at ω_2 and sharp edges at ω_1 and ω_3. The EPR line of TEMPO radical in a glassy matrix is the superposition of 3 such lines, each corresponding to one of the 3 magnetic states $m = -1, 0, +1$ of the nuclear spin of ^{14}N that has hyperfine interaction with the paramagnetic electron located at the N—O bond.

The lineshapes described above can be seen in solid samples at such low temperatures that the rotational motion of the paramagnetic molecules is slowed down well below the Larmor precession frequency. At high temperatures, and in particular in the liquid state, the EPR absorption signal becomes progressively narrower and more symmetric, as long as the dipole-dipole relaxation time T_2 remains small in comparison with the correlation time τ_c, defined as the time required for the rotational autocorrelation function to drop by $1/e$. If the g-anisotropy is not large, the centroid of the EPR absorption line approaches the average value g_{av}

$$g_{av} = \frac{1}{3}\left(g_{xx} + g_{yy} + g_{zz}\right) = \frac{1}{3}\left(2g_\perp + g_\parallel\right). \tag{3.19}$$

When the ratio τ_c/T_2 becomes small (compared with unity) at high temperature, the linewidth corresponds to the dipolar one with motional narrowing, and one may be able to resolve the hyperfine structure of the magnetic resonance if the concentration is low enough, usually well below 10^{17} spins/cm^3.

The free radicals and other paramagnetic molecules usually contain many electrons and atoms; the unpaired electron usually belongs to one of these atoms or ions, such as the Cr(V) ion in PD-Cr(V) complex or in Cr(V)-EHBA (see Chapter 7). In order to estimate the principal values of the g-tensor in such a real case, we must make simplifying assumptions in the treatment of the spin-orbit coupling. We have to ignore possible polarization of the electron pairs forming the molecular orbitals and assume that the total spin-orbit coupling operator is obtained by summing the terms similar to Eq. 3.2 written for each atom in the molecule. The calculation of the spin-orbit coupling constants requires additional assumptions on the representation of the molecular orbitals as a linear combination of the atomic orbitals, and on the interaction of the unpaired electron with the atomic orbitals which determine the electric field strength **E** near the nucleus. The detailed calculations are beyond the scope of this book, and numerical values are given for a large number of paramagnetic molecules in the literature.

If one electron is lacking in an otherwise filled shell of the atom, it can be regarded as a moving hole or positive charge in a filled shell. This changes the sign of Eq. 3.2 and therefore the sign of the spin-orbit coupling constant λ, which may be used to distinguish the radiolytic hole paramagnetic centers (V-centers) from the electron paramagnetic centers (F-centers). A way to understand this is to parametrize the g-shift as

$$g_{ii} = g_e \frac{a_{ii}\lambda}{\Delta E}. \tag{3.20}$$

Here a_{ii} describes the symmetry of the crystal field and its strength, and ΔE is the energy difference between the pair of electrons (with which the unpaired electron is associated) and the unpaired electron itself. This difference can be positive or negative, and consequently the principal values of the g-tensor can be shifted up or down from the free-electron value. It is, however, common that the shift is negative in stable free radicals, which can be understood by the requirement that the free electron be somewhat shielded from the outside of the molecule or ion to ensure chemical stability.

3.2.2 Hyperfine Splitting

The hyperfine interaction arises from the scalar magnetic coupling $\gamma_e\gamma_n\mathbf{S}\cdot\mathbf{I}$ between the magnetic moments of an unpaired electron and a nearby nucleus, as was described in Section 2.1.4. It results from the overlap of the wave functions and can be represented in the general form

$$\mathcal{H}_{hf} = \int dv\, |\psi(\mathbf{r})|^2\, \gamma_e\gamma_n\hbar^2 \left\{ \frac{8\pi}{3}\mathbf{I}\cdot\mathbf{S}\delta(\mathbf{r}) + \frac{3(\mathbf{I}\cdot\mathbf{r})(\mathbf{S}\cdot\mathbf{r})}{r^5} - \frac{\mathbf{I}\cdot\mathbf{S}}{r^3} \right\}, \qquad (3.21)$$

where the first term of the integrand can be derived from the Dirac equation and is non-zero for *s*-electrons (or when the electron wave function has *s*-component). The second and third terms, which arise from the dipolar coupling at distance *r*, is non-zero for electrons which are not in a pure *s*-state. The integration is over the electron coordinates. The expression is linear in I_x, I_y and I_z, and S_x, S_y and S_z. It can be diagonalized (by the appropriate choice of the principal axes) and represented in dyadic form by

$$\mathcal{H}_{hf} = \mathbf{S}\cdot\tilde{\mathbf{A}}\cdot\mathbf{I} \equiv A_x I_x S_x + A_y I_y S_y + A_z I_z S_z \qquad (3.22)$$

where

$$\tilde{\mathbf{A}} = \mathbf{i}A_x\mathbf{i} + \mathbf{j}A_y\mathbf{j} + \mathbf{k}A_z\mathbf{k}. \qquad (3.23)$$

For pure *s*-electrons the hyperfine tensor is isotropic. In complicated molecules the orbital of an unpaired electron may be represented by a linear combination of atomic orbitals in *s*- and *p*-states; the resulting hyperfine tensor in high field is

$$\tilde{\mathbf{A}} = \begin{bmatrix} a & 0 & 0 \\ 0 & a & 0 \\ 0 & 0 & a \end{bmatrix} + \begin{bmatrix} b_1 & 0 & 0 \\ 0 & b_2 & 0 \\ 0 & 0 & b_3 \end{bmatrix}, \qquad (3.24)$$

where $b_1 + b_2 + b_3 = 0$ and a is the isotropic hyperfine constant.

If the isotropic hyperfine energy dominates the anisotropic one, the spin Hamiltonian of the unpaired electron can be approximated by[3]

$$\mathcal{H} = \hbar\mu_B \mathbf{B}\cdot\mathbf{g}\cdot\mathbf{S} + \sum_i a_i \mathbf{I}_i\cdot\mathbf{S}, \qquad (3.25)$$

where the sum is over the hyperfine nuclei in the molecule.

Although there is no general reason for the principal axes of the anisotropic *g*- and hyperfine tensors to coincide, they often do so in relatively simple molecules. We shall assume this, and for simplicity let us also take an isotropic *g*-tensor (or alternatively consider an oriented single crystal) in the solid sample. The energy levels are then,

---

[3] We suppose that the orbital angular momentum is strongly quenched.

$$E = \left(g\mu_B B + \sum_i a_i m_i \right) m_S, \qquad (3.26)$$

because (to a good approximation in high field) $\mathcal{H}$ commutes with S_z and the eigenfunctions can be taken as those of S_z with $a_i S_z I_z$ only having diagonal matrix elements (in first order). The resulting energy level splitting can be best represented graphically; an example is shown in Figure 3.2. We note that for the case of N equivalent nuclei all a_i are equal and yield the split of the ESR spectrum into $2NI + 1$ lines. The intensity ratios for such allowed electron spin transitions $\Delta m_S = \pm 1$, $\Delta m_I = 0$ are given by the binomial coefficients (obtained from Pascal's triangle, for example). These ratios are valid for high temperatures, and follow from the number of possible combinations of the states m_i to get the same sum of all m_i:

$$\text{Intensity ratios}: 1 : M : \ldots : \binom{M}{k} : \ldots : M : 1,$$

where $M = 2NI$ is the row of Pascal's triangle and k its column; these intensity ratios are also called binomial coefficients.

When DNP changes significantly the polarization of the hyperfine nuclei, noticeable changes result in the intensity ratios of the hyperfine components in the ESR spectrum, if they are discernible in it. This will be discussed again in Chapters 4 and 5.

The number of equally split lines and their intensity ratios may allow to determine the nuclear spin I and the number of equivalent nuclei. This is often the case with protons, which produce large resolved hyperfine splittings owing to their large nuclear moment.

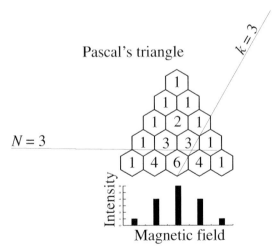

Figure 3.2 Hyperfine splitting due to N equivalent spins $I = \frac{1}{2}$ in the paramagnetic molecule with $S = \frac{1}{2}$, illustrated here for $N = 4$. The population ratios of the hyperfine levels are equal at the high temperature where the g-factor anisotropy is averaged out by motion in the liquid sample. The hyperfine splitting can be seen in the EPR spectra at 0.3 T when the sample is sufficiently diluted, typically $\leq 10^{17}$ spins/cm^3

3.2 EPR Absorption Spectrum

This facilitates the identification of the radicals produced by radiation damage in the target, if the experimental EPR spectrum can be determined at a temperature where the broadening due to the anisotropy of the g- and hyperfine tensors is absent because of motional narrowing, but the radicals survive long enough for the measurement.

The identification of the radical may be further facilitated if the nuclear spin transitions $\Delta m_i = \pm 1$, $\Delta m_S = 0$, can be observed by NMR. For $S = \frac{1}{2}$ (which is nearly always the case in polarized targets) and one nucleus with $I = \frac{1}{2}$, as illustrated in Figure 3.3A, the NMR line will be split in two equally intense components

$$\omega = \gamma_n B_0 + \frac{am_s}{\hbar} \qquad (3.27)$$

separated by $a/\hbar$ from each other, if $\hbar \gamma_n B_0 \gg a$. For $a \gg \hbar \gamma_n B_0$ the NMR is observed at frequencies

$$\omega = \frac{a}{\hbar} \pm \gamma_n B_0; \qquad (3.28)$$

however, this may be observed directly only at low field in some cases, but if the hyperfine coupling is anisotropic, the resulting large broadening requires the use of electron-nuclear double resonance (ENDOR) technique, which consists of observing the nuclear spin resonance by its effect on the electron spin resonance (see, for example, Ref. [13] pp. 266–268). This is a powerful technique that allows to measure hyperfine couplings with a high accuracy.

The second term in Eq. 3.21 gives rise to the anisotropic hyperfine tensor

$$\mathcal{H} = \mathbf{S} \cdot \tilde{\mathbf{b}} \cdot \mathbf{I} = \mathbf{S} \cdot \begin{bmatrix} b_1 & 0 & 0 \\ 0 & b_2 & 0 \\ 0 & 0 & b_3 \end{bmatrix} \cdot \mathbf{I}, \qquad (3.29)$$

where averaging over the electron spatial wave function is performed. In complex molecules summing must again be performed over all nearby nuclei. If the isotropic hyperfine coupling is zero (purely non-s unpaired electron, which is rare), the complete spin Hamiltonian, neglecting g-anisotropy and taking B along z, becomes

$$\mathcal{H} = g\mu_B B_0 S_z + \hbar \gamma_n \mathbf{B}_{\text{eff}} \cdot \mathbf{I} \qquad (3.30)$$

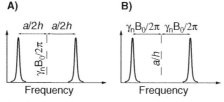

Figure 3.3 NMR line splitting for $I = \frac{1}{2}$ due to hyperfine interaction with nearby electron spin $S = \frac{1}{2}$. (a) $\hbar \gamma_n B_0 \gg a$; (b) $a \gg \hbar \gamma_n B_0$

where

$$\hbar\gamma_n \mathbf{B}_{\text{eff}} = S_z\left(b_1\mathbf{i} + b_2\mathbf{j} + b_3\mathbf{k}\right). \quad (3.31)$$

We may interpret this as an interaction of the nuclear magnetic moment with an internal field $\mathbf{B}_{\text{eff}}$, the magnitude of which depends on the projection S_z of the electron spin on the z-axis. The electron spin remains quantized along the main field, but the nuclear spin is quantized along the effective internal field $\mathbf{B}_{\text{eff}}$, the direction of which changes when the electron spin flips. The magnetic energy levels of the electron based on this hyperfine interaction are

$$E(m_S, m_I) = \left(g\mu_B B_0 + \sqrt{b_1 + b_2 + b_3}\, m_I\right) m_S. \quad (3.32)$$

The effect of Eq. 3.31 is to split the resonance of an electron with spin 1/2, to $2I + 1$ components for each nearby nucleus. This splitting depends on the orientation of the principal axis of the g-tensor with respect to the main field. A thorough discussion on the resulting complex behavior of the lineshape, including effects of the isotropic hyperfine tensor, can be found in Ref. [9], pp. 30–39.

In the case of axial symmetry often encountered (at least approximatively) in the case of more heavy nuclei such as ^{13}C and ^{14}N, the lineshape function for $S = 1/2$ coupled to one nucleus with spin $I = 1/2$ resembles Eq. 3.10, which resulted from g-factor anisotropy [9]:

$$f(\omega) = \frac{1}{2|m_I|\sqrt{6ab + 3b^2}} \cdot \frac{\omega - \omega_e}{\sqrt{(\omega - \omega_e)^2 - m_I^2(a-b)^2}}, \quad (3.33)$$

when $m_I^2(a-b)^2 \leq (\omega - \omega_e)^2 \leq (a+2b)^2$ and $f(\omega) = 0$ elsewhere. Here a is the isotropic hyperfine constant and $b = b_3 = -2b_1 = -2b_2$, and $\omega_e = g_e\mu_B B_0/\hbar$.

Again, if the nuclear polarization deviates significantly from zero, the electron lineshape function becomes asymmetric due to different populations of the levels m_I.

The complete spin Hamiltonian in a high field can be now written as

$$\mathcal{H} = g\mu_B \mathbf{B}\cdot\tilde{\mathbf{g}}\cdot\mathbf{S} + \sum_j \left(\hbar\gamma_j \mathbf{B}\cdot\mathbf{I}_j + \mathbf{I}_j\cdot\tilde{\mathbf{A}}_j\cdot\mathbf{S}\right), \quad (3.34)$$

neglecting the dipolar interactions among the electron spins themselves and with distant nuclear magnetic moments, and assuming no anisotropic hyperfine interaction. The summation is over the nuclei j, whose spins are I_j, gyromagnetic ratios γ_j and hyperfine tensors A_j. The NMR lines of randomly oriented diluted paramagnetic molecules, as calculated from Eq. 3.34, spread over many MHz for protons. The resulting NMR spectra will be discussed in Chapter 5.

Because the axes of quantization for I_j and S are not same, the axis of I_j changes when the electron spin flips. The state $|m_S, m_I\rangle$ before flip becomes $|m_S \pm 1, m_I'\rangle$ after flip; with $m_I \neq m_I'$. Such a transition represents the simultaneous flip of an electron and a nucleus.

However, these transitions do not in general polarize the hyperfine nuclei, because the new state represents an admixture of states m_I', referring to the new axis of quantization. This may be a handicap for DNP, since the resulting lower polarization for the hyperfine nuclei may cause a leakage of polarization for the matrix nuclei whose resonance is within the NMR line of the hyperfine nuclei.

For unpaired non-s electrons one might expect zero isotropic hyperfine interaction. This is, however, rarely entirely true because the unpaired electron may polarize the inner s-electrons in the molecule, causing isotropic splitting of the ESR line.

The anisotropic hyperfine interaction also often extends to the nuclei of the surrounding diamagnetic molecules. This is seen as a broadening of the hyperfine lines, which depends on the isotopic composition of the matrix molecules. If the matrix is a hydrocarbon glass, deuteration may lead to resolved hyperfine lines in the solid material.

The anisotropic hyperfine interaction as well as the g-tensor anisotropy average out in liquid samples of low viscosity, leaving a set of resolved lines due to the isotropic hyperfine interaction, shifted by the average of the g-factor of Eq. 3.19 and representing transitions between the energy levels

$$E = g_{av}\mu_B B_0 m_S + \sum_j a_j m_{I_j} m_S. \qquad (3.35)$$

These resolved lines can also be seen in solid samples at such elevated temperatures that the radical molecule rotates much faster than the hyperfine frequency $a/\hbar$. This happens, for example, with $\cdot NH_2$ in solid NH_3 at 78 K temperature. Below 4 K the rotation becomes so slow that the resolution is lost. The only radicals which are known to display resolved hyperfine structure below 1 K are $\cdot H$ and $\cdot D$. The doublet of $\cdot H$ and triplet of $\cdot D$ may be split, shifted or smeared further by interactions with the matrix or lattice, in particular due to the polarization of the inner shell electrons in the matrix molecules.

3.2.3 Dipolar Broadening of the EPR Line

DNP in solid target materials requires electron spin concentrations in the range of 10^{19}–10^{20} spins/cm^3. The dipolar interaction of the magnetic dipole moments of the unpaired electrons causes significant broadening of the ESR in solids at these concentrations. The dipolar interaction with nuclear spins also broadens the ESR line. The dipolar broadening is said to be homogeneous, because it results from fluctuating dipolar fields which average to zero over sufficiently long periods of time. We may thus imagine that the Larmor precession frequency of each electron spin fluctuates about a central value given by the average magnetic induction in the material.

The dipolar broadening sets the scale for two important parameters in polarized targets: the homogeneity of the static magnetic field over the target volume, and the range of microwave frequencies within which the maximum DNP can be expected.

Ignoring first the nuclear spins, the electron spin Hamiltonian is the sum of the Zeeman and dipolar spin-spin Hamiltonians. These spins are separated from each other by distance

r_{jk}, the vector $\mathbf{r}_{jk}$ making an angle of θ with the steady external field $\mathbf{B}_0$, and the Hamiltonian is obtained by rewriting Eq. 2.14 with $\gamma_j = g_j \mu_B/\hbar$ from Eq. 1.8:

$$\mathcal{H}_{tot} = \sum_j g_j \mu_B B_0 S_{jz} + \frac{\mu_0 \mu_B^2}{8\pi} \sum_{j \neq k} \frac{g_j g_k}{r_{jk}^3}\left(1 - 3\cos^2\theta_{jk}\right)\left(3S_{jz}S_{kz} - \mathbf{S}_j \cdot \mathbf{S}_k\right). \tag{3.36}$$

Here the scalar product of the spin vectors causes the simultaneous exchange of all the components of the two spins with identical g-factors (or when the g-factors are sufficiently near compared with the dipolar broadening of the resonance line). This exchange process conserves the energy and angular momentum in low-order perturbation theory and can be often neglected despite the fast cross-relaxation transitions and spin diffusion resulting from the term. In the case of a relatively large inhomogeneous broadening (also due to hyperfine interactions), the effect of the cross-relaxation in the linewidth is strongly suppressed. The remaining terms give the spin pair energy spectra

$$E_1 + E_2 = g\mu_B B_0 (m_1 + m_2) + \frac{\mu_0 g^2 \mu_B^2}{8\pi r_{12}^3}\left(1 - 3\cos^2\theta\right) m_1 m_2, \tag{3.37}$$

where m_1 and m_2 are the magnetic quantum numbers, i.e. eigenvalues of the spin operators S_{1z} and S_{2z}. We note that this simple system has four energy levels, among which there are two first-order dipole transitions in high field. These transitions are split by the effect of the equal but opposite local dipole field projections along the z-axis, B_{loc}:

$$2B_{loc} = \frac{\Delta E}{g\mu_B} = \frac{3g\mu_B}{2 r_{12}^3}\left(3\cos^2\theta - 1\right). \tag{3.38}$$

Depending on the orientation θ of the dipole pair and on their distance r, the total dipolar energy of a spin in a large assembly of other spins varies, causing smooth broadening of the resonance. This broadening depends in magnitude on the spin concentrations, and in shape on the spin distribution. The calculation of the dipolar width involves numerical summing over the orientations and distances of a relatively large number of neighboring spins, which yields the shape function of the envelope curve describing the sum of the resonance lines of Eq. 3.21. The symmetric bell-shaped curve has a definite full width at half maximum (FWHM) in the case of unpolarized spins; this width cannot be calculated analytically, but it may be estimated from the root-mean-square (RMS) width which is given by the second moment of the resonance line:

$$\Delta B^{FWHM} \cong 2\Delta B_{SS}^{RMS} = \frac{2\hbar}{g\mu_B}\sqrt{\langle \Delta^2 \rangle}, \tag{3.39}$$

where the second moment of Eq. 2.110 for like spins S can be written in the form

$$\langle \Delta^2 \rangle_{SS} = \left(\frac{\mu_0}{4\pi}\right)^2 \frac{g^4 \mu_B^4}{\hbar^2} S(S+1)\frac{3}{4}\sum_k \frac{\left(1 - \cos^2\theta_{jk}\right)^2}{r_{jk}^6}. \tag{3.40}$$

3.2 EPR Absorption Spectrum

To approximate our amorphous polarized target materials with dissolved paramagnetic free radicals or complexes, we may use the result of Van Vleck for a powder of cubic crystals

$$\left\langle \Delta^2 \right\rangle_{SS} = \left(\frac{\mu_0}{4\pi}\right)^2 \frac{g^4 \mu_B^4}{\hbar^2} S(S+1) \frac{5.1}{d^6}, \tag{3.41}$$

where the distance between the lattice points d can be approximated from $d^{-3} = n_S$ (the number density of the electronic spins). The often-used parameter 'dipolar frequency' is the square root of this and is for spin 1/2 and $g = 2$

$$D_{SS} = \sqrt{\left\langle \Delta^2 \right\rangle_{SS}} = 63.8 \cdot 10^6 \frac{n_S}{10^{20} \text{cm}^{-3}} \text{ rad/s.} \tag{3.41'}$$

The dipolar width can be written in the practical form

$$\Delta B_{FWHM}^{SS} = \frac{\mu_0}{4\pi} g\mu_B \sqrt{S(S+1)} \cdot 4.5 \, n_S. \tag{3.42}$$

The above result is valid for a spin system of like spins with low polarization and uniform distribution. If the polarization is nearly complete, the resonance line will be shifted to a higher frequency due to the magnetization of the polarized spins, by an amount comparable with the average magnetic induction

$$\Delta B_M = \mu_0 M = \mu_0 g \mu_B n_S. \tag{3.43}$$

Comparing with Eq. 3.42 we note that the dipolar FWHM width at low polarization amounts to about 31% of the static dipolar induction at complete polarization. The width arises mostly from contributions of the randomly oriented nearest neighbor spins which do not sum to zero because there are only a few neighbors. At full polarization the sums of the dipolar fields are equal at all sites of a simple cubic lattice, and therefore the dipolar broadening theoretically should vanish. Electronic spins in polarized targets, however, are always diluted and therefore a substantial fraction of the width of Eq. 3.42 remains at high polarization.

The above expressions for the width of the resonance line are valid only when there is no inhomogeneous broadening of the EPR, a condition which is met in LMN single crystals doped with Nd^{3+} ions, for example. In glassy solids the anisotropies of the g-factor and of the hyperfine tensor cause broadening, which dominates the dipolar one, and therefore the above equations have to be modified so that they will represent broadening of the spins S by unlike spins S', with the same density and magnetic moment. This is accomplished by multiplying the results simply by 2/3, as was described in Chapter 2. The result for the dipolar broadening of the resonance line is then

$$\Delta B_{SS'}^{FWHM} = 4.8 \cdot 10^{-24} \frac{n_S}{\text{cm}^{-3}} \text{ T (uniform isotropic distribution).} \tag{3.44}$$

For a random distribution of diluted spins the second moment becomes very large, because there may be spin pairs separated by a distance substantially smaller than the average distance. Using a Lorentzian lineshape model with cut-offs due to the closest possible approach of the diluted spins, Abragam [14] obtained the truncated Lorentzian FWHM

$$\delta = \frac{\pi}{2\sqrt{3}} \left(\frac{\langle \Delta^2 \rangle^2}{\langle \Delta^4 \rangle} \right)^{\frac{1}{2}} \sqrt{\langle \Delta^2 \rangle}. \tag{3.45}$$

which yields for diluted unlike spins $S = 1/2$ in a cubic lattice the width

$$\Delta B_{SS'}^{\text{FWHM}} = \frac{\mu_0}{4\pi} g\mu_B 5.9 \cdot n_S. \tag{3.46}$$

If the distribution of the spins is completely random, a statistical theory gives the fully Lorentzian shape with

$$\delta = \frac{2\pi^2}{3\sqrt{3}} \frac{\mu_0}{4\pi} \frac{(g\mu_B)^2}{\hbar} n_S \tag{3.47}$$

for like spins, and width (for unlike spins)

$$\Delta B_{SS'}^{\text{FWHM}} = \frac{\mu_0}{4\pi} g\mu_B 5.1 \cdot n_S. \tag{3.48}$$

The agreement of Eqs. 3.46 and 3.48 is fairly good and yields, based on the latter, the FWHM

$$\Delta B_{SS}^{\text{FWHM}} = 9.4 \cdot 10^{-24} \frac{n_S}{\text{cm}^{-3}} \text{ T (random isotropic distribution)}. \tag{3.49}$$

We note that the FWHM is nearly twice larger for the random isotropic distribution compared with the uniform isotropic distribution.

For very large dilutions, with spin densities well below 10^{19} cm^{-3}, the distribution is often random isotropic, and Eq. 3.49 describes well the dipolar broadening. At spin densities well above 10^{19} cm^{-3}, the uniform distribution may be a much better description, in particular if the paramagnetic molecules are large, which prevents random close pairs. However, if the electron spin polarization is very high, the lineshape in this case again tends towards Lorentzian in the central part, because of the random distribution of the very dilute spins with opposite polarization.

The electron spin densities in polarized target materials are up to $n_S \approx 10^{20}$ cm^{-3}. The dipolar widths of EPR in polarized targets therefore range up to 0.5 mT (= 5 G); the broadening due to anisotropic g-factor and hyperfine splitting always dominate this dipolar broadening. The dipolar broadening is then seen as a rounding of the edges of the distribution function describing the effects of anisotropy, which will be discussed later in this chapter.

The dipolar interaction of the electrons with nuclear spins also leads to the broadening of the electron resonance line. Because of their low precession frequency, the cross-relaxation transitions among the nuclear spins are so slow that the nuclear dipolar field can be regarded as static compared with the Larmor precession frequency of the electrons.

In the case where the method of moments is justified, the RMS dipolar width is calculated from the square root of the second moment of the resonance line. Approximating the nuclei as those of a powder of simple cubic crystals, the second moment is obtained from Eq. 3.41 by multiplying with 4/9 and changing the magnetic moment of the second spin species to that of the nucleus:

$$\langle \Delta^2 \rangle_{IS} = \frac{4}{9} \left(\frac{\mu_0}{4\pi} \right)^2 \gamma_I^2 g^2 \mu_B^2 I(I+1) \frac{5.1}{d^6}. \tag{3.50}$$

The FWHM linewidth becomes

$$\Delta B_{IS}^{FWHM} = \frac{2\hbar}{g\mu_B} \sqrt{\langle \Delta^2 \rangle} = \frac{\mu_0}{4\pi} \hbar \, \gamma_I \sqrt{I(I+1)} \, 3.01 \cdot n_I \tag{3.51}$$

in the case of nuclear spins I with low polarization, in contact with electron spins S which can be thought to be so dilute that their own dipolar width is negligible; the electrons may be fully polarized, however. The approximation of a powder of simple cubic crystals is rather good for glassy materials where the distribution of the nuclear spins is rather uniform around the paramagnetic electrons and the orientations of their position vectors are random.

Choosing the canonical density of protons in frozen butanol target material, $n_p = 0.79 \times 10^{23}$ cm^{-3}, we obtain their contribution to the dipolar width and dipolar frequency of the electrons:

$$\Delta B_{I_p S}^{FWHM} = 0.58 \text{ mT} \tag{3.52}$$

$$D_{I_p S} = 51.0 \cdot 10^6 \text{ rad/s}, \tag{3.53}$$

which is seen to dominate the width due to the dipolar coupling among the electrons themselves if their concentration is below 10^{20} cm^{-3} and they are uniformly distributed. At lower concentration the broadening due to protons dominates, and when the protons get polarized during DNP, the electron line will get narrowed.

In the case of completely deuterated butanol, the contribution of the deuteron spins to the electron linewidth is

$$\Delta B_{I_d S}^{FWHM} = 0.178 \text{ mT}, \tag{3.54}$$

$$D_{I_d S} = 15.7 \cdot 10^6 \text{ rad/s}. \tag{3.55}$$

Other nuclei such as ^{13}C and ^{17}O contribute negligibly because of their low magnetic moment and concentration.

The polarization of the electrons due to the low temperature and high field, and that of the nuclei due to high DNP, may reduce the above widths by as much as a factor of 1/2. The narrowing at high polarization, however, results in a lineshape which is closer to Lorentzian than Gaussian near the center of the line. Absorption at deviation larger than one FWHM from the center (for a narrow line) or from the edge (for a line inhomogeneously broadened by the anisotropy of the g-factor, for example) may thus be less sensitive to the polarization of the various spin species in the material.

Appendix A.6 describes a normalized lineshape function, which can be changed from Lorentzian to a shape close to Gaussian by the variation of one parameter. Such a function might be useful for the phenomenological description of simple narrow EPR lineshapes at high polarization.

The total RMS width due to the dipolar interactions with various spin species is obtained quite accurately by summing the second moments and taking the square root of the sum. The FWHM is close to one obtained from the second moments if all major contributions produce a shape close to a Gaussian lineshape, with a fourth moment about three times the square of the second moment. This condition is not very well satisfied for highly polarized spins, diluted spins and random arrays of spins, but the error in the procedure of using the root of the sum of squares of the individual contributions is not very significant in view of the need for the knowledge of this parameter. The total dipolar width is mainly needed for getting an idea of the scale in which the magnetic field homogeneity and microwave frequency must be optimized in order to obtain maximum DNP. It is also important for the cross-relaxation transitions, which play a vital role in DNP and nuclear spin relaxation.

In order to see how the criteria for the field homogeneity and microwave frequency range should be set, let us consider the shape of the absorption line for electrons with both homogeneous and inhomogeneous broadening. The inhomogeneous broadening, for simplicity, is assumed to be due to the anisotropy of the axially symmetric g-tensor described by Eq. 3.10, and the homogeneous broadening is due to dipolar interactions. The absorption line is the convolution of the two lineshapes, which yields the shape shown in Figure 3.4 featuring edges rounded by the dipolar broadening. On the low-frequency side, the rounding follows rather close the complementary error function (erfc), whereas on the high-frequency side it comes closer to the Gaussian shape. We need to know these rounding functions roughly because it is important to understand how the microwave power absorption varies as a function of frequency in the range where the target material is transparent to the microwave radiation. At the electron spin densities of interest for polarized targets, the material is almost perfectly black to the microwaves at the center of the absorption line, so that only a thin layer on the surface of the target then absorbs power. The penetration improves at the edges of the line until the material is sufficiently transparent and yields a fairly uniform transverse field distribution throughout the volume of the target.

The region of interest for DNP begins where the absorption function has dropped by about one order of magnitude from maximum. In this region the absorption drops steeply when going further away from the center of the line; one dipolar FWHM deviation further drops the absorption by about two orders of magnitude. It is clear then that field uniformity

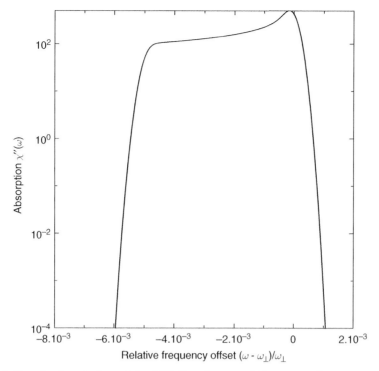

Figure 3.4 Gaussian broadening of the ESR line due to axially symmetric g-tensor, shown in logarithmic scale to emphasize the fall of the absorption signal just outside the endpoints of the unbroadened spectrum

must be much better than the dipolar FWHM, which is around 0.7 mT for polarized proton targets with paramagnetic center density of $10^{20}\,\mathrm{cm}^{-3}$, and about 0.3 mT for a deuterated material with $0.5 \times 10^{20}\,\mathrm{cm}^{-3}$ spin density. In the latter case, the field uniformity should clearly be better than 0.03 mT if highly uniform polarization is required. For a field of 2.5 T a relative uniformity of about 10^{-5} is therefore necessary if a high and homogeneous deuteron polarization is required. For proton targets the requirement is less critical, because the FWHM is twice larger, and because a variation of the spin temperature in the volume entails a much smaller variation on the proton polarization when it is already close to ± 1.

The optimization of the microwave frequency and power, to be discussed in more detail in Chapters 4 and 10, is related both with the field uniformity and with the dipolar width of the spin system. If the magnetic induction over a small sample is perfectly homogeneous, absorption of the microwave power by the spin system varies by about two orders of magnitude within one FWHM. Changing the frequency by the equivalent of one FWHM, however, does not result in such a large power variation, because the transverse field is increased when the magnetic absorption decreases. Therefore, the speed of polarization is not very sensitive to the variation of the microwave frequency, as long as it does not exceed 10^{-4}, which might bring it beyond the cut-off frequency.

Because at high field the dipolar broadening does not depend on the steady field value, the requirements for the relative homogeneity of the field and the time stability of the field and the frequency are stricter at higher fields, as will be shown also in Chapters 4 and 11.

3.2.4 Exchange Narrowing

At high concentration, when the distance between molecules with unpaired electron is about 0.5 nm (concentration approaching 10^{22} spins/cm^3), the electrons may exchange their positions due to the uncertainty principle. This leads to a narrowing effect somewhat similar to the motional narrowing; however, the second moment of the line remains constant because the narrowing of the center of the line is associated with slight extension of the wings. The width at half maximum is reduced and the shape approaches to a Lorentzian rather than the Gaussian-like shape resulting from pure dipolar interactions.

The exchange Hamiltonian (in analogy to the dyadic notation used in Eqs. 3.5 and 3.23)

$$\mathcal{H}_{exch} = \mathbf{S}_i \cdot \tilde{\mathbf{J}}_{ij} \cdot \mathbf{S}_j \tag{3.56}$$

can be thought to consist of isotropic and anisotropic parts. For like spins the resulting local field fluctuation has the rate $|J|/h$, which gives the reduction of the dipolar broadening. For unlike spins (different g-factor or hyperfine splitting, even because of their anisotropy), complicated effects may arise, varying from the broadening of the hyperfine lines to the fusion and sometimes to the narrowing of the hyperfine lines or g-anisotropy broadened line.

Electron spin concentrations above 10^{21} cm^{-3} tend to give poor results for DNP; the highest recorded optimum concentration is 1.6×10^{20} cm^{-3} in PD-Cr(V). The cause of this could be exchange effects in clusters of two or more paramagnetic molecules, which are also the supposed cause of increased nuclear spin-lattice relaxation [15], to be discussed in Section 3.4. A magnetic transition (ferro- or antiferromagnetic) in larger clusters might be another cause of these. The optimization of the electron spin concentration will be discussed further in Chapter 7.

Radiolytic radicals are often formed in pairs with short distance. Such pairs have a large dipolar interaction in addition to the exchange term. No narrowing of resonance line is expected, because there is no fluctuating field from many dipoles [9].

3.3 Saturation of Electronic Spin System

The power absorbed from a transverse linearly polarized field $2B_1$ per unit volume of target material was calculated in Section 2.2.1; we shall rewrite Eq. 2.58 here:

$$\frac{\dot{Q}}{V} = \frac{1}{2}\omega\mu_0^{-1}B_1^2\chi''(\omega), \tag{3.57}$$

recalling that the linearly polarized field was composed of counter-rotating components with magnitude B_1. Only the field that rotates in the same sense as the spins causes spin

3.3 Saturation of Electronic Spin System

transitions, while the other component is 'sterile'. By expressing the absorption part of the complex susceptibility in the terms of the normalized lineshape function of Eq. 2.63, and by using relation 1.66 between the static susceptibility and the spin polarization, the power can be put in the form

$$\frac{\dot{Q}}{V} = \frac{\pi}{4}\hbar\omega n_s S_e P_0 \omega_1^2 f(\omega), \tag{3.58}$$

where P_0 is the electron polarization in equilibrium with the lattice. This can be further developed by using the transition probability of Eq. 1.58 for spin $S_e = 1/2$, and by defining the dimensionless saturation function

$$s(\omega) = W(\omega)T_{1Z} = \frac{\pi}{2}\omega_1^2 f(\omega) \cdot T_{1Z}, \tag{3.59}$$

where T_{1Z} is the Zeeman spin-lattice relaxation time (relaxation time of the angular momentum or longitudinal magnetization). By inserting this, the power per unit volume reads, in the linear response approximation

$$\frac{\dot{Q}}{V} = \frac{\hbar\omega n_s}{4T_{1Z}} P_0 s(\omega). \tag{3.60}$$

We see here that the power absorbed is linearly proportional to the spin density and polarization, magnetic field strength, inverse relaxation time and the saturation function, which itself is proportional to the lineshape function.

When the strength of the transverse oscillating field is increased beyond the linear response region, the linear response theorem is no longer valid. The power absorbed by the spin system is then obtained by using the Provotorov equations, which yield for the spin susceptibility of Eq. 3.57 a different value given by Eq. 2.142. Using this and the saturation function defined above leads to

$$\frac{\dot{Q}}{V} = \frac{\hbar\omega n_s}{4T_{1Z}} P_0 \frac{s(\omega)}{1 + s(\omega)(1 + a\Delta^2/D^2)}, \tag{3.61}$$

where the usual definitions are taken for the angular frequency offset Δ and the dipolar frequency D, and $a = T_{1D}/T_{1Z} = 1/3$ as was stated in Section 2.2.4. We see here that additional requirements are needed in the case of long irradiation times for the linear response theorem to be valid: we must have $s(\omega) \ll 1$, and $s(\omega)a\Delta^2/D^2 \ll 1$ in order that the experimental susceptibility would follow the lineshape function obtained in the linear response approximation.

The Provotorov saturation equations deviate from the phenomenological model of Bloembergen, Purcell and Pound [16] by the term $s(\omega)a\Delta^2/D^2$ in the denominator; this term follows from the inclusion of the dipolar interactions and dipolar temperature in the quantum statistical treatment of the problem.

The above power absorption 3.61 is valid in the limit of the high-temperature approximation for a dipolar lineshape; these were the assumptions made by Provotorov when deriving the equations. There is no similar quantum statistical treatment for the case of high-electron polarization which applies for the polarized target operating conditions. Phenomenologically at least the following modifications can be suggested for the extension of Eq. 3.61 towards polarizations which do not satisfy the conditions of the high-temperature approximation:

(1) The dipolar frequency becomes smaller and depends both on polarization and on saturation.
(2) The dipolar relaxation time becomes longer because of smaller cross-relaxation rate; this rate scales with $1 - P_0^2$.
(3) The lineshape function changes from a broader near-Gaussian shape towards a narrower truncated Lorentzian shape.
(4) At high spin density and polarization the effective field, the polarization and the magnetization have a large transverse component which adds on the external RF field; this amplifies the effective field and makes the response of the spin system increasingly non-linear, compared with the usual Provotorov equations.

The item 1 can be demonstrated in the central part of the NMR signal of protons which are highly polarized, by turning the microwave saturation of the electron spins on and off. The proton absorption signal gets narrower and higher when the electron spins are saturated; this is a sudden change if b_0 and/or B_1 is large, in accordance with Eq. 2.198. When the microwave power is reduced, the proton resonance signal gets broader and lower in a slower time scale, measuring the parameter T_{1D} of Eq. 2.198 alone. The demonstration can be made without signal averaging, if the electron spin density is high, and therefore its dipolar interaction with the proton spin system is large.

It would be tempting to add the above phenomenological features 1–4 by hand in Eq. 3.61 in order to gain at least qualitative understanding of the low-temperature saturation problem. The usefulness of this approach, however, is questionable because the lineshape of the diluted paramagnetic system is not amenable to precise numeric calculations even in the high-temperature approximation. We therefore proceed to make some numeric estimates directly with Eq. 3.61.

Let us assume electron spins 1/2 with $g = 2$ and density $n_e = 5 \times 10^{19}$ cm^{-3} which give $D = 60.2$ Mrad/s taking into account broadening by electrons themselves, Eq. 3.41', and by protons, Eq. 3.53. At 2.5 T field we have $\omega = 2\pi \cdot 70$ GHz and take $a = 3$ and $P_0 \approx 1$. The spin-lattice relaxation time is $T_{1Z} = 38$ ms for PD-Cr(V) at low temperatures and 2.5 T field. The magnetic power dissipation is then

$$\frac{\dot{Q}}{V} = \dot{q}(0) \frac{s(\Delta)}{1 + s(\Delta)(1 + a\Delta^2/D^2)}. \tag{3.62}$$

where

$$\dot{q}(0) = \frac{\hbar \omega n_s}{4T_{1Z}} P_0 = 96 \frac{mW}{cm^3}. \quad (3.63)$$

For the sake of a qualitative discussion, we assume that the lineshape is Gaussian defined by the dipolar width so that the saturation function is

$$s(\Delta) = \frac{\pi \omega_1^2 T_{1Z}}{2\sqrt{2\pi} D} \exp(-\Delta^2 / 2D^2) = s(0) \exp(-\Delta^2 / 2D^2), \quad (3.64)$$

where

$$s(0) = \frac{\pi \omega_1^2 T_{1Z}}{2\sqrt{2\pi} D} = \left(\frac{\omega_1}{5.0 \cdot 10^4 \, s^{-1}} \right)^2. \quad (3.65)$$

In the high-temperature approximation, the power absorption as a function of frequency deviation can then be plotted for different transverse field strengths which correspond to different values of $s(0)$; such plots are shown in Figure 3.5. We assume that the microwave power fed into the cavity is absorbed predominantly by non-resonant losses and that the spin system absorbs 96 mW/cm³ at exact resonance for $s(0) = \infty$, as in the

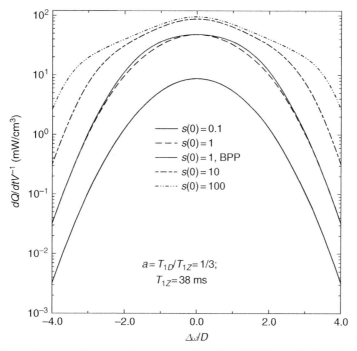

Figure 3.5 Magnetic power absorption due to saturation of an ESR line with Gaussian broadening, with $a = 3$, $P_0 \approx 1$ and $T_{1Z} = 38$ ms. For comparison we show the model of Bloembergen, Purcell and Pound labelled BPP [16]

above example for PD-Cr(V) at 2.5 T field. We see that the saturation function $s(\Delta)$ varies by a factor 500 when the frequency offset varies from $4D$ to $2D$. This means that the microwave power absorption varies by almost the same factor in this frequency range, if $s(0) \leq 10$.

The above can be used for evaluating the uniformity of the microwave power absorption in a real magnetic field, which is slightly non-uniform. If the magnetic field varies within the volume of the material by an amount equivalent to D and the average frequency offset is $3D$, the power absorption in the material varies by a factor of about 30 when the transverse field strength corresponds to $s(0) \leq 10$. It is evident from this that the relative field uniformity must be much better than D/ω_0 because uniform power absorption is required for a uniform polarization.

It is clear that at low temperature and high polarization there must be some quantitative changes in the above parameters, although the qualitative conclusions may be valid as such. The dipolar widths become smaller and the lineshape becomes narrower in the central part; these suggest that the variation of the saturation becomes a steeper function of frequency. The dipolar relaxation time becomes longer and therefore the power absorption becomes smaller, particularly for large values of $s(\Delta)$. It is therefore evident that at high polarization the magnetic field must be even more uniform, and that the optimum frequency at which DNP is made becomes restricted to a narrower range. The above considerations will be used for estimating the requirements of the magnetic field uniformity for polarized targets in Chapter 9.

The saturation of an inhomogeneously broadened EPR line must include a treatment of the cross-relaxation between spin packages with different Larmor frequencies. In Chapter 2 we discussed the cross-relaxation between two distinct lines under saturation of one of them. This treatment, however, cannot be extended to a continuous spectrum of lines without complications. Although these complications could be numerically handled, it would be difficult to judge the quality of the outcome. Therefore, we prefer to resort to a phenomenological approach guided by Provotorov's result only for the line with dipolar broadening.

3.4 Relaxation of Electron Spins

3.4.1 Electron Spin-Lattice Relaxation

The success of DNP using paramagnetic impurity spins depends critically on their spin-lattice and spin-spin relaxation, in addition to the width and shape of the resonance line. The Zeeman and dipolar spin-lattice relaxation times T_{1Z} and T_{1D}, together with the spin density n_s, determine the power absorbed by the material under saturating microwave irradiation, as was discussed in Section 3.3. The cross-relaxation of the electron spins determines the cooling rate of the nuclear spins. The same electron spins also cause the nuclear spin-lattice relaxation, unless some unwanted impurities cause faster relaxations. In good target materials the unwanted impurities must be reduced to a minimum.

3.4 Relaxation of Electron Spins

In high field the spin-lattice relaxation establishes the thermal equilibrium between the Zeeman energy reservoir $\langle \mathcal{H}_z \rangle = \hbar \gamma_e B_0 \langle I_z \rangle$ and the lattice, by interchange of energy. In steady state the interchange of energy results in equal flow of energy to and from each magnetic level. Considering spin 1/2 and denoting the upper and lower levels by subscripts + and −, respectively, this detailed balance is written $N_+ W_{+-} = N_- W_{-+}$, or

$$\frac{N_-}{N_+} = \frac{W_{+-}}{W_{-+}}. \qquad (3.66)$$

The medium where the electron spins are bound can be represented by a continuum of energy levels related with the excitations of the medium. The excitations which couple with the electron spins are phonons, discussed in Section 2.4.5, which obey Bose–Einstein statistics so that their number density is from Eqs. 2.157 and 2.171

$$p_{ph} \Delta \omega = \frac{\sigma(\omega) \Delta \omega}{V} p_{ph} = \Sigma(\omega) \frac{1}{\exp(\hbar \omega / kT) - 1}, \qquad (3.67)$$

which is the product of the phonon density of states $\Sigma(\omega)$ in the frequency interval $\Delta \omega$, and their population number p_{ph}. We recall that

$$\Sigma(\omega) = \frac{3\omega^2 \Delta \omega}{2\pi^2 v^3}. \qquad (3.68)$$

The transition probabilities among the magnetic levels are those of stimulated emission and absorption of phonons given by

$$W_{+-} = w\Sigma(\omega) p_{ph};$$
$$W_{-+} = w\Sigma(\omega)(p_{ph} + 1), \qquad (3.69)$$

where w is a constant which depends on the coupling between the lattice and the effective spin of the electron, to be discussed below. These give immediately

$$\frac{N_-}{N_+} = \frac{W_{+-}}{W_{-+}} = \frac{p_{ph}}{p_{ph} + 1} = e^{-\frac{\hbar \omega}{kT}}. \qquad (3.70)$$

The levels are therefore populated according to the Boltzmann statistics, which is the consequence of the Bose–Einstein statistics of the phonon system. This can be easily generalized to any number of magnetic levels as long as the number is finite and their total spectrum is narrow in comparison with the spectrum of phonons.

At high temperatures the approach of the population ratio towards thermal equilibrium with lattice can be described often by a single time constant T_1, whereas at low temperatures it is the inverse spin temperature which approaches its equilibrium value exponentially, as was discussed in Section 2.4.5. The population ratio or polarization therefore is not expected to change exponentially during relaxation at low temperatures, but the deviation is quite small and can be only seen as a lengthening of the time constant close to thermal equilibrium with lattice. The deviation is therefore difficult to observe.

The spin-lattice interaction generally takes place via the spin-orbit coupling, the strength of which is modulated by the lattice phonons causing changes in the electric field configuration of the electron orbital. The paramagnetic spins used in polarized targets have small deviations only from the $g = 2.0023$ free-electron value; the g-factor can therefore be modulated only slightly, resulting in a relatively small relaxation rate. The relaxation rates due to the single-phonon (direct) process and the two-phonon Raman process were derived in Chapter 2. An additional two-phonon process is also possible where the paramagnetic ion or molecule is excited to a low-lying orbital level by the absorption of the first phonon, followed by the emission of the second phonon which brings the ion or molecule to the ground-state orbital level with different magnetic state.

In order to obtain order-of-magnitude estimates for the matrix elements F_n of the lattice operators, Orbach [17, 18] expanded the lattice electric potential in the terms of the powers of the stress ε induced by the phonons, by

$$V = V^{(0)} + \varepsilon V^{(1)} + \varepsilon^2 V^{(2)} + \cdots \qquad (3.71)$$

where the first term is the static potential giving rise to the static crystal field, and the following terms represent the additional potential generated by the strain, in increasing order of the perturbing stress

$$\varepsilon = \frac{\rho_{ph} d\omega}{2\rho v_a}. \qquad (3.72)$$

Here the matrices $V^{(n)}$ are of the same order of magnitude, and rapid convergence is obtained at low temperatures because of the smallness of ε.

Direct Process

In the case of electron spin $S = 1/2$, symmetry under time reversal demands that the matrix elements of an electric potential $V^{(1)}$ vanish in zero magnetic field [19]. However, the presence of an external magnetic field produces admixtures of excited states of order $\hbar\omega/\Delta$ where Δ is a splitting of the crystal field. This yields the rough estimate

$$F_1 \approx \frac{\hbar\omega}{\Delta} \frac{e|V^{(0)}|}{\hbar}, \qquad (3.73)$$

where $\Delta/\hbar$ is of the order 3×10^{11} s for rare-earth ions [8]. Substituting this in Eq. 2.181 yields

$$\frac{1}{T_{1e}^{direct}} \approx 9\Omega_D \left(\frac{\omega_0}{\Omega_D}\right)^5 \left(\frac{e|V^{(0)}|}{\Delta}\right)^2 \frac{\hbar\Omega_D}{mv_a^2} \coth\left(\frac{\hbar\omega_0}{2kT}\right). \qquad (3.74)$$

In this expression the term

$$\left(\frac{e\left|V^{(0)}\right|}{\Delta}\right)^2 \tag{3.75}$$

must be understood as a very roughly estimated constant. Its definition is particularly unclear in glassy materials where the meaning of the crystal field is vague, because the wave function of the unpaired electron may be overlapping with a large fraction of the paramagnetic molecule. The constant is therefore best determined from experimental data, provided that the data shows the characteristic features of the direct process which are the frequency (and field) dependence ω^5, and absence of temperature dependence in the region where the coth-term approaches unity.

It should be noted that for spin higher than 1/2 (non-Kramers doublet) the magnetic states are not symmetric under time reversal, because the ground states generated by the static crystal potential may have different orbital states between which the matrix elements of $V^{(1)}$ are non-zero and independent of frequency [8]. In this case, which is uncommon in polarized targets, the relaxation rate varies as ω^3.

It is notable that in the relaxation due to the direct process the temperature dependence vanishes at very low temperatures, where relaxation becomes dominated by the spontaneous emission of phonons. All other known processes due to phonons, to be discussed below, have a fairly steep temperature dependence which makes their rate drop below that of the direct process at high field.

Another point to be noted is that the term of Eq. 3.75, the static part of the electric potential in the lattice, can be strongly influenced by the static stress in the lattice created by radiation damage, which then can control the electron spin lattice relaxation time due to the direct process. This may explain why the radiation damage first improves DNP in deuterated ammonia targets after annealing [20].

Raman Processes

Two types of Raman processes can be distinguished [8]; these are called first- and second-order processes depending on whether the phonons are real (of frequency within the phonon spectrum) or virtual (of frequency above the phonon spectrum).

The matrix elements of $V^{(2)}$ also vanish between the magnetic states of spin $S = 1/2$ because of symmetry under time reversal but, as above, the symmetry is removed by the application of a magnetic field. The matrix element of the lattice operator F_2 is analogously estimated roughly as

$$F_2 \approx \frac{\hbar\omega_0}{\Delta'} \frac{e\left|V^{(0)}\right|}{\hbar}, \tag{3.76}$$

where Δ' is the energy of the excited intermediate state which is not necessarily the same as Δ. The relaxation rate for the first-order Raman process, using Eqs. 2.188 and 2.190, can now be written as

$$\frac{1}{T_{1e}^{\text{Raman}}} = \frac{9^2}{\pi}\Omega_D \left(\frac{\hbar\omega_0}{mv_a^2}\right)^2 \left(\frac{e|V^{(0)}|}{\Delta'}\right)^2 \left(\frac{T}{\Theta_D}\right)^7 P_6(x_0) I_6, \qquad (3.77)$$

which exhibits strong temperature dependence roughly proportional to T^7 and moderate field dependence roughly proportional to B^2.

In the second type process, the phonons are virtual and may therefore have frequencies beyond the Debye cut-off. The matrix element of the lattice operator F_2 is due to an interference term between the absorbed and emitted waves and involves in first order the phonon frequency; it is estimated roughly as

$$F_2 \approx \frac{\omega'}{2} \left(\frac{e|V^{(0)}|}{\Delta''}\right)^2. \qquad (3.78)$$

The integration over the phonon spectrum analogously to Eq. 2.188 yields

$$\frac{1}{T_{1e}^{\text{Raman}}} = \frac{9^2}{\pi}\Omega_D \left(\frac{\hbar\Omega_D}{mv_a^2}\right)^2 \left(\frac{T}{\Theta_D}\right)^9 \int_0^{X_D - x_0} \frac{x^5 (x+x_0)^3 e^{x+x_0}}{(e^x - 1)(e^{x+x_0} - 1)} dx, \qquad (3.79)$$

with X_D and x_0 defined in Eq. 2.187. At low temperatures the integral can be extended to infinity and it converges to

$$\int_0^{X_D - x_0} \frac{x^5 (x+x_0)^3 e^{x+x_0}}{(e^x - 1)(e^{x+x_0} - 1)} dx \cong \int_0^\infty \frac{x^5 (x+x_0)^3 e^{x+x_0}}{(e^x - 1)(e^{x+x_0} - 1)} dx \cong P_8(x_0) \int_0^\infty \frac{x^8 e^x}{(e^x - 1)^2} dx = P_8(x_0) I_8,$$

$$(3.80)$$

where the integral has the numeric value

$$I_8 = \int_0^\infty \frac{x^8 e^x}{(e^x - 1)^2} dx = 40484.4 \qquad (3.81)$$

and the function $P_8(x_0)$ is described by the polynomial

$$P_8(x_0) \cong 1 + 0.3743 x_0 + 5.3881 \cdot 10^{-2} x_0^2 + 3.0124 \cdot 10^{-3} x_0^3 \qquad (3.82)$$

to an accuracy better than 1% for $x_0 \leq 10$, as shown in Figure 3.6.

Similar to the direct process, in the Raman processes the static lattice stress may strongly influence the relaxation rates via terms 3.76 and 3.78.

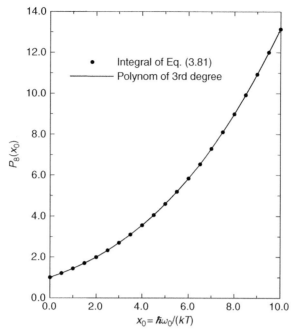

Figure 3.6 Polynomial fit to the temperature and field dependence of the integral of Eq. 3.79 approximated by Eq. 3.80

Orbach Process

The Orbach relaxation process [17, 18] involves two virtual phonons which are within the available spectrum but not necessarily thermally populated. The first phonon is absorbed exciting the ion or molecule to the energy Δ above the magnetic levels whose splitting is much smaller. The system subsequently decays, by emitting a second phonon, back to the ground state with roughly equal probability of ending to any of the two magnetic levels (supposing spin 1/2). The resulting electron spin-lattice relaxation rate is [8]

$$\frac{1}{T_{le}^{\text{Orbach}}} \approx 9\Omega_D \frac{e^2 \left|V^{(1)}\right|^2}{\hbar\Omega_D m v_a^2} \left(\frac{\Delta}{\hbar\Omega_D}\right)^3 \frac{1}{e^{\Delta/kT} - 1}, \qquad (3.83)$$

This rate has no field dependence and a steep temperature dependence at such low temperatures that $\Delta/kT \gg 1$. In polarized targets this rate is therefore visible only if $\Delta/k < 5\,\text{K}$.

A variant called Orbac–Aminov process is also independent of magnetic field; this has an additional term that is temperature independent when $\hbar\omega_e/kT \gg 1$ [21]. All Orbach processes are also susceptible to be influenced by the static stress of the lattice due to its effect on $V^{(1)}$.

Other Spin-Lattice Relaxation Mechanisms

Blume and Orbach [22] describe a low-temperature mechanism which may also dominate the direct process at low fields. In their case the ground state consists of multiplets with splittings of the same order as the magnetic level spacing. In two-phonon processes the splitting can then be neglected with respect to the phonon energies, allowing to approximate the relaxation rate by

$$\frac{1}{T_{1e}^{Blume}} = \frac{9^2}{\pi} \Omega_D \left(\frac{e|V^{(0)}|}{mv_a^2} \right)^2 \left(\frac{e|V^{(0)}|}{\hbar \Omega_D} \right)^2 \left(\frac{T}{\Theta_D} \right)^5 \int_0^\infty \frac{x^4 e^x}{(e^x - 1)^2} dx. \tag{3.84}$$

This rate could also dominate the rate due to the direct process at low field and moderately low temperatures.

Waugh and Slichter [23] have suggested that a slight change in the electron spin quantization axis, due to the flip of a hyperfine nucleus, could give the required degree of freedom for the relaxation of the nuclear spins by paramagnetic molecules at very low temperatures, where the probability of electron spin flips becomes extremely small. At low field or at low effective field, this could also entail electron spin relaxation.

Experimental Spin-Lattice Relaxation Times

Very few direct electron spin-lattice relaxation time measurements of polarized target materials have been made. The only systematic study is that of Ruby, Benoit and Jeffries [24] who measured at 0.25 T field and 0.3 K to 4.3 K temperatures the spin-lattice relaxation time of Nd^{3+} in LMN in several samples with concentrations from 1% to 5%. No concentration dependence was observed, in agreement with the theoretical models discussed above. The relaxation rate was observed to obey

$$\frac{1\,s}{T_{1e}} = 300 \left(\frac{B}{1\,T} \right)^5 \coth \frac{\hbar \omega_e}{kT} + 6.3 e^{-(47.6\,K)/T} \tag{3.85}$$

in good agreement with the rates due to the direct and Orbach processes. de Boer [25] measured, using direct pulse-recovery method, the relaxation times of PD-Cr(V) and porphyrexide in PD and butanol-water, respectively, at 2.5 T field and 0.5 K temperature with the results

$$T_{1e}(Cr(V)) = 38 \pm 2 \text{ ms}; \tag{3.86}$$

$$T_{1e}(PX) = 3.3 \text{ ms}. \tag{3.87}$$

These values correspond to the direct process which can be expected to dominate in these conditions. This has been verified by measuring the temperature and field dependences of the nuclear spin lattice relaxation, to be discussed in Chapter 5.

Indirect evidence from proton spin-lattice relaxation measurements in PD-Cr(V) suggests that at low fields and temperatures the direct process may be dominated by a relaxation rate which is field independent and has the temperature dependence

$$\frac{1}{T_{1e}^{\text{Low-field}}} \approx d e^{-\Delta/kT}, \qquad (3.88)$$

where $\Delta/k = 0.5\,\text{K}$ and d is a constant. A similar term was found by Harris and Yngvesson [26] in the electron spin relaxation in iridium salts; they could interpret it by a two-phonon process causing transitions between the magnetic levels of exchange-coupled clusters. This clearly requires a rather high spin concentration, unless there is a mechanism of obtaining clustering due to a non-uniform distribution of the paramagnetic molecules. Such clustering could take place if the material is not a good glass-former, for example. This point will be discussed in Chapters 5 and 7. Other measurements of proton spin relaxation as a function of Cr(V) concentration also suggest the occurrence of clusters [27] with exchange coupling at high concentration.

The irradiation of materials sometimes tends to yield pairs of paramagnetic centers, whose exchange coupling produces effects in the lineshape and relaxation; this was discussed in the Second Polarized target Workshop at Rutherford Laboratory [28, 29].

Electron spin-lattice relaxation of two trityl radicals, d24-OX063 and Finland trityl, were studied under conditions relevant to their use in dissolution DNP [30]. The dependence of relaxation kinetics on temperature up to 100 K and on concentration up to 60 mM was obtained at X- and W-bands (0.35 T and 3.5 T fields, respectively). The relaxation is quite similar at both bands and for both trityl radicals. At concentrations typical for DNP, relaxation is mediated by excitation transfer and spin-diffusion to fast-relaxing centers identified as clusters of 3 trityl radical molecules that spontaneously form in the frozen samples. These centers relax by an Orbach–Aminov mechanism and determine the relaxation, saturation and electron spin dynamics during DNP.

The measurement of the electron spin-lattice relaxation time in situ in a polarized target cannot be made by the usual saturation technique, because of the poor signal-to-noise ratio with the large untuned cavity and large target. Furthermore, it is difficult to saturate the resonance significantly because of the large amount of power required, and because of the large electron concentration, the saturation in the material tends to be inhomogeneous in the center of the resonance. The best in situ technique, although rarely used so far, is the nucleus electron double resonance (NEDOR) method (see Ref. [31] p. 402), which will be discussed in Section 3.5.2.

Indirect determinations of electron spin-lattice relaxation times are based on the experimental temperature and field dependences of nuclear spin lattice relaxation, which always proceeds predominantly via the electron spin system at low temperatures. This will be discussed in Chapter 5.

3.4.2 Phonon Bottleneck

In some crystalline materials the phonon system may not be able to transmit the Zeeman heat due to spin relaxation at the speed of the relaxation rate from the spins to the phonons. In this case the phonon system is not in internal thermal equilibrium, and the relaxation rate will become dependent on the heat transmission capability of the phonon system. This is characterized by a strongly non-exponential relaxation, demonstrated by the phonon avalanche effect [32]. Here the electron spin population is first inverted by adiabatic fast passage. Immediately after the passage phonons are emitted spontaneously if $h\nu > kT_L$. These heat up the phonon 'bath' by increasing the phonon density at frequency ν above their initial thermal density. This, in turn, increases the rate of the phonon emission by the process of induced emission. These cumulate and result in a 'phonon avalanche' which can be detected even directly, in addition to the observation of the anomalous relaxation of the electron spins.

Phonon bottleneck effects are also seen in the grain-size dependence of relaxation [33], because the phonon relaxation time depends on the crystal size. The bottleneck can best be observed in fairly faultless single crystals or large crystallites where there exist ballistic phonons. These are long-wavelength phonons which are predominantly scattered by the walls of the crystal, so that their decay time becomes

$$\tau_{ph} \cong \frac{d}{v_a}, \tag{3.89}$$

where d is the characteristic size of the crystal and v_a is the acoustic speed. This follows from the fact that the phonon frequency may only be converted by inelastic wall collisions. For the ballistic phonons the elastic wall collisions are more frequent than the inelastic ones, so that Eq. 3.57 must be then regarded as a lower limit for the phonon decay time constant. The ballistic phonons, furthermore, have a large acoustic mismatch with the helium coolant, resulting in their large probability of reflection and therefore in the Kapitza thermal boundary resistance, which will be discussed in Chapter 8.

Ballistic phonon effects are usually seen only below 20 K temperature, because phonons with energy higher that $k_B \cdot (20\,\text{K})$ undergo inelastic collisions with other phonons. Furthermore, above 20 K the dominant phonon wavelength is so short that the wall effects play a minor part in the phonon decay and thermalization, and also the Kapitza resistance is dominated by other effects which limit the heat transfer through the solid-helium boundary.

Let us consider a system of N electron spins $S = 1/2$ in a high field so that the magnetic levels are populated in thermal equilibrium with the lattice at temperature T_L by numbers N_a and N_b. These numbers are determined by the equilibrium phonon distribution function 3.67 denoted here by p_0 so that

$$\frac{N_a}{N_b} = \frac{p_0 + 1}{p_0} = \exp\left(\frac{\hbar\omega}{kT_L}\right). \tag{3.90}$$

3.4 Relaxation of Electron Spins

If this equilibrium is perturbed so that the level populations become n_a and n_b, with $n_a + n_b = N_a + N_b = N$, the microscopic differential equations which determine the approach back to equilibrium are [8]

$$-\dot{n}_a = +\dot{n}_b = B'\Sigma(\omega)\left[n_a p_{ph} - n_b\left(p_{ph}+1\right)\right], \quad (3.91)$$

$$\dot{p}_{ph} = \frac{1}{\tau_{ph}}\left(p_0 - p_{ph}\right)B'\Sigma(\omega)\left[n_a p_{ph} - n_b\left(p_{ph}+1\right)\right], \quad (3.92)$$

where p_{ph} is the instantaneous phonon distribution function which yields the correct ratio of the probability of phonon emission of a photon to that of the absorption of a phonon by the spin system as $(p_{ph}+1)/p_{ph}$. The rate constant B' [8] is related to the electron spin lattice relaxation time without phonon bottleneck by

$$B'\Sigma(\omega)(2p_{ph}+1) = \frac{1}{\tau_{1e}} \quad (3.93)$$

and

$$\Sigma(\omega) = \frac{3\omega^2}{2\pi^2 v_a^3}\Delta\omega \quad (3.94)$$

is the number of phonon modes per unit volume, within the frequency spectrum of the electron spin resonance. The rate equations can be simplified by defining the variables

$$x = \frac{n_a - n_b}{N_a - N_b}, \quad y = \frac{p_{ph} - p_0}{p_0 + \frac{1}{2}}, \quad (3.95)$$

which yields the non-linear equations

$$\dot{x} = \frac{1}{\tau_{1e}}(1 - x - xy), \quad (3.96)$$

$$\dot{y} = -\frac{1}{\tau_{ph}}y + b\dot{x}, \quad (3.97)$$

where

$$b = \frac{N/V}{\Sigma(\omega)(2p_0+1)^2} = \frac{N/V}{\Sigma(\omega)}\tanh^2\left(\frac{\hbar\omega}{2kT_L}\right) = \frac{C_S}{C_{ph}}, \quad (3.98)$$

is the ratio of the heat capacity of the spin system to that of the part of the phonons which are resonant with the spin system.

Equations 3.96 and 3.97 have been discussed by Faughnan and Strandberg [34]. The recovery of the electron spin polarization after a microwave pulse, for example, is initially rapid until the phonons are heated up to the electron spin temperature. The approach to this happens according to the rate equation

$$\dot{P}_e = -\frac{1}{\tau_{1e}} \frac{P_e - P_0}{1 + \sigma P/P_0}, \qquad (3.99)$$

where

$$\sigma = \frac{\tau_b}{\tau_{1e}} = 1 + (1+b)\frac{\tau_{ph}}{\tau_{1e}} \cong \frac{b\tau_{ph}}{\tau_{1e}} = \frac{C_S \tau_{ph}}{C_{ph}\tau_{1e}} \qquad (3.100)$$

and τ_b is defined as the time constant by which the spin and phonon systems then evolve together at a slower speed, approaching

$$\tau_b = \tau_{1e} + (1+b)\tau_{ph}. \qquad (3.101)$$

At low temperatures where b becomes large the term $b\tau_{ph}$ may dominate the relaxation time due to the direct process. The effective relaxation rate is then approximately

$$\frac{1}{\tau_b} = \frac{1}{\tau_{ph}} \frac{3\omega^2 \Delta\omega}{2\pi^2 v_a^3 N/V} \coth^2\left(\frac{\hbar\omega}{2kT_L}\right). \qquad (3.102)$$

The phonon bottleneck does not influence the Raman process directly because the phonons generated in the relaxation process cover a wide spectrum. However, if the direct process happens in parallel and if there is inelastic phonon scattering which converts the narrow-band phonons into the wide spectrum, the Raman relaxation may also be influenced by the non-thermal phonons even if the direct process does not dominate the relaxation rate.

The phonon bottleneck has consequences in DNP and nuclear spin lattice relaxation; these will be discussed in Chapters 4 and 5. The strength of the phonon bottleneck increases with decreasing phonon heat capacity and increasing heat capacity of the spins. It is therefore stronger in the conditions which are favorable for obtaining high nuclear polarization by DNP, that is high field and low temperature. The strength σ also increases with the Debye temperature Θ_D, which is typically high for good and hard crystals, whereas soft materials such as organic glasses have a low Θ_D. Furthermore, a high electron spin concentration increases the strength of the phonon bottleneck.

The stationary saturation of EPR in dilute paramagnets at helium temperatures has been described by the spin temperature model by Kochelaev [35], while taking into account the phonon bottleneck. This explains the results obtained experimentally in crystalline materials; the model has not been applied to glassy materials, where ballistic phonons may have a short mean free path.

Microwave frequency modulation during DNP might alleviate the adverse effects of the phonon bottleneck, because a wider range of the phonon frequencies will be available for the direct process; this will be also discussed in Chapter 4.

3.4.3 Cross-Relaxation

With cross-relaxation we mean transitions where two nearby spins with opposite orientations undergo a simultaneous transition so that angular momentum is conserved. In the case of spin 1/2 this exchanges the orientations of the spins, and for higher spin the cross-relaxation transitions take place so that $\Delta(m_1 + m_2) = 0$. With identical spins these transitions conserve both energy and angular momentum and they are therefore quite fast. The origin of these is the term B in Eq. 2.11b of the two-spin dipolar Hamiltonian; this term is therefore called the 'flip-flop' term. These flip-flops enable the relaxation of the dipolar energy and the establishment of the dipolar temperature, which in equilibrium is equal to the Zeeman temperature. Under saturation these temperatures are given by the Provotorov equations in the rotating frame, where they are close to each other if the spin-lattice relaxation is not too fast. The cross-relaxation is therefore very important for nuclear polarization, which uses the cooling of the electron dipolar temperature as a principal mechanism.

In Chapter 2 we reviewed the formal quantum statistical theory of the cross-relaxation between two spin species whose Larmor frequencies are close to each other but whose absorption spectra have no significant overlap. In the rather complicated dynamic equations there appears a long time constant which determines the rate of approach of the Zeeman temperatures of the two species to a common equilibrium temperature. This time constant is close to

$$\frac{1}{T_{12}} = \frac{2\pi}{\hbar^2} \frac{\text{Tr}\left\{\left(\mathcal{H}_D^+\right)_{-\Delta_{12}} \left(\mathcal{H}_D^-\right)_{\Delta_{12}}\right\}}{\text{Tr}\left\{I_z^2\right\}}, \quad (3.103)$$

where the terms appearing under the trace are the dipolar terms B of Eq. 2.11b

$$\mathcal{H}_{CR}^+ = -\frac{1}{4}\sum_{i<k} a_{ik} I_i^+ S_k^-, \quad (3.104)$$

$$\mathcal{H}_{CR}^- = -\frac{1}{4}\sum_{i<k} a_{ik} I_i^- S_k^+ \quad (3.105)$$

transformed to the rotating frame. These are the flip-flop terms causing cross-relaxation. The dipolar temperature of the spin system approaches equilibrium, even in the absence of an RF field, at a faster rate

$$\tau_D = \frac{T_{12}}{1 + \delta + \varepsilon \Delta_{12}^2 \hbar^2}, \quad (3.106)$$

where ε is roughly the inverse of the square of the total dipolar energy, and $\delta = 1$ when both spins are 1/2.

These equations are valid under strong saturation, but they are derived under the assumption of a high temperature. The time constants have not been evaluated in a more explicit form, and therefore we shall also discuss a more phenomenological approach for estimating the cross-relaxation rate in spin systems, due to Abragam [14].

We are interested first in the rate W_{ff} of the flip-flop transitions per spin in an electronic system with only one spin species so that all electrons have the same Larmor frequency; their linewidth is given by a normalized lineshape function $f(\omega)$, which has the second moment of Eq. 3.41' and roughly Gaussian shape so that

$$f(\omega_0) = \frac{1}{\sqrt{2\pi}} \frac{1}{\langle \Delta^2 \rangle^{\frac{1}{2}}}. \tag{3.107}$$

The oscillating component of the dipolar field arising from the term B has the mean square value

$$\langle \Delta_B^2 \rangle = \langle \Delta^2 \rangle - \frac{4}{9} \langle \Delta^2 \rangle = \frac{5}{9} \langle \Delta^2 \rangle, \tag{3.108}$$

which may be assumed to be the effective RF field which causes the flip-flop transitions at the rate

$$W'_{ff} = \frac{\pi}{2} \omega_1^2 f(0) = \frac{\pi}{2} \frac{5}{9} \langle \Delta^2 \rangle \frac{1}{\sqrt{2\pi}} \frac{1}{\langle \Delta^2 \rangle^{\frac{1}{2}}} = \frac{5}{18} \sqrt{\frac{2\langle \Delta^2 \rangle}{\pi}} \cong \frac{1}{4.5} \langle \Delta^2 \rangle^{\frac{1}{2}} \tag{3.109}$$

if the frequency and amplitude of the field would not fluctuate.

Equation 3.109 ignores the fact the frequency ω_1 fluctuates about the Larmor frequency with the same spread as the dipolar lineshape function. This requires that the normalized lineshape be convoluted by itself in evaluating the flip-flop rate, and leads to a rate which is about 10 times lower for a powder of cubic crystals [14]:

$$W_{ff} = \left(\frac{5}{18}\sqrt{\frac{2}{\pi}}\right)^3 \frac{\int g^2(\omega) d\omega}{g(\omega_0)} \langle \Delta^2 \rangle^{\frac{1}{2}} \cong \frac{1}{46} \langle \Delta^2 \rangle^{\frac{1}{2}}, \tag{3.110}$$

which is based on the approximation of the lineshape function by a Gaussian. Using the linewidth of Eq. 3.49 with spin density of $10^{19}\,cm^{-3}$ we find the flip-flop rate of about 15 kHz for the electrons with g-factor 2.

For unlike spins S_1 and S_2 the overlap of the frequency spectra of the dipolar field of species 1 and the absorption of species 2 reduces the above rate by

$$\frac{1}{2} \frac{\int g_1(\omega - \omega_{0,1}) g_2(\omega - \omega_{0,2}) d\omega}{g_1(\omega_{0,1})} \cong \exp\left[-\frac{(\omega_{0,1} - \omega_{0,2})^2}{2(\langle \Delta_1^2 \rangle + \langle \Delta_2^2 \rangle)}\right] \tag{3.111}$$

for Gaussian shapes [14]. This lowers the mutual flip-flop rate considerably if the Larmor frequencies are separated by more than the sum of the linewidths of the two species. There are, however, flip-flop processes where an even number of spins undergo a simultaneous transition so that the change in the total Zeeman energy is approximately conserved, and the small difference is compensated by a change in the dipolar energy. This may increase the flip-flop rate by several orders of magnitude, and it has been observed with nuclear spins.

An inhomogeneously broadened electron spin resonance line is described by Abragam and Borghini [36] as an assembly of 'spin packets' with distinct Larmor frequencies and dipolar widths estimated by the unlike spin approximation. This makes a continuum of frequencies among which the flip-flop processes produce spectral diffusion, when only one part of the spectrum is perturbed by RF saturation. Because the dipolar width of the spin packets depends linearly on the spin density, the cross-relaxation rate can be also expected to increase with the spin density, probably even faster than linearly. The number density of the spins therefore controls the dynamic behavior of the spin system under RF saturation, changing from inhomogeneous to homogeneous behavior when the density varies from a certain lower limit to an upper limit. Evidence from polarized target materials where a large variation of density is possible, such as PD-Cr(V), suggests that in this material at 2.5 T field the lower limit is around 10^{18} spins/cm^3, and the higher limit is about 10^{20} spins/cm^3. The spread of Larmor frequencies in this case is about 350 MHz and the dipolar frequencies due to the electrons themselves are

$$D = \sqrt{\langle \Delta^2 \rangle_{ss}} = \frac{\mu_0}{4\pi} \frac{g^2 \mu_B^2}{\hbar} \sqrt{\frac{3}{4}} \sqrt{5.1} n_s = 2\pi \frac{n_s}{10^{20} \text{cm}^3} \cdot 10.2 \text{ MHz}, \quad (3.112)$$

which gives $2\pi \cdot 0.1$ MHz and $2\pi \cdot 10$ MHz, for the lower and upper limits, respectively. The contribution of the nuclear spins is 8.1 MHz for protons in normal butanol, and 2.5 MHz for fully deuterated butanol; these dominate the dipolar width in the lower range of concentrations. The nuclear spins are therefore expected to play an important role in the cross-relaxation of electrons at high field.

As the rate at which the dipolar temperature is established in the spin system depends on the cross-relaxation rate, the spin density can control the relative importance of the mechanisms which contribute in the DNP. This will be discussed in more detail in Chapter 4.

3.5 EPR Spectroscopy

3.5.1 EPR Spectroscopy in Liquid and Solid State

X-Band EPR Spectrometer

The X-band[4] EPR spectrometer consists of a TE$_{102}$ single-mode cavity, iris coupled to an X-band waveguide in a uniform field of an electromagnet with an H-yoke. The frequency of operation is usually 9 GHz to 10 GHz and the field can be swept around the value at or above 0.3 T. The microwaves are produced by a klystron or Gunn diode source feeding the microwave bridge followed by a Schottky diode detector. Derivative absorption spectra detected by the signal of the diode are recorded by modulating the magnetic field using a modulation coil set while scanning the main power supply of the magnet. The modulation

---
[4] See Appendix A.8 for standard microwave frequency band definitions.

frequency and amplitude, and the field scan rate can be adjusted for the type of material; these parameters control the resolution of amplitude and linewidth under the various situations resulting from the sample properties [12]. The derivative signals can be integrated once or twice to obtain the lineshape and/or its integral. The details of the technique of EPR spectroscopy are beyond the scope of this book, and we therefore describe here only the applications relevant to polarized targets.

EPR Spectrum

In liquid or gaseous state the linear and rotational motion of the molecules causes averaging of the anisotropies in the g-shift and hyperfine splittings, and dipolar fields may also average out. This allows high-resolution studies of the isotropic part of the hyperfine structure, provided that the radical life is long enough at the temperature required. The hyperfine splitting constants have been measured usually in this way.

Light radical molecules may also rotate in the solid matrix even below 4 K temperature; this accounts for the possibility of resolving the hyperfine structure of radicals such as $\cdot NH_2$ in polycrystalline NH_3 at liquid nitrogen temperature, when the radical concentration is low enough.

In polarized targets the desired electron spin concentration is usually above $10^{19}\,cm^{-3}$ at which dipolar broadening makes the resolution of the hyperfine lines impossible; EPR studies and radical identification then require dilution of the liquid substance. The response of the EPR spectrometer also becomes non-linear in this range of concentrations, so that the signal size is not a good measure of the concentration. EPR spectra of diluted ethanediol (ED)-Cr(V) are shown in Figure 3.7; the first shows $2NI + 1 = 9$ hyperfine lines due

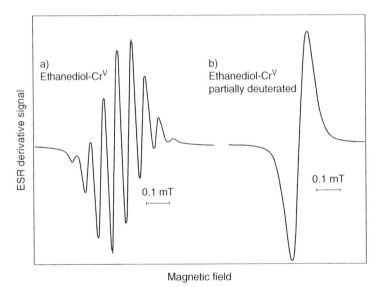

Figure 3.7 EPR derivative spectra of diluted ED-Cr(V) complexes; (a) normal ED, (b) partially deuterated ED (with the OH radicals undeuterated)

to eight equivalent protons, with intensity ratios 1:8:28:56:70:56:28:8:1 (note that the first and last lines are not resolved in the vertical scale of the figure). These lines are separated by 0.03 mT. The second spectrum is that of partially deuterated ED-Cr(V), where hyperfine structure is not resolved because of the much smaller splitting by the 17 lines due to the hyperfine coupled deuterons. When the spectra are taken with a wider field scan, four weak lines separated by 1.7 mT are seen; these are due to the hyperfine interaction of the electron with ^{53}Cr nuclei whose spin is 3/2 and natural abundance 9.55%. It is therefore clear that the unpaired electron wave function is concentrated on the Cr nucleus and it spills only slightly onto the surrounding protons of the molecule.

The undiluted Cr(V) complexes in PD and ED, for example, exhibit one broad line at room temperature at concentrations above 10^{19} cm^{-3}. The width of the line depends on the concentration; this feature can be used for its determination in such materials. Glättli [37] obtained for ED-Cr(V) results, which can be approximately expressed as

$$\frac{n_s}{10^{19}\text{cm}^3} = -7.7684 + 3.0422\frac{\Delta H}{\text{Oe}} - 1.6787\cdot 10^{-2}\left(\frac{\Delta H}{\text{Oe}}\right)^2, \quad (3.113)$$

where ΔH is the separation of the peaks of the derivative spectrum at 3 kOe (Oe = 10^{-4} A/m). de Boer [15] similarly obtained for PD-Cr(V)

$$\frac{n_s}{10^{19}\text{cm}^3} = -6.3221 + 3.1040\frac{\Delta H}{\text{Oe}} - 4.3164\cdot 10^{-4}\left(\frac{\Delta H}{\text{Oe}}\right)^2. \quad (3.114)$$

These are plotted in Figure 3.8. The scale in both cases was determined by diluting both samples sufficiently and comparing the signal size with that of a known solution of DPPH.

The frequency in an EPR spectrometer is accurately measured, but the field cannot be known so precisely. The mean g-factor g_{av} is therefore measured by comparing with a standard sample such as DPPH ($g = 2.0036$), diphenyl nitrogen oxide ($g = 2.0065$), Mn^{2+} ions in magnesium oxide ($g = 2.0023$) or Cr$_2$O$_3$·4H$_2$O ($g = 1.95$). The standard sample thus calibrates the field at one point. For a higher accuracy two standards are used, because the field of an iron-core magnet is not accurately proportional to the current in the coil.

Relaxation Time Measurement

The spin lattice relaxation time of the paramagnetic electrons is important for the DNP and experimental values are needed at 1 K and in the magnetic field of operation, usually 2.5 or 5 T; this will be discussed in detail in Chapter 4. The relaxation times obtained by the EPR spectrometer operating at RT or 78 K cannot be extrapolated to low temperatures.

In solid samples at low temperatures, the relaxation times can be determined from the measured saturation curves, which were discussed in Section 3.3. For broad lines the spin-lattice relaxation time T_1 can be determined from the dependence of the signal height on the saturation constant. A transmission-type saturation recovery measurement in situ was described by de Boer [38]; this requires a fast ferrite modulator and yields direct measurement of the recovery of the absorption part of the susceptibility. For lines with

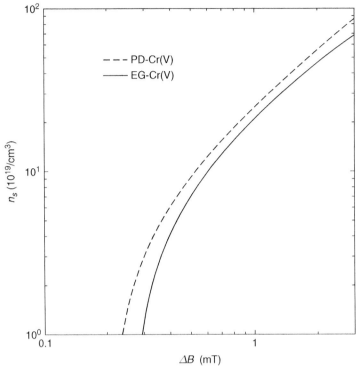

Figure 3.8 The Cr(V) concentration as a function of the unresolved EPR linewidth in 0.3 T field at RT, for high spin concentrations that are of interest to polarized targets. Based on data from Ref. [15], figure reproduced with permission from Elsevier

dominant dipolar broadening, the dipolar relaxation time T_{1D} can also be determined from the dependence of the lineshape on the saturation constant, as was also discussed above. Furthermore, the lineshape in a solid sample allows to estimate the anisotropies of the g-tensor and of the hyperfine tensor.

Indirect experimental values can be obtained from the temperature and field dependences of the nuclear spin lattice relaxation times. This will be discussed in Chapter 5.

High-Field EPR Spectrometers

The range of commercial EPR spectrometers extends up the Q-band with 1.22 T electromagnet and 34 GHz source. With superconducting magnets equipment is available for W-band spectroscopy at 3.35 T and 94 GHz, while microwave bridges extend to over 200 GHz frequency.[5] However, samples of polarized target materials cannot be easily mounted inside a single mode cavity because of its small dimensions at these mm-wave

---

[5] Bruker Corporation.

frequencies. Therefore, open resonators are used, in particular various types of Fabry–Perot interferometers.

A particularly elegant and efficient V-band[6] EPR spectrometer using an open cavity has been developed by the Polarized Target Group of Ruhr University at Bochum. The spectrometer operates down to 1 K at 2.5 T field; this is the first time to be able to measure EPR spectra under the conditions in which polarized targets operate [39]. The adjustable semiconfocal Fabry–Perot interferometer is mounted vertically in a vertical cryostat that can operate at 1 K, 78 K and RT. The waveguide is coupled through the upper mirror which is flat, and the sample droplet is placed on the lower spherical mirror having a radius of about 1 m. This geometry does not require extremely precise control of the mirror alignment, in contrast with the interferometers with flat mirrors. Also, the sample does not have to be optically flat. The irradiated samples which can be only handled under LN_2 were attached to the lower mirror with a film of PTFE.

The team of Ruhr University at Bochum made a series of EPR measurements on several successful polarized deuterated target materials and the spectra were analyzed with commercial software to extract the hyperfine structure and the width due to the anisotropy of the g-factor [40]. The derivative spectra were measured at 1 K in 2.5 T field, and these derivative signals were integrated in order to detect the edges of the shapes due to g-anisotropy. These results will be summarized in Table 3.1. The linewidth ΔB is the FWHM width of the integrated signal, used for a theoretical DNP model to compare with a series of DNP results [40] (not shown here).

Table 3.1 g-factor anisotropies, FWHM linewidths ΔB and main HFS anisotropies of several paramagnetic centers measured using a V-band Fabry–Perot interferometer at 1 K temperature in 2.5 T magnetic field, from Ref. [40]. The parameter ΔB corresponds to the spectral width as given by the commercial software package. The irradiated d-butanol sample contains mainly the alcoxy radicals (see Section 3.6.4).

Matrix material	Paramagnetic center	$\Delta g/g$ 10^{-3}	ΔB (mT)	A_{xx} (mT)	A_{yy} (mT)	A_{zz} (mT)
d-Butanol	EDBA	5.89	12.3			
d-Butanol	TEMPO	3.61	5.25	0.67	0.69	3.65
d-Butanol	PX	4.01	5.20	0	0	2.4
d-Butanol	Irrad.	1.25	3.10			
d-Butanol	Finland-d36	0.50	1.28			
d-PD	OX 063	0.28	0.86			
ND_3	$\cdot ND_2$	2.3	4.80			
^6LiD	F-center	0.0	1.80			

---

[6] See microwave band definitions in Appendix A.8.

3.5.2 EPR Spectroscopy in situ in a Polarized Target

Bolometric Detection

The cavity of a polarized target cannot be tuned accurately to the frequency of operation around 70 GHz, because it is usually very large in comparison with the 4 mm wavelength. The targets themselves are substantially larger in all dimensions compared with the wavelength. The cavity is always filled with addenda such as wires, thermometers, NMR coils, helium coolant and a cell confining the target beads. Often a separate chamber confines the liquid helium, because it has also been found that it is advantageous to separate the cavity from the chamber so that the cavity losses can be absorbed at a temperature higher than that of the target. Furthermore, the coupling between the waveguide and the cavity is very good; often the guide is gradually tapered to the size of the cavity in order to maximize the transmission of power into the target. The guide itself is quite lossy because it includes thermally isolating sections made of poorly conducting metals such as cupronickel or stainless steel. It is clear that accurate spectroscopy is impossible under these circumstances.

It has been, however, noticed that the resistance thermometers, inside the mixing chamber of a dilution refrigerator which cools the target, are very sensitive to the microwave field in the cavity. The thermometers then operate as a bolometer. This gives the possibility of observing the electron spin resonance because of the strong absorption, which lowers the microwave field strength and causes a substantial cooling of the bolometer. The response of such a bolometer is not linear, but its power dependence can be estimated from the thermal contact to the helium bath which obeys the Kapitza conductance law:

$$\dot{Q}_{Bol} = \alpha \sigma \cdot \left(T_{Bol}^4 - T_{He}^4 \right); \tag{3.115}$$

this will be discussed also in Chapter 8. Here α is a constant and σ is the surface area of the bolometer in contact with the liquid coolant. The power dissipated in the bolometer is proportional to the square of the electromagnetic field strength at its location. Measurement of the temperature of the bolometer therefore allows to determine the field strength at a given location in the cavity. The calibration of the bolometer can be made at a known power off-resonance, if the Q-factor of the cavity is known or can be estimated. The power itself can be estimated off-resonance by using the refrigerator as a calorimeter; this requires the prior measurement of its cooling power as a function of temperature. A small bolometer is also sensitive to the mode structure in the cavity, if there are standing waves causing nodes to appear; the bolometer calibration then depends on the frequency and on the dispersion of the target material which changes the electrical dimensions of the cavity when the magnetic field is changed close to the resonance.

Figure 3.9 shows EPR signals of deuterated PD-Cr(V) at 70 GHz in a dilution refrigerator at about 0.1 K temperature [41]. The frequency is constant and the bolometer reading is plotted during the sweep of the magnetic field. The upper signal (A) is obtained at the optimum microwave frequency for positive polarization, and the lower signal (B) at the optimum microwave frequency for negative polarization. The horizontal axis is labelled

3.5 EPR Spectroscopy 141

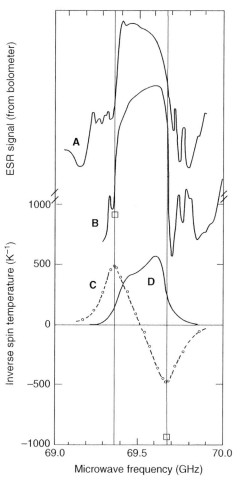

Figure 3.9 Bolometric EPR spectra of deuterated PD-Cr(V) in situ in CERN frozen spin-polarized target [41], compared with the inverse spin temperature and the EPR line obtained by microwave transmission techniques. See the text for the explanation of the different curves

in the terms of microwave frequency normalized to 2.5 T field in order to facilitate comparison with the lower curves, which show the microwave frequency scan of inverse spin temperature by DNP for the same material in a ^{3}He refrigerator (C), and the rough shape of the EPR absorption measured with a very small sample in the ^{3}He refrigerator (D) for normal PD-Cr(V) with 1.6×10^{20} spins/cm^3 [42].

The small-sample EPR signal exhibits the shape expected for axially symmetric g-tensor with $\Delta g/g \approx 4 \times 10^{-3}$, whereas the curves obtained with the bolometer show only a strong absorption feature within the range of Larmor frequencies. This absence of the absorption lineshape is interpreted to be due to the fact that the target material absorbs at resonance practically all incident waves with a transverse field component, so that the material is not transparent to such radiation.

The cavity used in Figure 3.9 has the diameter of 4 cm and length of 20 cm; the target is 1.8 cm in diameter and 12 cm in length. The coupling between the waveguide and the cavity is made by cutting a tapered slot to the guide all along the target. Such a cavity has a large number of overlapping modes so that the intensity recorded by the bolometer can be expected to depend strongly on the applied frequency and on the cavity tune, which is influenced by all materials in it. As a consequence of this, the bolometric signal is surrounded by additional peaks which are understood to be due to the tuning of the cavity by the dispersion of the target material at frequencies off-resonance. It can be seen that these features do not reproduce at different frequencies; this confirms that they are due to the cavity rather than the target material, which does not have such anomalies in the paramagnetic susceptibility.

The asymmetry of the bolometric signals, reversed for the two frequencies, is due to the inhomogeneity of the magnetic field when it is offset from the value of 2.5 T. At this high field the cobalt-iron pole pieces and part of the iron yoke are magnetized to saturation, so that the shimming can only be performed at one field value. The inhomogeneity of the field therefore 'smears' the absorption signal in proportion to the deviation from the value of 2.5 T.

In smaller samples mounted in small cavities, the bolometric signal features related with the dispersion are much less pronounced, as shown in Figure 3.10. The irradiated ammonia sample of about 1 cm^3 is mounted in a cavity about 2 cm long and 2 cm in diameter, and the spectrum is recorded at 140 GHz frequency in a 5 T superconducting magnet. Due to the low spin density of about 10^{19} cm^{-3}, some details of the spectrum can now be resolved so that the three main hyperfine lines are distinguishable; these are due to the two equivalent

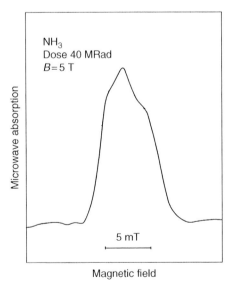

Figure 3.10 Bolometric EPR spectrum of the ·NH$_2$ radical in irradiated solid NH$_3$ in 5 T field below 500 mK temperature. The concentration is $\approx 10^{19}$ spins/cm^3

protons of the ·NH$_2$ radical. The main causes of the line broadening are hyperfine and dipolar interactions, because the width of the spectrum is identical at 2.5 T.

3.5.3 Effect of Frequency Modulation in the EPR Spectrum

When the microwave frequency is modulated, the EPR spectrum recorded by a bolometer in a large cavity changes substantially in the regions outside the absorption band. Spectra recorded with and without modulation are shown in Figure 3.11 for a 60 cm long target 5 cm in diameter mounted in a cavity 21 cm in diameter and 100 cm in length [43]. The target material is deuterated butanol with 6.4×10^{19} cm^{-3} electron spins of the deuterated complex EDBA-Cr(V). Modulation of the microwave frequency causes the flattening of the peaks outside the absorption band, and the increase of the absorption near the main band. This has been studied from the first principles [43] and it has been concluded that there are two leading mechanisms which cause these phenomena. These are due to the optical properties of the cavity which are varied by the effect of the magnetic field on the complex propagation constant in the target material, and to the slow cross-relaxation of the electronic spins whose frequency is shifted outside the main band due to the hyperfine interaction with the ^{53}Cr nuclear spins and with the nearby deuteron spins. Slow dipolar relaxation may also contribute to the latter, as will be discussed below.

The complex propagation constant k of microwave radiation is [44]

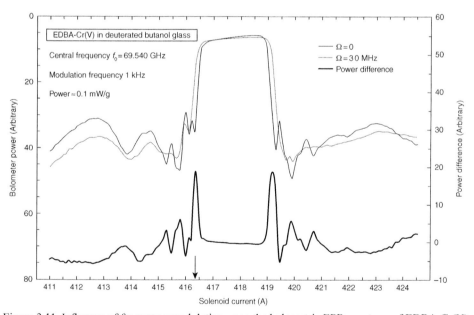

Figure 3.11 Influence of frequency modulation upon the bolometric EPR spectrum of EDBA-Cr(V) in deuterated butanol glass with 5% deuterated water. The bottom trace is the difference of the modulated and unmodulated curves; modulation increases the power absorption significantly at the edges of the EPR absorption line

$$k^2 = \varepsilon\mu = \frac{\varepsilon'\omega^2}{c^2}\left(1 + \chi' - i\chi''\right)\left(1 - i\frac{\varepsilon''}{\varepsilon'}\right),\tag{3.116}$$

where $c^2 = \left(\varepsilon_0 \mu_0\right)^{-1}$ is the squared speed of light in vacuum. Assuming weakly magnetic medium so that $|\chi'| \approx |\chi''| < 0.08$ (average absorption in the main line) and that the absorption part of the dielectric constant has no strong resonances and is much smaller than the dispersion so that $\varepsilon''/\varepsilon' \ll 1$, we may write the propagation constant in the approximate form

$$k = \alpha - i\beta \cong \frac{\sqrt{\varepsilon'}\omega}{c}\left[1 + \frac{\chi'}{2} - \frac{i}{2}\left(\chi'' + \frac{\varepsilon''}{\varepsilon'}\right)\right],\tag{3.117}$$

where α is the propagation constant and β is the attenuation constant. This approximate formula shows that the dispersion in the target material can change the speed of propagation by about 4%, which is enough for shifting several tens of nodes of standing waves to the microwave bolometer located in the central part of the cavity.

If the microwave frequency is changed by a few MHz, the bolometric pattern of the features outside the main spectrum changes considerably. It can be expected that if the microwave frequency is continuously modulated during the field scan with an amplitude of, say, 10 MHz at a frequency which is higher than the inverse of the response time of the bolometer (100 ms), averaging of the features would result. This is what is observed, as shown by Figure 3.11, apart from the region adjacent to the strong absorption line where a substantial additional absorption is evident, as shown by the subtracted spectrum. This additional absorption is visible already with a modulation amplitude of a few MHz and is close to saturation above 6 MHz which is less than the dipolar frequency estimated at about 10 MHz. The band of the additional absorption is about 60 MHz wide, twice larger than the peak-to-peak modulation amplitude $\Omega = 30$ MHz for this plot. The width of the additional absorption band is quite insensitive to the modulation amplitude, but it is less strong for amplitudes below 6 MHz.

The EIO tube microwave source has an emission bandwidth of about 100 kHz, measured with a spectrum analyzer equipped with a harmonic mixer downconverter. This is much less than the dipolar width, and in the absence of cross-relaxation a hole would be 'burned' in the inhomogeneously broadened spectrum; the width of the hole would be close to the dipolar frequency 10 MHz. The appearance of the spectral hole can be explained by considering the hyperfine interaction of the Cr(V) paramagnetic spin with the ^{53}Cr nucleus and with nearby deuterons, which causes additional broadening in the range of 5 MHz to 10 MHz, visible also in the NMR spectrum of the hyperfine nuclei (see Chapter 5). Saturation with a narrow emission band depopulates the hyperfine lines in the band and increases the population of the hyperfine lines outside the band; absorption therefore becomes very small, i.e. a hole is burned in the line. A change in the frequency repopulates the lines, and a continuous modulation at a frequency higher than the induced transition rate will yield a larger absorption of microwaves. This has also a remarkable effect on the speed of DNP, as will be discussed in Chapter 4.

3.5.4 NEDOR Spectroscopy

The NEDOR method [31, 45] is based on the use of the NMR signal for detecting the EPR. This is not to be confused with the ENDOR technique where the electron spin resonance is used for observing the NMR of hyperfine nuclei, for example. In NEDOR the NMR lock-in signal, or preferably the dispersion signal, is observed at its central region while saturating the EPR line or while recovering from a saturating microwave pulse. As the saturation changes the electron magnetization, the internal field shifts roughly by the amount of the static magnetization of Eq. 1.64

$$\Delta B = \mu_0 \Delta M = \mu_0 g \mu_B n_S S \Delta P(S) = 1.165 \Delta P(S) \frac{n_S}{10^{20} \text{cm}^{-3}} \text{ mT}, \qquad (3.118)$$

which can be almost as much as the NMR linewidth for protons in the case of a high electron spin density. The shift is accompanied by a change of the shape of the NMR line which is visible to naked eye in the case of a spin density of 10^{20} cm^{-3} in a PD-Cr(V) sample, even when saturating at the frequency of DNP. This can be tested by observing the NMR line center jump up a few percent when turning the microwave power on, and to note the height decay back in about a tenth of a second (at the rate of the electron spin-lattice relaxation) after turning the microwaves off.

At low spin density this simple observation is more difficult to make because the contribution of the electrons to the NMR lineshape is much smaller than the dipolar width due to the nuclei alone. Therefore, the shift of the NMR line center is a more sensitive, and also a more quantitative measure of the NEDOR signal.

In quantitative measurements we must take into account the microscopic distributions of the nuclear and electron spins, and the shape of the sample. As was discussed in Chapter 2, the first requires the summing and averaging of the dipolar fields from nearby electrons on protons at various lattice sites, and the second is given by an integral over the distant electrons up to the sample boundary. In glassy materials and in dilute crystalline paramagnets a model must be made to estimate the nearby dipolar field, whereas the remote dipolar field is given by integral which has position-independent values for elliptic sample shapes. For a sphere and cubic symmetry of the electrons surrounding the nuclei, the change in the internal field becomes exactly zero. For elongated (cigar) shape with long axis parallel to the field the first moment the field changes by

$$\Delta B = \frac{2\pi}{3} \mu_0 g \mu_B n_S S \Delta P(S), \qquad (3.119)$$

whereas if the long axis is perpendicular to the field, the field shift seen by the nuclei is

$$\Delta B = -\frac{\pi}{3} \mu_0 g \mu_B n_S S \Delta P(S). \qquad (3.120)$$

The highest sensitivity is therefore reached with a cylindrical sample parallel to the main field.

Polarized targets are usually made of spherical beads which have a very small field shift. The net shift therefore arises from neighboring beads and from the overall shape of the target container. The filling factor of the beads, usually around 0.6, must be taken into account in the electron spin density n_S.

In large targets the NEDOR technique enables to measure the electron spin-lattice relaxation time in situ by using the equipment readily available. The only small modification is to change the phase angle of the NMR circuit by 90° so as to observe the dispersion signal rather than the absorption, and to set the frequency of the RF source to the center of the NMR line. The microwave power is then pulsed by the appropriate control of the source, while recording the NMR signal transient. Averaging is useful for small signals which result from the use of small microwave power necessary at low temperatures. The record of the exponential recovery of the longitudinal magnetization gives directly the relaxation time T_{1Z}.

The NMR signal height is sensitive to the dipolar interactions between the electrons and the nuclear spins. The record of the recovery of the height is therefore likely to have a major contribution from the dipolar relaxation time T_{1D} of the electrons.

The in situ tests are best made at the frequencies of DNP which allows to work with large and steady NMR signal. The penetration of the microwave power inside a large target is then ensured, and one can work with buried NMR coils. Furthermore, if the microwave pulse is allowed to change the nuclear polarization, the control of the systematics of the NEDOR technique is difficult, because the nuclear polarization contributes substantially to the internal field shift.

When working with small samples, the NEDOR technique can be used for the measurement of the EPR line. It is essential that the coil is outside the sample because the microwave power does not penetrate the sample entirely in the central part of the absorption line at high spin density. The microwave power must be very small in order that the change in the nuclear polarization remains small while scanning the microwave frequency. The scan is best performed up and down in frequency, in order to average out systematics due to the changes in the nuclear polarization.

Because the measurement of the shift of the NMR line is quantitative and accurate, it can be used for determining the electron spin density in irradiated samples where there is no other direct measure available for this parameter. This requires, however, the complete saturation of the EPR line, and therefore it is mainly limited to small samples. The NMR signal shift is obtained accurately by calibrating the dispersion signal with a controlled change in the RF source frequency, or by a change in the applied field measured by another NMR signal.

It is advantageous to use a cylindrical sample with axis parallel to the static field, in order to maximize the NEDOR signal size. For spherical samples the NEDOR signal is small. This is due to the external term in the first moment which vanishes for a spherical sample, leaving alone the internal sum which is small.

The NEDOR techniques can be used for testing the various hypotheses underlying the frequency modulation effects in EPR and DNP, notably the effects of dispersion and of the repopulation of the hyperfine lines.

3.6 Paramagnetic Compounds and Radiolytic Centers Used in Polarized Targets

There is a vast number of know paramagnetic compounds containing one or more unpaired electrons, such as

- Free radicals
- Metallo-organic compounds
- Transition metal ions in crystalline lattice
- Radiolytic radicals and F-centers

In these categories there are good examples of stable substances that can be used for DNP, and we shall briefly discuss them below. A longer listing of potentially useful paramagnetic substances is given in Appendix 2.

3.6.1 Free Radicals

The structure of free radicals is studied by EPR spectroscopy because their molecular form can be inferred from the hyperfine spectrum obtained in liquid state and from the g-tensor seen in the spectrum in solid state.

A radical is an atom, a molecule or an ion that has an unpaired valence electron; this makes radicals highly chemically reactive. While most organic radicals have short lifetimes, there are notable exceptions such as the hydroxyl radical (HO·), a molecule that has one unpaired electron on the oxygen atom. Two other examples are triplet oxygen and triplet carbene ($:CH_2$), which have two unpaired electrons.

The prime example of a stable radical is molecular dioxygen (O_2). Triplet oxygen, $^3O_2$, refers to the $S = 1$ electronic ground state of molecular oxygen; this is the most stable and common allotrope of oxygen. Molecules of triplet oxygen contain two unpaired electrons, making triplet oxygen an unusual example of a stable and commonly encountered biradical. Atmospheric oxygen naturally exists as a biradical in its ground state as triplet oxygen. The low reactivity of atmospheric oxygen is due to its biradical state. The biradical state also results in the paramagnetic state of liquid oxygen, which is demonstrated by its attraction to an external magnet that is well known in cryogenics. Paramagnetic oxygen dissolved in a glassy matrix has a broad EPR line and short spin-lattice relaxation time; such a line cannot be used for DNP. Furthermore, it causes fast 'leakage' relaxation of nuclear spins, and therefore the polarized target materials are fabricated in such a way that contact with atmospheric oxygen is completely eliminated.

Biradicals are molecules containing two radical centers; they can also occur in metal-oxo complexes, lending themselves for studies of spin forbidden reactions in transition metal chemistry. Very large biradicals have been developed for DNP enhanced magic angle spinning NMR studies of chemical and biochemical structures; an example is 1-(TEMPO-4-oxy)-3-(TEMPO-4-amino)propan-2-ol (TOTAPOL) [46].

Another common example is nitric oxide (NO·). There are also many thiazyl radicals, which show low reactivity and remarkable thermodynamic stability.

Persistent radical compounds are those whose longevity is due to a molecular cage around the radical center, which makes it physically difficult for the radical to react with another molecule. Examples of these include

- Gomberg's triphenylmethyl radical, abbreviated as trityl, and its derivatives;
- Fremy's salt (potassium nitrosodisulfonate $(KSO_3)_2NO \cdot$);
- aminoxyls (general formula $R_2NO \cdot$, known as nitroxyl radicals and nitroxides) such as TEMPO, TEMPOL;
- nitronyl nitroxides;
- azephenylenyls and radicals derived from PTM (perchlorophenylmethyl radical);
- tris(2,4,6-trichlorophenyl)methyl radical TTM.

Many such free radicals can be precipitated from solutions by crystallization and therefore they can be dosed accurately when preparing them into a glassy matrix for DNP.

In 2004 Bunyatova [47] reviewed the list of free radicals that have been successfully used for the DNP of polarized target materials, in particular glassy hydrogen-rich diols and alcohols, but also some polymers that include scintillating (detector) materials. The list includes

- diphenyl picryl hydrazyl (DPPH);
- bisdiphenylallyl (BDPA);
- pycril-N-aminocarbazyl (PAC);
- porphyrexide (PX);
- porphyrindene (PB);
- α, γ-bisdiphenylene-b-phenylallyl (BDPA);
- aminoxyl radicals (TEMPO, OH-TEMPO, TEMPOL or 4-hydroxy-TEMPO, oxo-TEMPO etc.).

To this list we may add the trityl radicals such as triphenylmethyl and its numerous derivatives, which were found to yield the highest deuteron polarizations as of today [48]. Triphenylmethyl is the first free radical discovered by M. Gomberg of the University of Michigan in 1900. The radical was used for the development of ESR spectroscopy; its spectrum of 196 lines arises from a 1:3:3:1 quartet splitting of 2.86 G from the 3 para-protons, a 1:6:15:20:15:6:1 septet of 2.61 G from the 6 ortho-protons and a further 1.14 G septet from the six meta-protons. When enriched by ^{13}C in the radical center, the splitting by the ^{13}C in the methyl position is 26 G and the g-factor is 2.0026.

The trityl radicals OX063 and Finland-D36 have very large molecular cages around the central carbon atom and therefore small g-shifts and g-asymmetries. These radicals were developed by Amersham Health[7] in their project of MRI signal enhancement by the DNP of injectable contrast agents.

Section A.2 reproduces a list of free radicals with the total width of their hyperfine structure, compiled by Borghini in 1966 [49].

---

[7] Formerly Nycomed, now GE Healthcare.

3.6.2 Metallo-Organic Compounds

The most successful transition metal ion for DNP is pentavalent chromium Cr(V) that contains one unpaired electron in the inner 3d shell. Such compounds can be obtained directly by reacting hexavalent potassium or sodium dichromate with a diol such as ED or PD; in these cases the reaction proceeds close to RT. The procedure, first time described by Garif'yanov et al. [50], is explained in Chapter 7. These compounds, however, cannot be extracted and purified from the diol solution. The fabrication of PT materials based on ED-Cr(V) and PD-Cr(V) will be discussed in Chapter 7.

By reacting potassium bichromate with various hydroxy acids, a series of pure Cr(V) compounds have been produced by Krumpolc and Rocek [51–55]. Most DNP results have been obtained with EHBA-Cr(V) and BHHA-Cr(V), to be discussed in better detail in Chapter 4. The synthesis of these compounds will be briefly described in Chapter 7.

Other metal ions that have shown potential as a paramagnetic complex active for DNP are Ti(III), Gd(III) and Mn(II). Titanium(III) complexes with oxalate Ti(III)-ox and with urea Ti(III)-urea were prepared and studied by Nakasuka [56]; these are stable and soluble and show EPR lines just above 10 mT width (at 77 K and 0.34 T field) that may be suitable for DNP.

Gd^{3+} and Mn^{2+} bound in complexes of the chelators DOTA and DTPA were first introduced as high-spin metal ions for MAS DNP by Corzilius et al. in 2011 [57]. A very narrow EPR central transition linewidth of 29 MHz at 5 T was observed for Gd-DOTA complex, which allowed induction of solid effect with an initially reported proton enhancement factor of about 12 in a urea model sample.

3.6.3 Ions in Crystalline Materials

Beyond the well-known lanthanum magnesium double nitrate LMN, Nd^{3+} has been used as a paramagnetic ion for the DNP of in $LaAlO_3$ [58], where ^{139}La reached polarizations of ±20%. The same ion was also tried in LaF_3.

Tm^{2+} ions in CaF_2 were used for DNP of ^{19}F by Urbina and Jacquinot [59] to reach 90% polarization in the course of their studies of nuclear magnetism.

3.6.4 Radiolytic Radicals and F-Centers

Glassy Organic Materials

Solid materials irradiated at low temperature accumulate radiolytic impurities, many of which are paramagnetic. The primary result of ionizing collision is often a free electron ejected from a molecule AB:

$$AB \rightarrow (\cdot AB)^+ + e^-. \quad (3.121)$$

The chemistry of the solid phase may ensue in several ways that were comprehensively treated by M. Symons in the Second Workshop of Polarized Target Materials [28]. Here we shall follow his presentation in an abbreviated way, to describe the three possible scenarios:

(1) the free electron may be trapped in a cavity; in glassy materials there may be several types of trapping cavities;
(2) the electron may be trapped by binding to another molecule AB: $AB + e^- \rightarrow (\cdot AB)^-$;
(3) the electron may be trapped by another solute molecule X: $e^- + X \rightarrow \cdot X^-$.

The consequent chemical reactions depend on the material, on the temperature and also on the amount of accumulated damage. We are here interested both in irradiation taking place at liquid N_2 or Ar temperature around 78 or 90 K, and in irradiation in situ during a scattering experiment at $T < 1$ K. The former corresponds to pre-irradiation of a target material and to annealing after radiation damage during an experiment.

In pure alkane hydrocarbons RH the electron is not trapped and the molecule therefore returns to an electrically excited neutral state which can either decay back to the initial ground state, or it can split in two radicals:

$$(RH)^* \rightarrow \cdot R + \cdot H. \qquad (3.122)$$

Depending on the temperature, atomic hydrogen $\cdot H$ may migrate away or stay close to $\cdot R$, which is called pair trapping. In such a case, electron spin exchange may happen that is visible in the EPR spectrum and that may lead to poor DNP. Upon annealing, molecular hydrogen will be formed.

In aryl hydrocarbons benzene (C_6H_6) and toluene ($C_6H_5CH_3$), the radical yields are reduced because the free electrons are easily transferred back. Benzene does not form a glass, whereas toluene does form a good glass and has a greater radical yield than benzene.

In alcohols R_2CHOH the glassy matrix can trap electrons in a cavity but also the alcohol molecule can react with an electron:

$$R_2CHOH \rightarrow (R_2CH\dot{O}H)^+ + e^- \qquad (3.123)$$
$$(R_2CH\dot{O}H)^+ + R_2CHOH \rightarrow R_2CH\dot{O} + (R_2CH\dot{O}H_2)^+$$

where $R_2CH\dot{O}$ is the alcoxy radical, and the free electron will be either trapped or it may react by generating another radical and a negative OH^- ion:

$$e^- + R_2CHOH \rightarrow R_2\dot{C}H + (OH)^-. \qquad (3.124)$$

The alcoxy radicals are orbitally degenerate but hydrogen bonding lifts this so that we get $g_\parallel \approx 2.08$ and $g_\perp \approx 2.002$; these electrons relax relatively quickly because of the anisotropy of the g-factor in the right order of magnitude. The electrons are trapped in shallow centers below 4 K and therefore absorb light in the infrared band. Annealing at 78 K allows the molecules to reorient which turns the traps deep, so that the irradiated material absorbs light in the visible spectrum; this permits the bleaching of the glass with visible light, which has been observed in butanol glass, for example. The bleaching mainly yields the two radicals of the reactions 3.123 and 3.124. The remaining deeply trapped electrons relax very slowly in the absence of cross-relaxation.

The $R\dot{O}$ radicals may also be formed but they are very reactive and yield the alcoxy radicals $R_2CH\dot{O}$:

$$\dot{RO} + R_2CHOH \rightarrow ROH + R_2\dot{CHO}, \tag{3.125}$$

which have $g \approx 2.003$ with proton hyperfine couplings with the hydrogens in the R groups. These are the main free radicals after bleaching with visible light at 78 K.

Irradiated solid NH_3 and ND_3 will be discussed in detail in Chapter 7, where the EPR spectra, DNP and relaxation will also be treated. Irradiation near 78 K yields $\cdot NH_2$ radicals that can rotate in solid NH_3 at that temperature, so that the isotropic hyperfine constants $a_H = (20$ to $24.5)$ mT, $a_N = (14.5$ to $15)$ mT, $g_{av} = (2.003$ to $2.00481)$ can be seen in a noble gas matrix, for example.

Inorganic Crystalline Materials

Here we are interested in simple cubic ionic crystalline materials and follow the paper of Henderson [29] in the Second Workshop of Polarized Target Materials. Such materials include notably alkali hydrides (LiH, etc.), LiF, CaF_2 and $Ca(OH)_2$. In these cases irradiation by ionizing radiation below 20 K produces only F-centers and X_2^--centers (called also H-centers). The F-center is a vacancy left by displaced anion (Li^+ ion) where a single electron is trapped. The H-center is an interstitial molecular ion such as F_2^- occupying a single anion site in the case of LiF. These are produced as F-center/H-center pairs at relatively modest recoil energies. The F-center wave function extends over several neighboring atoms and is visible as rich hyperfine structure of the EPR spectrum. The X_2^--ion is covalently bonded to two other X^- ions. Both centers are paramagnetic with effective spin $S = \frac{1}{2}$ at low temperatures.

The F-center EPR spectra are rather complex, although the g-factors are isotropic and close to the free-electron value. This follows from the extension of the wave functions that spread over many neighboring nuclei. As was discussed in Section 3.2.2, when the overlap of the electron wave function covers N equivalent spin $I = 1$ nuclei in a shell, for example, 6 $^7Li^+$ ions with in the first shell in ^7LiF or in ^7LiH, there will be $2NI + 1 = 19$ lines with intensity distributions determined from Pascal's triangle. The second shell contains $N' = 12$ F or H nuclei with spin $\frac{1}{2}$ and gives rise to $2N'I + 1 = 13$ lines. The net result is an inhomogeneously broadened EPR line, with T_1 in the range (10 to 100) ms at helium temperatures.

The X_2^- centers have relatively simpler EPR spectra which show g-tensor anisotropy and hyperfine structure. There are discrete hyperfine splitting lines due to the dominant interaction with the two nuclei of the X_2^- radical. Each of these lines is inhomogeneously broadened by weaker interactions with the neighboring ions. The spin-lattice relaxation time in the range of (0.1 to 1) ms is shorter than that of the F-centers because of the anisotropy of the g-tensor that improves the coupling with the lattice phonons.

The radiolysis producing the F-center/H-center pairs is efficient: only about 5 eV to 10 eV energy deposit is required for one pair. This implies that both UV and X-ray irradiation are efficient. Electrons and protons will also produce defects by the ionization mechanisms. However, protons and neutrons also damage by direct displacement of ions in the anion and cation sublattices, and the ensuing defects are more complicated due to the cascade of recoils. Such irradiation is less efficient for the production of isolated F-centers and F-center/H-center pairs.

References

[1] B. Odom, D. Hanneke, B. d'Urso, G. Gabrielse, New measurement of the electron magnetic moment using a one-electron quantum cyclotron, *Phys. Rev. Lett.* **97** (2006) 030801.

[2] T. Kinoshita, B. Nizic, Y. Okamoto, Eighth-order QED contribution to the anomalous magnetic moment of the muon, *Phys. Rev. D* **41** (1990) 593–610.

[3] T. Kinoshita, W. B. Lindquist, Theory of the anomalous magnetic moment of the electron – numerical approach, *Phys. Rev. Lett.* **47** (1981) 1573.

[4] S. J. Brodsky, S. D. Drell, Anomalous magnetic moment and limits on fermion substructure, *Phys. Rev. D* **22** (1980) 2236–2243.

[5] S. J. Brodsky, V. A. Franke, J. R. Hiller, et al., A nonperturbative calculation of the electron's magnetic moment, *Nuclear Physics B* **703** (2004) 333–362.

[6] O. V. Lounasmaa, *Experimental Principles and Methods below 1 K*, Academic Press, New York, 1974.

[7] D. S. Betts, *An Introduction to Millikelvin Technology*, Cambridge University Press, Cambridge, 1989.

[8] A. Abragam, B. Bleaney, *Electron Paramagnetic Resonance of Transition Ions*, Clarendon Press, Oxford, 1970.

[9] S. Y. Pshetzhetskii, A. G. Kotov, V. K. Milinchuk, V. A. Roginskii, V. I. Tupikov, *EPR of Free Radicals in Radiation Chemistry*, John Wiley & Sons, New York, 1974.

[10] V. A. Roginskii, V. I. Tupikov, *EPR of Free Radicals in Radiation Chemistry*, Wiley, New York, 1974.

[11] J. E. Wertz, J. R. Bolton, *Electron Spin Resonance: Elementary Theory and Practical Applications*, McGraw-Hill, New York, 1972.

[12] G. R. Eaton, S. S. Eaton, D. P. Barr, R. T. Weber, *Quantitative EPR*, Springer Verlag, Wien, 2010.

[13] C. P. Slichter, *Principles of Magnetic Resonance*, 3rd ed., Springer-Verlag, Berlin, 1990.

[14] A. Abragam, *The Principles of Nuclear Magnetism*, Clarendon Press, Oxford, 1961.

[15] W. de Boer, High proton polarization in 1,2-propanediol at ^{3}He temperatures, *Nucl. Instr. and Meth.* **107** (1973) 99–104.

[16] N. Bloembergen, E. M. Purcell, R. V. Pound, Relaxation effects in nuclear magnetic resonance absorption, *Phys. Rev.* **73** (1948) 679–712.

[17] R. Orbach, Spin-lattice relaxation in rare-earth salts, *Proc. R. Soc. A* **264** (1961) 458–484.

[18] R. Orbach, Spin-lattice relaxation in rare-earth salts: field dependence of the two-phonon process, *Proc. R. Soc. A* **264** (1961) 485–495.

[19] J. H. Van Vleck, Time reversal symmetry, *Phys. Rev.* **57** (1940) 426.

[20] D. G. Crabb, D. B. Day, The Virginia/Basel/SLAC polarized target: operation and performance during experiment E143 at SLAC, in: H. Dutz, W. Meyer (eds.) *7th Int. Workshop on Polarized Target Materials and Techniques*, Elsevier, Amsterdam, 1994, pp. 11–19.

[21] L. K. Aminov, On the theory of spin-lattice relaxation in paramagnetic ionic crystals, *Soviet Physics JETP* **15** (1962) 547–549.

[22] M. Blume, R. Orbach, Relaxation by two phonons in ground-state multiplets, *Phys. Rev.* **127** (1961) 1787.

[23] J. S. Waugh, C. P. Slichter, Mechanism of nuclear spin-lattice relaxation in insulators at very low temperatures, *Phys. Rev.* **B37** (1988) 4337.

[24] R. H. Ruby, H. Benoit, C. D. Jeffries, Electron relaxation in LMN(Nd^{+++}), *Phys. Rev.* **127** (1962) 51.

[25] W. de Boer, *Dynamic Orientation of Nuclei at Low Temperatures*, CERN Yellow Report CERN 74–11, 1974.

[26] E. A. Harris, K. S. Yngvesson, Relaxation by exchange interaction, *J. Phys. C (Proc. Phys. Soc.)* **1** (1968) 990, 1011.

[27] V. A. Atsarkin, P. E. Budkooskiy, G. A. Vasneva, V. V. Demidov, Proton spin relaxation with exchange-coupled clusters, in: *Proc. 17th Coll Ampere, Turku*, 1972.

[28] M. Symons, Radiation induced paramagnetic centers in organic and inorganic materials, in: G. R. Court, et al. (eds.) *Proc. of the 2nd Workshop on Polarized Target Materials*, SRC, Rutherford Laboratory, 1980, pp. 25–28.

[29] B. Henderson, Inorganic materials, in: G. R. Court, et al. (eds.) *Proc. Second Workshop on Polarized Target Materials*, SRC, Rutherford Laboratory, 1980, pp. 29–32.

[30] H. Chen, A. G. Maryasov, O. Y. Rogozhnikova, et al., Electron spin dynamics and spin–lattice relaxation of trityl radicals in frozen solutions, *Physical Chemistry Chemical Physics* **18** (2016) 24954–24965.

[31] A. Abragam, M. Goldman, *Nuclear Magnetism: Order and Disorder*, Clarendon Press, Oxford, 1982.

[32] W. J. Brya, P. E. Wagner, Paramagnetic relaxation to a bottlenecked lattice: Development of the phonon avalanche, *Phys. Rev. Lett.* **14** (1965) 431.

[33] P. L. Scott, C. D. Jeffries, Spin-lattice relaxation in some rare-earth salts at helium temperatures; observation of the phonon bottleneck, *Phys. Rev.* **127** (1962) 32–51.

[34] B. W. Faughnan, M. W. P. Strandberg, The role of phonons in paramagnetic relaxation, *J. Phys. Chem. Solids* **19** (1961) 155.

[35] B. I. Kochelaev, Spin temperature and non-equilibrium phonons, in: M. Goldman, M. Porneuf (eds.) *NMR and More: Scientific Day in Honour of Anatole Abragam*, Les Editions de Physique Les Ulis, Saclay, 1994, pp. 279–291.

[36] A. Abragam, M. Borghini, Dynamic polarization of nuclear targets, in: C. J. Gorter (ed.) *Progress in Low-Temperature Physics*, Interscience Publ., 1964, pp. 384–449.

[37] H. Glättli, Organic materials for polarized proton targets, in: G. Shapiro (ed.) *Proc. 2nd Int. Conf. on Polarized Targets*, LBL, University of California, Berkeley, Berkeley, 1971, pp. 281–287.

[38] W. de Boer, Dynamic Orientation of Nuclei at Low Temperatures, PhD, 1974, Delft University of Technology

[39] J. Heckmann, S. Goertz, W. Meyer, E. Radtke, G. Reicherz, EPR spectroscopy at DNP conditions, *Nucl. Instrum. Methods Phys. Res.* **A526** (2004) 110.

[40] J. Heckmann, W. Meyer, E. Radtke, G. Reicherz, S. Goertz, Electron spin resonance and its implication on the maximum nuclear polarization of deuterated solid target materials, *Phys. Rev. B* **74** (2006).

[41] T. O. Niinikoski, Polarized targets at CERN, in: M. L. Marshak (ed.) *Int. Symp. on High Energy Physics with Polarized Beams and Targets*, American Institute of Physics, Argonne, 1976, pp. 458–484.

[42] W. de Boer, Dynamic orientation of nuclei at low temperatures, *J. Low Temp. Phys.* **22** (1976) 185–212.

[43] Spin Muon Collaboration (SMC), B. Adeva, E. Arik, et al., Large enhancement of deuteron polarization with frequency modulated microwaves, *Nucl. Instrum. and Methods* **A 372** (1996) 339–343.

[44] Y. F. Kisselev, T. O. Niinikoski, *Frequency Modulation Effects in EPR and Dynamic Nuclear Polarization*, Preprint CERN-PPE/96–146, 1996.

[45] A. Abragam, M. Goldman, Dynamic nuclear polarisation, *Rep. Prog. Phys.* **41** (1978) 395.
[46] A. S. Lilly Thankamony, J. J. Wittmann, M. Kaushik, B. Corzilius, Dynamic nuclear polarization for sensitivity enhancement in modern solid-state NMR, *Progress in Nuclear Magnetic Resonance Spectroscopy* **102–103** (2017) 120–195.
[47] E. I. Bunyatova, Free radicals and polarized targets, *Nuclear Instruments and Methods in Physics Research* **A 526** (2004) 22–27.
[48] S. T. Goertz, J. Harmsen, J. Heckmann, et al., Highest polarizations in deuterated compounds, *Nuclear Instruments and Methods in Physics Research* **A 526** (2004) 43–52.
[49] M. Borghini, *Choice of substances for polarized proton targets*, CERN Yellow Report CERN 66-3, 1966.
[50] N. S. Garif'yanov, B. M. Kozyrev, V. N. Fedotov, Width of the EPR line of liquid solutions of ethylene glycol complexes for even and odd chromium isotopes, *Sov. Phys. Dokladyy* **13** (1968) 107.
[51] M. Krumpolc, J. Rocek, Stable chromium(V) compounds, *J. Am. Chem. Soc.* **98** (1976) 872–873.
[52] M. Krumpolc, J. Rocek, Three-electron oxidations. 12. chromium(V) formation in the chromic acid oxidation of 2-hydroxy-2-methylbutyric acid, *J. Am. Chem. Soc.* **99** (1977) 137–143.
[53] M. Krumpolc, B. G. DeBoer, J. Rocek, A Stable Cr(V) compound. synthesis, properties, and crystal structure of potassium bis(2-hydroxy-2-methylbutyrato)-oxochromate(V) monohydrate, *J. Am. Chem. Soc.* **100** (1978) 145–153.
[54] D. Hill, R. C. Miller, M. Krumpolc, J. Rocek, A new CrV doping agent for polarized targets, *Nuclear Instruments and Methods* **150** (1978) 331–332.
[55] M. Krumpolc, J. Rocek, Synthesis of stable chromium(V) complexes of tertiary hydroxy acids, *Journal of the American Chemical Society* **101** (1979) 3206–3209.
[56] N. Nakasuka, Some metal complexes as free radicals for polarized targets, in: E. Steffens, et al. (eds.) *Proc. 6th Workshop on Polarized Solid Targets*, Springer Verlag, Heidelberg, 1991, pp. 344–346.
[57] B. Corzilius, A. A. Smith, A. B. Barnes, et al., High-field dynamic nuclear polarization with high-spin transition metal ions, *Journal of the American Chemical Society* **133** (2011) 5648–5651.
[58] T. Adachi, K. Asahi, M. Doi, et al., Test of parity violation and time reversal invariance in slow neutron absorption reactions, *Nucl. Phys.* **A577** (1994) 433 c–442 c.
[59] C. Urbina, J. F. Jacquinot, Low field behavior of Tm^{2+} in CaF_2 at ultra-low nuclear spin temperature, *Physica B+C* **100** (1980) 333–342.

4
Dynamic Nuclear Polarization

There are two main categories of methods of dynamic nuclear polarization (DNP) in solid materials:

- methods based on the saturation of weak two-spin transitions which are calculable in the second-order perturbation theory;
- methods based on the off-resonance saturation of one-spin transitions calculable in the first-order perturbation theory.

In both cases, quantum statistics is needed in order to understand theoretically the behavior of nuclear polarization under the various experimental situations: external field, lattice temperature and microwave frequency and power. The first category, however, is simpler to model because rate equations can be written directly for the magnetic level populations regardless of the lattice temperature. In the second case a theoretically correct model involves the use of the density matrix, which can be simplified to an analytically calculable form only in the high-temperature approximation, a regime that is not of main interest in polarized targets.

The density matrix was introduced in Section 1.2.2 and its application for resonance saturation was discussed in Section 2.2.4, for the case of low polarization of the electron spins (i.e. high-temperature approximation).

At high polarization the density matrix must be kept in its exponential form and therefore theoretical estimates are limited to cases where other simplifications can be made. The object of the calculations is the spin temperature, which is a constant of motion, and therefore the second category methods are said to be based on *dynamic cooling of the spin-spin interactions*.

Historically, the first category was experimentally and theoretically understood well before the second, and therefore most discussions of DNP usually begin with introducing the so-called solid effect. In polarized targets, however, the dynamic cooling has gained a much wider use than the solid effect, and therefore we shall begin here with it, after a phenomenological introduction.

A third category of DNP methods is based on the Overhauser effect, which works in metals and liquids. This was the earliest one to be discovered, but it is of limited interest in polarized targets, because only low nuclear polarizations can be obtained. On the other hand, DNP at higher temperatures met in magic angle spinning NMR spectroscopy is reported to be effective also in solids at high magnetic fields up to 18 T [1]. We shall therefore treat the Overhauser methods only briefly at the end of this chapter.

Since the first book by Jeffries in 1963 [2], DNP has been reviewed by Abragam and Goldman in 1978 [3], Jeffries in 1991 [4], Atsarkin in 2011 [5], Slichter in 2014 [6] and Wenckebach in 2016 [7]. A review of DNP and polarized targets in general was presented by Crabb and Meyer in 1997 [8].

4.1 Phenomenology of Dynamic Cooling of Nuclear Spins

Here we shall discuss the cooling of nuclear spins to temperatures well below that of the lattice. We shall do it before discussing the relaxation of the nuclear spins, which will be done only in Chapter 5, and rely on the fact that the nuclear spins have a contact with the lattice predominantly via the electronic spin system. In other words, if the part of the electronic system which makes the contact with nuclei is cooled or heated, the nuclear spin temperature follows this with almost no direct contact with the lattice.

The elements of quantum statistics were discussed in Section 1.2; this underlies the thermodynamics of spin systems for which equilibrium can be reached under favorable conditions. The transport of magnetic energy (=heat) by relaxation and RF saturation was discussed in Chapters 2 and 3.

The spin system of polarized target materials is usually presented in the thermodynamic terms using heat reservoirs as shown in Figure 4.1. The various nuclear spin species have distinct Zeeman energy reservoirs which couple with the energy reservoir of the electron dipolar spin-spin interactions; this, in turn, is coupled with the electron Zeeman interaction reservoir that is dynamically cooled in the rotating frame by applying a strong microwave field. Although the existence of these reservoirs cannot be directly proven, there is good experimental evidence in favor of such a phenomenological model, which appears to be well obeyed under certain conditions. This evidence and the conditions will be discussed in this first part of the chapter.

Let us assume a paramagnetic electron spin system with a Hamiltonian

$$\mathcal{H} = \mathcal{H}_z + \mathcal{H}_1, \tag{4.1}$$

which is split here in the Zeeman and non-Zeeman parts. We assume furthermore such a high field B_0 that the Zeeman part dominates the non-Zeeman part:

$$\left|E_m - E_{m+1}\right| = \hbar\omega_L = \hbar g\mu_B B_0 \gg \left\langle \mathcal{H}_1 \right\rangle. \tag{4.2}$$

We shall not specify the exact character of $\mathcal{H}_1$ but assume that it causes a broadening of the magnetic energy levels roughly symmetrically about the unperturbed ones as shown in Figure 4.2, with total widths of about $\Delta E > \left\langle \mathcal{H}_1 \right\rangle$. This is in particular due to the inhomogeneous character of the broadening arising from the anisotropies of the g-factor and the hyperfine tensor, which cause energy level shifts exceeding the homogeneous broadening due to the dipolar interactions. These interactions were discussed in Chapters 2 and 3. We

4.1 Phenomenology of Dynamic Cooling of Nuclear Spins

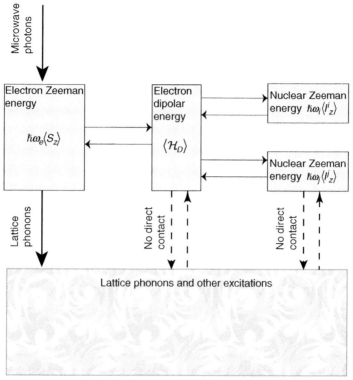

Figure 4.1 Thermal block diagram model of DNP by cooling of the electron spin-spin interactions with off-resonance microwave irradiation. The success of the model requires that the spin-lattice relaxation of the electrons establishes rapidly thermal equilibrium in their Zeeman energy reservoir, and that the temperature of their dipolar energy reservoir follows the Zeeman temperature closely; this requires a sufficient spin density. Moreover, a good thermal contact is needed between the nuclear spins and the electron dipolar reservoir which is assured by cross-relaxation transitions. The nuclear spin Zeeman energy baths are in good contact only with the electron spin dipolar bath, and in negligible contact directly with the lattice

also assume that thermal distribution among the 'magnetic sublevels' will be reached with a characteristic time constant T_{cross}, which is rather fast because the required transitions conserve energy and angular momentum; this relaxation time is not very much longer than the dipolar relaxation time T_{1D} but these are related as was discussed in Section 3.4.3.

Relaxation towards the lattice temperature T_L due to the spin-lattice interactions will establish a Maxwell–Boltzmann distribution of the magnetic level populations, as was discussed in Section 2.2.5; such a population distribution is shown by the shaded areas and the curve in Figure 4.2a. The relaxation is characterized by a roughly exponential decay with a time constant T_{1Z}, and after several relaxation times the average electron spin polarization will be, according to Eq. 1.67

$$P_0 = \tanh\left(\frac{1}{2}\alpha_0\bar{\omega}\right). \tag{4.3}$$

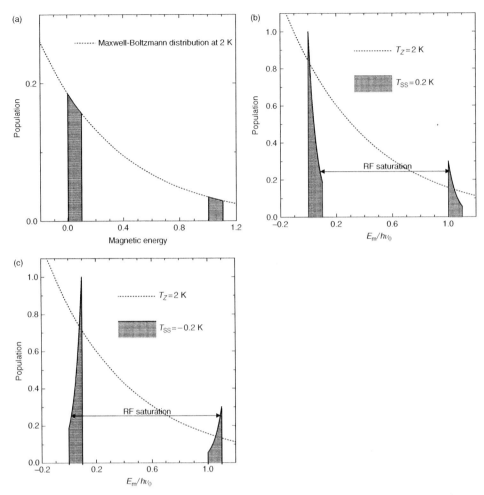

Figure 4.2 'Toy model' simulation of DNP with broad inhomogeneous 'box distribution' of g-factors in 2.5 T field. (a) Population distribution of the magnetic energy levels in thermal equilibrium with the lattice at 2 K temperature; (b) population distribution in the static frame when the spin system is saturated by microwaves yielding +0.2 K spin temperature in the rotating frame; (c) population distribution at a microwave frequency that yields –0.2 K spin temperature in the rotating frame

where $\bar{\omega}$ is roughly defined as a mean frequency difference between the distributions of the two levels. It is clear that at very low spin temperatures the population of the 'magnetic sublevels' is not constant within each level, as can be seen in Figure 4.2a. This point is very important for the correct calculation of the spin temperature during DNP, as will be discussed later in this chapter. The curve in Figure 4.2a illustrates the Maxwell–Boltzmann distribution function for electrons with $\bar{g} = 2$ and $\Delta g / \bar{g} = 0.1$ in 2.5 T field at the temperature of 2 K. The widths of the magnetic levels are thus exaggerated in order to be able to visualize the variation of the populations within the magnetic sublevels.

Let us now apply a transverse oscillating field, which is strongly saturating so that the spins which satisfy the resonance condition will be quantized along the rotating effective field aligned perpendicular to the main field. Their z-components average to zero in the laboratory frame, which means that both magnetic levels are equally populated for those spins that are in resonance. The saturation conditions were discussed in Section 2.4.4 where the Provotorov equations were introduced. The Zeeman and dipolar temperatures are in equilibrium for all times exceeding

$$t_{\exp} \gg \frac{D^2}{\pi \omega_1^2 \Delta^2 g(\Delta)}, \tag{4.4}$$

and the spin-lattice relaxation does not significantly reduce the inverse Zeeman temperature in dynamic equilibrium when

$$T_{1Z} \gg \frac{1}{\pi \omega_1^2 g(\Delta)}. \tag{4.5}$$

Two cases of particular interest are graphically illustrated in Figures 4.2b,c: saturation off-resonance at frequencies just below and just above the absorption line, as shown by the horizontal arrows. The fact that cross-relaxation transitions cause a spectral redistribution faster than the spin-lattice relaxation entails now a new distribution function within the magnetic sublevels, characterized by a Maxwell–Boltzmann distribution with a different temperature. This temperature can be much lower than that of the lattice, if the condition of Eq. 4.5 is satisfied. In Figure 4.2b the frequency is close to the lower absorption edge, which yields a positive spin temperature of 0.2 K, whereas in Figure 4.2c the RF frequency is close to the upper absorption edge; this forces the distribution so that it can be characterized only by a negative temperature of –0.2 K.

Figure 4.2 shows that in the two conditions illustrated, the temperature of the heat reservoir that can be associated with $\langle \mathcal{H}_1 \rangle$ will be lowered roughly by

$$|T_{SS}| \cong \frac{\Delta E}{\hbar \overline{\omega}} T_L, \tag{4.6}$$

if the system is such that it can be characterized by a thermalized statistical distribution. There are no solid theoretical arguments either in favor or against such an assumption, and the experimental evidence that this happens at least sometimes is indirect. We shall return to this question later in this chapter. We just note here that if the spectral width of the electron resonance line is of the order of a few ‰ of the mean resonance frequency, the spin interaction temperature may be lowered by a factor of several hundreds. If the lattice temperature is a few hundred millikelvin, the spin interactions may be cooled to about 1 mK. In terms of nuclear polarizations in a field of 2.5 T, this corresponds to a nearly complete proton polarization and even the deuterons may attain a polarization of ±0.8.

Equation 4.6 expresses the dynamic cooling of heat reservoirs in a model that was illustrated in Figure 4.2. The RF quanta are absorbed by the Zeeman interaction reservoir that is in contact with the spin-spin interaction reservoir. The strength of the saturation also determines the strength of the contact between the two reservoirs; this influences the speed at which the dynamic equilibrium will be reached. The speed can be calculated in many cases using the Provotorov equations, which were introduced in Section 2.4.4 and which will be further discussed later in this chapter. Figures 4.2b and c show how nuclear spins are then cooled by contact with the electron spin interaction reservoir. The strength of this coupling also depends on the saturation, which is more difficult to estimate, but in many cases is not the main bottleneck in the cooling speed.

The most convincing experimental evidence for the cooling mechanism of DNP comes from the fact that all different nuclear spin species reach the same spin temperature in steady state, and that their spin temperatures also evolve along a common curve during polarization growth and decay. This was first time proposed by Borghini in 1968 [9] and was shown in glassy butanol doped with porphyrexide in 1971 [10]. Moreover, using a partially deuterated ethanediol-Cr(V) sample, it was established that if a nuclear spin species was heated up by RF irradiation after freezing the polarizations obtained by DNP, the subsequent relaxation of all nuclear species happened towards a common spin temperature, and that the new equilibrium spin temperature was reached at a rate that depends linearly on the applied microwave power [11]. Figure 4.3 illustrates the transient towards common spin temperature when about 300 µW microwave power was applied to the 1 g sample in which proton spin polarization was zeroed by RF saturation. The transient was three times faster when three times more power was applied [11].

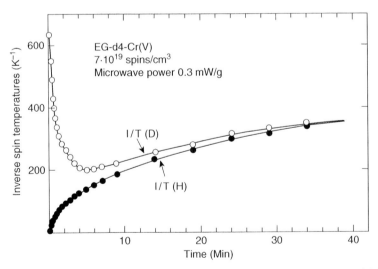

Figure 4.3 Transient recovery of equal spin temperatures for protons and deuterons with 300 µW/g microwave power at optimum frequency, after high DNP of protons was selectively zeroed by RF saturation. Reprinted, by permission from Springer Nature Customer Service Center GmbH, from Ref. [11]

The spin-spin interaction temperature resulting from dynamic cooling is calculable in many cases under certain assumptions and approximations. We are interested in the conditions where high polarizations are obtainable, which include high electron spin density, high field and low lattice temperature. In this case numerical solutions have been obtained for EPR lines where the main cause of broadening is inhomogeneous, i.e. caused by a spectral distribution of Larmor frequencies due to the anisotropy of the g-tensor or the hyperfine tensor. In the high-temperature approximation more cases are calculable and even analytical results have been obtained.

4.2 Dynamic Cooling of Electron Dipolar Interactions: High Temperatures

Let us first consider an electronic spin system with no other than dipolar broadening of the resonance line. Ignoring also the nuclear spins first, the Hamiltonian in the rotating frame is

$$\mathcal{H}^* = \hbar \Delta S_z + \hbar \omega_1 S_x + \mathcal{H}_{ss}^{\prime *}, \tag{4.7}$$

where $\Delta = \omega_S - \omega$, $\omega_1 = g\mu_B B_1/\hbar$ is the strength of the rotating field expressed in angular frequency units, and $\mathcal{H}_{ss}^{\prime *}$ is the electron dipolar Hamiltonian, truncated after the terms A and B as in Eq. 2.14, and transformed into the rotating frame.

The density matrix before the application of the transverse oscillating field is

$$\rho_0 = \frac{e^{-\mathcal{H}/kT_L}}{\mathrm{Tr}\{e^{-\mathcal{H}/kT_L}\}} \cong 1 - \frac{\mathcal{H}}{kT_L} \tag{4.8}$$

assuming the high-temperature approximation.

The Provotorov equations 2.141 and 2.142 describe the time evolution of the Zeeman and dipolar temperature reservoirs of the electrons during saturation. We rewrite them here using the saturation function $s(\Delta)$ of Eq. 3.59 from Section 3.3

$$\dot{\alpha}(t) = -\frac{1}{T_{1Z}} s(\Delta)\left[\alpha(t) - \beta(t)\right] - \frac{1}{T_{1Z}}\left[\alpha(t) - \frac{\omega_0}{\Delta}\alpha_L\right]; \tag{4.9}$$

$$\dot{\beta}(t) = \frac{1}{T_{1Z}} s(\Delta)\frac{\Delta^2}{D^2}\left[\alpha(t) - \beta(t)\right] - \frac{1}{T_{1D}}\left[\beta(t) - \beta_L\right]. \tag{4.10}$$

In steady state the two time derivatives are zero and we may solve the inverse temperatures; their difference is obtained immediately

$$\alpha - \beta = \beta_L \frac{\omega}{\Delta} \frac{1}{1 + s(\Delta)\left(1 + \frac{T_{1D}}{T_{1Z}}\frac{\Delta^2}{D^2}\right)}, \tag{4.11}$$

which is seen to become small when s increases. Here it was assumed that the relaxation of the dipolar and Zeeman temperatures in the rotating frame take place so that they feel the same lattice temperature, i.e. $\alpha_L = \beta_L$. The inverse dipolar temperature is then

$$\beta = \beta_L \frac{\omega}{\Delta} \frac{s(\Delta)\dfrac{T_{1D}}{T_{1Z}}\dfrac{\Delta^2}{D^2}}{1+s(\Delta)\left(1+\dfrac{T_{1D}}{T_{1Z}}\dfrac{\Delta^2}{D^2}\right)} + \beta_L \qquad (4.12)$$

and the inverse Zeeman temperature is

$$\alpha = \beta_L \frac{\omega}{\Delta} \frac{1+s(\Delta)\dfrac{T_{1D}}{T_{1Z}}\dfrac{\Delta^2}{D^2}}{1+s(\Delta)\left(1+\dfrac{T_{1D}}{T_{1Z}}\dfrac{\Delta^2}{D^2}\right)} + \beta_L. \qquad (4.13)$$

These equations relate the spin temperature to the microwave power and EPR lineshape. The nuclear spins that are in contact with the electron spin-spin interaction reservoir cool rapidly to the same temperature. The contact is promoted greatly by the microwave transitions off-resonance, which substantially increase the probability of thermal mixing.

We note immediately that if the saturation is very weak, the dipolar temperature of Eq. 4.12 does not differ significantly from the lattice temperature, but the Zeeman temperature can be much colder. This is easily understood from the fact that the cooling 'power' available between the two reservoirs depends on the microwave power, and when this is very small, the dipolar relaxation ensures that the dipolar temperature stays closer to that of the lattice than that of the Zeeman reservoir in the rotating frame. The situation then corresponds to the conditions of validity of the linear response theory.

In the limit of very high microwave power we have $s(\Delta) \gg 1$, and the two temperatures tend to the same asymptotic value

$$\alpha = \beta = \beta_L \frac{\omega \Delta}{\Delta^2 + \dfrac{T_{1Z}}{T_{1D}} D^2}, \qquad (4.14)$$

which has a minimum and a maximum of

$$\beta = \pm \frac{\beta_L}{2\sqrt{\dfrac{T_{1Z}}{T_{1D}}}} \frac{\omega}{D} \qquad (4.15)$$

at the frequencies

$$\Delta = \mp \sqrt{\dfrac{T_{1Z}}{T_{1D}}} D. \qquad (4.16)$$

As the ratio ω/D may be several thousand in a high field, the dipolar temperature and the Zeeman temperature in the rotating field can be three orders of magnitude lower than the lattice temperature. If the nuclei relax efficiently towards the electron temperature, very high nuclear polarizations can theoretically be obtained at the spin temperatures below 1 mK when the lattice is at a temperature below 1 K.

In practice, the rate at which the nuclear spins cool to the temperature of the electron dipolar interactions can be high only when the nuclear Larmor frequency is of the same magnitude or smaller than the dipolar width of the EPR line. For a large nuclear moment, a high concentration of paramagnetic spins is therefore required, or else the effect can be seen only at a relatively low magnetic field. For a low electron spin concentration, the cooling of the nuclear spins by the homogeneously broadened EPR line requires a very low magnetic field or a low nuclear dipole moment.

Examples of cases where the nuclear spins can be cooled by an electron spin system with predominantly dipolar broadening are deuterons in LMN doped with Nd^{3+} [12], and deuterons in glassy m-xylene-D6 doped with the free radical BDPA [13].

The rate at which the inverse dipolar and Zeeman temperatures approach each other is approximately

$$\tau^{-1} \cong 2\frac{s(\Delta)}{T_{1Z}}\left(1+\frac{\Delta^2}{D^2}\right), \tag{4.17}$$

whereas the rate at which the Zeeman temperature approaches a steady value is

$$\tau_z^{-1} \cong \frac{1}{T_{1Z}}\left(1+s(\Delta)\right). \tag{4.18}$$

These rates are equal when

$$s(\Delta) = \frac{1}{1+2\Delta^2/D^2}; \tag{4.19}$$

at a lower saturation the Zeeman temperature approaches faster an equilibrium value, whereas at higher saturation the equilibrium between the dipolar and Zeeman temperatures is reached faster than the final equilibrium spin temperature. It is the latter situation that is desirable for DNP. These considerations neglect, however, the thermal load coming from the cooling of the Zeeman energy reservoir of the nuclear spins, which can be phenomenologically visualized in Figure 4.1.

In Section 4.2 the high-temperature approximation of the spin Hamiltonian was assumed; in the following sections this is not made unless stated explicitly.

4.3 Dynamic Cooling with Inhomogeneously Broadened Resonance Line

The DNP by dynamic cooling of electron spin-spin interactions is also called DNP by thermal mixing; both names refer to the same mechanism in which the nuclear spins couple strongly with the electron spin-spin interaction reservoir which is cooled in the rotating frame due to saturating off-resonance microwave transitions.

4.3.1 Spin Hamiltonian under Strong Saturation

Let us consider the description of the saturation of a spin system in a strong magnetic field aligned along the z-axis $\mathbf{B}_0 = B_0 \mathbf{e}_z$. The most general spin Hamiltonian is

$$\mathcal{H}_{tot} = \mathcal{H}_Z + \mathcal{H}_{HF} + \mathcal{H}'_D + D + \mathcal{H}_{SL} + \mathcal{H}_{RF}, \tag{4.20}$$

where the Zeeman (Z), hyperfine (HF) dipolar (D), spin-lattice (SL) and near-resonance oscillating field (RF) components are expressed using the following notation. The Zeeman component is the interaction energy with the static field of the N_e electrons with spin $\mathbf{S}_i$ and g-tensor $\mathbf{g}_i$, and the N_n nuclei with spin $\mathbf{I}_j$ and Larmor frequency ω_j:

$$\mathcal{H}_Z = -\sum_{i=1}^{N_e} \mu_B \mathbf{B}_0 \cdot \tilde{\mathbf{g}}_i \cdot \mathbf{S}_i - \sum_{j=1}^{N_n} \hbar \omega_j I_{jz}, \tag{4.21}$$

where μ_B is the Bohr magneton and the sum over the electrons involves terms written in the dyadic notation of Eq. 3.5. We shall assume that the g-tensor is axially symmetric, with an anisotropy of the order of 0.5%. When we shall study the steady-state saturation of the electron spin system (dynamic equilibrium), we may drop the nuclear Zeeman term in the treatment, justified by the facts that the RF frequency is far from the nuclear resonance and that in steady state the nuclear polarization is constant and therefore contributes nothing in the energy balance. The Zeeman term due to the nuclei having hyperfine interaction with the electrons is included in the nuclear part of Eq. 4.21.

The hyperfine interaction of the electrons, each having contact with N_{hf} nearby hyperfine nuclei $\mathbf{I}_k$, can be described in high field by

$$\mathcal{H}_{HF} = \hbar \sum_{i=1}^{N_e} \sum_{k=1}^{N_{hf}} \mathbf{S}_i \cdot \mathbf{A}_{ik} \cdot \mathbf{I}_k, \tag{4.22}$$

where $\mathbf{A}_{ik}$ is the hyperfine tensor of the i-th electron, written out in Eq. 3.24. On the other hand, the hyperfine term results in the inhomogeneous broadening of the electron resonance line, and it can be treated together with the g-tensor broadening as was done in the original paper of Borghini [9].

$\mathcal{H}'_D$ describes the secular parts of the dipolar spin-spin interactions among the spins S, among the nuclear spins I, and between S and I:

$$\mathcal{H}'_D = \mathcal{H}'_{SS} + \mathcal{H}'_{II} + \mathcal{H}'_{SI}. \tag{4.23}$$

The form of these terms is given in Eq. 2.14. The first and last terms on the right side of Eq. 4.23 are very important for the success of the DNP, because they provide the necessary relaxation mechanisms for the rapid thermalization in the electronic system, and for the good heat transport between the nuclear Zeeman system and the electron dipolar spin-spin interaction system. However, in order to be able to obtain estimates for dynamic cooling, these terms must be omitted when the inhomogeneous broadening dominates the homogeneous one.

Neglecting the electron-electron dipolar term in the calculation of the steady-state spin temperature is likely to produce a small error, which we shall discuss also in the end of this section in the light of the rate equations derived under the high-temperature approximation. The part of the dipolar Hamiltonian which describes the interaction between the nuclear and electron spins is important for the rate of DNP 'cooling speed' of the nuclei, but has less important effect on the steady-state ultimate spin temperature in the case of negligible 'leakage' relaxation of the nuclear spins that might be caused by electron spins whose resonance absorption line has no overlap with the line which is being saturated.

The interaction between the electronic spin system and the 'lattice' is due to $\mathcal{H}_{SL}$ and it arises mainly from the modulation of the g-factor by the thermal lattice excitations, resulting in energy exchange and therefore spin relaxation; this was described in Chapter 3. This term sets the scale for the speed of the DNP in most cases of interest for the polarized targets, and its introduction will be discussed after the treatment of the problem with reduced contact to the lattice (phonon bottleneck).

There could, in principle, also exist direct 'leakage' relaxation of the nuclear spins towards the lattice temperature, caused by non-resonant electronic impurities or impurities with very broad line, such as paramagnetic O_2, or by relaxation resulting from contact with hyperfine nuclei having faster spin lattice relaxation. No such effects are seen in good materials, and we shall present some quantitative evidence for the small size of such terms at the end of this section.

The last term $\mathcal{H}_{RF}$ in Eq. 4.20 describes the interaction of the electronic spins with an oscillating microwave field $2B_1 \cos\omega t$ applied perpendicular to the main field at a frequency ω, close to the average Larmor frequency $\bar{\omega}$ of the electron paramagnetic resonance:

$$\mathcal{H}_{RF} = -\mu_B \sum_{i=1}^{N_e} \mathbf{B}_1 \cdot \mathbf{g}_i \cdot \mathbf{S}_i e^{i\omega t}, \tag{4.24}$$

The Hamiltonian, which adequately describes our spin system in high field under saturating RF irradiation, is now

$$\mathcal{H}_{tot} = -\sum_{i=1}^{N_e} \mu_B \mathbf{B}_0 \cdot \tilde{\mathbf{g}}_i \cdot \mathbf{S}_i + \hbar \sum_{i=1}^{N_e} \sum_{k=1}^{N_{hf}} \mathbf{S}_i \cdot \mathbf{A}_{ik} \cdot \mathbf{I}_k + \mathcal{H}'_D + \mathcal{H}_{SL} - \mu_B \sum_{i=1}^{N_e} \mathbf{B}_1 \cdot \mathbf{g}_i \cdot \mathbf{S}_i e^{i\omega t}. \tag{4.25}$$

Various solutions to the steady-state DNP have been worked out using the above Hamiltonian, under different approximations. Most of these can be solved only in the case of the high-temperature approximation, under which the spin temperature and polarization can be obtained using approximate analytical equations. In the limit of low temperatures there is no general solution and numerical methods must be used for obtaining the spin temperatures and polarizations.

4.3.2 Energy Conservation

In dynamic equilibrium all parts of the system are in internal thermal equilibrium and all temperatures are steady; some of the temperatures may be equal but this need not be the case. The constants of motion (spin temperatures) can then be solved from the equations that describe the conservation of energy. To derive the relevant equation, we shall follow the treatment of Borghini [14].

The total time derivative of the total energy described by Eq. 4.20 follows from the explicit variations due to the spin-lattice interactions, to the applied RF field, and to the spin-spin interactions, so that

$$\frac{d}{dt}\langle \mathcal{H}_{tot}\rangle = \left[\left.\frac{\partial}{\partial t}\right|_{SL} + \left.\frac{\partial}{\partial t}\right|_{RF} + \left.\frac{\partial}{\partial t}\right|_{SS}\right]\langle \mathcal{H}_{tot}\rangle. \tag{4.26}$$

On the other hand, we have in steady state

$$\frac{d}{dt}\langle \mathcal{H}_{tot}\rangle = 0. \tag{4.27}$$

If there is no exchange of energy between the spin system and the rest of the sample in the processes due to spin-spin interactions, we have also

$$\left.\frac{\partial}{\partial t}\right|_{SS}\langle \mathcal{H}_{tot}\rangle = 0, \tag{4.28}$$

which is certainly true in pure spin flip-flop processes. Small deviations from this could be due to certain rare relaxation phenomena, but experimental evidence tells that these cause negligible contribution to the energy balance.

In high field, the part of the total Hamiltonian that is subject to the RF field is then that describing the electron Zeeman interaction, so that

$$\left.\frac{\partial}{\partial t}\right|_{RF}\langle \mathcal{H}_{tot}\rangle = \hbar\omega\left.\frac{\partial}{\partial t}\right|_{RF}\langle S_z\rangle, \tag{4.29}$$

where S_z is the projection of the total electron spin onto the z-axis. This follows simply from the fact that the emission or absorption of a photon of energy $\hbar\omega$ happens each time when an electron spin flips, because the RF field is much smaller than the dipolar and hyperfine fields so that the z-axis remains the axis of quantization. This is true for transitions involving any order in perturbation theory in their description. Furthermore, because the spin-spin processes involve the exchange of orientation of two spins, we have

$$\left.\frac{\partial}{\partial t}\right|_{SS}\langle S_z\rangle = 0. \tag{4.30}$$

Because $\langle S_z\rangle$ is constant in steady state, we have now

$$\frac{\partial}{\partial t}\langle S_z\rangle = \left(\left.\frac{\partial}{\partial t}\right|_{SL} + \left.\frac{\partial}{\partial t}\right|_{RF} + \left.\frac{\partial}{\partial t}\right|_{SS}\right)\langle S_z\rangle = 0, \tag{4.31}$$

which becomes

$$\left.\frac{\partial}{\partial t}\right|_{RF}\langle S_z\rangle = -\left.\frac{\partial}{\partial t}\right|_{SL}\langle S_z\rangle, \tag{4.32}$$

after inserting Eq. 4.30.

4.3 Dynamic Cooling with Inhomogeneously Broadened Resonance Line

Combining Eqs. 4.26, 4.28 and 4.32 with 4.27 yields now

$$\left.\frac{\partial}{\partial t}\right|_{SL} \langle \mathcal{H}_{tot} - \hbar\omega S_z \rangle = 0, \tag{4.33}$$

which is valid in all high-field situations and at any temperature. This equation that states that energy and angular momentum are conserved has been used for the calculation of steady-state spin temperatures and polarizations for several types of spin systems. The limitation of this equation, as Borghini has pointed out [14], is that it only gives the polarization in the limit of strong saturation, and does not give the dependence of polarization on the intensity of the RF field. This dependence can only be treated in the high-temperature approximation, which is not of great interest for polarized targets because only low polarizations are then obtained (by definition of the high-temperature approximation).

Given the fact that we must have $s(\omega) \gg 1$ for the model to work at the optimum frequency of DNP does not mean that high DNP is impossible even if the electron spin-lattice relaxation time is short and therefore the power dissipation is high. Each electron spin system has optimizable parameters such as the spin density n_e and the microwave frequency and power that have optimum values; moreover, the T_{1e} can be controlled by the magnetic field on which it depends strongly when the direct process dominates the other spin-lattice relaxation mechanisms.

4.3.3 DNP with Inhomogeneously Broadened EPR Line

In most free radicals, the EPR line is much more broadened by the anisotropy of the g-tensor and of the hyperfine tensor than by the dipolar spin-spin interactions. In a magnetic field of 2.5 T, for example, the Cr(V) complexes with diols, and the stable compounds, such as EHBA-Cr(V), have spectral linewidths of 300 MHz to 400 MHz while the dipolar broadening at spin density of 10^{20} cm^{-3} is 0.5 mT or 14 MHz as was calculated in Chapter 3, Eq. 3.44 for like spins. As the g-tensor broadens the line so much that the spin packets do not entirely overlap, the broadening, in fact, should be predominantly by unlike spins, which was taken into account in Eq. 3.44. At high polarization the dipolar broadening can be even much less than 14 MHz, but this depends on the way how the paramagnetic molecules are distributed in the material.

Provotorov's derivation [15] of cross-relaxation for inhomogeneous spin systems, using the high temperature approximation and assuming strongly saturating RF field, has been extended to show that the spin system can be described as having a single temperature in a suitable reference frame, when the cross-relaxation is faster than spin-lattice relaxation. This condition is satisfied at low temperatures when the spin density is high enough, as was shown in Chapter 3.

We shall express now the Hamiltonian in angular frequency units in order to simplify the subsequent formulas. With anisotropic g-tensor only and assuming a high steady field superimposed by transverse RF field oscillating at frequency ω, the simplified Hamiltonian is obtained from Eq. 4.25:

$$\hbar\mathcal{H} = -\sum_{i=1}^{N_e} \mu_B \mathbf{B}_0 \cdot \tilde{\mathbf{g}}_i \cdot \mathbf{S}_i - \mu_B \sum_{i=1}^{N_e} \mathbf{B}_1 \cdot \mathbf{g}_i \cdot \mathbf{S}_i \, e^{i\omega t}. \tag{4.34}$$

This can be rewritten as

$$\hbar\mathcal{H} = -\sum_{i=1}^{N_e} \hbar\omega_i S_i - \hbar\omega_1 e^{i\omega t} \sum_{i=1}^{N_e} S_i, \tag{4.35}$$

where $\omega_1 = \overline{g}\mu_B B_1 / \hbar$.

By recalling the definitions of Section 1.2.2, and by removing the main time dependence of the Hamiltonian using transformation to the rotating frame by

$$U = \exp\left(-i\omega t \sum_i S_z^i\right),$$

we get the Hamiltonian in the frame rotating at frequency ω:

$$\hbar\mathcal{H}^* = -\sum_{i=1}^{N_e} \hbar(\omega_i - \omega) S_i. \tag{4.36}$$

Before turning the microwave field on the density matrix is

$$\rho_0 = \frac{\exp(-\alpha_0 \mathcal{H})}{\mathrm{Tr}\{\exp(-\alpha_0 \mathcal{H})\}}, \tag{4.37}$$

where $\alpha_0 = \hbar/(kT_0)$ is the inverse spin temperature close to that of the lattice; after reaching the steady state the density matrix in the rotating frame becomes

$$\rho^* = \frac{\exp(-\alpha \mathcal{H}^*)}{\mathrm{Tr}\{\exp(-\alpha \mathcal{H}^*)\}}, \tag{4.38}$$

where $\alpha = \hbar/(kT_S)$ is the inverse spin temperature in dynamic equilibrium. In this new equilibrium in the rotating frame, we may now write the expectation values of Eq. 4.33 in the form

$$\mathrm{Tr}\{\rho^* \mathcal{H}^*\} = \mathrm{Tr}\{\rho_0 \mathcal{H}^*\}, \tag{4.39}$$

which can be expressed in the terms of the sums

$$\sum_i \Delta_i P_i = \sum_i \Delta_i P_i^0, \tag{4.40}$$

or

$$\sum_i \Delta_i (P_i - P_i^0) = 0, \tag{4.41}$$

where

$$\Delta_i = \omega_i - \omega, \quad \bar{\omega} = \sum_i \omega_i, \quad P_i = \tanh(\tfrac{1}{2}\alpha\Delta_i), \quad P_0^i = \tanh(\tfrac{1}{2}\alpha_0\omega_i). \quad (4.42)$$

These give

$$\sum_i \Delta_i \left[\tanh(\tfrac{1}{2}\alpha\Delta_i) - \tanh(\tfrac{1}{2}\alpha_0\omega_i) \right] = 0. \quad (4.43)$$

We note immediately that at low temperatures the hyperbolic tangents are both close to unity, which requires high numeric accuracy for the solution of α from the equation; their difference deviates substantially from zero only when Δ_i is close to zero. The iterative solution then works better if we subtract 1 from both tanh functions, and develop Eq. 4.43 into frequency-dependent and frequency-independent parts by substituting $\omega_i = \omega + x_i$ and $\Delta_i = \Delta + x_i$. The frequency-independent hyperbolic tangents can then be separated out of the sum:

$$\tanh(\tfrac{1}{2}\alpha_0\bar{\omega}) - \tanh(\tfrac{1}{2}\alpha\Delta) = \sum_i \left(1 + \frac{x_i}{\Delta}\right) \times T(x_i, \alpha, \Delta), \quad (4.44)$$

where $T(x_i, \alpha, \Delta)$ is

$$T(x_i, \alpha, \Delta) = \frac{\tanh(\tfrac{1}{2}\alpha x_i)\left(1 - \tanh^2(\tfrac{1}{2}\alpha\Delta)\right)}{1 + \tanh(\tfrac{1}{2}\alpha\Delta)\tanh(\tfrac{1}{2}\alpha x_i)} - \frac{\tanh(\tfrac{1}{2}\alpha_0 x_i)\left(1 - \tanh^2(\tfrac{1}{2}\alpha_0\bar{\omega})\right)}{1 + \tanh(\tfrac{1}{2}\alpha_0\bar{\omega})\tanh(\tfrac{1}{2}\alpha_0 x_i)}. \quad (4.45)$$

The right-hand side of Eq. 4.44 sums to a value $\ll 1$ at low temperatures, which makes the iteration fast. The inverse spin temperature α can now be numerically solved for each lineshape function $f(\omega)$ by iteration of the integral equation

$$\tanh(\tfrac{1}{2}\alpha_0\bar{\omega}) - \tanh(\tfrac{1}{2}\alpha\Delta) = \int_{-\infty}^{\infty} f(x_i)\left(1 + \frac{x_i}{\Delta}\right) T(x_i, \alpha, \Delta) dx_i. \quad (4.46)$$

In practice the numeric integration needs to be carried out over the absorption lineshape function only in each iteration cycle.

The hyperbolic function $T(x_i, \alpha, \Delta)$ approaches at low temperatures the Heaviside step function, with a step from -2 to 0 occurring at $x_i = \Delta$ (for positive polarization).

As an example of the application of Eqs. 4.40–4.46 let us consider deuterated propanediol (PD-d8) with PD-Cr(V) complexes obtained by reaction of PD-d8 with potassium bichromate. This gives an axially symmetric g-factor with an EPR line width of about 0.4% at 2.5 T field for which the theoretical lineshape is shown in Figure 3.1 (with different width of 0.5%). Figures 4.4a, b and c show the results obtained by solving the integral equation 4.46 at different inverse lattice temperatures α_0 and at different microwave frequency offsets Δ. The resulting inverse spin temperature α is shown in Figure 4.3 as a function of microwave frequency offset Δ from the center of the EPR line for the lattice temperatures of 300 mK, 500 mK and 700 mK. The maximum positive and negative α are

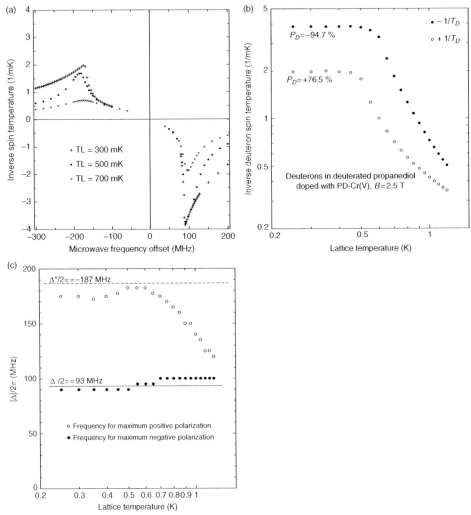

Figure 4.4 Simulation of DNP at 2.5 T with axially symmetric g-factor giving rise to inhomogeneously broadened 280 MHz wide EPR line, similar to that of perdeuterated PD-Cr(V). (a) Steady-state inverse spin temperature as a function of microwave frequency offset for three lattice temperatures; (b) maximum inverse temperature as a function of lattice temperature; (c) optimum frequency offset as a function of lattice temperature

more sharply defined at the negative values, as can be expected for the lineshape function that has a steep edge and a sharp peak at $\Delta/2\pi = 93$ MHz. The maximum positive polarizations are broader and they are obtained at frequencies just above the lower frequency edge of the EPR line at $\Delta/2\pi = -187$ MHz.

The maxima of the positive and negative inverse spin temperatures are plotted as a function of the lattice temperature in Figure 4.4b. Both of these increase steeply down to $T_L = 0.6$ K and then level off below $T_L = 0.5$ K. As can be expected from the asymmetric lineshape of the EPR line, the negative maxima of $\alpha = -3.85$ (mK)$^{-1}$ are higher than the positive ones around $\alpha = +1.98$ (mK)$^{-1}$; these correspond to the deuteron polarizations of -94.7% and $+76.5\%$, respectively.

The frequency offsets $|\Delta^\pm|/2\pi$ at which the maxima are obtained are plotted in Figure 4.4c as a function of the lattice temperature. While the frequency for the negative maximum decreases by about 10 MHz from 1.2 K to 250 mK, the frequency offset for the positive maximum increases by about 50 MHz when the lattice temperature decreases from 1.2 to 0.6 K. Below this temperature the offset decreases by about 10 MHz.

At the time of performing the above simulations in the 1980s, the best deuteron polarizations of PD-Cr(V) at 2.5 T field were $+44\%$ and -47% in a dilution refrigerator running around 300 mK coolant temperature. While it is clear that the simulations assume ideal conditions of DNP and full saturation, it is nevertheless interesting to try to understand the leading reasons for the difference between the experimental and theoretical numbers. The main reasons are known today to be due to both experimental and theoretical problems:

- The lattice temperature during DNP was not measured and it was substantially higher than the measured coolant temperature.
- The EPR saturation was incomplete.
- The nuclear spin polarization was measured using an RF field that was too high.
- The effect of the polarization of the hyperfine nuclei on the EPR line was not included in the simulation.
- The effect of dipolar broadening was not included in the simulation.

In the following we shall briefly discuss the above items in more detail.

Temperature of the Lattice

The heat transfer between the target material and the helium coolant will be discussed in Chapter 8. Specifically, using the thermal boundary conductance of Section 8.1.1 for similar glassy solids, we may take for PD-d8

$$\frac{\dot{Q}}{A_{beads}} = C_K \left(T_L^4 - T_{He}^4 \right),$$

where $C_K = 30$ W/(K^4 m^2) is the Kapitza conductance, and using an estimated microwave power absorption of 0.2 mW/cm^3 and surface-to-volume bead diameter $d = 0.2$ cm, with a filling factor $f = 0.6$, we may determine the lattice temperature

$$T_L = \left(T_{He}^4 + \frac{\dot{Q}}{C_K V_t} \frac{V_t}{A_{beads}} \right)^{1/4} = \left(T_{He}^4 + \frac{\dot{Q}}{C_K V_t} \frac{d}{6f} \right)^{1/4} = \left(T_{He}^4 + (0.246\,\text{K})^4 \right)^{1/4}. \quad (4.47)$$

At coolant temperature of 250 mK these yield the lattice temperature $T_L = 295$ mK. The lattice temperature is therefore about 50 mK higher than that of the dilute solution cooling the target. With a much higher power density, the lattice temperature may become almost entirely determined by it, when the target is cooled by a dilution refrigerator.

EPR Saturation

By examining the saturation function of Figure 3.5 plotted using the spin-lattice relaxation time 38 ms of PD-Cr(V) and the spin density $n_e = 5 \times 10^{19}$ cm^{-3}, we see that the power density of 0.2 mW/cm^3 corresponds to $s = 0.1$–1.0 when the frequency deviates around $3.0D$ from the Larmor frequency. This validates marginally the hypothesis of strong saturation required for the simplified steady-state Eq. 4.33 for the Hamiltonian of Eq. 4.25. Given that one must shift the microwave frequency further away from the Larmor frequency spectrum in order to get $s \gg 1$, one must choose to work with frequency deviations between $3D$ and $4D$; in our example this equals 45–60 MHz offsets from the main line. The transition rate then yields a low cooling rate for the dipolar reservoir, which, in turn, may become insufficient in comparison with the direct lattice contact for the nuclear spins via impurity spin interactions.

The high-temperature approximation of Eq. 4.11 yields the steady-state difference between the inverse temperatures of the electron Zeeman and dipolar reservoirs in the case of incomplete saturation:

$$\alpha - \beta = \beta_L \frac{\omega}{\Delta} \frac{1}{1 + s(\Delta)\left(1 + \frac{T_{1D}}{T_{1Z}} \frac{\Delta^2}{D^2}\right)}. \tag{4.48}$$

Based on this we can qualitatively state that the electron dipolar reservoir will remain significantly hotter than the electron Zeeman reservoir when $s < 0.1$.

Saturation of Nuclear Spins by NMR Measurement

In a very good target material the number of harmful impurity spins can be very low, which is witnessed by nuclear spin-lattice relaxation measurements. Then the leading loss of nuclear spin polarization takes place via RF saturation during NMR polarization measurement. In Section 6.2.5 a lengthy discussion will be presented on the effects of saturation of the deuteron NMR line during polarization measurement using the Liverpool Q-meter with a standard coil current of 0.3 mA; Eq. 6.72 yields an estimate of about 180 h for the saturation time constant of the deuteron NMR signal. This is comparable with the DNP time constant at low s, and it was shown that almost twice higher deuteron polarization was obtained when the duty cycle of the NMR measurement was substantially reduced [16].

Neglect of Electron Dipolar Interactions

Given the facts of finite saturation required by low lattice temperature, and given that this leads to optimum frequencies in the dipolar wings of the EPR line, it is tempting to include the dipolar broadening in the simulation of DNP by the cooling of the dipolar interaction reservoir. This should be done from the first principles by including the effects of cross-relaxation. However, if the cross-relaxation is sufficiently fast due to the high electron spin density, one may estimate the effect of dipolar broadening from ad hoc arguments by noticing that the frequency offsets grow up to 60 MHz in order to get far enough from the main Larmor frequency spectrum. This is 30% to 40% of the optimum frequency offsets without dipolar interactions, which reduce the ultimate inverse spin temperatures by the same amount.

Neglect of Hyperfine Interactions

The hyperfine nuclei are also polarized during DNP and they have multiple effects on the spin temperature. A high polarization of the solvent nuclei also imposes a high polarization of the hyperfine nuclei, which changes the shape of the EPR line, thus complicating the simulation. The effect will be estimated in Section 4.6 for the crystalline LMN doped with Nd^{3+}; this, unfortunately, cannot be extended to glassy materials doped with free radicals. We only know that in the latter case the hyperfine nuclei get the same sign of polarization as the solvent nuclei (see Chapter 5), but quantitative measurements are needed for the better understanding of their role and impact in DNP [17].

Comparison of Experimental Frequency Dependence of DNP with Simulation

Based on the above arguments, in particular with the strong frequency dependence of the saturation in the region of optimum DNP, it is not surprising that the agreement between the experimental frequency dependence and simulation is rather qualitative than quantitative. However, the various theoretical estimates have led to improved understanding of the phenomenology which, in turn, has had an impact on the experimental techniques and choices of target materials. For the future choices, it is clear that better measurements are needed in the understanding of the EPR spectrum (see Refs. [18, 19]) and in the knowledge of the electron spin-lattice and cross-relaxation. Moreover, better experimental values are desirable for the thermal boundary conductance between the glassy solids and liquid helium coolants.

Comparison with Other Simulations

Similar substantial deviations between simulated and experimental spin temperatures have been seen earlier by de Boer, for example [20]. These deviations were made smaller by introducing a leakage factor in Eq. 4.40 that was adjusted so as to get a better theoretical agreement. Phenomenologically the leakage factor may be justified if paramagnetic impurities such as oxygen cause nuclear spin-lattice relaxation with a rate comparable to the rate of DNP. Such a leakage, however, should be visible in the spin-lattice relaxation of the nuclear spins, and it should also make the temperatures of the various nuclear spin species unequal, both in steady state and during transients. No evidence for these have been found in good target materials.

In the light of much improved deuteron polarizations beyond 60% in a large deuterated target using EDBA-Cr(V) complex [21], and around 80% in smaller samples doped with a trityl complex [22], we may suggest that during DNP the leakage of nuclear spin polarization is small if not negligible in good target materials.

4.3.4 Time Evolution of Spin Temperatures during DNP at Low Temperatures

The above spin temperature model due Borghini allows one to determine the steady-state spin temperature after thermal equilibrium has been reached between the various heat reservoirs. It also requires that the saturation of the microwave transitions is high. Thus, we have only rough estimates of the speed at which nuclear spin polarization is obtained, and how far from thermal equilibrium the various nuclear spin systems are.

To improve on this, Wenckebach has extended Provotorov's time-dependent treatment of DNP and cross-relaxation to low temperatures where the density matrix must be kept in its exponential form [23]. In his treatment the only assumption is that spectral diffusion in the EPR spectrum is faster than any other processes. The resulting two equations for the spin temperatures cannot be solved analytically even approximatively. Wenckebach proceeded then to solve them numerically for the EPR line of the free radical TEMPO [23]. It is interesting that in 3.4 T field and at 0.75 and 1.5 K temperature the ultimate spin temperatures at saturation parameters $s = 1$ and $s = 10$ are almost identical, whereas at $s = 0.1$ and below the maximum inverse temperature drops rapidly. Similarly, the speed at which the DNP is obtained increases little above $s = 1$ but drops rapidly around $s = 0.1$.

4.4 Solid Effect

Let us consider a narrow electron spin resonance line with width $\Delta\omega_e \ll \omega_n$. In such a case the first-order forbidden transitions, where one electron and one nuclear spin flip simultaneously, can be well resolved from the first-order allowed electron spin transition. Such a weak narrow line can be observed when the electron spin density is fairly low, which means the range around or below 10^{19} cm^{-3}. In addition, it is required that there is no substantial anisotropy in the g-tensor and the hyperfine tensor. Alternatively this can happen in a single crystal that is suitably oriented with respect to the magnetic field.

In discussing the 'solid effect' below we shall follow the presentations of Schmugge and Jeffries [12] and of Abragam and Goldman [24]. The semi-phenomenological model is based on probabilities of transitions induced by the microwave field and by the lattice interactions, and allows to include effects due to the nuclear polarization 'leakage' via unwanted paramagnetic impurities, and due to the nuclear spin diffusion barrier.

The energy levels and state vectors for an isolated pair of electron and nuclear spins are shown in Figure 4.5. The mixing constants depend on the distance between the electron and the nucleus, and on the polar angles between the main field and the connecting vector. In the case that the electron and nuclear wave functions do not overlap (no hyperfine interaction), they are given by Eqs. 2.10 and 2.11c – f:

$$q = \frac{3}{2}\frac{\hbar\gamma_S\gamma_I}{\omega_I r^3}\sin\theta\cos\theta \cdot e^{i\phi} \ll 1;$$

$$p = \sqrt{1-qq^*} \cong 1.$$

(4.49)

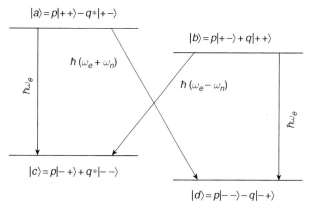

Figure 4.5 Eigenvalues of a pair of electron and nuclear spins ½ coupled by dipole-dipole interaction in a high field

4.4.1 Resolved Solid Effect

Level Populations, Transition Rates and Rate Equations

The populations of the four levels are p_a, p_b, p_c and p_d with $p_a + p_b + p_c + p_d = 1$. The electron and nuclear polarizations are then

$$P_e = (p_a + p_b) - (p_c + p_d) \tag{4.50}$$

and

$$P_n = (p_a + p_c) - (p_b + p_d). \tag{4.51}$$

The population ratios satisfy at all time during microwave irradiation

$$\frac{p_a}{p_c} = \frac{p_b}{p_d}, \tag{4.52}$$

which presupposes that the relaxation transition probabilities can be written

$$W_{|a\rangle \to |c\rangle} = W_{|b\rangle \to |d\rangle} = W_L, \tag{4.53}$$

$$W_{|c\rangle \to |a\rangle} = W_{|d\rangle \to |b\rangle} = rW_L, \tag{4.54}$$

$$W_{|a\rangle \to |d\rangle} = W_{|b\rangle \to |c\rangle} = U_L, \tag{4.55}$$

$$W_{|d\rangle \to |a\rangle} = W_{|c\rangle \to |b\rangle} = rU_L, \tag{4.56}$$

with

$$r = \exp(-\hbar\omega_e / k_B T); \; U_L = \overline{4qq^*} W_L. \tag{4.57}$$

The averaging must be understood over those $n = N_n/N_e$ nuclei which surround the electron but have no hyperfine interaction with it:

$$\overline{4qq^*} = \frac{4}{n}\sum_i q_i q_i^* = \frac{N_e}{N_n}\alpha = C\alpha, \tag{4.58}$$

where

$$\alpha \cong \frac{8\pi}{20}\frac{\gamma_S}{\gamma_I}\left(\frac{\Delta B_n}{B_0}\right)^2 \ll 1 \tag{4.59}$$

follows from averaging of the polar angles

$$\langle \cos^2\theta \sin^2\theta \rangle = \frac{2}{15}. \tag{4.60}$$

The populations of the levels can then be solved in terms of polarizations:

$$\begin{aligned}
p_a &= \frac{1}{4}(1+P_e)(1+P_n); \\
p_b &= \frac{1}{4}(1+P_e)(1-P_n); \\
p_c &= \frac{1}{4}(1-P_e)(1+P_n); \\
p_d &= \frac{1}{4}(1-P_e)(1-P_n).
\end{aligned} \tag{4.61}$$

The RF field applied along the main field at a frequency $\omega_e + \omega_n$ or $\omega_e - \omega_n$ causes transitions at a rate which is calculated using perturbation theory to the second order; the result normalized per nucleon is

$$W_{RF}^{(2)} = \overline{4qq^*}W_{RF}^{(1)} = C\alpha W_{RF}^{(1)}, \tag{4.62}$$

where $W_{RF}^{(1)}$ is the first-order transition probability of Eq. 1.58 for the electron alone in the same transverse field.

Rate Equations and Dynamic Equilibrium Polarization

Using the expressions above we may now write the rate equations for the change of the electron and nuclear polarizations during the saturation of the first-order forbidden resonance at $\omega_e - \omega_n$. After quite some algebra these read

$$\begin{aligned}
\dot{P}_e &= -\frac{1}{C}W_{RF}^{(2)}(P_e - P_n) - (1+r)(W_L + U_L)(P_e - P_0); \\
\dot{P}_n &= -W_{RF}^{(2)}(P_n - P_e) - C(1+r)U_L P_n(1 - P_e P_0).
\end{aligned} \tag{4.63}$$

4.4 Solid Effect

The electron concentration appears here because the rates were normalized per nucleon, and there are C electrons per nucleon. In a high field qq^* is very small and therefore $U_L \ll W_L$ can be neglected in the first of these equations. Furthermore, from Chapter 3 we recall that

$$(1+r)W_L = \frac{1}{\tau_{1e}} \tag{4.64}$$

and

$$(1+r)U_L = \frac{\alpha}{\tau_{1e}}. \tag{4.65}$$

With these Eq. 4.63 can be put to the form

$$\dot{P}_e = -\frac{W_{RF}^{(2)}}{C}(P_e - P_n) + \frac{1}{\tau_{1e}}(P_e - P_0);$$

$$\dot{P}_n = -W_{RF}^{(2)}(P_n - P_e) - \frac{C\alpha}{\tau_{1e}} P_n(1 - P_e P_0). \tag{4.66}$$

By defining the second-order saturation constant per electron

$$s = s_e^{(2)} = \frac{W_{RF}^{(2)} \tau_{1e}}{C}, \tag{4.67}$$

these can be rewritten

$$\dot{P}_e = -\frac{1}{\tau_{1e}}\left[s(P_e - P_n) + (P_e - P_0)\right];$$

$$\dot{P}_n = -\frac{C\alpha}{\tau_{1e}}\left[\frac{s}{\alpha}(P_n - P_e)\right] - P_n(1 - P_e P_0). \tag{4.68}$$

Because of the smallness of α, the rate of change of the electron polarization is much faster than that of the nucleon, and the electrons are therefore at all times at a quasi-equilibrium obtained by solving the first equation with its left-hand side put to zero:

$$P_e = \frac{sP_n + P_0}{1+s} \cong P_0 - s(P_0 - P_n). \tag{4.69}$$

This approximation is justified because the second-order transition probability is so small that the saturation constant is well below 1 for all conceivable microwave field strengths.

Assuming now that we are close to dynamic equilibrium so that $P_n \approx P_0$, and that the saturation constant is so small that we may write $P_e = P_0$, the second equation becomes linear and can be written as

$$\dot{P}_n = -\frac{C\alpha}{\tau_{1e}}\left[\left(\frac{s}{\alpha} + 1 - P_0^2\right)P_n - \frac{s}{\alpha}P_0\right]. \tag{4.70}$$

The nuclear polarization then approaches the equilibrium value

$$\dot{P}_n^{max} = \frac{P_0}{1 + \frac{\alpha}{s}(1 - P_0^2)} \qquad (4.71)$$

with the rate

$$\frac{1}{\tau_{pol}} = \frac{C\alpha}{\tau_{1e}}\left(\frac{s}{\alpha} + 1 - P_0^2\right) = \frac{C}{\tau_{1e}}\left[s + \alpha(1 - P_0^2)\right]. \qquad (4.72)$$

We see that the rate is linear with the electron concentration and spin-lattice relaxation rate, and grows linearly with the saturation parameter if we have $s > \alpha$, which can be reasonably achieved at high field where $\alpha \approx 10^{-4}$–10^{-3}. The nuclear polarization can therefore be expected to saturate just below the static electron polarization P_0. The polarization time constant is of the order of $\tau_{1e}/(Cs) = 10^2$ s for $Cs = 10^{-6}$ and $\tau_{1e} = 10^{-4}$ s, typical for LMN targets.

When saturating the second-order transition a–d at the frequency $\omega_e + \omega_n$ the only change in Eqs. 4.63 is to replace P_n by $-P_n$. The results above are therefore valid also in this case, with only the change of sign in Eq. 4.71.

It has turned out in practice that the maximum nucleon polarization does not reach as high values as P_0. This may be partly explained by the fact that high saturation also means dielectric heating of the crystal immersed in the helium coolant, and therefore P_0 is lowered because the lattice temperature is increased. As an example let us assume that $s = 10^{-3}$ and $\alpha = 10^{-4}$. If the difference between the electron and nuclear polarizations is 0.1, the magnetic power dissipation is

$$\frac{\dot{Q}_{mag}}{V} = \frac{s}{\tau_{1e}}(P_e - P_n)\frac{N_e}{V}\hbar\omega_e \cong 0.5\frac{\text{mW}}{\text{cm}^3} \qquad (4.73)$$

at 70 GHz frequency and with the electron density of $10^{19}/\text{cm}^3$ and spin-lattice relaxation time of 0.1 ms. The saturation constant $s = 10^{-3}$ requires

$$\omega_1^2 \cong \frac{2s}{\alpha}\frac{\Delta\omega_e}{\pi\tau_{1e}}, \qquad (4.74)$$

which is equivalent of the transverse field strength of

$$B_1 = \frac{\omega_1}{\omega_0}B_0 = 34\ \mu\text{T} \qquad (4.75)$$

when the line width of the allowed transition is 35 MHz and the static field strength is 1.8 T, as for Nd^{3+} ions in a LMN crystal. The dielectric losses are then

$$\frac{\dot{Q}_{diel}}{V} \cong \delta \cdot v_e \frac{B_1^2}{2\mu_0} = 3\ \frac{\text{mW}}{\text{cm}^3} \qquad (4.76)$$

for the loss tangent $\delta = 10^{-4}$. Such a power can be absorbed at 1 K without large temperature drop at the interface between the solid and the coolant, but increasing the saturation parameter by several orders of magnitude would turn out problematic. The crystal overheating, however, does not explain in this case the discrepancy between the observed maximum nuclear polarizations of about 80% and the electron polarization of nearly 100%. The heating of the crystal, however, is a critical factor because if the temperature is raised sufficiently, the Orbach process may start to dominate in the spin-lattice relaxation, which becomes quickly substantially faster because of the steep temperature dependence. This results in larger magnetic power absorption and ultimately in a thermal run-off.

The cavity losses for the Q-factor $Q_{cav} = 10^3$ are similarly

$$\frac{\dot{Q}_{cav}}{V_{cav}} \cong \frac{1}{Q_{cav}} \cdot v_e \frac{B_1^2}{2\mu_0} = 30 \frac{mW}{cm^3} \qquad (4.77)$$

and they are much higher than the magnetic and dielectric losses, particularly because the cavity volume must be substantially higher than that of the crystal, in order to ensure good cooling and to provide space for the NMR coil for polarization measurement.

Leakage Factor

Another factor which may reduce the nuclear polarization below the value predicted by Eq. 4.71 is relaxation by impurity electronic spins other than those that are used for the DNP by the solid effect. Assuming that these cause nuclear spin-lattice relaxation at the rate

$$\dot{P}_n^{(imp)} = -\frac{P_n}{\tau'_{1n}}, \qquad (4.78)$$

Eqs. 4.63 are easily modified by adding this term to the second equation. This causes a straightforward modification into the results: the equilibrium polarization becomes

$$\dot{P}_n^{max} = \frac{P_0}{1 + \frac{\alpha}{s}\left(1 - P_0^2\right) + \frac{f\alpha}{s}}, \qquad (4.79)$$

where

$$f = \frac{\tau_{1e}}{\tau'_{1n} C\alpha} \qquad (4.80)$$

is called the leakage factor. The rate of nuclear polarization growth is also slightly changed and Eq. 4.72 becomes

$$\frac{1}{\tau_{pol}} = \frac{C}{\tau_{1e}}\left[s + \alpha\left(1 - P_0^2\right) + f\alpha\right]. \qquad (4.81)$$

We notice that if $\alpha/s = 0.1$, f needs to be around 1 in order to explain a 10% discrepancy between the experimental DNP and Eq. 4.71. Variation of the microwave power thus allows to control the leakage term and to measure the leakage factor, if all other parameters are known.

At this point it is useful to consider a practical example of an LMN crystal with 1% of the La atoms replaced with Nd [12]. The concentration is then roughly $C = 4 \times 10^5$ and $\alpha = 8 \times 10^{-5}$. The nuclear spin relaxation time in an undoped crystal may be 10^5 s at 1 K, and the electron spin relaxation time is roughly 10^{-4} s at 2 T field. These give the leakage factor $f = 0.3$, which is too small to explain the experimental nuclear polarization of 70% at 1.2 K temperature with high microwave power.

A possible explanation is that the microwave field also induces transitions in the impurity electron spin system and therefore increases the polarization leakage. This can be expected if the impurities have such broad resonance lines that they cannot be observed. Alternatively, the phonon system may be overheated at frequencies close to the electron spin resonance, thus causing faster electron spin-lattice relaxation and therefore larger polarization leakage through nuclear spin-lattice relaxation.

Diffusion Barrier

Nuclei nearest to the electronic spin have their resonance frequency shifted significantly from its average value in the material, because of the static part of the dipolar field of the electron, Eq. 2.4. This is

$$B_z^{static} = -\frac{\mu_0}{4\pi} \frac{\hbar \gamma_e S_z}{r^3} \left(1 - 3\cos^2\theta\right) \qquad (4.82)$$

which has the value of 1.76 mT at a distance of 1 nm from an electron spin $S = 1/2$ with a g-factor 2. This may be compared with the dipolar field of protons with the density 0.79×10^{23} cm^{-3}, Eq. 2.7, which is 0.4 mT. Protons within the radius of about 1 nm from the electron have their frequency shifted more than the dipolar width of the bulk material; their number is above 300 in this volume. If the electron density is 10^{19} cm^{-3}, about 4% of the protons have a resonance frequency shifted out from the resonance line of the bulk material. These protons, however, are most effectively polarized by the solid effect, and we must ask the question whether the spin diffusion may be limited so that it will dominate the speed of the spread of the polarization into the bulk material.

Experimentally the question is best studied by observing nuclear spin relaxation limited by the diffusion barrier radius b, defined as

$$b^3 \cong \frac{4\pi}{\mu_0} \frac{\Delta\omega_n}{\hbar \gamma_e \gamma_n}, \qquad (4.83)$$

where $\Delta\omega_n$ is the nuclear dipolar linewidth. Within the barrier relaxation is fast and it is limited by spin diffusion outside the barrier. The decay of average polarization then becomes non-exponential and can be analyzed in the terms of an appropriate model. These will be discussed in the next chapter.

In solid effect the diffusion barrier does not seem to limit the polarization growth speed at the usual paramagnetic center densities around 10^{19} cm^{-3}. This may be due to the fact that during saturation of the forbidden resonance line the electron spins have significant probability of finding themselves in a state with zero z-component, during which time the diffusion barrier vanishes and nuclei within the diffusion barrier may cross-relax so that their polarization can diffuse outwards.

Phonon Bottleneck

The phonon bottleneck was seen to cause non-exponential spin-lattice relaxation of the electron spins in Chapter 3, with the approach of equilibrium at the time constant of $\sigma\tau_{1e}$ where σ is the phonon bottleneck constant defined by Eq. 3.100. A strong phonon bottleneck causes the phonon system to appear as heated up to a higher temperature, although in fact their distribution function may be not the equilibrium Maxwell–Boltzmann distribution. The electronic spin system is warmed up to this temperature, which leads to the replacement of Eqs. 4.68 by

$$\dot{P}_e = -\frac{1}{\tau_{1e}}\left[s(P_e - P_n) + \frac{P_e - P_0}{P_0 + \sigma P_e}P_0\right], \tag{4.84}$$

while the second equation remains the same because the bottleneck only influences the first-order relaxation transitions but not the much less frequent crossed relaxation transitions. This situation has been analyzed carefully by Abragam and Goldman [3].

In the absence of nuclear polarization leakage, the main effect of the phonon bottleneck is to lengthen the polarization build-up time. The phonon bottleneck has little effect in the DNP by solid effect if the leakage factor f is small, but it begins to reduce the polarization when $\sigma > (f\alpha)^{-1}$, which requires a rather strong bottleneck coefficient [3].

4.4.2 Differential Solid Effect

If the resonance line of the electron spin system is not narrow in comparison with the nuclear Larmor frequency, it is possible that both crossed transitions are saturated at the same time. If the electron line is completely inhomogeneous with negligible cross-relaxation, the rate equations 4.68 can be modified under the condition that the saturation by allowed transitions causes negligible change in the electron polarization, at least in dynamic equilibrium. Such a situation may arise if the electron resonance line is composed of many hyperfine lines, for example. It may also happen whenever a 'hole' can be burned in the electron resonance line due to the absence of cross-relaxation.

The steady-state nuclear polarization then becomes

$$P_n = P_0 \frac{s^+ - s^-}{s^+ + s^- + \alpha(1 - P_0^2)}, \tag{4.85}$$

where the second-order transition saturation factors are

$$s^{\pm} = \frac{\pi}{2}\alpha\tau_{1e}\omega_1^2 g(\omega \pm \omega_n), \tag{4.86}$$

and $g(\omega)$ is the normalized lineshape function of the electron resonance line. If the electron frequencies are spread over a spectrum much wider than that of the nuclear Larmor frequency, the nuclear polarization as a function of microwave frequency will resemble the derivative of the electron absorption spectrum. This is called the 'differential solid effect' [2].

4.5 Cross-Effect
Phenomenological Model

Hwang and Hill [25] introduced in 1967 a phenomenological model to describe their previous results [26] of DNP in polystyrene doped with Ley's radical, obtained at 4.2 K temperature and 2.5 T field. The model described the microwave frequency dependence of the dynamic enhancement of proton polarization in three samples with different radical concentrations, but required the introduction of three adjustable parameters obtained by fitting the experimental frequency dependences of the nuclear spin polarization enhancements.

In the experimental conditions of Ref. [26], the EPR line was thought to be primarily due to hyperfine structure and therefore consist of many narrow spin packets broadened slightly by dipolar interactions among the packets and with nuclear spins. This picture of the inhomogeneously broadened EPR line composed of 'spin packets' was originally introduced by Portis [27] in 1953 to describe the saturation of F-centers created by irradiation, with a low density.

The spin packets were thought to be saturable almost independently, but unlike the differential solid effect, the DNP was based mainly on allowed transitions followed by stimulated emission and multispin flip-flops between the spin packets interleaved spatially. The nuclear spin polarization then followed the derivative of the EPR lineshape, because the crossed transitions involving nuclear and electron spins had asymmetric rates when the microwave irradiation was off-resonance.

The cross-effect might describe correctly spin systems with such low density that the dipolar energy reservoir is very small, so that the spin packets are almost independent. The rates of transitions producing DNP are then rather slow. In systems with such high density that DNP is fast, the nuclear spins couple with the electron dipole-dipole energy reservoir with a much faster rate than with individual spin pair or multispin transitions, independent of whether stimulated emission or virtual photons are evoked. In any case, the proponents of the cross-effect failed to model the contribution of virtual photons in their off-resonance saturation, and ignored that the differential solid effect transitions require such a high microwave power that the cross-effect transitions would be oversaturated.

The three parameters obtained by fitting the experimental frequency dependences of the nuclear spin polarization enhancements were as follows:

- the width of the saturating 'ensemble' (probably meaning the width of the 'hole burnt' in the EPR line – the original paper [25] is not very clear at this point);
- the saturation parameter s defined by Eq. 3.59;

- an ad hoc parameter relating the RF-induced transition rates with those caused by spin-lattice relaxation.

The fit yields saturation parameters suggesting that at low electron spin concentration the DNP might be dominated by the solid-effect transitions, while the high-concentration sample could be dominated by spin-temperature effects. This conclusion is also supported by the variation of the ad hoc parameter with the electron spin concentration yielded by the fits. The hole widths yielded by the fits suggest that the lower concentration samples saw a much higher RF field strength, as can be expected.

Our conclusion is that the cross-effect might be observed in materials with low concentration of electronic impurities, which is not of greatest interest for polarized targets. On the other hand, applications in chemistry and biology may be satisfied by more modest polarization enhancements and with low concentration of paramagnetic impurities; in dissolution DNP and magic angle spinning MAS-DNP the cross-effect may therefore offer a good model for DNP [28].

The success of the DNP by cross-effect at high temperatures (>10 K) and low radical concentrations is amplified by the ingenious development of biradicals that naturally create pairs of cross-relaxing electron spins in a matrix where the spin concentration is otherwise so low that cross-effect is less probable that the differential solid effect [29].

Extension of the Cross-Effect Model to Low Temperatures

In order to develop a theoretically sound model for the cross-effect that is valid at low temperatures, Wenckebach first built a quantum statistical model to describe the EPR line in the density domain where the spectral diffusion due to cross-relaxation transitions is not fast enough to obtain a constant Zeeman temperature of the electrons in the rotating frame [30]. Instead then, the Zeeman temperature of each spin packet will evolve individually and reach an equilibrium value that is different for each packet.

Rather than going into mathematical details, we shall here only briefly overview the physical principles and the main results of the model of Wenckebach.

The EPR line was thought to consist of these spin packets with width of 10 MHz, much broader than the dipolar width which was in the range around 0.3 MHz. He then described the saturation of this line by defining the saturation parameter s of Eq. 3.59 and a cross-relaxation parameter $s_{ff} = W_{ff} T_{1S}$ where W_{ff} describes the rate of the cross-relaxation and T_{1S} the electron spin-lattice relaxation time [30]. The Larmor frequencies of spin packets do not overlap, and the spectral diffusion occurs only due to the dipolar widths of the packets. The rate of cross-relaxation is approximately $W_{ff} = \pi M_2 D/(2\Delta_p)$ where M_2 is of the order of the second moment of the homogeneous line broadening, D is the homogeneous line width and Δ_p is the width of the spin packet.

The spectral diffusion is then described using the saturation and cross-relaxation parameters s and s_{ff}, by writing the simultaneous rate equations for the polarizations and non-Zeeman energies of all spin packets. For $2n$ spin packets this involves the solution of $2n + 2$ differential equations, which yield the $2n$ Zeeman temperatures of the spin packets and their common non-Zeeman (spin-spin interaction) temperature.

In the derivation of the above equations, the energy balance was taken into account in the same way as in the model of Borghini that covers the case of fast spectral diffusion, as was described in Section 4.3.2. By adjusting the cross-relaxation rate parameter s_{ff} high enough, the model of Wenckebach coincides with that of Borghini [9], which gives confidence for the soundness of the principle and approximations that needed to be made [30]. On the other hand, if the temperature is assumed to be so high that a linear approximation can be made in the density matrix and polarization, the equations reduce to those of Provotorov [15] for cross-relaxation.

Using his model Wenckebach simulated the influence of the cross-relaxation parameter on the shape and width of the hole burnt in the EPR spectrum of TEMPO radical, with the result that at $s = 10$ and $s_{ff} < 10^2$ a hole is burnt in the spectrum, while at $s_{ff} > 10^4$ the cross-relaxation is sufficient to establish a common Zeeman temperature in the entire spectrum [30]. The simulation was made for 3.4 T field and 1.5 K lattice temperature.

The polarization of the nuclear spins in the cross-effect is obtained mainly by the simultaneous flip of one nuclear spin and the flip-flop of a pair of opposite electron spins whose Larmor frequencies differ by the Larmor frequency of the nuclear spin; this triple flip conserves energy and is therefore frequent. In the absence of microwave irradiation, these transitions bring the nuclear spin system to the lattice temperature. When the off-resonance microwave irradiation is turned on, it creates a hole in the EPR spectrum and a steep gradient in the spectral polarization of the electrons; this difference in the electron spin polarization is then transferred to the nuclear spin polarization [30]. The simulation in 3.4 T field and 1.5 K temperature of a glass with protons and ^{13}C spins doped with TEMPO shows that at $s_{ff} > 10^4$ the different nuclear species reach practically the same spin temperature, while at $s_{ff} < 10^3$ a substantially lower inverse spin temperature results and the ^1H and ^{13}C spin temperatures are clearly different [30]. As the parameter s_{ff} depends on the electron spin density and spectral width, these results are directly relevant for the choice of the free radical and density of doping.

4.6 DNP of Hyperfine Nuclei

The nuclei in paramagnetic atoms, molecules and the atoms just around the paramagnetic centers in crystals and other solids may have their wave function overlap with that of the paramagnetic electronic system. This leads to the interaction of Eq. 3.21, which has two terms: the scalar or Dirac term and the dipolar term leading to anisotropy. The scalar term alone splits the narrow EPR line into narrow hyperfine components, as was discussed in Section 3.2.2. The selective saturation of one or more of these lines will cause the population ratios of the hyperfine magnetic levels change, and leads to DNP of the hyperfine nuclei.

4.6 DNP of Hyperfine Nuclei

Let us assume a system of dilute paramagnetic atoms or ions in a single crystal where the axial symmetry leads to the Hamiltonian

$$\mathcal{H}_{HF} = A\hbar S_z I_z + \frac{1}{2}b\hbar\left(I_+ S_- + I_- S_+\right), \tag{4.87}$$

where $S = \frac{1}{2}$ is the electron spin and, for simplicity, also the nuclear spin is $I = \frac{1}{2}$. In a high field the energy levels are then admixtures given by Eq. (4.87) as shown by Figure 4.6. In high field the electron Zeeman splitting is larger than $A\hbar$, whereas the nuclear Zeeman splitting is often (but not always) smaller than $A\hbar$ (see Figure 3.3). If the anisotropic term is smaller than the isotropic one, the electron resonance line is split by the hyperfine interaction with this single nucleus to $2I + 1 = 2$ resolved components which can be saturated individually. The magnetic energy levels of this simplified system are, from Eq. 3.26,

$$E(m_S, m_n) = (g\mu_B B_0 + A m_n)m_S; \tag{4.87a}$$

these are illustrated in Figure 4.6.

Let us denote the level populations by p_i, $i = a, b, c, d$, normalized so that

$$p_a + p_b + p_c + p_d = 1. \tag{4.88}$$

If the transition a–c is saturated by a transverse microwave field, the populations of the corresponding energy levels are equal so that

$$p_a = p_c. \tag{4.89}$$

On the other hand, relaxation transitions tend to establish the population ratios determined by the Boltzmann factors

$$\frac{p_b}{p_c} = \frac{p_a}{p_d} = \exp\left(-\frac{\hbar\omega_e}{kT}\right) = B^{-1} \tag{4.90}$$

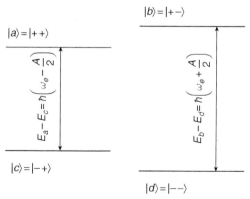

Figure 4.6 Magnetic energy levels of hyperfine-coupled pair of electron and nuclear spins 1/2, together with the state vectors and energy differences of the allowed RF transitions that generated DNP

and

$$\frac{p_b}{p_d} = \exp\left(-\frac{\hbar(\omega_e + A/2)}{kT}\right) = B'^{-1}. \quad (4.91)$$

By solving the four equations we obtain the hyperfine nuclear polarization

$$P_n^* = p_a + p_c - p_b - p_d = \frac{2B - B' - 1}{2B + B' + 1}. \quad (4.92)$$

If the temperature is low so that $B \gg 1$, the maximum nuclear polarization becomes

$$P_n^* = \frac{2 - B'/B - B^{-1}}{2 + B'/B + B^{-1}} \cong \frac{1}{3}, \quad (4.93)$$

assuming that $B'/B \approx 1$.

When the transition b–d is saturated, the equations to be solved are

$$p_b = p_d, \quad (4.94)$$

$$\frac{p_a}{p_d} = \frac{p_b}{p_c} = \exp\left(-\frac{\hbar\omega_e}{kT}\right) = B^{-1}, \quad (4.95)$$

and

$$\frac{p_a}{p_c} = \exp\left(-\frac{\hbar(\omega_e - A/2)}{kT}\right) = B''^{-1}. \quad (4.96)$$

In this case the nuclear polarization becomes

$$P_n^* = \frac{B + B/B'' - 2}{B + B/B'' + 2}. \quad (4.97)$$

If again the temperature is low so that $B \gg 1$ and it is assumed that $B''/B \approx 1$, the maximum nuclear polarization reaches

$$P_n^* = \frac{1 + B''^{-1} - 2B^{-1}}{1 + B''^{-1} + 2B^{-1}} \cong \frac{1 - B^{-1}}{1 + 3B^{-1}} \cong 1. \quad (4.98)$$

In both of the above cases, the polarization of the hyperfine nucleus is positive; it is thus impossible to obtain negative nuclear polarization by saturating the first-order transitions of the hyperfine line of the electron. The correction due to the hyperfine nuclei to the polarization of the non-hyperfine nuclei, whether they are same nuclear species or background nuclei, is thus larger for negative polarization than for positive polarization. This correction, however, is usually smaller than 1% because the relative abundance of the hyperfine nuclei is at most the same as the abundance of the dilute electrons. Furthermore, the hyperfine nuclei can be diluted by using isotopic enrichment so that the electrons belong to isotopes with no nuclear spin. An example of this is the Nd^{3+} ions in LMN targets where the enrichment with Nd was used.

When the crossed transition a–d is saturated with a large microwave field parallel to the steady field, the hyperfine nuclear polarization is obtained similarly

$$P_n^* = \frac{1 - B'^{-1}B''^{-1}}{1 + 2B''^{-1} + B'^{-1}B''^{-1}} \cong 1, \tag{4.99}$$

where the numeric value refers again to the case of low temperatures. In the case that the transition b–c is saturated, requiring also a high microwave field parallel to the steady field, the polarization of the hyperfine nuclei reaches at low temperature the same value

$$P_n^* = \frac{1 - B'^{-1}B''^{-1}}{1 + 2B''^{-1} + B'^{-1}B''^{-1}} \cong 1. \tag{4.100}$$

The crossed transitions happen at the frequency of the electron spin resonance of the isotopic species with no nuclear spin. This transition leads to no nuclear polarization in materials such as LMN doped with Nd^{3+} ions, and is therefore of no major concern for DNP using the solid effect, if the first-order forbidden transitions are well resolved from the main resonance line of the electrons.

If the hyperfine nucleus has a spin higher than ½, the derivation of their polarization becomes arithmetically more complicated although straightforward. In all cases the polarization is positive, however.

In glassy materials doped with free radicals or metallo-organic complexes, the hyperfine Hamiltonians are generally anisotropic, in particular for protons or deuterons belonging to these molecules. The DNP of these hyperfine nuclei cannot then be simply obtained using the above expressions. The cases of EHBA-Cr(V) and BHHA-Cr(V) have been studied in deuterated butanol with about 2% unsubstituted hydrogen [31]. Their hydrogen NMR lines at high positive polarization are shown in Figure 5.7, which indicates that the polarization of all hyperfine nuclei corresponds to the same sign of spin temperature as that of the matrix protons. At negative polarizations the hyperfine nuclei are polarized negatively, which proves that they are cooled by the same mechanism as the matrix nuclei. Unfortunately there is no quantitative study of the magnitude of the polarization of hyperfine coupled nuclei, the main reason being that at the time of these studies the sweep widths and dynamic ranges of the RF sources and Q-meters were not well adapted to sweep width well in excess of 1 MHz.

The above discussion on the polarization of the hyperfine nuclei applies therefore to the case of well-resolved EPR lines, and for broader EPR lines each case should be studied individually.

4.7 The Overhauser Effect

Overhauser proposed in 1953 [32] that the saturation of the resonance of conduction electrons in a metal will lead to nuclear polarization, because the nuclear spin-lattice relaxation is very fast due to simultaneous flips of the electron and nuclear spins. Despite general scepticism among the leading experts, Carver and Slichter [33, 34] proved this experimentally in metallic lithium powder at 350 K temperature (!) and 3 mT field, where they reached an enhancement of the ^{7}Li spin polarization by a factor of 100. They also demonstrated similar results in liquid ammonia with dissolved sodium, where the sodium atoms dissociate into diamagnetic Na$^+$ ions and paramagnetic electrons, which move rapidly in the liquid and cause fast proton spin relaxation.

In the free-electron model of metals, the thermal motion excites electrons above the Fermi surface. This gives rise to a temperature-independent susceptibility due to the electrons, which have their energy in the band of about kT around the Fermi energy E_F. In this band the population ratio of the magnetic levels is given by the Boltzmann ratio B

$$\frac{N_+}{N_-} = e^{-\hbar\omega_e/kT} = B^{-1}, \tag{4.101}$$

while the remaining conduction electron spins are paired and therefore contribute nothing to the susceptibility.

In a high field the spin Hamiltonian of a single conduction electron and a nearby nucleus is

$$\mathcal{H} = g\mu_B B_0 S_z - \hbar\gamma_n B_0 I_z + A\mathbf{I}\cdot\mathbf{S} = \hbar\omega_e S_z + \hbar\omega_n I_z + A\mathbf{I}\cdot\mathbf{S}, \tag{4.102}$$

where the scalar term $A\mathbf{I}\cdot\mathbf{S}$ is due to the overlap of the electron and nuclear wave functions. Although the electron interacts simultaneously with several nuclei, we may derive the rate equations by treating one electron-nucleus pair at a time, and then allow A to have varying strengths when averaging over all spin pairs. In this scalar term only the part $AI_z S_z$ has diagonal elements so that to a good approximation the energy eigenvalues are

$$E = g\mu_B B_0 m_S - \hbar\gamma_n B_0 m_I + A m_I m_S. \tag{4.103}$$

In this system the electron and nuclear resonance frequencies are (instantaneously)

$$\omega_e = \gamma_e B_0 + \frac{A}{\hbar} m_I;$$
$$\omega_n = \gamma_n B_0 - \frac{A}{\hbar} m_S, \tag{4.104}$$

because the first-order electron spin transitions are those with $\Delta m_S = \pm 1$, $\Delta m_I = 0$, and the nuclear spin transitions with $\Delta m_S = 0$, $\Delta m_I = \pm 1$. For simplicity we shall assume that $S = 1/2$ and $I = 1/2$. We may then label the four instantaneous energy states as shown in Figure 4.7, which shows the first-order RF transitions and the dominant relaxation transitions with $\Delta m_S = \pm 1$, $\Delta m_I = 0$ and $\Delta m_S + \Delta m_I = 0$; the relaxation transitions $\Delta m_S + \Delta m_I = \pm 2$ and $\Delta m_S = 0$, $\Delta m_I = \pm 1$ are very slow in comparison with the first ones. It is this relaxation behavior which makes the Overhauser effect work.

Referring to the notation of Figure 4.7, the magnetic energy differences are

$$E_b - E_c = \hbar(\gamma_e + \gamma_n)B_0;$$
$$E_d - E_c = \frac{A}{2} + \hbar\gamma_n B_0; \tag{4.105}$$
$$E_a - E_c = \frac{A}{2} + \hbar\gamma_e B_0,$$

4.7 The Overhauser Effect

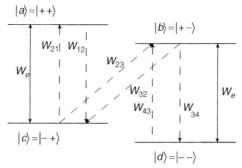

Figure 4.7 The Overhauser effect in a metal. The transitions W_{ij} are induced thermally and tend to establish thermal equilibrium; the thermal transitions with dashed arrows are fast, whereas those with $\Delta m_S + \Delta m_I = \pm 2$ and $\Delta m_S = 0$, $\Delta m_I = \pm 1$ are slow and they are not marked with arrows. The transverse RF field induces transitions at the rate W_e between the states a and c and/or between states b and d

and in thermal equilibrium without saturating microwave field, the population ratios of the eigenstates p_i, $i = a$ to d obey the Boltzmann ratios

$$\frac{p_a}{p_c} = \frac{p_b}{p_d} = e^{-(E_a - E_c)/kT} = e^{-(\hbar\omega_e + A/2)/kT} = B'^{-1};$$

$$\frac{p_b}{p_c} = e^{-\hbar\omega_e/kT} = B^{-1}.$$
(4.106)

If the transition a–c is strongly saturated by a resonant transverse field, the populations of these energy levels are equalized so that

$$p_a = p_c.$$

At the same time the fast spin-lattice relaxation transitions keep the Boltzmann ratios unchanged:

$$\frac{p_b}{p_c} = e^{-\hbar\omega_e/kT} = B^{-1};$$
(4.107)

$$\frac{p_b}{p_d} = e^{-(\hbar\omega_e + A/2)/kT} = B'^{-1}.$$
(4.108)

Because the sum of population numbers is $p_a + p_b + p_c + p_d = 1$, we may now solve these equations to get the nuclear polarization

$$P_n \equiv \frac{\langle I_z \rangle}{I} = \frac{1}{I}\sum_i p_i \langle i | I_z | i \rangle = p_a - p_b + p_c - p_d$$
$$= \frac{2B - B' - 1}{2B + 1 + B'}.$$
(4.109)

If also the transition b–d is saturated simultaneously, we have $p_a = p_c$, $p_b = p_d$ and $p_b = p_c/B$ and

$$P_n = \frac{B-1}{B+1} = P_e. \qquad (4.110)$$

In the limit of strong saturation at low temperatures, the nuclear polarization therefore becomes equal to the polarization of the conduction electrons (with opposite sign if the nuclear gyromagnetic moment is positive) that are thermally excited above the Fermi surface. Only one sign of polarization is obtained, however; if the sign of the nuclear gyromagnetic factor is negative, the sign of polarization is the same as that of the electrons.

In practice each unpaired electron couples with many nuclei, and the nucleus couples with many electrons simultaneously. This results in an electron spin resonance line where the two first-order RF transitions of Figure 4.7 cannot be resolved. This can be thought to result from the averaging of A over a range of values that depend on the overlapping of the nuclear and electron wave functions at particular instants and positions. Both first-order transitions are therefore always saturated and the electron resonance line is as if it were homogeneously broadened because of the hyperfine coupling with the nuclei.

In the high-temperature limit the strong saturation at the center of the electron absorption line therefore results in the enhancement of the nuclear polarization over its thermal equilibrium value by

$$\frac{P_n}{P_{n,0}} = \frac{\omega_e}{\omega_n} = \frac{\gamma_e}{\gamma_n}, \qquad (4.111)$$

which can be several hundred for protons in practice.

In the Overhauser effect it is important that nuclear relaxation takes place predominantly via the flip-flop transitions b–c and rarely via the double-flip transitions a–d.

In polarized targets the Overhauser effect has a limited applicability because a very high RF field strength and a low steady field are required. The metal must be finely powdered and the strong heating precludes cooling to low temperatures. As a result high nuclear polarizations are unattainable.

The Overhauser effect also works with paramagnetic ions or free radicals in a solution [35] (see also Section 7.5 of Ref. [36]) where the Brownian motion induces the nuclear relaxation via electron-nucleus flip-flop transitions with $\Delta m_S + \Delta m_I = 0$. Similarly, the Overhauser effect has also been observed in solids containing paramagnetic free radicals strongly coupled by the exchange interaction [37–39].

In liquids the heating by the microwave field is less strong, but because no other substances than helium are in liquid state at low temperatures, low nuclear polarizations are again obtained.

The Overhauser effect in liquid state could be useful for enhancing the signal size and the contrast in magnetic resonance imaging (MRI). The problem is that free radicals tend to be toxic in living organisms. One way to alleviate this would be to use short-lived excitations of diamagnetic molecules into a paramagnetic state, such as triplet states accessible by resonant optical absorption.

Tests in Bitter Magnet Laboratory of MIT have revealed that Overhauser effect also works in insulating solids such as polystyrene in which BDPA radical was dispersed [1]. These tests were carried out at high fields up to 18.8 T and at temperatures around 80 K, common in DNP magic angle spinning experiments. Microwave power variation studies showed that the Overhauser effect saturates at considerably lower power levels than the solid effect in the same samples. These results provide new insights into the mechanism of the Overhauser effect, and also provide a new approach to perform DNP experiments in chemical, biophysical and physical systems at high magnetic fields.

4.8 Frequency Modulation Effects in DNP

Microwave frequency modulation (FM) has been sometimes applied during DNP, in particular with LMN targets where the solid effect is used. In such cases the improvements of a few percent in the ultimate proton polarization has been explained by its compensating effect on the inhomogeneity of the magnetic field in the target volume. Other cases where FM is effective include the cross-effect (Section 4.5) and differential solid effect (Section 4.4.2). Because it was known that in EHBA-Cr(V) the dynamic cooling of the spin-spin interactions is effective, the dramatic improvement of the deuteron polarization of glassy deuterated butanol doped with EDBA-Cr(V) came as a surprise [21].

In Section 3.5.3 we discussed the effect of microwave FM on the in situ EPR spectra obtained by the bolometric technique. The experimental observations in a large deuterated target (see Figure 3.11) were clear and convincing, but they were not quantifiable theoretically. It was suggested that the dispersion part of the complex susceptibility of the paramagnetic dopant EDBA-Cr(V) can produce the effects seen when the frequency is not modulated, but this alone could not explain the observed enhancement of microwave power absorption with FM. It was therefore also suggested that when the frequency is not modulated and power is applied near the optimum frequencies, a hole is burned in the EPR spectrum because of the interaction of the paramagnetic electrons with nearby hyperfine nuclei that are polarized. When modulating the frequency with sufficient amplitude, all hyperfine lines within the modulated range share the power and, together with cross-relaxation, the entire EPR spectrum is responding collectively rather than only by the narrow part in which the hole was burnt. This qualitatively explains the higher power absorption and higher speed of DNP. The DNP speed enhancement by a factor up to 2 is shown by Figure 4.8 [21].

The hypothesis of slow cross-relaxation in EDBA-Cr(V) is not supported by the cross-relaxation measurement between protons and ^{13}C spins in undeuterated butanol with 5% water doped with EHBA-Cr(V) [40]. The polarization recovery time constant for ^{13}C spins below 100 mK was found to be in the order of 3,000 min, which can be compared with that measured between protons and deuterons in partially deuterated ethylenglycol

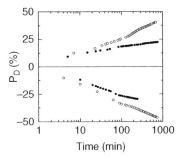

Figure 4.8 Deuteron polarization growth as a function of time in the two oppositely polarized halves of the SMC double-cell target made of deuterated butanol with 5% heavy water and 7×10^{19} spins/cm^3 of EDBA-Cr(V) [21]. Black dots: no FM; open circles: 20 MHz peak-to-peak FM at 1 kHz frequency. The natural emission bandwidth of the EIO source was 0.1 MHz. The target cells were 40 cm long and 5 cm in diameter, and the homogeneity of the 2.5 T superconducting magnet was 10^{-4}

(EG-d4) doped with EG-Cr(V) around 5,000 min. This explains why equilibrium could be obtained between the spin temperatures of the various nuclei after long DNP. On the other hand, when turning the microwaves on, the same time constant in EG-Cr(V) was 2 min with 0.3 mW/g power, and it was found to be inversely proportional to the microwave power with at least three times higher power [11]. This is in strong contrast with the results on EHBA-Cr(V) featuring recovery time constants of 295 min with 4.5 μW/g and 190 min with 13.5 μW/g [40]; this can be interpreted as a hole burned in the EPR line already at these low power densities.

The final deuteron polarization of the large deuterated target of SMC was higher by a factor up to 1.7 when applying FM [21]. In order to reach the maximum deuteron polarizations of +51% and −60%, the NMR polarization measurement duty cycle had to be reduced, because the Q-meter was found to cause some saturation in continuous measurement mode [41]. In the case of continuous NMR measurement, the saturation time constant was 180 h, i.e. just in the same range as the equilibrium time constant for different nuclei. This will be discussed in more detail in Section 6.2.5.

Small sample DNP tests with the same target material yielded better results without FM, but these tests were carried out with other microwave sources and perhaps with higher power density [42]. Also, the Q-meter coil had less target material in the proximity of the coil wire. The two large double-cell targets of SMC, with different magnetic field uniformities, showed identical improvements with FM; this excludes that the compensation of the magnetic field inhomogeneity could contribute significantly to the enhancement. Because of the differences of response to FM between small samples and the large targets, a study was carried out in order to see possible effects due to the electrodynamic response of the large volumes of the cavity and the target.

Firstly, the frequency f_m at which the full enhancement due to FM was reached was determined to be about 100 Hz, which is close to the inverse spin-lattice relaxation time of the EDBA-Cr(V). Modulation at a frequency higher than this brought little additional

4.8 Frequency Modulation Effects in DNP

absorption. Therefore, all subsequent power absorption tests were carried out at $f_m = 500$ Hz or 1 kHz, despite of the fact that at low power the additional absorption levelled off at a somewhat lower modulation frequency.

Secondly, the dependence of the additional power absorption on the amplitude of the FM (defined as the peak-to-peak frequency deviation) was studied using different input powers as shown in Figure 4.9. The absorbed microwave power was determined from the coolant temperature and the previously measured cooling performance of the dilution refrigerator. The additional absorption grows rapidly up to about 10 MHz p–p FM amplitude and then more slowly; when optimizing the p–p amplitude with regard to the resulting DNP, the highest values of positive and negative polarizations were obtained using 30 MHz p–p FM amplitude, at center frequencies of $f^+ = 69.070$ GHz and $f^- = 69.540$ GHz, in a magnetic field corresponding to 106.45 MHz proton NMR frequency.

Thirdly, saturation of the magnetic absorption was studied with and without FM, at the optimum frequency $f^- = 69.540$ GHz for the negative polarization and in the central part of the EPR spectrum [21]. The input power was measured by a calorimeter at room temperature, and the additional magnetic losses in the target were obtained from the reading of the in situ bolometers. The results are shown in Figure 4.10 where the magnetic losses, measured with the bolometers, are plotted as a function of input microwave power, measured using the calorimeter at room temperature. As can be expected, when the microwave

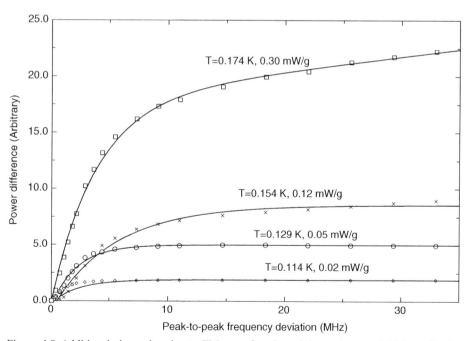

Figure 4.9 Additional absorption due to FM, as a function of the peak-to-peak FM amplitude, at different microwave powers, from Ref. [41]. Highest deuteron polarizations were reached at 30 MHz p–p FM amplitude. Reprinted from Ref. [41], with permission from Elsevier

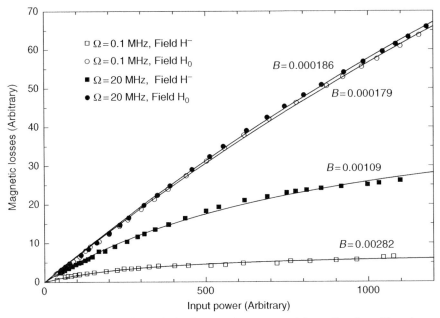

Figure 4.10 Relative magnetic losses in the deuterated target material, as a function of the microwave power. The small deviation B of the upper curves from straight lines is due to the small non-linearity of the calorimeter; here the frequency is tuned to the center of the EPR line. The lower curves describe power losses at the optimum frequency for negative DNP; in both of these the higher curvature B is due to increasing saturation of the resonant transitions at the semi-transparent edge of the EPR line. The difference of the two lower curves is due to the additional absorption caused by FM

frequency is tuned to the center of the EPR line, the power is insufficient to saturate the resonance, and therefore the FM has no effect in the relationship between the power in the input and in the cavity; the target is 'black' (using optical terminology). On the other hand, when irradiating the target at the frequency of optimum negative polarization, the target is semi-transparent, and the cavity power can be substantially altered by FM that influences the spectral spread of the power over the EPR line, and alters the standing waves in the cavity, thus making the power spread spatially more uniformly.

The improvement in the speed and final value of DNP with FM is demonstrated in Figure 4.11 where the time evolution of the deuteron polarization is plotted for the two opposite polarized cells of the SMC target [43]. On the left is shown the difference of the mean polarizations of the target cells, with no FM applied during the first 55 h of DNP. At time 55 h the FM with 20 MHz p–p FM was turned on in both EIO tubes feeding the two target cells; this is seen as a dramatic increase in the speed of DNP. On the right is shown the speed enhancement during the first 10 h of DNP, without and with 20 MHz p–p FM, for positively and negatively polarized cells.

Ajoy et al. have modified the FM techniques by applying swept microwave frequency combs [44] in a 7 T field. They observe that ^{13}C DNP enhancement in their microdiamond

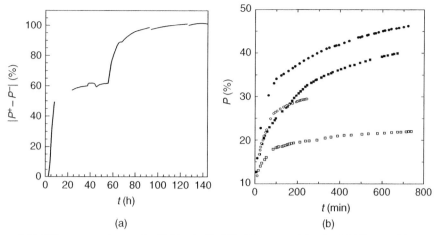

Figure 4.11 Effect on turning on the FM in the BuOH-d10 target of the SMC [41]. (a) Difference of the opposite polarizations of the two target cells, when 20 MHz FM was turned on at time 55 h. (b) Initial 10 h of DNP growth without FM (open symbols) and with 20 MHz p–p FM (black symbols). The circles refer to negative polarizations and squares to positive polarizations. The microwave power and frequency were continuously optimized during data taking. Reprinted from Ref. [41], with permission from Elsevier.

sample increases from 30 to 100 when applying the modulation. Furthermore, they speculate that applying this technique to samples doped with TEMPO with low density could produce gains better than one order of magnitude in the enhancement. A microwave frequency comb is obtained by the modulation of the amplitude and/or phase of the microwave source at a suitable radiofrequency; in the case of DNP using hyperfine broadened EPR spectrum, the suitable RF frequency should be equal to the hyperfine splitting. In the case of ^{6}LiD, for example, this splitting is about 10 MHz.

When using solid effect for DNP with a hyperfine split EPR line such as that of ^{169}Tm^{2+} ($I = \frac{1}{2}$) in CaF$_2$, Abragam and coworkers note that the saturation of the satellite of only one of the EPR lines results in its rapid depopulation, which can be cured by saturating the other line EPR during 4% of the time. This repopulates the original line and entails much faster DNP with the original satellite line that is saturated 96% of the time, as is described on p. 361 of Ref. [24]. This is a clean mechanism for the benefit of FM in the DNP by solid effect.

Noda and Koizumi used microwave FM with peak-to-peak width of 100 MHz in 3.35 and 6.7 T field in their DNP apparatus [45]. At 1.2 K and 6.7 T, they achieve high proton spin polarizations of +76% and −84% with short build-up time constants of 3.2 and 4.1 min in a polystyrene film doped with 50 mM of TEMPO. The gain in polarization is by a factor of about 1.2 in a field of 3.35 T and the speed of DNP is practically unchanged. In contrast, the gain in the speed of DNP is remarkable in 6.7 T field, because the DNP becomes very slow without FM, for both concentrations of TEMPO. The authors suggest that the line width at higher field causes slower spectral diffusion and therefore a much slower DNP, when FM is not applied.

References

[1] T. V. Can, M. A. Caporini, F. Mentink-Vigier, et al., Overhauser effects in insulating solids, *The Journal of Chemical Physics* **141** (2014) 064202–064208.

[2] C. D. Jeffries, *Dynamic Nuclear Orientation*, Wiley-Interscience Publications, Hoboken, NJ, 1963.

[3] A. Abragam, M. Goldman, Principles of dynamic nuclear polarisation, *Reports on Progress in Physics* **41** (1978) 395–467.

[4] C. D. Jeffries, History of the development of polarized targets, *High Energy Spin Physics*, Springer, Berlin, Heidelberg, 1991, pp. 3–19.

[5] V. A. Atsarkin, Dynamic nuclear polarization: yesterday, today, and tomorrow, *Journal of Physics: Conference Series* **324** (2011)012003.

[6] C. P. Slichter, The discovery and renaissance of dynamic nuclear polarization, *Reports on Progress in Physics* **77** (2014) 072501.

[7] W. T. Wenckebach, *Essentials of Dynamic Nuclear Polarization*, Spindrift Publications, The Netherlands, 2016.

[8] D. G. Crabb, W. Meyer, Solid polarized targets for nuclear and particle physics, *Annu. Rev. Nucl. Part. Sci.* **47** (1997) 67–109.

[9] M. Borghini, Spin-temperature model of nuclear dynamic polarization using free radicals, *Phys. Rev. Lett.* **20** (1968) 419–421.

[10] M. Borghini, K. Scheffler, Experimental evidence for dynamic nuclear polarization by cooling of electron spin-spin interactions, *Phys. Rev. Letters* **26** (1971) 1362–1365.

[11] W. de Boer, M. Borghini, K. Morimoto, T.O. Niinikoski, F. Udo, Dynamic polarization of protons, deuterons and carbon-13 nuclei: thermal contact between nuclear spins and electron spin-spin interaction reservoir, *J. Low Temp. Phys.* **15** (1974) 249–267.

[12] T. J. Schmugge, C. D. Jeffries, High dynamic polarization of protons, *Phys. Rev.* **138** (1965) A1785–A1801.

[13] M. Borghini, W. de Boer, K. Morimoto, Nuclear dynamic polarization by resolved solid-state effect and thermal mixing with an electron spin-spin interaction reservoir, *Phys. Lett.* **48A** (1974) 244–246.

[14] M. Borghini, Mechanisms of nuclear dynamic polarization by electron-nucleus dipolar coupling in solids, in: G. Shapiro (ed.) *Proc. 2nd Int. Conf. on Polarized Targets*, LBL, University of California, Berkeley, Berkeley, 1971, pp. 1–32.

[15] B. N. Provotorov, A quantum statistical theory of cross relaxation, *Soviet Phys.- JETP* **15** (1962) 611–614.

[16] Spin Muon Collaboration (SMC), B. Adeva, S. Ahmad, et al., Measurement of the deuteron polarization in a large target, *Nucl. Instr. and Meth. in Phys. Res.* **A349** (1994) 334–344.

[17] T. O. Niinikoski, Dynamic nuclear polarization with the new complexes, in: G. R. Court, et al. (eds.) *Proc. of the 2nd Workshop on Polarized Target Materials*, SRC, Rutherford Laboratory, 1980, pp. 60–65.

[18] J. Heckmann, S. Goertz, W. Meyer, E. Radtke, G. Reicherz, EPR spectroscopy at DNP conditions, *Nucl. Instrum. Methods Phys. Res.* **A 526** (2004) 110–116.

[19] J. Heckmann, W. Meyer, E. Radtke, G. Reicherz, S. Goertz, Electron spin resonance and its implication on the maximum nuclear polarization of deuterated solid target materials, *Phys. Rev.* **B 74** (2006) 134418.

[20] W. de Boer, Dynamic orientation of nuclei at low temperatures, *J. Low Temp. Phys.* **22** (1976) 185–212.

[21] Spin Muon Collaboration (SMC), B. Adeva, E. Arik, et al., Large enhancement of deuteron polarization with frequency modulated microwaves, *Nucl. Instrum. and Methods* A **372** (1996) 339–343.

[22] S. T. Goertz, J. Harmsen, J. Heckmann, et al., Highest polarizations in deuterated compounds, *Nuclear Instruments and Methods in Physics Research* A **526** (2004) 43–52.

[23] W. T. Wenckebach, Dynamic nuclear polarization via thermal mixing: beyond the high temperature approximation, *J. Magn. Res.* **277** (2017) 68–78.

[24] A. Abragam, M. Goldman, *Nuclear Magnetism: Order and Disorder*, Clarendon Press, Oxford, 1982.

[25] C. F. Hwang, D. A. Hill, Phenomenological model for the new effect in dynamic polarization, *Phys. Rev. Letters* **19** (1967) 1011–1014.

[26] C. F. Hwang, D. A. Hill, New effect in dynamic polarization, *Phys. Rev. Letters* **18** (1967) 110–112.

[27] A. M. Portis, Electronic structure of F centers: saturation of the electron spin resonance, *Phys. Rev.* **91** (1953) 1071–1078.

[28] A. S. Lilly Thankamony, J. J. Wittmann, M. Kaushik, B. Corzilius, Dynamic nuclear polarization for sensitivity enhancement in modern solid-state NMR, *Progress in Nuclear Magnetic Resonance Spectroscopy* **102–103** (2017) 120–195.

[29] C. Song, K.-N. Hu, C.-G. Joo, T. M. Swager, R. G. Griffin, TOTAPOL: a biradical polarizing agent for dynamic nuclear polarization experiments in aqueous media, *Journal of the American Chemical Society* **128** (2006) 11385–11390.

[30] W. T. Wenckebach, Spectral diffusion and dynamic nuclear polarization: beyond the high temperature approximation, *Journal of Magnetic Resonance* **284** (2017) 104–114.

[31] T. O. Niinikoski, NMR of the paramagnetic compound molecules in a deuterated matrix, in: G. R. Court, et al. (eds.) *Proc. of the 2nd Workshop on Polarized Target Materials*, SRC, Rutherford Laboratory, Abingdon, UK, 1980, pp. 62–65.

[32] A. W. Overhauser, Polarization of nuclei in metals, *Phys. Rev.* **92** (1953) 411–415.

[33] T. R. Carver, C. P. Slichter, Experimental verification of the Overhauser nuclear polarization effect, *Phys. Rev.* **102** (1956) 975–980.

[34] T. R. Carver, C. P. Slichter, Polarization of nuclear spins in metals, *Phys. Rev.* **92** (1953) 212–213.

[35] A. Abragam, Overhauser effect in nonmetals, *Phys. Rev.* **98** (1955) 1729–1735.

[36] C. P. Slichter, *Principles of Magnetic Resonance*, 3rd ed., Springer-Verlag, Berlin, 1990.

[37] H. G. Beljers, L. van der Kint, J. S. van Wieringen, Overhauser effect in a free radical, *Phys. Rev.* **95** (1954) 1683.

[38] A. Abragam, A. Landesman, J. M. Winter, Overhauser effect with exchange coupling?, *C. R. Acad. Sci.* **246** (1958) 1849.

[39] R. H. Webb, Dynamic polarization anomalies in organic free radicals, *Phys. Rev. Lett.* **6** (1961) 611–613.

[40] S. Bültmann, G. Baum, C. M. Dulya, et al., Cross-relaxation between protons and ^{13}C nuclei, in: H. Dutz, W. Meyer (eds.) *Proc. 7th Int. Workshop on Polarized Target Materials and Techniques*, North-Holland, 1995, 106–107.

[41] Spin Muon Collaboration (SMC), D. Adams, B. Adeva, et al., The polarized double-cell target of the SMC, *Nucl. Instr. and Meth. in Phys. Res.* A **437** (1999) 23–67.

[42] Y. F. Kisselev, The modulation effect on the dynamic polarization of nuclear spins, in: H. Dutz, W. Meyer (eds.) *Proc. 7th Int. Workshop on Polarized Target Materials and Techniques*, Elsevier, Amsterdam, 1995, 99–101.

[43] Spin Muon Collaboration (SMC), D. Adams, B. Adeva, et al., The polarized double-cell target of the SMC, *Nucl. Instr. and Meth.* A **437** (1999) 23–67.

[44] A. Ajoy, R. Nazaryan, K. Liu, et al., *Enhanced dynamic nuclear polarization via swept microwave frequency combs*, Proceedings of the National Academy of Sciences, 2018.

[45] Y. Noda, S. Koizumi, Dynamic nuclear polarization apparatus for contrast variation neutron scattering experiments on iMATERIA spectrometer at J-PARC, *Nuclear Instruments and Methods in Physics Research Section A: Accelerators, Spectrometers, Detectors and Associated Equipment* **923** (2019) 127–133.

5
Nuclear Magnetic Resonance and Relaxation

We shall first discuss the origin of the spins and magnetic dipole moments of the nucleons and nuclei (Section 5.1). The nuclear magnetic resonance (NMR) lineshape in solids will then be reviewed in Section 5.2 in general theoretical terms first, before turning to the microscopic sources of line broadening that are valid for solid materials only. Here frequent reference is made to Chapters 1–3. The relaxation mechanisms of nuclear spins will then be described in Section 5.3.

5.1 Nuclear Spins and Magnetic Moments

5.1.1 Nucleon Dipole Moments

A nucleus with a non-zero spin I has a magnetic dipole moment defined by Eq. 1.11

$$\hat{\vec{\mu}} = \gamma \hbar \hat{\mathbf{I}} = g \mu_N \hat{\mathbf{I}}, \tag{5.1}$$

where γ is the nuclear gyromagnetic ratio and

$$\mu_N = \frac{\hbar e}{2 m_p} \tag{5.2}$$

is the nuclear magneton which is defined identically to the Bohr magneton of Eq. 1.9 with proton mass m_p replacing the electron mass. Equation 5.1 also defines the nuclear g-factor whose experimental values for the proton, neutron and heavier nuclei deviate substantially from the Dirac value 2, a fact which suggests that nucleons and nuclei have a rich structure.

The nuclei consist of protons and neutrons, the nucleons, which themselves are the simplest nuclei with spin 1/2 and g-factors of 5.587 and -3.826, respectively. In the naïve non-relativistic quark model the nucleon has three quarks arranged in a symmetric S-state, and the nucleon magnetic moment operator is

$$\hat{\vec{\mu}}_n = \sum_i \frac{g_i q_i}{2 m_i} \hbar \hat{\mathbf{I}}_i, \tag{5.3}$$

where the sum is over the constituent quarks with $i = u, d$ labelling the quark flavors. The proton consists of two u quarks with charge $q_u = 2e/3$ and one d quark with charge $q_d = -e/3$, whereas the neutron is made of one *u* quark and two *d* quarks. Assuming that the quarks are pointlike and have *g*-factors close to 2, and that all quarks have the same mass m_q, the expectation value of Eq. 5.3 yields the magnetic moment of $(m_p/m_q)\mu_N$ and $-2/3(m_p/m_q)\mu_N$ for the proton and the neutron, respectively, using angular momentum algebra. The ratio of these magnetic moments is -1.5, very close to the experimental value -1.46. This was one of the first successes of the simple quark model of hadrons.

Using similar arguments and the magnetic moment of the Λ particle which contains one *s* quark, the naïve quark model also predicts values for the magnetic moments of all baryons containing *u*, *d* and *s* quarks. Such predictions are surprisingly close to the experimental magnetic moments. In view of these and other successes of the model to predict the static properties of the ground states of baryons, it was mysterious that deep inelastic scattering of polarized muons on polarized protons and deuterons yielded spin-dependent structure functions of the nucleons whose integrals could not be understood without evoking models which involve only a small (30%) contribution of the quark spins to the nucleon spin, the rest being contributed by the orbital motion and by the gluons. [1, 2]. The dilemma has been theoretically resolved by taking into account the axial anomaly [3] that adds a gluonic component to the helicity of the proton. Furthermore, following the measurements of the deuteron spin asymmetry and experimental tests of the Bjorken sum rule, the strong coupling constant could be determined with improved accuracy and the nuclear spin decomposition could be theoretically understood [4]. Perturbative quantum chromodynamics (QCD) corrections played an essential role in reconciling the interpretations of the data taken using different polarized targets [5].

Deur, Brodsky and de Téramond review the present theoretical understanding of the nucleon spin in the light of the deep inelastic scattering of polarized muons and electrons on polarized targets [6]. They make the strong point that the nucleon spin provides a critical window for testing QCD, the gauge theory of the strong interactions, since it involves fundamental aspects of hadron structure which can be probed in detail in these experiments. This was one of the main motivations for developing in the 1970s and 1980s the large twin polarized target techniques for the polarized muon beams, and radiation-hard polarized target materials for polarized electron beams.

5.1.2 Nuclear Moments

Magnetic Moment

Table A3.1 shows the spins, magnetic moments and electric quadrupole moments of selected stable nuclei and some radioactive ones. The magnetic moments, defined by Eq. 5.1, are expressed in Table A3.1 scaled by the nuclear magneton μ_N of Eq. 5.2, which is an analogy with the Bohr magneton of Eq. 1.9.

The magnetic dipole operator for a nucleus is a sum of two terms originating from orbiting protons and from the intrinsic spins of the nucleons [7]:

5.1 Nuclear Spins and Magnetic Moments

$$\frac{\hat{\vec{\mu}}}{\mu_N} = \sum_{k=1}^{A} g_L^{(k)} \hat{\mathbf{L}}^{(k)} + \sum_{k=1}^{A} g_I^{(k)} \hat{\mathbf{i}}^{(k)}, \qquad (5.4)$$

where $g_L^{(k)}$ and $g_I^{(k)}$ are known as the orbital and spin g-factors. For protons the orbital g-factor is 1 and for neutrons it is 0, because neutral particles contribute no orbital magnetic moment. The expectation value of the z-component of this operator is the magnetic moment, which can be formally expressed using Glebsch–Gordan coefficients and the reduced matrix element $\langle I \| \hat{\vec{\mu}} \| I \rangle$ describing the dynamics of the nuclear ground state with spin I:

$$\mu = \left(\frac{I}{I+1}\right)^{\frac{1}{2}} \langle I \| \hat{\vec{\mu}} \| I \rangle. \qquad (5.5)$$

The deuteron is experimentally known to have spin 1 and magnetic moment $\mu_d = 0.857438 \mu_N$ [7]. If the two nucleons were bound only by central forces, the ground state of deuteron would be $^3S_1$ with $L=0$ and $S=1$, and the magnetic moment would be the sum of the moments of the proton and the neutron, which is 2.5% higher than the experimental value. Corrections due to a tensor force between the nucleons lead to an admixture of the $^3D_1$ state in the deuteron ground state wave function and the magnetic moment [7]

$$\mu_d = \mu_p + \mu_n - \frac{3}{2}\left(\mu_p + \mu_n + \frac{1}{2}\right) P_D + \Delta\mu_D^{MEC}, \qquad (5.6)$$

where $P_D = 4.81\%$ is the D-state probability and $\Delta\mu_D^{MEC} \approx 0.02 \mu_N$ is a theoretical correction due to meson-exchange current [7].

The ^{3}H and ^{3}He nuclei are isospin mirrors and angular momentum algebra yields the magnetic moments:

$$\begin{aligned} \mu(^3\text{H}) &= \mu^{(0)}(A=3) + \mu^{(1)}(A=3), \\ \mu(^3\text{He}) &= \mu^{(0)}(A=3) - \mu^{(1)}(A=3), \end{aligned} \qquad (5.7)$$

where $\mu^{(0)}(A=3)$ and $\mu^{(1)}(A=3)$ are isoscalar and isovector contributions given by [7]:

$$\begin{aligned} \mu^{(0)}(A=3) &= \frac{1}{2}(\mu_p + \mu_n)(P_S + P_{S'} - P_D) + \frac{1}{2} P_D + \Delta\mu_0^{MEC}, \\ \mu^{(1)}(A=3) &= \frac{1}{2}(\mu_p - \mu_n)\left(P_S - \frac{1}{3}P_{S'} + \frac{1}{3}P_D\right) - \frac{1}{6} P_D + \Delta\mu_1^{MEC}, \end{aligned} \qquad (5.8)$$

and the probabilities P_i are those of the principal S-state $^2S_{1/2}$ (0.897), the mixed-symmetry state $^2S'_{1/2}$ (0.017) and the three $^4D_{1/2}$-states (0.086). The probabilities of the three P-states are small and are ignored. The meson-exchange term is small and positive for the isoscalar case, similar in magnitude to that of the theoretical correction for the deuteron, but it is somewhat more significant for the isovector case. The model gives a rather precise result for the isoscalar moment, but the isovector moment has a larger error due to the fact that the corrections are more dependent of the model.

Among heavier nuclei ^{6}Li has spin 1 and a magnetic moment 4.1% smaller than that of deuteron; the nucleus can be understood as a deuteron loosely bound to a ^{4}He nucleus. Corrections from higher-state admixtures to the ground S-state are therefore small, and the ^{6}Li nucleus can therefore be used in a polarized neutron target with corrections which are not substantially larger than those applied for the deuteron.

Other slightly heavier nuclei that are simple and accurate to describe theoretically belong to the category of odd-mass nuclei; the reason for this is the Pauli exclusion principle, which tends to lead to the pairing of the equal nucleons in light nuclei. For odd-proton nuclei the magnetic moment often is close to [7]

$$\mu = \left(j - \frac{1}{2}\right)\mu_N + \mu_p, \qquad j = l + \frac{1}{2};$$
$$\mu = \frac{j}{j+1}\left[\left(j + \frac{3}{2}\right)\mu_N - \mu_p\right], \qquad j = l - \frac{1}{2};$$
(5.9)

the experimental values of all odd-proton nuclei fall between these two predictions. Examples of the case $j = l + \frac{1}{2}$ are ^{19}F ($l = 0$) and ^{7}Li ($l = 1$) whose magnetic moments are 0.16 and 0.54 nuclear magnetons lower, respectively, than the values predicted by Eq. 5.9. For the case $j = l - \frac{1}{2}$, $l = 0$ a good example is ^{15}N whose magnetic moment is -0.28298 μ_N while the prediction above gives -0.2645 μ_N.

Similarly, the magnetic moment of odd-neutron nuclei fall between the predictions

$$\mu = \mu_n, \qquad j = l + \frac{1}{2};$$
$$\mu = -\frac{j}{j+1}\mu_n, \qquad j = l - \frac{1}{2}.$$
(5.10)

A light nucleus whose magnetic moment falls 0.1 μ_N below the prediction of the second equation 5.10 is ^{13}C, which has $l = 1$, whereas ^{17}O ($l = 2$) has a magnetic moment that is less than 0.1 μ_N above the first of the two equations 5.10.

The magnetic structures of other light nuclei and of all heavier nuclei are more complicated, with the exception of ^{14}N. The shell model of the ^{14}N is a zero-spin ^{12}C core with the remaining proton and neutron in a $P_{1/2}$ state. It can be shown that their spins have probability 1/3 to be oriented parallel to the ^{14}N spin and a probability 2/3 to be oriented antiparallel to it [8]. Thus, the polarization of the ^{14}N spin corresponds to 1/3 of the proton and neutron polarizations in this nucleus. The proton and neutron together are assumed to behave like a deuteron, and therefore produce a scattering asymmetry of

$$A(^{14}\text{N}) = -\frac{1}{3}A(^2\text{H}).$$
(5.11)

Quadrupole Moment

The quadrupole moments of the nuclei are sensitive to the probabilities of the D- and higher states, and theoretical models describe only the lighter nuclei with reasonable accuracy.

Table A3.1 lists the electric quadrupole moments in addition to the spins and magnetic moments of selected stable nuclei and some radioactive ones. The magnetic moments, defined by Eq. 5.1 $\mu = \gamma\hbar I = g\mu_N I$, are expressed in Table A3.1 scaled by the nuclear magneton μ_N of Eq. 5.2, which is an analogy with the Bohr magneton of Eq. 1.9. The magnetic moments of nuclei are obtained mostly from NMR measurements in diamagnetic liquids, where chemical shifts (to be discussed later in this section) up to several hundred ppm put limits to the uncertainty of their value; we note that the zeros in the end of the numbers given are not significant. The gyromagnetic ratios are related with the given experimental magnetic moments by Eq. 2.5:

$$\gamma = \frac{\mu}{\hbar I}. \tag{5.12a}$$

Table A3.1 also gives the NMR frequencies of the free nuclei at 2.5 T magnetic field; this number is practical for finding the approximate frequency in any weakly magnetic material and in any field by

$$f = \frac{B}{2.5\,\text{T}} f_{2.5}. \tag{5.12b}$$

The natural abundances are given with 10 ppm resolution, but the last zeros are not significant.

We emphasize that the resonance frequencies of the nuclei in Table A3.1 refer to free nuclei, a condition which can be met with in atomic beam experiments. The NMR frequency in solids is shifted due to interactions with surrounding electrons and other nuclei, because solids have a high density. The resonance signal will then be shifted, broadened and possibly split depending on the detailed physics and chemistry of the substance. These and the resulting frequency dependence of the transverse RF susceptibility will be discussed in the following sections. The NMR signal measurement techniques, suitable for determining the nuclear polarization in polarized targets, are the subject of Chapter 6.

Nuclei with spin $I \geq 1$ have a quadrupole moment, which is also listed in Table A3.1. The quadrupole frequency and splitting depends on the strength of the electric field gradient tensor at the location of the nucleus. This, in turn, depends on the chemical binding and environment of the atom in the material of interest, so that no universal frequency may be associated with the quadrupole moment. The influence of the quadrupole moment on the NMR absorption spectrum will be discussed with examples in Section 5.2.2.

5.2 Transverse Magnetic Susceptibility and NMR Absorption Signal

5.2.1 NMR Linewidth in Linear Response Theory

The NMR technique of nuclear spin polarization measurement, which will be discussed in Chapter 6, relies on the accurate recording of the absorption lineshape. The lineshape depends on the various interactions of the nuclear moments among themselves, with moments of other spin species, and with electromagnetic fields due to the structure of the solid material. These interactions were already discussed in Chapter 2, and many results which are directly related with lineshapes were derived. In this section we shall discuss the lineshapes themselves in the limit of very low saturation where the linear response theory is valid and where the complications due to the spin dynamics can therefore be avoided. We recall here Eq. 2.140 which gives the condition to the RF field strength for the applicability of the linear response theory

$$\omega_1 = \gamma B_1 \ll \sqrt{\frac{1}{\pi T_2 t_{\text{exp}}}}, \tag{5.13}$$

where $1/T_2 \approx D$ is defined by

$$D^2 = \frac{\text{Tr}\{\mathcal{H}_D^2\}}{\hbar^2 \text{Tr}\{I_z^2\}} \tag{5.14}$$

and is related to the dipolar line width of the absorption signal; t_{exp} is the time during which the susceptibility is measured. In swept-frequency techniques a safe limit is obtained by replacing t_{exp} by the time required for scanning the FWHM of the absorption lineshape when the main cause of line broadening is the dipolar interaction.

If the frequency sweep is produced by a stepped-frequency digital synthesizer, the step interval is the relevant t_{exp}, which must be longer than D^{-1} so that the transverse magnetization has time to stabilize after each step.

The requirement of Eq. 5.13 is most difficult to satisfy for nuclei with large magnetic moment in materials leading to a relatively narrow dipolar linewidth. As an example, protons in normal butanol or ammonia may be assumed to have $D = 2\pi \cdot 30$ kHz, which leads to the requirement

$$B_1 \ll \frac{1}{\gamma}\sqrt{\frac{1}{\pi T_2 t_{\text{exp}}}} \approx \frac{D}{\sqrt{2\pi}\gamma} = 0.3 \text{ mT}. \tag{5.15}$$

This field limit is obtained at 1 mm radius from a wire carrying a current of 1.5 mA, and may be compared with the usual current of about 0.3 mA, which is normally applied in the proton NMR coil in a series tuned Q-meter circuit. It might be of interest to lower the current to 0.1 mA or below if a high precision (e.g. 1%) of the NMR signal shape or integral is needed. The lineshape distortion resulting from not obeying well the requirement of Eq. 5.13 can be obtained from the Provotorov equations, which were discussed in Section 2.4.4.

5.2.2 General Features of the Complex Susceptibility

Let us briefly summarize here the general characteristics of the lineshape functions and the complex transverse susceptibility, which were obtained in Sections 2.4.1 and 2.4.2 under the assumptions of linear response and high temperature for fairly narrow lineshapes. The lineshape functions for the absorption and dispersion were Fourier transforms of the pulse response $G(t)$

$$g(\Delta) = \frac{1}{\pi} \int_0^\infty G(t) \cos \Delta t \, dt \tag{5.16}$$

and

$$g'(\Delta) = \frac{1}{\pi} \int_0^\infty G(t) \sin \Delta t \, dt. \tag{5.17}$$

The integral of the absorption lineshape is normalized to 1

$$\int_{-\infty}^\infty g(\Delta) d\Delta = 1 \tag{5.18}$$

and the absorption lineshape is symmetric in the case of dipolar interactions only at high temperatures.

The derivatives of $G(t)$ immediately after the pulse at $t = 0$ are related with the even moments of the absorption lineshape function

$$M_{2n} = (-1)^n \left(\frac{d^{2n} G(t)}{dt^{2n}} \right)_{t=0} \tag{5.19}$$

and the odd moments are zero if the lineshape is symmetric, which can be shown to be the case for dipolar interactions. For $n = 0$ we get $M_0 = G(0) = 1$, which agrees with the definition of $G(t)$.

The dispersion lineshape can be expressed as a series expansion of $1/\Delta$

$$g'(\Delta) = \frac{1}{\pi \Delta} \sum_{n=0}^\infty \frac{M_{2n}}{\Delta^{2n}}, \tag{5.20}$$

which converges for values of Δ for which the second and higher terms in the sum are much smaller than 1; in practice this requires that $\Delta^2 > 2 M_2$ in the case of dipolar lineshapes. For an infinitely narrow resonance, the dispersion lineshape is

$$g'(\Delta) = \frac{1}{\pi \Delta} M_0 = \frac{1}{\pi \Delta}, \tag{5.21}$$

which agrees with that of Eq. 2.83 obtained directly from the Kramers and Krönig relationship if we recall that the absorption line has antisymmetric components about zero frequency, and that the dispersion part of the susceptibility is related to its lineshape function by

$$\chi'(\Delta) = -\frac{\pi}{2}\omega_0\chi_0 g'(\Delta). \tag{5.22}$$

The absorption part of the susceptibility is similarly related to its lineshape function by Eq. 2.85, which we rewrite here

$$\chi''(\Delta) = \frac{\pi}{2}\omega_0\chi_0 g(\Delta). \tag{5.23}$$

We recall that these formulae assumed the absorption line to be narrow and that the polarization to be low. The deviations due to nearly complete proton polarization, however, are rather small and can be often neglected if the frequency shift due to the internal magnetization is taken into account in the magnetic induction. At high polarization, however, the absorption line narrowing can be substantial, and the lineshape is different from that at low polarization; furthermore, the line often becomes asymmetric. Deviations for lines which are broadened by quadrupole interaction can be quite substantial, in particular at high polarization and at low field.

Equation 5.20 is practical for fitting experimental dispersion signals to theoretical expressions because the drift of the NMR circuit forces to have a reliable model for the signal shape at frequencies far from the Larmor frequency. Because the moments can be determined accurately from the absorption signal, these can be used for the calculation of the dispersion signal shape which enables a fit to be made to eliminate the circuit drift effects. This will be discussed in better detail in Chapter 6.

5.2.3 Broadening by Dipolar Interactions in Solids

Low Polarization

At high temperature the method of Van Vleck can be used for the calculation of the moments of the dipolar lineshape resulting from the truncated Hamiltonian of Eq. 2.100. For like spins the dipole-dipole interaction Hamiltonian was written in the form

$$\mathcal{H}_D = \frac{\mu_0 \hbar^2 \gamma^2}{2}\sum_{j\neq k}\frac{1}{r_{jk}^3}\left(1-3\cos^2\theta\right)\left(3I_{jz}I_{kz}-I_j\cdot I_k\right). \tag{5.24}$$

and the second moment was given by

$$\left\langle\Delta^2\right\rangle_{\parallel} = \frac{1}{3}\mu_0^2\hbar^2\gamma^4 I(I+1)\sum_k b_{jk}^2, \tag{5.25}$$

where the lattice sum is over the square of the function of distances and direction cosines between the nucleus j and its neighbors k

$$b_{jk} = \frac{3}{2}\frac{1-3\cos^2\theta_{jk}}{r_{jk}^3}. \tag{5.26}$$

5.2 Transverse Magnetic Susceptibility and NMR Absorption Signal

The fourth moment for like spins is by a factor of about three times the square of the second moment.

For a powder of cubic crystals the sum can be evaluated yielding the second moment

$$\langle \Delta^2 \rangle_{II} = 5.1 \mu_0^2 \hbar^2 \gamma^4 I(I+1) \frac{1}{d^6} \tag{5.27}$$

where d is the lattice spacing between the like nuclei. This equation is also useful for glassy materials where the lattice spacing must be replaced by the average spacing between the like nuclei $d = n^{1/3}$.

For unlike nuclei which have gyromagnetic ratios γ_I and γ_S, and corresponding Larmor frequencies so far apart that the resonance lines are well resolved, the dipolar Hamiltonian must be rewritten

$$\mathcal{H}_D = \frac{\mu_0 \hbar^2 \gamma_I \gamma_S}{2} \sum_{j \neq k} \frac{1}{r_{jk}^3} (1 - 3\cos^2\theta) \cdot 2 \cdot I_{jz} I_{kz} \tag{5.28}$$

because the term B due to mutual flip-flop transitions does not contribute, as was discussed in Section 2.3.1. The resulting second moment of spin I is 4/9 times that of Eq. 5.25

$$\langle \Delta^2 \rangle_{IS} = \frac{4}{27} \mu_0^2 \hbar^2 \gamma_I^2 \gamma_S^2 S(S+1) \sum_k b_{jk}^2. \tag{5.29}$$

In evaluating the sum for the case where the spins S are those of dilute electronic impurities one must approximate the distribution of distances and direction angles as random, which leads to the approximate expression for the second moment contribution.

High Polarization

The above formulae are only valid at high temperatures, which is not the domain of main interest in polarized targets. Abragam and Goldman have derived general equations for the lineshape functions without resorting to the high-temperature approximation:

$$f(\Delta) = \frac{\omega_1}{4} \tanh \frac{\beta \omega_0}{2} \int_{-\infty}^{+\infty} \left[H(\beta,t) \cos(\Delta t) + F(\beta,t) \sin(\Delta t) \right] dt, \tag{5.30}$$

where the functions H and F represent the real and imaginary parts of the expectation value of the anticommutator of the transverse spin components:

$$H(\beta,t) + iF(\beta,t) = \langle \{I^-, \tilde{I}^+(t)\} \rangle_0, \tag{5.31}$$

where the subscript 0 refers to unperturbed conditions, and

$$\tilde{I}^+ = e^{+i\mathcal{H}_D t} I^+ e^{-i\mathcal{H}_D t}. \tag{5.32}$$

The function $H(\beta,t)$ is even in β and t, and $F(\beta,t)$ is odd in β and t. Analytical expressions cannot be obtained for the general lineshape functions directly from these equations, but they can be used for numerical evaluations in principle. Some simple results can be obtained, however, directly from the symmetry of the functions. Notably, this leads to

$$f(\Delta,\beta) = -f(-\Delta,-\beta) \tag{5.33}$$

i.e. by inversing the spin temperature the lineshape function is antisymmetric with respect to the Larmor frequency at low polarization.

Moments of Dipolar Lineshapes

Abragam and Goldman [9] derived general expressions for the moments of dipolar lineshapes at high polarization. The zeroth moment is

$$M_0 = \int_{-\infty}^{+\infty} f(\Delta) d\Delta = \frac{\pi}{2}\omega_1 \langle [I_-, I_+] \rangle_0 = -\pi\omega_1 \langle I_z \rangle_0, \tag{5.34}$$

which is a more general proof of the linear relationship between the absorption lineshape function and the polarization, valid also at low spin temperatures only in the case of dipolar interactions.

We quote here similarly the results of Abragam and Goldman [9] for the higher moments. The nth moment, if defined as

$$M_n = \frac{(-1)^n \int_{-\infty}^{+\infty} \Delta^n f(\Delta) d\Delta}{\int_{-\infty}^{+\infty} f(\Delta) d\Delta}, \tag{5.35}$$

leads to the first moment

$$M_1 = \frac{\langle [\mathcal{H}'_D, I_+] I_- \rangle_0}{\langle I_+ I_- \rangle_0}, \tag{5.36}$$

which can be evaluated similarly to Van Vleck's method of Section 2.4.3 for spin i

$$M_1(P) = \frac{3}{2} a_i I P. \tag{5.37}$$

Here

$$a_i = \frac{\mu_0}{4\pi} \hbar \gamma_I^2 \sum_j \frac{1 - 3\cos^2\theta_{ij}}{r_{ij}^2} \tag{5.38}$$

is the Weiss field factor for spin i which may vary depending on its location in the sample. For a spherical sample of spins in crystallographically equivalent positions, all a_i are equal. If this is not the case, average must be taken over all positions i, which can be carried out by dividing the sample into two regions. The inner region is a sphere centered at i of diameter much larger than the lattice spacing, in which the Weiss field can be summed as above, and the outer region limited by the surface of the inner sphere and by the sample boundaries, over which an integral can be taken disregarding the crystal structure. The procedure is rather tedious, and because polarized targets are usually composed of small spherical beads, we need here only the Weiss field result. For targets made of irregularly shaped pieces, averaging must be made also over the various possible shapes; the result should intuitively converge towards that of a sphere if the NMR coil is in contact of a sufficiently large number of such pieces and they are randomly oriented.

The factor 3/2 in the first moment comes from the effect of the resonant transverse field of the identical spins and must be removed when calculating the Weiss field effect of one spin species to another.

The polarization dependencies of the second and higher moments for spin 1/2 are [9]

$$M_2(P) = M_2(0)(1-P^2) \tag{5.39}$$

in the case of spherical sample and simple cubic lattice with no other spins contributing. It should be noted that the centroid for the second moment is taken to be the point about which the first moment is zero, and not the Larmor frequency at low polarization. In glassy materials the second moment cannot be expected to go to zero at $P^2 = 1$, but a significant narrowing and shape difference of the proton absorption line is always observed.

In polarized targets, however, the assumption of a single spin species is never the case because electronic spins are needed. The additive contribution of the electronic spins must therefore be calculated separately; this results in a non-zero residual width at $P^2 = 1$.

For the conditions of simple cubic symmetry and spherical sample, the third and fourth moments of a single spin 1/2 system read [10]

$$M_3(P) \cong -0.39 \left[M_2(0) \right]^{3/2} (1-P^2) P; \tag{5.40}$$

$$M_4(P) \cong 2.18 \left[M_2(0) \right]^2 (1-P^2)(1-0.42P^2). \tag{5.41}$$

The non-zero values of the third moment for $0 < P^2 < 1$ shows that the absorption line is asymmetric; the asymmetry is likely to be highest around $P^2 = 1/3$ where the third moment has maximum value.

The ratio

$$\frac{M_4(P)}{M_2^2(P)} \cong 2.18 \frac{1-0.42P^2}{1-P^2} \tag{5.42}$$

diverges at $P^2 = 1$, which indicates that the lineshape approaches the Lorentzian shape. This can be understood by considering almost all spins pointing in the same direction except for a few randomly placed spins in opposite orientation. This random array should give a Lorentzian lineshape with cut-offs due to the nearest approach limited by the lattice spacing, similar to the shape obtained for highly diluted spins [11].

In real target materials, the contributions of all spins must be summed for the first and higher moments. This leads to much less pronounced effects of nuclear polarization on the moments, because the electron spin polarization is almost complete and therefore constant. The following qualitative changes, however, can be observed:

(1) The shift of the centroid of all lines is proportional to P at low polarization; at high polarization the shifts scale with a weighted average of the polarizations of the different nuclei.
(2) All lines become narrower; the narrowing scales with the root of the second moment at low polarization where the shape is close to a Gaussian, whereas at high polarization the Lorentzian-shaped lines have a narrowing which scales with $1 - |P|$.
(3) The absorption lines become asymmetric at high polarization. In glassy materials, however, the dipolar line broadening is to an increasing extent due to the random orientations of the direction vectors of the spin pairs, and therefore the line asymmetry is not very pronounced and may be hidden by the asymmetry due to the circuit used for the NMR signal measurement.

5.2.4 Quadrupolar Interactions and Lineshapes

In the following we shall discuss the lineshape resulting from the Hamiltonian of Eq. 2.19, rewritten here explicitly in the coordinate system where the electric field gradient tensor is diagonal

$$\mathcal{H}_Q = \hbar\omega_Q \left[(3c^2 - 1)(3I_z^2 - I(I+1)) + \frac{\eta}{2}(I_+^2 + I_-^2) \right], \quad (5.43)$$

where $c = \cos\theta$. In the case of spin 1 and $\eta = 0$, this was transformed to the coordinate system $0x'y'z'$ in which the magnetic field is along z'

$$\mathcal{H}_Q = \hbar\omega_Q \left[(3c^2 - 1)(3I_{z'}^2 - 2) + 3c\sqrt{1-c^2}\{I_{z'},(I_+ + I_-)\} + \frac{\sqrt{1-c^2}}{2}(I_+^2 + I_-^2) \right]. \quad (5.44)$$

The asymmetry parameter η is rather small in the presently known target materials and its influence on the lineshape will be only briefly discussed below.

In high magnetic field, the quadrupole shift of the energy levels can be obtained precisely from first-order perturbation expansion, allowing one to obtain the powder lineshape functions in analytical form. This will be discussed in detail for spin $I = 1$ below. We shall then describe how non-zero asymmetry parameter changes the lineshape in high field. Finally we shall show that when the quadrupole energy is not very small compared with the Zeeman energy, the lineshapes become distorted so that only numerically simulated results are available; recent work on ^{14}N in NH_3 will be reviewed below to illustrate such methods.

High Field

Let us assume that the magnetic field is so high that the second-order terms in Eq. 2.22 are much smaller than the first-order terms of Eq. 2.21, which is sufficiently well satisfied (in comparison with the accuracy of signal measurement) when

$$\varepsilon = \frac{\omega_Q}{\omega_L} < 10^{-2}. \tag{5.45}$$

This ratio is about 0.13×10^{-2} at 2.5 T field for deuterons covalently bonded to carbon chains in deuterated hydrocarbons. In such cases the use of first-order theory is well justified, and the relatively small linewidth also allows to measure the absorption signal at constant magnetic field using the frequency-sweep method. The case where $\varepsilon > 10^{-2}$ will be discussed later in this section, with an example of ^{14}N in NH$_3$ with $\varepsilon = 0.0514$.

The first-order energy levels for spin 1 are, from Eqs. 2.20 and 2.21

$$E_m = -\hbar\omega_L m + \hbar\omega_Q(3\cos^2\theta - 1)(3m^2 - 2); \tag{5.46}$$

here m are the magnetic quantum numbers of the unperturbed Hamiltonian, and the shifted energy levels are shown in Figure 5.1 for $\cos^2\theta = 1$ and $\cos^2\theta = 0$. The angular frequency ω_Q is defined by Eq. 2.23 as

$$\omega_Q = \frac{e^2 qQ}{8\hbar}. \tag{5.47}$$

The energy differences for the transitions between the two lower levels ($m = 1, 0$) and the two higher levels ($m = 0, -1$), labelled by subscripts + and –, respectively, are

$$\Delta E_\pm = \hbar\left[\omega_L \mp \omega_Q(3\cos^2\theta - 1)\right]. \tag{5.48}$$

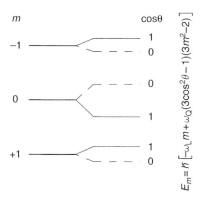

Figure 5.1 The magnetic energy level shifts for spin $I = 1$ by quadrupole interaction, for two polar angles producing the extreme values of the frequency spectra of the two magnetic transitions. The illustration is schematic

The frequency spectra for the two transitions are obtained analogously to Eqs. 3.7–3.9 by considering the uniform distribution over the solid angle $\Omega = 2\pi$ of the upper-hemisphere polar angles θ between the direction of the electric field gradient and the magnetic field vectors. Only positive values of $\cos\theta$ need to be taken because of the symmetry properties of the quadrupole Hamiltonian. The frequency spectrum is therefore

$$f(\omega)d\omega = \frac{d\Omega}{2\pi} = \frac{1}{2\pi}2\pi\sin\theta\,d\theta = d(\cos\theta). \qquad (5.49)$$

This gives the number density of nuclei within the frequency interval $d\omega$; the relation between the polar angle and the angular frequency is obtained directly from Eq. 5.48

$$\Delta E_\pm = \hbar\omega_\pm = \hbar\left[\omega_L \mp \omega_Q(3\cos^2\theta - 1)\right], \qquad (5.50)$$

which yields

$$\cos\theta = \sqrt{\frac{1 \mp x}{3}} \qquad (5.51)$$

where

$$x = \frac{\omega - \omega_L}{3\omega_Q}. \qquad (5.52)$$

Inserting this into Eq. 5.49 results in

$$f_\pm(\omega) = \frac{d}{d\omega}\cos\theta = \frac{dx}{d\omega}\frac{d}{dx}\cos\theta = \frac{1}{3\sqrt{3\omega_Q}}\sqrt{\frac{1}{1 \mp x}} \qquad (5.53)$$

for the two transitions. These two shape functions are shown in Figure 5.2, together with their sum broadened by Gaussians with two values of σ.

In the foregoing the lineshape function was obtained without resorting to formal expressions of susceptibility, thus ignoring any spin-spin interactions. This approach of can be justified when the spread of the resonance frequencies due to the quadrupole interactions is much larger than that due to the dipolar interactions. A system of spins obeying such a condition behaves under strong RF field in a different way compared with a system with dominant homogeneous broadening, as was already discussed in Chapter 4. In particular, the absorption spectrum is obtained by summing over the absorption spectra of the 'spin packages', even at low temperatures. In doing so care must be taken to use the right population ratios between the magnetic levels. This ratio

$$r(\theta) = \frac{p_m}{p_{m-1}} = e^{-(E_m - E_{m-1})/kT} \qquad (5.54)$$

is not independent of the polar angle θ because the energy differences, given by Eq. 5.50, depend on it. The ratio, however, is constant to a good approximation if the maximum variation of the exponent above is much less than 1, i.e.

5.2 Transverse Magnetic Susceptibility and NMR Absorption Signal

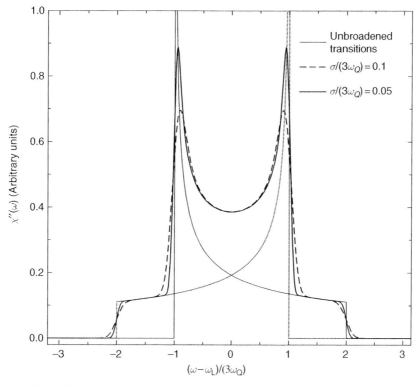

Figure 5.2 Absorption signal shapes for spin $I = 1$ with quadrupole broadening of the two magnetic transitions

$$\left|\frac{3\hbar\omega_Q}{kT}\right| \ll 1. \tag{5.55}$$

Taking the value $3\omega_Q = 2\pi \cdot 65$ kHz (C–D bond in butanol) the requirement becomes

$$|T| \gg 3.1\,\mu\text{K}. \tag{5.56}$$

It should be noted that this requirement is independent of the value of the magnetic field, and that it is easily satisfied in polarized targets where spin temperatures may reach values not much less than ±1 mK. Lower spin temperatures may be obtained by demagnetizing to a low field, but then there will be additional complications in the lineshape arising from the low field value itself. These will be discussed later in this Section.

If the condition above holds, the absorption signal arising from the two transitions can be described as a superposition of the two functions 5.53, weighted by the population ratio r which is assumed to be constant:

$$f(x) = \frac{r-1}{1+r+r^2}\{rf_+(x) + f_-(x)\} = \frac{1}{3\sqrt{3}\omega_Q}\frac{r-1}{1+r+r^2}\left\{r\sqrt{\frac{1}{1-x}} + \sqrt{\frac{1}{1+x}}\right\}. \tag{5.57}$$

The multiplier of the shape function which has a pole at $x = 1$ is thus growing faster with inverse spin temperature than the other, producing an asymmetric lineshape. The multiplier of the function with a pole at $x = -1$ reaches a maximum at $r = 1 + \sqrt{3}$ and then decreases to zero at $P = 1$. At negative spin temperatures the ratio r is smaller than 1, and the asymmetry is reversed. If the asymmetry is determined precisely, for example, by fitting a model shape to the NMR signal data with r as one of the free parameters, polarization can be determined from this by

$$P = \frac{r^2 - 1}{1 + r + r^2}, \tag{5.58}$$

provided that the polarization sampled by the NMR probe is very uniform, and that the experimental signal distortions caused by the resonant circuit are corrected for, either by including the circuit effects in the model of the signal shape, or by separate circuit modelling. If the polarization is not uniform or there are no strong reasons to believe so, the variation of the polarization can be estimated by comparing r_p calculated from P obtained from the integral of the NMR signal, with r determined from the signal asymmetry. This will be discussed in more detail in Chapter 6.

If the quadrupole energy and therefore the angular frequency of Eq. 5.47 has positive sign, the pole at x is at the high-frequency side of the centroid of the combined lineshape function of Eq. 5.57. This allows one to determine the sign of qQ from the polarized deuteron signal asymmetry. If the quadrupole moment and its sign are known, the polarized NMR signal gives the sign and magnitude of the electric field gradient tensor. The method has been used more often in the opposite sense, to determine signs of quadrupole moments of nuclei from signals arising from compounds where the electric field gradient is known.

When the asymmetry parameter η is non-zero, the derivation of the lineshape function becomes tedious, and we shall quote the results only here. The first-order energy levels for spin 1 are

$$E_m = -\hbar\omega_L m + \hbar\omega_Q \left(3\cos^2\theta - 1 + \eta \sin^2\theta \cos 2\varphi\right)\left(3m^2 - 2\right), \tag{5.59}$$

where θ and φ are the Euler angles defining the relative orientations of the magnetic field and the principal axis of the electric field gradient tensor. The shape function can now be expressed as

$$f(z,\eta) = \frac{1}{\pi\sqrt{\eta(1-z)}} K(y) \quad \text{for} \quad -\frac{\eta}{2} - \frac{1}{2} \leq z \leq \frac{\eta}{2} - \frac{1}{2};$$

$$f(z,\eta) = \frac{2}{\pi\sqrt{(3-\eta)(\eta+1-2z)}} K(y^{-1}) \quad \text{for} \quad \frac{\eta}{2} - \frac{1}{2} < z \leq 1; \tag{5.60}$$

where

$$y = \frac{(3-\eta)(\eta+1+2z)}{4\eta(1-z)} \tag{5.61}$$

and K is the complete elliptic integral of first kind defined by

$$K(y) = \int_0^{\pi/2} \frac{1}{\sqrt{1-y\sin^2\theta}} d\theta. \tag{5.62}$$

The function K is tabulated in most mathematical tables of functions; for numerical evaluations it can be calculated to a precision of 3×10^{-5} from the approximation [12]

$$K(y) = a_0 + a_1(1-y) + a_2(1-y^2) - \left[b_0 + b_1(1-y) + b_2(1-y^2)\right]\ln(1-y), \tag{5.63}$$

where

$$\begin{aligned} a_0 &= 1.3862944 \\ a_1 &= 0.1119723 \\ a_2 &= 0.0725296 \\ b_0 &= 0.5 \\ b_1 &= 0.1213478 \\ b_2 &= 0.0288729 \end{aligned} \tag{5.64}$$

This approximate function has been used successfully for fitting model lineshapes with experimental data of deuteron NMR signals [13].

Low-Field or Large Quadrupole Interaction

At zero field the Zeeman term vanishes in Eq. 5.46 and the energy eigenvalues are in the case of axially symmetric electric field gradient tensor ($\eta = 0$)

$$E_m = \frac{e^2qQ}{4I(2I-1)}\left(3m^2 - I(I+1)\right). \tag{5.65}$$

The states $m = \pm n$ are thus degenerate, and there is only one resonance frequency for spin 1

$$\omega_q(1) = \frac{3}{4}\frac{e^2qQ}{\hbar}. \tag{5.66}$$

This resonance can be seen using a coil in a circuit identical to that used for NMR measurements. Although the term 'quadrupole resonance' is often used for the observable signal, the transition is due to the magnetic interaction of the dipole moment of the nucleus with the RF magnetic field.

At zero external field, the spin polarization is zero, because there is no direction in space on which the spin vector has a non-zero projection. At low temperatures and zero field, however, a large resonance signal may be observed, due to the uneven populations of the quadrupole energy levels; these populations obey Boltzmann statistics. This clearly violates the statement that the integral of the NMR absorption signal is proportional to the vector polarization and indicates that at low field values there may also be an error in such a statement. We shall therefore discuss below the quadrupole broadened absorption signal at such field values where the inequality 5.45 is not well satisfied. This situation arises, for example, for ^{14}N in NH$_3$ at 2.5 T field, where $\varepsilon = 5.14 \times 10^{-2}$, and therefore clearly observable effects can be seen in the population of the magnetic states and in the relation between the population ratios and the vector polarization. We should also note that the quadrupole broadened absorption line spans the frequency range of $6\omega_Q$, which is equal to the Larmor frequency when $\varepsilon = 1/6$. Such broad lines cannot be measured at constant field, which complicates the analysis of experimental signals measured piecewise at different field values SMC [8]. This adds to the difficulty that the amplitudes of such broad lines are extremely small and can be measured only at relatively high polarization.

At intermediate fields between zero and high values, the projections of the spin vectors on the axis along the magnetic field become gradually larger with the field, and the state vectors gradually change from completely mixed states towards pure magnetic states. The state vectors $|m\rangle$ are obtained by solving the energy eigenvalues directly from the Schrödinger equation

$$\mathcal{H}|m\rangle = E_m|m\rangle, \tag{5.67}$$

where the Hamiltonian is the sum of the Zeeman term and the quadrupole term of Eq. 5.44:

$$\mathcal{H} = \mathcal{H}_Z + \mathcal{H}_Q = \hbar\omega_L I_{z'} + \mathcal{H}_Q. \tag{5.68}$$

The explicit matrix representation of the Hamiltonian is obtained by inserting in Eq. 5.44 the matrix representations of the components of the spin 1 vector given by Eq. 1.32. This yields

$$\frac{\mathcal{H}}{\hbar\omega_L} = I_{z'} + \frac{\mathcal{H}_Q}{\hbar\omega_L} = \begin{pmatrix} 1 & 0 & 0 \\ 0 & 0 & 0 \\ 0 & 0 & -1 \end{pmatrix} + \begin{pmatrix} A & B & C \\ B & -2A & -B \\ C & -B & A \end{pmatrix}, \tag{5.69}$$

where

$$\begin{aligned} A &= \varepsilon(3c^2 - 1); \\ B &= 3\varepsilon c\sqrt{2(1-c^2)}; \\ C &= 3\varepsilon(1-c^2) \end{aligned} \tag{5.70}$$

and

$$\varepsilon = \frac{\omega_Q}{\omega_L}; \quad c = \cos\theta. \tag{5.70a}$$

The eigenvalues are the solutions of

$$\det(\mathcal{H} - E_m\mathbf{1}) = 0 = -\hbar\omega_L \begin{vmatrix} 1+A-e_m & B & C \\ B & -2A-e_m & -B \\ C & -B & -1+A-e_m \end{vmatrix}, \tag{5.71}$$

where

$$e_m = -\frac{E_m}{\hbar\omega_L}. \tag{5.72}$$

The roots of Eq. 5.71 are the solutions of the third-order polynomial

$$e_m^3 - e_m(1 + 12\varepsilon^2) - 2\varepsilon(3c^2 - 1) + 16\varepsilon^3 = 0. \tag{5.73}$$

It can be seen immediately that if $\varepsilon = 0$ the roots are ± 1 and 0, as expected for the pure Zeeman states. Figure 5.3 shows the polynomial plotted for three values of ε and for the values 0, 0.5 and 1 of c. There are three roots for each case and it can be seen that the largest value of ε requires treatment beyond the first-order perturbation method, because of the large asymmetry of the polynomial. The first- and second-order energy levels were already given in Section 2.1.2, Eqs. 2.21 and 2.22, and we shall examine here the accuracy of the

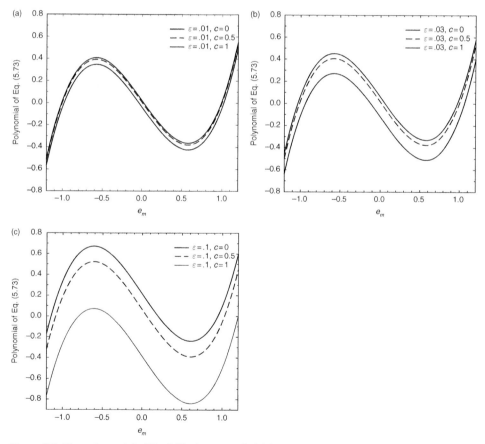

Figure 5.3 The polynomial of Eq. 5.73, the roots of which are the solutions of the eigenvalue equation of Eq. 5.71. The polynomial is plotted for (a) $\varepsilon = 0.01$, (b) $\varepsilon = 0.03$, (c) $\varepsilon = 0.1$, for three values of $\cos\theta$

second-order treatment by comparing with the exact solution in the case of ^{14}N in solid NH$_3$ in a 2.5 T field, in which case we have $\varepsilon = 0.0514$.

The general solutions of the third-order polynomial 5.73 are

$$e_m = 2\sqrt{\frac{1+12\varepsilon^2}{3}}\cos\left[\frac{\phi}{3} + \frac{\pi}{3}(4 - m - 3m^2)\right], \qquad (5.74)$$

where m has the values of -1, 0 and $+1$, and

$$\phi = \arccos\left[\frac{3\varepsilon(3c^2 - 1 - 8\varepsilon^2)}{1+12\varepsilon^2}\sqrt{\frac{3}{1+12\varepsilon^2}}\right]. \qquad (5.75)$$

These energies are plotted in Figure 5.4a as a function of $c = \cos\theta$ for ^{14}N in solid NH$_3$ in a 2.5 T field; in this scale the energies obtained from the first- and second-order perturbation

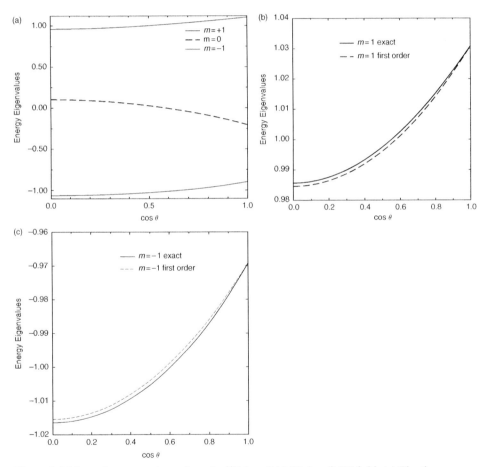

Figure 5.4 Magnetic energy eigenvalues for ^{14}N in solid NH$_3$ in a 2.5 T field. (a) The three energy eigenvalues of Eq. 5.74 as a function of $\cos\theta$. (b) and (c) Comparison of the first-order and exact energy eigenvalues for the states $m = +1$ and $m = -1$

treatment are indistinguishable. The plots of Figures 5.4b and c zoom up on the energy level differences between the exact solution and the first-order treatment in the case of the magnetic states $m = -1$ and $m = +1$; the energy from the first-order treatment of $m = 0$ practically coincides with the exact solution and is not shown. Figure 5.4 suggests that the first-order expressions for the NMR signal are likely to yield vector polarizations that may have a few percent systematic errors.

The second-order perturbation theory gives energy levels which deviate only slightly from the exact solution as shown by Figure 5.5. The deviation is zero at $\cos\theta = 0$ and at $\cos\theta = 1$, and it peaks just above $\cos\theta = 0.8$ for all three magnetic states. Therefore, for the practical purposes, the second-order expression is sufficiently accurate for the numeric evaluation of the NMR signal shape and its relation with the vector polarization.

The exact solution, however, is equally easily amenable to the numerical calculation of lineshapes and vector polarization as a function of the spin temperature. These calculations require the knowledge of the state vectors which are obtained from Eq. 5.67 for each energy eigenvalue. The explicit matrix equation is

$$\mathcal{H}|m\rangle = \hbar\omega_L \begin{pmatrix} 1+A & B & C \\ B & -2A & -B \\ C & -B & -1+A \end{pmatrix} \begin{pmatrix} a_m \\ b_m \\ c_m \end{pmatrix} = E_m|m\rangle = \hbar\omega_L e_m \begin{pmatrix} a_m \\ b_m \\ c_m \end{pmatrix}. \quad (5.76)$$

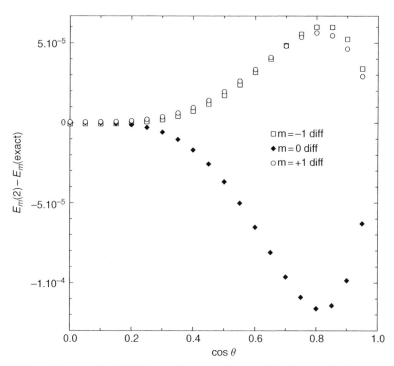

Figure 5.5 Magnetic energy level differences between the second-order perturbation expression and the exact solution, for ^{14}N in solid NH_3 in a 2.5 T field

This set of three linear equations has the solutions

$$|m\rangle = \begin{pmatrix} a_m \\ b_m \\ c_m \end{pmatrix} = N \begin{pmatrix} (1+A+C-e_m)^{-1} \\ B\left[C(2A+e_m)-B^2\right]^{-1} \\ (1-A-C+e_m)^{-1} \end{pmatrix}. \qquad (5.77)$$

The normalizing constant N is found by requiring that $\langle m|m\rangle = 1$, because the set of equations give only two of the ratios of the components b_m/a_m and c_m/b_m, which define a_m/c_m. The signs are known therefore relative to each other and one must resort to the original equations to fix the sign of the major component. This is best done by writing the state vector in the form

$$|m\rangle = \begin{pmatrix} a_m \\ b_m \\ c_m \end{pmatrix} = N' \begin{pmatrix} \sqrt{1-c^2}\left[1+e_m-2\varepsilon\right](e_m-2\varepsilon) \\ c\sqrt{2}\left[1-(e_m-2\varepsilon)^2\right] \\ \sqrt{1-c^2}\left[1-(e_m-2\varepsilon)\right](e_m-2\varepsilon) \end{pmatrix} \times \text{sign}(m) \qquad (5.78)$$

using relations 5.70. This form is practical for numeric computations because it avoids divergences of normalization at all values of c except $m = 0$, $c = 0$ and $m = \pm 1$, $c = 1$. The divergence must be properly accounted for in numeric calculations. Another method, which is more correct theoretically but somewhat cumbersome in programming, is to fix the value of the leading component to 1 and determine the remaining two components from their ratios with the fixed one; these ratios do not diverge at any values of m and c.

The components of the state vectors of Eq. 5.78 are needed for the calculation of the NMR signal shapes as a function of spin temperature.

The vector polarization as a function of the spin temperature can be calculated by using the density matrix techniques of Section 1.22, with the density matrix of Eq. 1.81 having the diagonal elements of Eq. 1.83:

$$P_I = \frac{\langle I_z \rangle}{I} = \frac{\sum_{i=-1}^{+1} m_i e^{-\frac{\langle m|\mathcal{H}|m\rangle}{k_B T}}}{\sum_{i=-1}^{+1} e^{-\frac{\langle m|\mathcal{H}|m\rangle}{k_B T}}}, \qquad (5.79)$$

and by integrating over the polar angle θ. Here $m_{-1} = c_{-1}$, $m_0 = b_0$, $m_1 = a_1$ are obtained from the components of the state vector $|m\rangle$ of Eq. 5.78. For ^{14}N in NH_3 in 2.5 T field the expression of Eq. 5.79 deviates by no more than a few % from the simple integral of the first-order energy levels as a function of θ. Numeric integration of the exact simulated Eq. 5.79 was compared with the first-order signal expression and it was found that at low polarization the integrated experimental signal underestimates the polarization more for negative than positive P, while at 50% polarization of ^{14}N the integrated NMR signal overestimates the polarization [14]. The study also confirms that the second-order perturbation calculates the energy levels and state vectors almost identically with the exact expressions, and because these are simpler to use, there is no need to apply the exact solution for the correction of the integrated absorption signal when determining the polarization of ^{14}N.

The SMC collaboration performed a study on the contribution of the ^{14}N polarization on the longitudinal muon-proton deep inelastic scattering asymmetry from their polarized NH$_3$ target [15]. Using cross-calibration techniques at a constant frequency of 6.47 MHz, they established that the equilibrium polarizations of ^{14}N and protons corresponded to equal spin temperatures over a broad range of polarizations, and that the thermal equilibrium was established with a time constant of about 25 min when microwaves were turned on for DNP. For the cross-calibration techniques corrections due to Eqs. 5.74 and 5.78 played a minor role because the nitrogen NMR signal peak measurements were made at 2.45 and 1.68 T fields. Because the proton signal measurement was performed at 0.15 T, care had to be taken to avoid thermal mixing and superradiance during the field ramps down and up.

Quadrupole Coupling Constants for Common Polarized Target Materials

The NMR absorption signal shapes that are quadrupole broadened are similar to the smooth curves of Figure 5.2 in the case of glassy deuterated solids and in microcrystalline ammonia obtained by rapid freezing. In slowly frozen and crushed ammonia, the crystals are large and therefore the NMR probe coil in contact with some tens of crystals may sense the material with higher sensitivity than the rest of it. This will result in a signal that looks rough because some polar angles θ are favored over other ones. In such a case, the NMR signal will be difficult to fit with theoretical expressions based on the uniform distribution of the polar angles.

Table 5.1 shows the quadrupole coupling parameters for deuterons in butanol and ammonia and for ^{14}N in ammonia. ^{6}Li is the only other spin-1 nucleus as shown in Table 5.1. In

Table 5.1 Total spectral width $12\omega_Q/2\pi$, dipolar broadening σ, and asymmetry parameter η for some solid polarized target materials at low temperature and 2.5 T field. Measurement at zero field for ^{14}N in NH$_3$ gives an extrapolated value of $12\omega_Q/2\pi = 4747.5 \pm 4.5$ kHz [16] at $T = 0$ K.

Material (chem. bond)	$12\omega_Q/2\pi$ (kHz)	σ (kHz)	η	Reference
ND$_3$ (N–D)	335.6	19.4	0.1263	[13]
NH$_3$ (H–^{14}N)	4744.0		0	[16]
d-Butanol (C–D)	258.96	4.0	0	[17]
d-Butanol (O–D)	319.2	4.0	0.15	[17]
D6-propanediol (C–D)	234			

^{6}LiD, however, the structure has a cubic symmetry that has zero electric field gradient so that both nuclei show no quadrupole broadening in the NMR signal. For half-integer spin nuclei 3/2 and higher, the quadrupole broadening is complex.

5.2.5 Molecular Spin Isomers and NMR Line Due to Indirect Spin Interactions

Nuclear spin influences the electron spin wave function and may produce effects which range from dramatic symmetry properties to small shifts in the effective field value at the nuclear sites. The symmetry effects result in molecular energy levels which depend on the total nuclear spin and rotational quantum numbers; of these hydrogen (H_2) and methane (CH_4) are good examples. Other materials where this happens include water (H_2O) and acetylene (C_2H_2). The exclusion principle and symmetry laws determine the ground state total nuclear spin, which is $I = 0$ in H_2, H_2O and C_2H_2, and $I = 2$ in CH_4, and in all these cases there is good experimental evidence that conversion to the ground state takes place in the molecular crystals of these materials [18]. It can then be concluded that only solid methane is a potentially good material for polarized targets. A more detailed discussion of the symmetry properties of molecules and brute-force polarizable materials will be given in Chapter 10.

In the molecular crystals discussed above, the binding forces between the molecules are weak, and therefore they preserve their rotational freedom even at low temperatures. However, if these molecules are strongly bound by electrostatic forces in a solid, the protons may regain their individuality. A clear example of this is water in an ionic crystal such as the lanthanum magnesium nitrate $La_2Mg_3(NO_3)_{12} \cdot 24H_2O$ (LMN), which has been used as a polarized proton target in the past. Another example is water in glassy solids such as 1-butanol with 5% H_2O. The absence of spin conversion to the ground state is easily verified by following the evolution of the NMR lineshape and the growth of the thermal equilibrium NMR signal integral at constant temperature. The absence of molecular rotation is also evident if the proton spin-lattice relaxation time becomes very long at low temperatures, because in the case that J is a good quantum number, the molecules in the $J = 1$, $I = 1$ state provide a relaxation mechanism which is much faster than that due to the paramagnetic impurities at low temperatures and high field.

When a molecule is strongly bound by electrostatic forces in the lattice of a crystal or in a glass matrix, the positions and axes of the molecule are well defined. Then the rotational moment, which is a conjugate variable to the orientation of the axes, is not defined at all because of the uncertainty principle. Consequently J is not a good quantum number and there is no correlation between the total nuclear spin value and the quantum state of the molecule [18]. This is very important for the measurement of polarization based on the calibration of the absorption signal integral at a known temperature, because the method relies on the knowledge of the thermal equilibrium polarization based on individually behaving spins.

Hydrogen

A free molecule of H_2 has a total wave function which is antisymmetric under exchange of the proton variables:

$$\psi(1,2) = -\psi(2,1), \tag{5.80}$$

because the proton is a fermion. It is important to note that this can only hold if the molecule does not experience external binding forces which are comparable with respect to the internal forces of the molecule. The total wave function in these conditions can be approximated by the factored form

$$\psi = \psi_e \psi_v \psi_r \psi_s, \tag{5.81}$$

where ψ_e, ψ_v, ψ_r and ψ_s are the electronic, vibrational, rotational and nuclear spin wave functions, respectively. The factorization is a valid approximation when the electronic and vibrational energy levels are very high with respect to the rotational and nuclear spin levels. The molecule is then in the ground state with respect to these at low temperatures. As the electronic and rotational wave functions are symmetric under exchange of the protons, we must have

$$\psi_r(1,2) \psi_s(1,2) = -\psi_r(2,1) \psi_s(2,1). \tag{5.82}$$

The rotational energy levels are

$$E_J = BJ(J+1), \tag{5.83}$$

where J is the rotational quantum number and $B/k = 86\,K$. For even values of J the rotational part of the wave function is symmetric, and for odd values it is antisymmetric:

$$\psi_r(1,2) = (-1)^J \psi_r(2,1). \tag{5.84}$$

From Eqs. 5.82 and 5.84 it follows that ψ_s must be antisymmetric for even values of J, and it must be symmetric for odd values of J. As the two proton spins add either to a total nuclear spin $I = 0$ or $I = 1$ and these give antisymmetric and symmetric ψ_s, respectively, it follows directly that for even J we must have $I = 0$ and for odd J we have $I = 1$.

Molecules with $I = 0$ are called para-hydrogen and those with $I = 1$ ortho-hydrogen. The ground state of the molecule is the para state with $J = 0$ and $I = 0$, which cannot be polarized because the total nuclear spin is zero. Labelling the energy levels by $E(J,I)$ we have then the energies of the four lowest rotational levels from Eq. 5.83: $E(0,0) = 0\,K$, $E(1,1) = 0\,172\,K$, $E(2,0) = 516\,K$ and $E(3,1) = 1,032\,K$.

In the gaseous state, after preparation by chemical means, the statistical weights are equal to $2I + 1$, which yields 3/4 of ortho-hydrogen molecules and 1/4 of para-hydrogen molecules. Collisions that change J between even and odd values are very rare because they involve the simultaneous change of I, so that the two molecular species behave as a mixture of two non-converting gases. The conversion between the two species can take place in contact with walls, and this is speeded up by magnetic materials.

Hydrogen solidifies at about 14 K where practically all para molecules are in the state $J = 0$ and all ortho molecules are in the state $J = 1$, because the higher rotational states are much above the two lowest states. Because the binding forces between the molecules are weak, J remains a good enough quantum number which allows to distinguish the two molecular isomers. This is not generally true for other solids where binding forces in solid state are larger and the excited molecular levels are not much above the rotational levels.

The ortho molecules, however, convert slowly in solid into the para form at a speed which depends on their molar fraction:

$$\dot{c} = -3c^2 \frac{1}{\text{hour}}. \tag{5.85}$$

The original $c = 0.75$ becomes 0.40 in roughly two days in a very pure sample. The conversion releases heat per unit mass by

$$\frac{\dot{Q}}{m} = \frac{-\dot{c}kB}{2m_p} = 3c^2 \frac{kB}{2m_p} \frac{1}{\text{hour}}, \tag{5.86}$$

which is 1.93 mW/g for $c = 0.75$. This is a very substantial amount of heat in view of the thermal conductivity of hydrogen and the Kapitza resistance to cooling medium at low temperatures. Brute-force polarized solid ortho-hydrogen has not been used as a polarized target, because the removal of the conversion heat would require large amounts of thermal contact material relative to the mass of hydrogen. Its use as a source of polarized hydrogen gas has been contemplated, however.

Because ortho-hydrogen has the ground state spin $I = 1$, its polarization is given for the Brillouin function 1.63′ for spin 1, which is 4/3 higher than that of spin 1/2 for low values of B/T. At 10 T field and 20 mK spin temperature, the polarization of ortho-hydrogen is 57%. Pure ortho-hydrogen can be prepared by selective adsorption on alumina powder, for example.

Methane

The methane molecule CH_4 can be modelled by the shape of an equilateral tetrahedron with hydrogen atoms at the four corners and the carbon atom at the center. Because of the symmetry of the molecule, the rotational levels can also be expressed by Eq. 5.83 with $B/k = 8$ K. The total nuclear spin can have the values of $I = 0, 1, 2$; the selection rules for the rotational and nuclear spin states are more complicated to find out and are given in the Table 5.2 where the allowed combinations are marked.

In the ground state, the methane molecule thus has the maximum nuclear spin $I = 2$, which is of interest in polarized targets. The first rotational level is at $E/k = 16$ K so that its equilibrium population is negligible below 1 K; however, in pure methane the conversion speed is very slow between the rotational states at low temperatures. The $J = 1$ rotational level has the statistical population of 9/16 after fabrication; its influence on the proton spin lattice relaxation is substantial.

Solid methane containing oxygen impurity on the level of 0.1–1% has much faster conversion speed in the range of hours [19] and the measured nuclear susceptibility shows that

Table 5.2 Combinations of the rotational and total nuclear spin states J and I allowed by the symmetry and exclusion principles in the methane molecule in solid methane.

J	$I=2$	$I=1$	$I=0$
0	Allowed		
1		Allowed	
2		Allowed	Allowed
3	Allowed	Allowed	
4	Allowed	Allowed	Allowed
5		Allowed	Allowed

all molecules have the total spin $I = 2$ and therefore the nuclear polarization is twice higher than in the case that the nuclear spins would be independent. The nuclear spin-lattice relaxation in the ground state is controlled by the paramagnetic oxygen impurity and is slowly increased to 1.35 s at 1 K and 2.5 T after 4 days of conversion [19].

Ammonia

An important material to understand from the point of view of nuclear spin states is ammonia, which is used in polarized proton and deuteron targets. The relatively high melting point of 195.4 K is a sign of strong binding in the solid, and this is confirmed by the large shift of 15% in the ^{14}N quadrupole coupling constant between the gaseous and solid states [16]. Studies of the width and shape of the NMR absorption signal show a transition in the solid at 65 K where the half-width, measured from the inflection points of the line, changes from a plateau of 21 kHz up to 50 K to a plateau of 16.5 kHz from 75 K upwards [20]. This transition is believed to be due to an onset for thermally activated rotation about the C_3 axis (symmetry axis of the molecule perpendicular to the triangle of the protons), with the experimental activation energy of 9.62 ± 0.17 kJ/mol. The rotation of the molecule is thus hindered below 50 K, and it may be expected that at low temperatures the second moment should be that due to a rigid lattice which amounts to 0.4607 mT². This is contradicted by the value measured at 2 K of 0.175 mT² [20], and was first interpreted as rotation by quantum mechanical tunnelling about the C_3-axis.

It has been later recognized that the NMR spectrum in ammonia and several other substances such as ammonium halides is more complex due to second-order couplings where the diamagnetic atomic electrons are involved. These effects were independently discovered by Hahn and Maxwell [21] and Gutowsky, McCall and Slichter [22] in substances such as PF_3, in which the NMR lines were split into equidistant components independent of field and temperature. The authors pointed out that the interaction which results in such a splitting cannot be due to the chemical shift because of the absence of the field dependence; it can only be of the form

$$\mathcal{H}_{1,2} = A_{1,2} \mathbf{I}_1 \cdot \mathbf{I}_2, \tag{5.87}$$

which produces a splitting similar to the isotropic hyperfine interaction and depends only on the relative orientations of the nuclei but not on the orientation of the molecule.

Ramsey and Purcell [23] proposed a model where the two nuclei interact with the nearby electrons and where the electron spins were also included. These ideas were extended to solids by Bloembergen and Rowland [24] and Ruderman and Kittel [25]. The interaction is a second-order effect due to the polarization of the electrons by the nuclear spin; the electron polarization is seen by all other nuclei in the molecule. These indirect nuclear spin interactions are called pseudo-dipolar and pseudo-exchange interactions, because their effective Hamiltonians have spin dependences identical to the dipolar and exchange interactions, respectively. An introductory discussion on these is given in Section 4.9 of Ref. [26].

In solids the pseudo-dipolar interaction causes broadening which is indistinguishable from ordinary dipolar broadening. The pseudo-exchange interaction of Eq. 5.87 can be observed in solids if the linewidth is narrow enough to enable resolving the resulting splittings.

An example where molecular isomerism is observed is the ammonium group in ammonium halides. In NH_4I the conversion between the nuclear spin isomers makes the lineshape evolve after quenching the sample to helium temperature [27]. This was later confirmed by observing the evolution of the NMR absorption signal integral after quenching to 4.2 K; the relative size grew from 4.64 after 0.5 h to 26.61 after 8 h [28].

In solid NH_3 below 1 K, the NMR line is split into three components which are not very well resolved, as can been seen in Figure 5.6 [29]. It was first thought that this could be

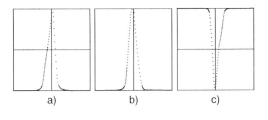

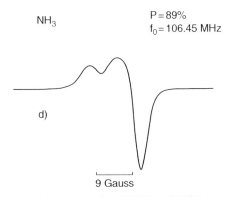

Figure 5.6 NMR absorption signals for protons in solid NH_3 at 2.502 T magnetic field at different values of DNP: (a) +90%, (b) 0.3% and (c) −93%, with 400 kHz frequency sweep about 106.45 MHz center frequency. The curve (d) is the first derivative of the curve (a). The NMR line consists of three equidistant unresolved lines that behave as if caused by isotropic hyperfine splitting by the term of Eq. 5.87

due to molecular spin isomerism but could equally well be caused by the dipolar coupling between the protons of the equilateral triangle of the molecule, as has been shown numerically [30]. No sign of conversion between nuclear spin isomers has been observed by following the evolution of the integrated absorption signal during TE calibration process, and it is therefore concluded that if the spin isomers do exist, molecular rotation is hindered, J is not a good quantum number at 1 K and the proton spins behave as independent.

5.2.6 Interactions with Electron Spins

Dipolar and Hyperfine Interactions in Dielectric Solids

The hyperfine and dipolar interaction between the nucleus and nearby paramagnetic electron was briefly discussed in Section 2.1.4. It arises from the scalar magnetic coupling $\gamma_e \gamma_n \mathbf{S} \cdot \mathbf{I}$ between the magnetic moments as shown by Eq. 2.44.

The dipolar terms lead to the broadening that can be evaluated by applying the Van Vleck formulas for unlike spins of Section 2.1.1.

The Fermi contact interaction term gives rise to the isotropic hyperfine tensor, which manifests itself as a splitting of the magnetic levels without broadening. The other terms depend on the relative orientations of $\mathbf{r}$ and $\mathbf{B}$ and lead to an anisotropic hyperfine tensor which splits and broadens the resonance lines.

Expression 2.44 is linear in I_x, I_y and I_z and in S_x, S_y and S_z and it can be shown that it can be reduced to the general form

$$\mathcal{H}_{hf} = \mathbf{S} \cdot \tilde{\mathbf{A}} \cdot \mathbf{I} \equiv A_x I_x S_x + A_y I_y S_y + A_z I_z S_z. \tag{5.88}$$

If there are several hyperfine nuclei in the molecule, very complicated EPR line structures may arise. In the case of only isotropic couplings, the magnetic levels are

$$E = \left(g\mu_B B + \sum_i a_i m_i \right) m_S. \tag{5.89}$$

In such cases the isotropic hyperfine constants a_i may be resolved in dilute liquid samples. The EPR spectra with hyperfine interactions were discussed in Chapter 3.

The hyperfine interaction will also be seen in the resonance of the hyperfine nuclei, which are split in the same way. The resonance frequencies are, in the isotropic case and in high field, for a given nucleus i,

$$\omega_i = \gamma_i B_0 + \frac{a_i m_S}{\hbar}, \tag{5.90}$$

where m_S is the projection of the electron spin (assuming $S = \frac{1}{2}$) on the direction of the magnetic field. Such spectra are often spread over several MHz and they are beyond accurate direct NMR spectroscopy in solids. These may be observable by the ENDOR techniques, which consist of measuring the effect of nuclear resonance saturation on the electron resonance signal. However, ENDOR cannot be easily applied in situ in a polarized target, because the multimode cavity around the target is not suitable for this purpose, and because the NMR probes usually sample only a small fraction of the target material.

The NMR spectra of the protons of paramagnetic molecules EHBA-Cr(V) and BHHA-Cr(V) have been measured in dynamically polarized deuterated glassy butanol with 5% deuterated water, as shown in Figure 5.7, which was measured at 5 T field and 0.3 K temperature. The glassy matrix of the butanol-water solvent contained 2% of unsubstituted protons that give rise to the narrow central signal in both cases. Judging from the chemical composition of the complex molecules and taking into account the undeuterated water of crystallization that is likely to be distributed uniformly in the glassy matrix, it can be deduced that the hyperfine proton polarization is not different from that of the solvent protons. This is also to be expected because the NMR resonance lines overlap so that cross-relaxation transitions can establish rapidly a uniform spin temperature within the spectra.

It remains to be measured how quickly the hyperfine protons reach common spin temperature with that of the matrix nuclei. Furthermore, it will be interesting to measure the spectra with negative polarization, because some of the hyperfine protons may be oppositely polarized due to the effects discussed in Section 4.6 in connection with DNP in crystalline materials with resolved hyperfine interactions.

Chemical Shift

The NMR techniques to measure accurate chemical shifts has become an important tool for analytical chemistry. These involve free induction decay NMR signal techniques with sophisticated pulse sequences and offline signal treatment for noise reduction. Sample

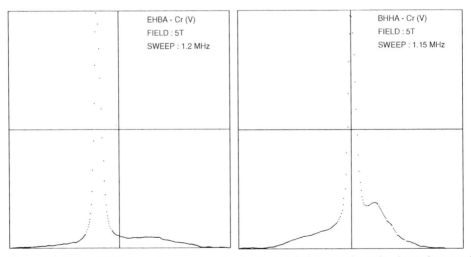

Figure 5.7 Proton NMR absorption spectra of undeuterated Cr(V) complexes in glassy deuterated butanol with 5% deuterated water, at 5 T magnetic field and 0.3 K temperature, at 90% proton polarization. Left: EHBA-Cr(V); Right: BHHA-Cr(V); concentration 5×10^{19} spins/cm^3

spinning compensates to a large extent most broadening mechanisms and permit high-resolution spectroscopy also in solid samples. This, combined with DNP to enhance the NMR signal size, has made it possible to study the chemistry of elements with rare nuclear spins.

As was discussed in Section 2.1.5, chemical shifts in diamagnetic materials are expressed as

$$\omega = (1-\sigma)\omega_L, \tag{5.91}$$

where ω_L is the Larmor frequency of the bare nucleus. However, as the field cannot be measured with high absolute accuracy, usually the shift is defined with respect to a known resonance line whose shift is thus arbitrarily defined to be zero. For protons and ^{13}C the usual reference is the Larmor frequency of protons in tetramethylsilane $Si(CH_3)_4$, abbreviated TMS. With reference to this, the shifts for protons in organic substance are up to 12 ppm.

The range of the chemical shift increases with the atomic weight; for ^{13}C this ranges up to 200 ppm in organic substances, and for fluorine the shift ranges 600 ppm.

The chemical shifts are anisotropic so that they can be best measured in liquid state, where pulsed NMR techniques produce very narrow resonance lines, because the rapid motion of molecules averages out entirely the dipolar broadening, and partly the quadrupole and chemical shift anisotropy broadenings. Progress in high-resolution pulsed NMR, however, with magic angle spinning (MAS) techniques, has enabled the measurement of chemical shifts also in solid samples. In the MAS technique the solid sample is rotated inside the NMR probe, driven by a miniature gas turbine, at speeds ranging from 1 kHz to 130 kHz, about an axis that makes the magic angle $\theta_m = 54.74°$ ($\cos2\theta_m = 1/3$) with respect to the steady magnetic field.

Such MAS-NMR measurements in dynamically polarized materials have made rare nuclei accessible for the MAS-DNP experiments in materials with suitable paramagnetic impurities [31]. In these experiments, performed at fields in excess of 5 T, the sample can be rotated down to 30 K temperature, using ^{4}He as drive gas [32].

Knight Shift

The Knight shift is due to the conduction electrons in metals, as was discussed in Section 2.1.5. These produce an effective field at the nuclear site, due to the spin orientations of the electrons thermally excited conduction electrons close to their Fermi surface, in the presence of an external field. This is responsible for the shift observed in the NMR and for the temperature-independent paramagnetic susceptibility. The shift comes from Pauli spin susceptibility, in addition to the electron s-component wave functions at the nucleus. In cubic metals the shift is isotropic, and in non-cubic metals the Knight shift depends on the relative orientations of the crystalline axes and the applied field. Then a tensor part must be added to the constant K of Eq. 2.52, for each spin species i:

$$\mathcal{H}_K = -\sum_i \gamma_i \mathbf{I}_i \cdot \mathbf{K}_i \cdot \mathbf{B}. \tag{5.92}$$

In axially symmetric crystals such as those with tetragonal structure, the tensor causes broadening similar to the g-shift in electron resonance. The broadening is smaller than the shift itself, which is around 1,000 ppm for ^{23}Na, for example.

The interaction between the conduction electrons and the nuclear spins also leads to a fast spin-lattice relaxation in metals, which was mentioned in Section 1.3.5 and which will be discussed further in Section 5.3.4.

DNP and NMR in metals are limited by the penetration of RF magnetic field (skin depth) and by eddy current heating; these force the studies to be limited to surface layers, thin films or fine powder samples.

5.3 Nuclear Spin Relaxation and Diffusion

The relaxation of the spin magnetization in dielectric solids was discussed in general terms in Section 2.3, where its relationship with phonons and other excitations was derived, in the terms of the lattice stress tensor coefficient F_1 for one-phonon (direct) process and F_2 for two-phonon (Raman) processes. These and the Orbach process were further elaborated for electron spins in Section 3.4. It was shown that below 1 K the direct Zeeman relaxation of nuclear spins was very slow, so that their only contact with the lattice remained via their interaction with the paramagnetic electrons.

5.3.1 Relaxation via Paramagnetic Electrons

The relationship of the spin-lattice relaxation of electrons and nuclei was studied in LMN doped with Nd^{3+} ions by Schmugge and Jeffries in 1965 [33]. As their treatment is equally valid in glassy hydrocarbons and irradiated materials with dilute paramagnetic spins, we shall briefly outline their theoretical arguments and results.

In their shell-of-influence model all nuclear spins within the shell $r_1 < r < r_2$ belong to the paramagnetic spin at $r = 0$. Here r_1 is the minimum distance between the nuclear spin and the electron spin, and r_2 is roughly half the average spacing between the paramagnetic spins

$$r_2 = \left(\frac{3}{4\pi n_e}\right)^{1/3}. \tag{5.93}$$

A single shell is then considered to represent the bulk of the material, in which the nuclear spins flip together with the electrons spins at the rate

$$\frac{1}{T_{1n}} = \bar{\sigma}\left(1 - P_e P_{e0}\right)\frac{1}{T_{1e}}, \tag{5.94}$$

where

$$\bar{\sigma} = \left\langle \frac{3}{10}\left(\frac{g\beta}{r^3 H}\right)^2 \right\rangle = \frac{3}{10}\left(\frac{g\beta}{H}\right)^2 \frac{1}{r_1^3 r_2^3} \tag{5.95}$$

is the mean ratio of the electron-lattice relaxation rate and electron-nucleus simultaneous relaxation rate within the shell; P_e is the instantaneous electron polarization and P_{e0} is the electron polarization in thermal equilibrium with the lattice. In the absence of microwave transitions $P_e \rightarrow P_{e0}$ rapidly, and the shell-of-influence model predicts

$$\frac{1}{T_{1n}} = \frac{3}{10}\left(\frac{g\beta}{H}\right)^2 \frac{1}{r_1^3 r_2^3} \text{sech}^2\left(\frac{\hbar\omega_0}{2kT_0}\right)\frac{1}{T_{1e}}$$

for the relationship between the two relaxation rates.

As was discussed in Section 3.4.1, Eq. 3.74 yields the electron spin-lattice relaxation below 1 K and at high magnetic field dominated by the direct process

$$\frac{1}{T_{1e}^{direct}} \approx 9\Omega_D \left(\frac{\omega_0}{\Omega_D}\right)^5 \left(\frac{e|V^{(0)}|}{\Delta}\right)^2 \frac{\hbar\Omega_D}{mv_a^2}\coth\left(\frac{\hbar\omega_0}{2kT_0}\right), \qquad (5.96)$$

where Ω_D is the Debye frequency, v_a is the velocity of the acoustic phonons and m is the mass of the atoms in the unit cell of the lattice in the case of crystalline materials. In glassy materials the meaning of these parameters remains vague, although the specific heat measurements [34] and electron spin-lattice time measurements [35] can be interpreted in the terms of the Debye model standing behind Eq. 5.96. These terms can therefore be interpreted as scaling parameters permitting rough comparison between different materials and, above all, as means of extrapolating the field and temperature dependences of the relaxation time. In this view it is noteworthy that the electron spin-lattice relaxation has no temperature dependence when the coth term approaches unity below 1 K, while T_{1n} increases steeply due to the sech² term when the temperature is lowered below 1 K. Also, the relaxation times scale with the magnetic field as $T_{1e} \approx B^{-5}$ and $T_{1n} \approx B^{-3}$. These temperature and field dependences were confirmed by the proton relaxation time measurements in propanediol-Cr(V) [36] shown in Figure 5.8, where the lines obey

$$\frac{1}{T_{1p}} = \frac{1}{AT_{1e}}\frac{h}{B^2}\frac{1\,T^2}{B^2}\text{sech}^2\left(\frac{\hbar\omega_0}{2kT_0}\right) + (1\,h)\cdot\left(3.96\frac{(B/1T)^{1.32}}{(T/1K)^{1.64}}\right)^{-1} \qquad (5.97)$$

with

$$\frac{1}{AT_{1e}} = \frac{1}{0.225}\left[\left(\frac{B}{1T}\right)^5 \coth\left(\frac{\hbar\omega_0}{2kT}\right) + 6.75\exp\left(-\frac{0.5\,K}{T}\right)\right]. \qquad (5.98)$$

The second term in Eq. 5.97 is empirical and was added to follow the temperature dependence of the experimental points above 50 h relaxation time. The first term in Eq. 5.98 is due to the direct process that dominates the magnetic field dependence, and the second term was added to improve the field dependence of the fitted functions at low magnetic field values. Such a field-dependent term may be due to a two-phonon process relaxing exchange coupled pairs or clusters of electrons [37]. Other measurements of the electron

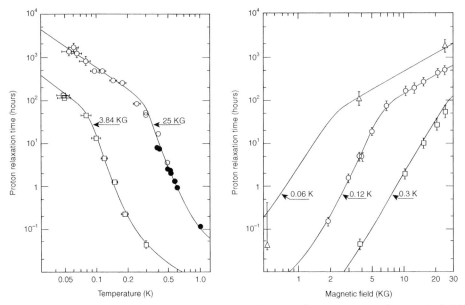

Figure 5.8 Proton spin-lattice relaxation time as a function of temperature and magnetic field in propanediol-Cr(V). Reprinted from Ref. [36], with permission from Elsevier.

spin density effects on the nuclear spin-lattice relaxation time also suggest that exchange coupling may be responsible for low-field transition rates exceeding that due to the direct process [38].

As was discussed in Section 3.4.1, the direct process of electron spin-lattice relaxation dominates the Orbach process and Raman process in polarized targets running in 2.5 T or 5 T magnetic field and at or below 1 K temperature. Frozen spin operation at a low field or with passages in a low field, however, requires the knowledge of nuclear spin-lattice and cross-relaxation phenomena; we recall here the low-field mechanisms of Blume and Orbach [39–41], the 'wobble' of Waugh and Slichter [42] and spin exchange of Harris and Yngvesson [37, 43] briefly discussed in Section 3.4.1.

5.3.2 Polarization Dependence of Nuclear Spin Relaxation

During the operation of CERN frozen spin polarized target in a 1 T magnetic field [44], it was noted that the positive proton polarization relaxed more slowly than the negative one. The target operation consisted of cycles of 24 h with about 3 h reversal procedure that included cooldown and transports between the polarizing 2.5 T field and 1 T holding field. The difference of the loss rates was substantial: 1.4±0.6%/d loss for +90% proton polarization and 4.3±2.0%/d loss for −90% polarization, based on the compilation of data from 60 days of operation. Superradiance was first ruled out during the transport of 0.3 T field minimum by performing sequences of successive transports; these yielded losses less than

0.1%/transport for both signs of polarization. It was also suspected that the optical spark chambers surrounding the target could interfere with the NMR Q-meter system and cause saturation that is higher with the negative polarization, but this was ruled out by finding no correlation between negative polarization loss and number of spark triggers [44].

The asymmetry in the relaxation rate was traced to originate from the asymmetry in the direction of energy flow between the proton spin system and the coolant dilute helium at about 30 mK temperature. The bottleneck of the heat transfer at this temperature is between the spin system reservoirs, shown in Figure 4.1, rather than between the lattice phonons and the coolant; the effect of the lattice heating or cooling only causes a secondary effect via the temperature of the electron spin-spin interaction reservoir that closely follows that of the proton spins, while the electron Zeeman temperature rises above or cools below that of the lattice phonons and produces the dependence of the relaxation rate on the nuclear spin polarization. Evidently the effect would be less significant for a deuterated target material because of the lower magnetic moment and therefore lower energy flow due to nuclear spin relaxation.

During operation in the beam the target beads of diameter $d = 1.5$ mm and bulk density ρ warm up slightly by energy deposited by the 6 GeV/c π^- beam spills of about 0.5 s duration every 3 s, with the mean flux $\dot{n}_b = 10^5$ s^{-1}cm^{-2}:

$$T_L^4 = T_{He}^4 + \frac{\dot{Q}_b / A_t}{\alpha} = T_{He}^4 + \frac{6\dot{n}_b}{\alpha \rho_t d} \frac{dE}{dx}. \tag{5.99}$$

Here $dE/dx = 2$ MeV cm^2/g is the minimum ionizing energy loss, α is the Kapitza conductance at the bead-helium interface, T_L is the lattice temperature and T_{He} is the coolant temperature. These yield [44]

$$T_L = \left[T_{He}^4 + \left(2.15 \text{ mK}\right)^4 \frac{\dot{n}_b}{\text{cm}^{-2}\text{s}^{-1}} \cdot \frac{d}{\text{cm}} \right]^{1/4} = 26 \text{ mK} \tag{5.100}$$

when the coolant temperature is 20 mK.

The thermal time constant of the propanediol-Cr(V) target bead is

$$\tau_L = \frac{c_L \rho d}{24 \alpha T_L^3} \cong 0.125 \text{ s} \cdot \left(\frac{10 \text{ mK}}{T_L}\right)^2, \tag{5.101}$$

where $c_L = 50 \times 10^{-7} \cdot T_L$ J/(gK^2) is the lattice specific heat of a similar amorphous solids [45]. This yields a time constant $\tau_L = 18$ ms at $T_L = 26$ mK, which means that the bead temperature follows the beam current closely, while averaging its faster time structure due to the bunches circulating in the machine during the slow extraction. By using the relaxation time of the positive polarization as a thermometer, the mean temperature of the beads was 28 mK during most of the operation [46]. This attenuates the time variation of T_L for both signs of polarization.

The thermal asymmetry in proton spin-lattice relaxation due to the Kapitza resistance can be estimated by solving T_L from

$$\alpha\left(T_L^4 - T_{He}^4\right) = \frac{\dot{Q}_b}{A_t} - n_p \hbar\omega_p \frac{V}{A_t} \frac{P_p}{2\tau_p(T_L)}, \qquad (5.102)$$

where $n_p = N_p/V$ is the number density of protons in the propanediol target, and $\tau_p(T_L)$ is the temperature-dependent proton spin-lattice relaxation time of Eq. 5.97). For 1 T field this yields the maximum asymmetry of 2×10^{-3} between +100% and −100% proton polarizations, and therefore rules out the bottleneck due to the Kapitza resistance.

In the above it was assumed that the electron Zeeman temperature T_Z was close to that of the lattice. If we now assume that T_Z may be heated or cooled by the nuclear spin system and that the nuclear spin-lattice relaxation depends directly on T_Z as shown by Eq. 5.97, we may equate the heat flow due to the nuclear spin relaxation with that of the electrons. The cases of high-positive and high-negative polarizations may need different treatments, because of the considerable difference in the electron Zeeman temperatures. In the case of negative proton polarization we have

$$N_e\left[P_e(T_L) - P_e(T_Z)\right]\frac{\hbar\omega_e}{\tau_{1e}} = N_p \frac{-P_p\hbar\omega_p}{2\tau_p(T_Z)}, \qquad (5.103)$$

where $T_Z \gg T_L$ and we may approximate $P_e(T_L) \approx 1$ to get

$$-\frac{\hbar\omega_e}{kT_Z} = \log\left(\frac{N_p}{N_e}\frac{\omega_p}{\omega_e}\frac{\tau_{1e}}{\tau_p(T_Z)}\frac{-P_p}{4}\right) \qquad (5.104)$$

from where we may solve T_Z by using again Eq. 5.97. To do this we also need Eq. 5.98 to scale the field dependence of the electron spin-lattice relaxation time from 2.5 T to 1 T field which yields $\tau_{1e} = 3.7$ s. These give the results $T_Z = 94$ mK and $\tau_p(T_Z) = 200$ h, which has the right order of magnitude given the large variance of the experimental loss of 4.3±2.0%/d.

In the case of positive polarization we might have $T_Z \ll T_L$ and could possibly solve T_Z now from

$$-\frac{\hbar\omega_e}{kT_L} = \log\left(\frac{N_p}{N_e}\frac{\omega_p}{\omega_e}\frac{\tau_{1e}}{\tau_p(T_Z)}\frac{P_p}{4}\right). \qquad (5.105)$$

However, in this case the extrapolation of Eq. 5.97 downwards to get $\tau_p(T_Z)$ is highly uncertain, because the relaxation mechanisms are unknown and there are no accurate experimental results below 50 mK temperature.

We may conclude that at a 1 T holding field the negative polarization loss time constant is always determined by the electron Zeeman temperature just below 100 mK, when the lattice temperature is well below this value and the holding field is 1 T. On the other hand, the much longer positive polarization loss time constant probably depends on a different mechanism and perhaps also on impurity electron spins.

The asymmetry in nuclear spin-lattice relaxation has been later confirmed by the results of the CERN Spin Muon Collaboration [8, 47].

5.3.3 Nuclear Spin Relaxation in Metals

Metals are important for brute-force polarized targets and for the nuclear demagnetization cooling systems, in addition to the research of nuclear magnetism which was discussed briefly in Chapter 1. The brute-force nuclear spin polarization will be discussed in Chapter 10. In metals the nuclear spin-lattice relaxation is due to their interaction with the conduction electrons that produces time constants in the range of milliseconds to seconds, in strong contrast with dielectric solids where phonons and paramagnetic impurities are at play.

The crystalline lattice of metals consists of positive ion cores of the atoms surrounded by a cloud of free or almost free conduction electrons; these are the valence electrons that are delocalized while the core ions are localized in their lattice positions. The conduction electrons are assumed to go into so-called Bloch states extended throughout the whole crystal. The free-electron wave functions are described by the Bloch functions

$$|\mathbf{k},s\rangle = \psi_{ks} = u_{\mathbf{k}}(\mathbf{r})e^{i\mathbf{k}\cdot\mathbf{r}}\psi_s, \qquad (5.106)$$

where ψ_s is the electron spin wave function and the plane wave part $\exp(i\mathbf{k}\cdot\mathbf{r})$ is modulated by the lattice function $u(\mathbf{r})$, which has the periodicity of the lattice and which peaks at the nuclei positioned at $\mathbf{r} = 0$. In the most simplified picture $u(\mathbf{r})$ is constant and the energy of the electron plane wave states is

$$E(\mathbf{k}) = \frac{\hbar^2 k^2}{2m_e}. \qquad (5.107)$$

In this case the constant energy surfaces are spheres in the space of the wave vector $\mathbf{k}$. The allowed values of $\mathbf{k}$ are distributed with density $V/8\pi^3$ in this space. For each value of $\mathbf{k}$ there are two electron states, of opposite spin according to the Pauli principle. Supposing that there are Z electrons per atom and 1 atom per unit cell of the lattice and N unit cells per unit volume, the electron states are filled up to a wave number k_F where

$$\frac{4}{3}\pi k_F^3 \frac{2}{8\pi^3} = ZN \qquad (5.108)$$

or

$$k_F = (3\pi^2 Z N)^{1/3} \qquad (5.109)$$

defines the radius of the Fermi sphere k_F (in our simplified model of free non-interacting electrons). The Fermi energy or Fermi level is then

$$E_F = \frac{\hbar^2 k_F^2}{2m_e}. \qquad (5.110)$$

This means that the wavelength of the electron near the Fermi level is comparable with the interatomic spacing:

$$\lambda_F = \frac{2\pi}{k_F} \cong \left(\frac{2}{Z}\right)^{1/3} 2.6 \cdot r_s, \qquad (5.111)$$

where r_s is the radius of the Wigner–Seitz sphere.[1]

In practice $u(\mathbf{r})$ is a smooth periodic function and the Fermi surface may have a complex shape that depends on the crystal structure and effective number of free electrons per unit cell. Many results can be, however, obtained using a much simplified model of the electron wave function, including the basic understanding of the nuclear spin-lattice relaxation. The effect of electrons on the relaxation of the nuclear spins can be understood as scattering of an electron with initial state $|\mathbf{k}s\rangle$ and final state $|\mathbf{k}'s'\rangle$ on a nucleus with initial state m and final state n. The detailed derivation, given in Section 5.3 of Ref. [26], is beyond our scope here, and we shall outline it only briefly.

Using the above notation and following Slichter [26], the rate for one nuclear spin transition is

$$W_{m\mathbf{k}s,n\mathbf{k}'s'} = \frac{2\pi}{\hbar} \left|\langle m\mathbf{k}s|V|n\mathbf{k}'s'\rangle\right|^2 \delta\left(E_m + E_{\mathbf{k}s} - E_n - E_{\mathbf{k}'s'}\right), \qquad (5.112)$$

where V is the interaction that provides the scattering and where it is assumed that the initial state is not empty and the final state is not already occupied. The total rate of transitions is obtained by adding up the rates from all initial and final states:

$$W_{mn} = \sum_{\mathbf{k}s;\mathbf{k}'s'} W_{m\mathbf{k}s,n\mathbf{k}'s'} f\left(E_{\mathbf{k}s}\right)\left[1 - f\left(E_{\mathbf{k}'s'}\right)\right], \qquad (5.113a)$$

which can be inserted in Gorter's formula for the relaxation time [49]

$$\frac{1}{T_1} = \frac{1}{2} \frac{\sum_{m,n} W_{mn}\left(E_m - E_n\right)^2}{\sum_n E_n^2}. \qquad (5.113b)$$

In Eq. 5.112 the probabilities of occupation of a given electron state $\mathbf{k}s$ are obtained from the Fermi distribution function

$$f\left(E_{\mathbf{k}s}\right) = \frac{1}{1 + \exp\left\{\left(E_{\mathbf{k}s} - E_F\right)/kT\right\}}. \qquad (5.114)$$

The dominant contribution to V of Eq. 5.112 is due to the s-state coupling in metals with a dominant s-character of the wave function at the Fermi surface:

$$V = \frac{8\pi}{3} \gamma_e \gamma_n \hbar^2 \mathbf{I} \cdot \mathbf{S} \delta(\mathbf{r}), \qquad (5.115)$$

---

[1] An introduction to the electrons and phonons in solids is given in Chapters 2 and 3 of Ref. [48].

5.3 Nuclear Spin Relaxation and Diffusion

where the nucleus is assumed at the origin. The initial and final wave functions are products of the spin function $|s\rangle$ and Bloch function of Eq. 5.106:

$$|m\mathbf{k}s\rangle = |m\rangle|s\rangle u_{\mathbf{k}}(\mathbf{r})e^{i\mathbf{k}\cdot\mathbf{r}}, \qquad (5.116)$$

$$|n\mathbf{k}'s'\rangle = |n\rangle|s'\rangle u_{\mathbf{k}'}(\mathbf{r})e^{i\mathbf{k}'\cdot\mathbf{r}}; \qquad (5.117)$$

which yield the matrix element

$$\langle m\mathbf{k}s|V|n\mathbf{k}'s'\rangle = \frac{8\pi}{3}\gamma_e\gamma_n\hbar^2 \langle m|\mathbf{I}|n\rangle \cdot \langle s|\mathbf{S}|s'\rangle u_{\mathbf{k}}^*(0)u_{\mathbf{k}'}(0). \qquad (5.118)$$

Inserting this into Eq. 5.112 yields for a single scattering event of an electron

$$W_{m\mathbf{k}s,n\mathbf{k}'s'} = \frac{2\pi}{\hbar}\frac{64\pi^2}{9}\gamma_e^2\gamma_n^2\hbar^4 \sum_{\alpha,\alpha'=x,y,z}\{\langle m|I_\alpha|n\rangle\langle n|I_{\alpha'}|m\rangle$$
$$\times \langle s|S_\alpha|s'\rangle\langle s'|S_{\alpha'}|s\rangle|u_{\mathbf{k}}(0)|^2|u_{\mathbf{k}'}(0)|^2 \qquad (5.119)$$
$$\times \delta(E_m + E_{\mathbf{k}s} - E_n - E_{\mathbf{k}'s'})\}.$$

Slichter substitutes this into Eq. 5.113a to compute W_{mn}, which can then be inserted into Eq. 5.113b to get the relaxation rate $(T_1)^{-1}$. The computation is tedious and requires several approximations in order to get usable results; we shall only list here the main steps and arguments:

- The sum of Eq. 5.113a over $\mathbf{k}$ and $\mathbf{k}'$ involves slowly varying functions and the sum can be performed by integrals using the density of states on the Fermi surface.
- The delta function of Eq. 5.119 simplifies the integrals.
- At such low temperatures that $kT \ll E_F$ the product of the Fermi distribution functions $f(E_{\mathbf{k}s})[1-f(E_{\mathbf{k}'s'})]$ resembles the delta function and simplifies further the sums and the integrals.
- The difference of the nuclear spin energy $E_m - E_n \ll kT$ which lets us ignore it in the Fermi functions.
- Because the density of states and the function $|u_{\mathbf{k}}(0)|^2$ are both slowly varying functions of the electron energy $E_{\mathbf{k}}$, the functions can be evaluated at the same value $E_{\mathbf{k}}$.
- The sums of the electron spin matrix elements yield $\delta_{\alpha\alpha'}/2$, because the integral over the Fermi distributions is independent of the electron spin states.

Using further the presentation of Slichter, Eq. 5.113a can be written

$$W_{mn} = a_{00}\sum_\alpha \langle m|I_\alpha|n\rangle;$$
$$a_{00} = \frac{64}{9}\pi^3\hbar^3\gamma_e^2\gamma_n^2\left\langle |u_{\mathbf{k}}(0)|^2\right\rangle_{E_F}^2 \rho^2(E_F)kT, \qquad (5.120)$$

where $\rho(E_F)$ is the density of states on the Fermi surface. This, however, describes the interaction of the electrons with a single nucleus at $r = 0$ and we must modify it because the electrons in fact see also the neighboring nuclei simultaneously, although with amplitudes that rapidly decrease with distance from $r = 0$. By summing up these Slichter arrives at a rate that is twice that of Eq. 5.113b:

$$\frac{1}{T_1} = a_{00} = \frac{64}{9}\pi^3\hbar^3\gamma_e^2\gamma_n^2\left\langle\left|u_\mathbf{k}(0)\right|^2\right\rangle_{E_F}^2 \rho^2(E_F)kT. \quad (5.121)$$

Here the lattice function $\left|u_{\mathbf{k}'}(0)\right|^2$ appears in the same way as in the Knight shift, Eq. 2.52, which we may use to relate it with the shift and spin susceptibility:

$$\left\langle\left|u_\mathbf{k}(0)\right|^2\right\rangle = \frac{3}{8\pi}\frac{1}{\chi_e^s}\frac{\Delta\omega}{\omega}, \quad (5.122)$$

to get finally the relation

$$T_1\left(\frac{\Delta\omega}{\omega}\right)^2 = \left(\frac{\chi_e^s}{\rho(E_f)}\right)^2 \frac{1}{\pi\hbar^3\gamma_n^2\gamma_e^2}\frac{1}{kT}. \quad (5.123)$$

If the free electrons can be assumed to form a non-interacting Fermi gas, it can be shown that its susceptibility is related with the density of states at the Fermi surface by

$$\chi_0^s = \frac{1}{2}\hbar^2\gamma_e^2\rho_0(E_F), \quad (5.124)$$

where the subscripts 0 label the approximate parameters valid for non-interacting electrons. In this approximation Eq. 5.123 becomes the well-known Korringa relation

$$T_1\left(\frac{\Delta\omega}{\omega}\right)^2 = \frac{\gamma_e^2}{\gamma_n^2}\frac{\hbar}{4\pi kT}. \quad (5.125)$$

The more accurate relation can be now written in the form [26]

$$T_1\left(\frac{\Delta\omega}{\omega}\right)^2 = \left(\frac{\chi_e^s}{\chi_0^s}\frac{\rho_0(E_f)}{\rho(E_f)}\right)^2 \frac{\gamma_e^2}{\gamma_n^2}\frac{\hbar}{4\pi kT}. \quad (5.126)$$

D. Pines has computed the correction terms for many metals for which the Korringa constant

$$\kappa = T_1 T \quad (5.127)$$

remains validated in a wide range of temperatures, for example, in aluminium where $\kappa = 1.85$ sK between the melting point near 1,000 K and superconducting transition temperature near 1 K.

Magnetic impurities and superconducting transition have a substantial effect on the conduction electrons and therefore on the Korringa law. In metals such as Cu and Ag annealing in low-pressure oxygen atmosphere neutralizes the magnetic impurities which is seen in the increased nuclear spin lattice relaxation time.

For copper $\kappa = 0.88$ sK near RT, increasing to 1.3 sK below 1 K and reaching at 0.2 K the relaxation time of about 2 h, i.e. $\kappa = 15$ sK. For silver $\kappa = 12$ sK and the relaxation time at 0.2 mK reaches 14 h.

In pure platinum we have a very low $\kappa = 0.03$ sK [50]; the Curie law of Pt has been used as a thermometer around and below 1 mK temperatures for the reason of this fast relaxation and because of the ability to use the Korringa constant as a method to calibrate the NMR absorption susceptibility at a temperature where the Korringa law is known to hold.

In a narrower range of temperatures the Korringa law holds for some intermetallic compounds such as $AuIn_2$ ($\kappa = 0.11$ sK), which is used for nuclear demagnetization refrigeration [50].

The Korringa constant is an important parameter to characterize the strength of the interaction between the conduction electron and the nuclei. For the materials where the coupling can be considered weak, the Korringa constant is larger than 1 Ks. As the spin lattice relaxation time T_1 becomes very long at temperatures below 1 mK, it is possible to cool only the nuclear spin system to nanokelvin or even picokelvin temperatures for a time long enough to perform nuclear magnetic investigations, while the conduction electrons stay at a much higher temperature ($T_e > 100$ μK). For systems such as these, nuclear magnetic ordering phenomena below the transition temperature T_c have been observed in silver ($T_c = 0.6$ nK) and copper ($T_c = 60$ nK) [50, 51]. It appears that the nuclear magnetic ordering temperatures roughly obey a simple rule

$$T_c \approx \frac{I(I+1)}{\kappa} \quad (5.128)$$

in many simple metals [51].

5.3.4 Cross-Relaxation and Diffusion of Nuclear Spins

Cross-Relaxation between Different Spin Species

In the absence of microwave-induced transitions the electron spin Zeeman temperature quickly relaxes close to that of the lattice, with the relaxation time of the direct process (which does not depend on temperature below 4 K). When the lattice is cooled below 1 K, the number of electron spin-spin transitions, however, is strongly reduced by the smallness of the factor $1 - P_e^2$ proportional to the exponentially reduced population of the electron spins in the upper energy state. This leads to the steep increase of the theoretical nuclear spin-lattice relaxation time of Eq. 5.97 that was experimentally validated by data of Figure 5.8.

It remained to be shown how the thermal equilibrium is reached between the different nuclear spin species and within the spectrum of inhomogeneously broadened NMR signal of deuterons, when the electron spin concentration is optimized for high DNP with fast build-up rate.

This is illustrated in Figure 5.9, which shows how the proton and deuteron spins approach equilibrium with the lattice, with each other, and within the quadrupole broadened deuteron spin system [52]. The partly deuterated d4-propanediol was chosen for this study because the Cr(V) electron spin density of 7×10^{19} cm^{-3} could be easily prepared in it; consequently DNP could be efficiently produced by microwave cooling of the electron spin-spin interaction reservoir. The measurements were made in a dilution refrigerator at 2.5 T magnetic field.

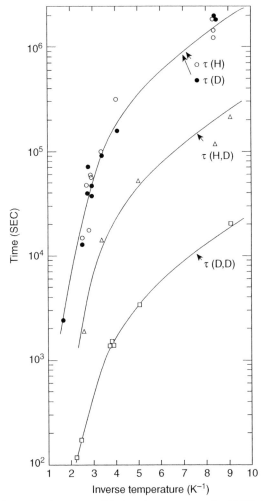

Figure 5.9 Spin-lattice and spin-spin equilibrium time constants in d4-ED-Cr(V): Upper curve and round symbols: proton and deuteron spin-lattice relaxation time; middle curve and triangle symbols: proton-deuteron spin-spin equilibrium time constant; lower curve and square symbols: recovery time constant for spin temperature in the deuteron NMR absorption spectrum after RF alignment. The curves are drawn to guide the eye. Reprinted, by permission from Springer Nature Customer Service Center GmbH, from Ref. [52].

After DNP the sample was cooled down in a few minutes to the minimum temperature below 100 mK and the various series of measurements were performed:

- In the first step, the spin-lattice relaxation times of the protons and deuterons were measured by following the evolution of their NMR signals and spin polarizations at various lattice temperatures. As can be expected, these relaxation times were not perfectly exponential, and the times shown in Figure 5.9 correspond to high positive and negative polarizations; the scatter is believed to be due to the differences of the polarizations between the data points.
- In the second step, after DNP and cooldown at 2.5 T field, the proton spin polarization was reduced close to zero by strong NMR saturation, and the evolution of the proton and deuteron NMR signals was recorded in order to determine the relevant time constant at various lattice temperatures.
- In the third step, after DNP and cooldown, the deuteron polarization was zeroed by strong RF saturation, and then the deuteron spin system was aligned by strong RF applied off-resonance of the proton NMR frequency. The recovery of the thermal equilibrium within the deuteron spectrum was followed by plotting the experimental quantity

$$X = A(D) - \frac{3}{4} P^2(D) \tag{5.129}$$

as a function of time.

The upper curve is quite similar to Eq. 5.97 shown in Figure 5.8; this shows that above 300 mK the direct process dominates in the temperature dependence of the nuclear spin relaxation, and that below 300 mK some less well-known mechanisms overtake in the evolution of the relaxation rate, similar to propanediol-Cr(V) with high spin density.

The fact that the middle curve has a shape similar to the upper one suggests that the temperature dependences arise from the same source, while the equilibrium time constants are different by about one order of magnitude. Referring to Figure 4.1, this might be explained, in relaxation between nuclear spin species, by the fact that little energy is required to pass to the lattice from the electron Zeeman energy reservoir while one nuclear spin system is cooling another via the electron spin-spin energy reservoir. Therefore both of the two terms of Eq. 5.97 appear to apply also to the cross-relaxation between the nuclear spin species in a high magnetic field.

Furthermore, the lower curve of Figure 5.9 also seems to have the same temperature dependence as the upper one, and we may draw the same conclusion as above: the cross-relaxation leading to the cooling of the dipolar spin temperature within the deuteron spectrum suggests that the two terms of Eq. 5.97 apply and limit the rate at which the deuteron spin system reaches the internal equilibrium. However, this happens again about ten times faster than the proton-deuteron relaxation, because a much lower quantity of energy needs to be transferred.

The electron spin concentration plays a decisive role in the above processes via Eqs. 5.94 and 5.95 that leads to linear relationship between the inverse nuclear spin relaxation time and the electron spin concentration.

Diffusion of Nuclear Spin Polarization

The diffusion of molecules in solids at low temperatures is hindered (apart from solid ^{3}He) and therefore polarization can diffuse only by cross-relaxation transitions that are effective within molecules and over molecular distances. Dipolar broadening provides a mechanism to conserve energy in spin flip-flop transitions between lines that have different chemical shifts and between hyperfine lines that have a small shift.

Close to a paramagnetic spin the dipole field increases causing a nuclear Larmor frequency shift that exceeds the dipolar frequency. Two-spin flip-flops are then slow and it has been speculated that the nuclei nearest to paramagnetic electrons, inside this diffusion barrier, are excluded from uniform DNP. This has been experimentally verified and it has been concluded that such a barrier is likely to be overcome by energy-conserving four-spin flip-flops.

The speed of polarization diffusion within paramagnetic molecules has been studied by time-resolved coherent neutron scattering [53]; this is described briefly in Section 11.3.2. The time scale for reaching uniform polarization in such molecules during DNP is tens of minutes, similar to the time scale of the DNP itself.

References

[1] European Muon Collaboration (EMC), J. Ashman, B. Badelek, et al., An investigation of the spin structure of the proton in deep inelastic scattering of polarized muons on polarized protons, *Nucl. Phys.* **B328** (1989) 1–35.

[2] European Muon Collaboration (EMC), J. Ashman, B. Badelek, et al., A measurement of the spin asymmetry and determination of the structure function g_1 in deep inelastic muon proton scattering, *Phys. Lett.* **B206** (1988) 364–370.

[3] R. D. Carlitz, J. C. Collins, A. H. Mueller, The role of the axial anomaly in measuring spin-dependent parton distributions, *Physics Letters B* **2014** (1988) 229–236.

[4] J. Ellis, M. Karliner, Determination of alpha_s and the nucleon spin decomposition using recent polarized structure function data, *Physics Letters B* **341** (1995) 397–406.

[5] E. Leader, *Spin in Particle Physics*, Cambridge University Press, Cambridge, 2001.

[6] A. Deur, S. J. Brodsky, G. F. de Teramond, The spin structure of the nucleon, *Reports on Progress in Physics* **82** (2019) 076201.

[7] B. Castel, I. S. Towner, *Modern Theories of Nuclear Moments*, Clarendon Press, Oxford, 1990.

[8] Spin Muon Collaboration (SMC), D. Adams, B. Adeva, et al., The polarized double-cell target of the SMC, *Nucl. Instr. and Meth. in Phys. Res.* **A437** (1999) 23–67.

[9] A. Abragam, M. Goldman, *Nuclear Magnetism: Order and Disorder*, Clarendon Press, Oxford, 1982.

[10] A. Abragam, M. Chapellier, J. F. Jacquinot, M. Goldman, Absorption lineshape of highly polarized nuclear spin systems, *J. Magn. Res.* **10** (1973) 322–346.

[11] A. Abragam, *The Principles of Nuclear Magnetism*, Clarendon Press, Oxford, 1961.

[12] M. Abramowitz, I. A. Stegun, *Handbook of Mathematical Functions with Formulas, Graphs and Mathematical Tables*, NIST, 1972.

[13] W. F. Kielhorn, A technique for measurement of vector and tensor polarization in solid spin one polarized targets, D. Ph., 1991, Physics, University of Texas in Austin.

[14] C. M. Dulya, J. Kyynäräinen, Influence of strong axial quadrupole interactions on the measurement of nuclear polarization, *Nuclear Instruments and Methods in*

Physics Research Section A: Accelerators, Spectrometers, Detectors and Associated Equipment **406** (1998) 6–12.

[15] Spin Muon Collaboration (SMC), D. Adams, B. Adeva, et al., The polarized double-cell target of the SMC, *Nucl. Instr. and Meth.* **A437** (1999) 23–67.

[16] S. S. Lehrer, C. T. O'Konski, Nuclear quadrupole resonance and bonding in crystalline ammonia, *J. Chem. Phys.* **43** (1965) 1941–1949.

[17] C. Dulya, D. Adams, B. Adeva, et al., A line-shape analysis for spin-1 NMR signals, *Nuclear Instruments and Methods in Physics Research Section A: Accelerators, Spectrometers, Detectors and Associated Equipment* **398** (1997) 109–125.

[18] M. Borghini, *Proton spin orientation*, CERN Yellow Report 68–32, 1968.

[19] Ö. Runolfsson, S. Mango, Nuclear magnetic resonance measurements of solid methane during the conversion to its ground state, *Phys. Lett.* **28A** (1968) 254–255.

[20] J. L. Carolan, T. A. Scott, A nuclear magnetic resonance study of molecular motion in liquid and solid ammonia, *J. Mag. Res.* **2** (1970) 243–258.

[21] E. L. Hahn, D. E. Maxwell, Spin echo measurements of nuclear spin coupling in molecules, *Phys. Rev.* **88** (1952) 1070–1084.

[22] H. S. Gutowsky, D. W. McCall, C. P. Slichter, Nuclear magnetic resonance multiplets in liquids, *J. Chem. Phys.* **21** (1953) 279–292.

[23] N. F. Ramsey, E. M. Purcell, Interactions between nuclear spins in molecules, *Phys. Rev.* **85** (1952) 143–144.

[24] N. Bloembergen, T. J. Rowland, Nuclear spin exchange in solids: Tl^{203} and Tl^{205} magnetic resonance in thallium and thallic oxide, *Phys. Rev.* **97** (1955) 1679–1698.

[25] M. A. Ruderman, C. Kittel, Indirect exchange coupling of nuclear magnetic moments by conduction electrons, *Phys. Rev.* **96** (1954) 99–102.

[26] C. P. Slichter, *Principles of Magnetic Resonance*, 3rd ed., Springer-Verlag, Berlin, 1990.

[27] Z. T. Lalowicz, J. W. Hennel, Evidence for ammonium group isomerism in NH_4I obtained by NMR, *Acta Phys. Pol.* **A40** (1971) 547–549.

[28] J. W. Hennel, Z. T. Lalowicz, Proton magnetic resonance intensity in ammonium salts at low temperatures, in: V. Hovi (ed.) *XVII Congress Ampere*, North-Holland, Turku, 1973, 217–218.

[29] T. O. Niinikoski, J.-M. Rieubland, Dynamic nuclear polarization in irradiated ammonia below 0.5 K, *Phys. Lett.* **72A** (1979) 141–144.

[30] E. R. Andrew, R. Bersohn, Nuclear magnetic resonance line shape for a triangular configuration of nuclei, *J. Chem. Phys.* **12** (1950) 159–161.

[31] T. Polenova, R. Gupta, A. Goldbourt, Magic angle spinning NMR spectroscopy: a versatile technique for structural and dynamic analysis of solid-phase systems, *Analytical chemistry* **87** (2015) 5458–5469.

[32] D. Lee, E. Bouleau, P. Saint-Bonnet, S. Hediger, G. De Paëpe, Ultra-low temperature MAS-DNP, *J. Magn. Res.* **264** (2016) 116–124.

[33] T. J. Schmugge, C. D. Jeffries, High dynamic polarization of protons, *Phys. Rev.* **138** (1965) A1785–A1801.

[34] C. Talón, M. A. Ramos, G. J. Cuello, et al., Low-temperature specific heat and glassy dynamics of a polymorphic molecular solid, *Phys. Rev. B* **58** (1998) 745–755.

[35] Y. Zhou, B. E. Bowler, G. R. Eaton, S. S. Eaton, Electron spin lattice relaxation rates for S = 1/2 molecular species in glassy matrices or magnetically dilute solids at temperatures between 10 and 300 K, *J. Magnetic Resonance* **139** (1999) 165–174.

[36] W. de Boer, T. O. Niinikoski, Dynamic proton polarization in propanediol below 0.5 K, *Nucl. Instrum. and Meth.* **114** (1974) 495–498.

[37] E. A. Harris, K. S. Yngvesson, Spin-lattice relaxation in some iridium salts I. Relaxation of the isolated $(IrCl_6)_2$-complex, *J. Phys. C (Proc. Phys. Soc.)* **1** (1968) 1011–1023.

[38] W. de Boer, High proton polarization in 1,2-propanediol at ^{3}He temperatures, *Nucl. Instr. and Meth.* **107** (1973) 99–104.

[39] R. Orbach, Spin-lattice relaxation in rare-earth salts, *Proc. R. Soc. A* **264** (1961) 458–484.

[40] R. Orbach, Spin-lattice relaxation in rare-earth salts: field dependence of the two-phonon process, *Proc. R. Soc. A* **264** (1961) 485–495.

[41] M. Blume, R. Orbach, Relaxation by two phonons in ground-state multiplets, *Phys. Rev.* **127** (1961) 1787.

[42] J. S. Waugh, C. P. Slichter, Mechanism of nuclear spin-lattice relaxation in insulators at very low temperatures, *Phys. Rev.* **B37** (1988) 4337–4339.

[43] E. A. Harris, K. S. Yngvesson, Spin-lattice relaxation in some iridium salts II. Relaxation of nearest-neighbour exchange-coupled pairs, *Journal of Physics C: Solid State Physics* **1** (1968) 1011–1023.

[44] T. O. Niinikoski, F. Udo, "Frozen spin" polarized target, *Nucl. Instr. and Meth.* **134** (1976) 219–233.

[45] R. B. Stephens, G. S. Cieloszyk, G. L. Salinger, Thermal conductivity and specific heat of non-crystalline solids: polystyrene and polymethyl methacrylate, *Phys. Lett.* **A 38** (1972) 215–217.

[46] T. O. Niinikoski, Polarized targets at CERN, in: M. L. Marshak (ed.) *Int. Symp. on High Energy Physics with Polarized Beams and Targets*, American Institute of Physics, Argonne, 1976, 458–484.

[47] Spin Muon Collaboration (SMC), D. Adams, B. Adeva, et al., Spin asymmetry in muon-proton deep inelastic scattering on a transversely-polarized target, *Phys. Lett.* **B336** (1994) 125–130.

[48] J. M. Ziman, *Principles of the Theory of Solids*, Cambridge University Press, Cambridge, 1965.

[49] C. J. Gorter, *Parametric Relaxation*, Elsevier, New York, 1947.

[50] W. Wendler, T. Herrmannsdörfer, S. Rehmann, F. Pobell, Electronic and nuclear magnetism in platinum-iron at ultralow temperatures, *Platinum Metals Rev.* **40** (1996) 112–116.

[51] A. S. Oja, O. V. Lounasmaa, Nuclear magnetic ordering in simple metals at positive and negative nanokelvin temperatures, *Rev. Mod. Phys.* **69** (1997) 1–136.

[52] W. de Boer, M. Borghini, K. Morimoto, T. O. Niinikoski, F. Udo, Dynamic polarization of protons, deuterons and carbon-13 nuclei: thermal contact between nuclear spins and electron spin-spin interaction reservoir, *J. Low Temp. Phys.* **15** (1974) 249–267.

[53] B. van den Brandt, H. Glättli, I. Grillo, et al., Time-resolved nuclear spin-dependent small-angle neutron scattering from polarised proton domains in deuterated solutions, *The European Physical Journal B – Condensed Matter and Complex Systems* **49** (2006) 157–165.

6
NMR Polarization Measurement

In this chapter we shall first briefly review the principles of the continuous-wave (CW) NMR techniques used for the measurement of the nuclear polarization in polarized targets. These principles were discussed extensively in Chapters 1, 2 and 5, and we shall refer back to them and remain brief here.

The circuit theory of the series-tuned Q-meter is then described in detail in Section 6.2, in view of calculating precisely the CW NMR absorption signal and its integral, the signal-to-noise ratio, the probe coupling and sampling and the signal saturation.

Optimization of the series-tuned Q-meter circuit is discussed in Section 6.3, on the basis of the above circuit theory.

Improved Q-meter circuits will be reviewed in Section 6.4. These include the capacitively coupled Q-meter, the crossed-coil NMR circuit and the introduction of quadrature mixer that enables the measurement of the real and imaginary parts of the radio frequency (RF) signal simultaneously.

Calibration and measurement of very small NMR signals then follows in Section 6.5. Here we shall also treat the signal-to-noise issues, the electromagnetic interferences (EMIs) and the NMR circuit drift issues.

6.1 Principles of the NMR Measurement of Polarization

6.1.1 Spin Polarization and Magnetization

The CW NMR method is certainly the most accurate polarization measurement technique today. It allows to determine the spin polarization of all nuclear species individually, and is not sensitive to impurities, contaminants or addenda other than those containing unpolarizable nuclei of the same species. Its principle relies on the relationship between the integral of the NMR absorption signal and the static magnetization due to the spin species of interest. The method is inherently accurate for nuclei with spin 1/2, but its accuracy can be shown to suffer little from the quadrupole interaction of higher spins, if the magnetic dipole interaction with the static field is substantially stronger than the quadrupole interaction.

The vector polarization $P(I)$ of the spin species I was defined by Eq. 1.60 as the expectation value of the component of the spin along the static field, divided by its maximum value:

$$P(I) = \frac{\langle I_z \rangle}{I}. \tag{6.1}$$

The spin polarization is associated with the static magnetization given by Eq. 1.64, which we also repeat here

$$M_z(I) = \hbar \gamma_I n_I I P(I). \tag{6.2}$$

Because dynamic polarization requires a rather large concentration of electronic spins with a large gyromagnetic factor, the determination of the nuclear spin polarization based on the measurement of the static magnetization in the material is subject to errors in the measurement of the electronic spin density. Although this, in principle, can be overcome by determining the magnetization from the frequency shift of a narrow NMR line between zero and finite polarizations of the given spins, with electronic polarization unchanged, there remains the problem of knowing the concentrations and polarizations of the contaminant nuclear spins in the material. This is why in all polarized targets today the magnetization and polarization are measured by magnetic resonance methods; their selectivity is based on the fact that the Larmor precession frequencies of the various nuclear spin species are usually very well resolved. It may be instructive, however, that even the resonance methods (basically) rely on the above *linear* relation of the static magnetization with respect to the polarization, which was pointed out already in Section 2.2.1.

This was seen by considering the static susceptibility[1] due to spins I

$$\chi_0 = \frac{\mu_0 M_z(I)}{B_0}, \tag{6.3}$$

which was also seen to be linear in $P(I)$ by inserting Eq. 6.2:

$$\chi_0 = \frac{\mu_0 \hbar \gamma_I n_I I}{B_0} P(I). \tag{6.4}$$

These equations are very general as they do not require the populations of the magnetic energy levels to be in thermal equilibrium (TE), although they were first discussed in Chapter 1, Eqs. 1.64–1.66, in the terms of Boltzmann distribution of the level populations.

6.1.2 Transverse Susceptibility

We shall now relate the static susceptibility to the generalized complex transverse susceptibility $\chi(\omega)$, which describes the response of the spin system to a time-dependent excitation $B_x = B_1 \cos \omega t$ of such small amplitude that the response of the system remains linear. This requires the amplitude B_1 to satisfy inequalities 5.14 and 2.139.

We remind that this frequency-dependent susceptibility has the real and imaginary parts

---

[1] In SI system, the usual definition is $\mathbf{B} = \mu_0(\mathbf{H} + \mathbf{M}) = (1 + \chi_0)\mu_0 \mathbf{H}$, which deviates slightly from the definition of Eq. 1.65. The difference is insignificant in dilute paramagnetic systems and totally negligible in nuclear magnetism, even at high polarizations.

6.1 Principles of the NMR Measurement of Polarization

$$\chi(\omega) = \chi'(\omega) - i\chi''(\omega) \tag{6.5}$$

called dispersion and absorption parts, respectively, because they describe the dispersion and absorption of the electromagnetic wave which excites the magnetization of the spin system in the plane perpendicular to the static field B_0; the absorption and dispersion of the wave were discussed in Section 2.2.1. In a linear system, these parts are related with each other by the Kramers–Krönig equations introduced in Section 2.2.2:

$$\chi'(\omega) - \chi'(\infty) = \frac{1}{\pi} P \int_{-\infty}^{+\infty} \frac{\chi''(\omega')}{\omega' - \omega} d\omega', \tag{6.6}$$

$$\chi''(\omega) = -\frac{1}{\pi} P \int_{-\infty}^{+\infty} \frac{\chi'(\omega') - \chi'(\infty)}{\omega' - \omega} d\omega', \tag{6.7}$$

which give restrictions and extremely general and useful relationships on the NMR signal by adding few other assumptions or observations. The only additional assumptions, which we require, are that

(1) the spin absorption line be a single narrow function of frequency;
(2) the dispersion due to the spin system should tend to zero at infinite frequency (which in practice means that the particular spin must not possess higher magnetic resonance frequencies due to the inner structure of the nucleon).

It is rather easy to show that the second assumption follows the first one, if the deeper structure of the nucleon can be regarded independent of the external spin degrees of freedom, which is well satisfied at field values obtainable in the laboratory.

Because the response of the spin system to a linearly polarized transverse oscillating magnetic field should be totally independent of the sign of the frequency of oscillation, the dispersion and absorption must be symmetric and antisymmetric functions, respectively, about the zero frequency, i.e. $\chi'(\omega) = \chi'(-\omega)$ and $-\chi''(\omega) = \chi''(-\omega)$. This also implies $\chi''(0) = 0$, i.e. that absorption be zero with a static transverse field, a result which can be understood even without linear response theorems. The same is true with dispersion at zero frequency: supposing that the transverse field $H_x = H_x(0)$ is steady and very much smaller than the steady main field $H_z = H_0$, its effect is to rotate the field and magnetization slightly so that

$$\frac{M_x(0)}{M_z} = \frac{H_x(0)}{H_z} \tag{6.8}$$

and the magnitude of the magnetization along the z-axis remains unchanged. We have then from Eq. 1.65

$$M_x(0) = \frac{\mu_0 M_z}{B_0} H_x(0) = \chi_0 H_x(0). \tag{6.9}$$

On the other hand, the definition of the susceptibility is just

$$M_x(0) = \chi'(0) H_x(0), \qquad (6.10)$$

which proves that

$$\chi'(0) = \chi_0. \qquad (6.11)$$

On the other hand, we may express $\chi'(0)$ using the Kramers–Krönig relation 6.6 at zero frequency

$$\chi_0 = \chi'(0) = \frac{1}{\pi} P \int_{-\infty}^{+\infty} \frac{\chi''(\omega')}{\omega'} d\omega'. \qquad (6.12)$$

As the absorption part of the susceptibility is antisymmetric with respect to 0, the integration can be performed from 0 to $+\infty$ and multiplied by 2; inserting now the static susceptibility from Eq. 5.6 gives

$$\chi_0 = \frac{\mu_0 \hbar \gamma_I n_I I}{B_0} P(I) = \frac{2}{\pi} \int_0^{+\infty} \frac{\chi''(\omega)}{\omega} d\omega, \qquad (6.13)$$

which yields for P after expressing the magnetic field in the terms of the Larmor frequency

$$P(I) = \frac{2}{\pi \mu_0 \hbar \gamma_I^2 n_I I} \int_0^{+\infty} \chi''(\omega) \frac{\omega_0}{\omega} d\omega. \qquad (6.14)$$

Assuming that the absorption signal is very narrow and in addition symmetric about its centroid, one may replace ω in the denominator of the integrand by the Larmor frequency ω_0, which yields the usual linear relationship of Eq. 2.65 between the integrated absorption signal and the polarization of the material.

In the case of deuterons and many other nuclei with $I \geq 1$, the quadrupole broadening may be so large that the above statements do not hold well enough. This happens in particular at low magnetic field (say, 0.5 T for deuterons), at high polarizations when the absorption lineshape becomes very asymmetric, and especially for ^{14}N. The usual formula must then be replaced by Eq. 6.14, which we shall always use unless stated differently.

At high field (say, 2.5 T), the difference between this expression and the more customary formula, with the integral of the absorption curve only, remains smaller than 1% at $P_D = 0.5$, and can easily be corrected for. Such corrections, however, have seldom been made, because in the experimental absorption signal there are always several other distortions which may cause even larger errors. These come from the measurement circuitry on the one hand, and on the other hand from the microscopic description of the non-saturating passage of the deuteron resonance line at very low spin temperature.

The integral of Eq. 6.14 accurately gives the polarization of the spins I, once the parameters in front of the integral have been determined experimentally. In practice, this is achieved by measuring the real part of the NMR signal, known to be a good representation of the imaginary part of the transverse susceptibility, under conditions in which the polarization is accurately known. This calibration process yields the missing parameters, and it is usually performed at 1 K temperature. At this temperature, the Curie law of Eq. 1.70 gives a very accurate value to the polarization. This and the calibration procedure for very small signals will be discussed further in Section 6.5.1.

6.2 NMR Signal Measurement Using a Series-Tuned Q-Meter

Two main types of Q-meter have been used for NMR signal acquisition and spin polarization measurement in polarized targets: the parallel-tuned circuit and the series-tuned circuit. Both circuits feature an inductive circuit element placed in or around the target, and they differ obviously by the placement of the tuning capacitor, either in parallel or in series with the inductor.

At resonance the parallel-tuned circuit has maximum impedance, whereas the series-tuned circuit has minimum impedance. For a linear response the coupling resistances (or impedances) should be substantially higher than the resonator impedance; in the series-tuned circuit this is easily realizable, while in the parallel-tuned circuit the resistances must be so high that the parasitic capacitances begin to influence the circuit tune significantly. Moreover, in the series-tuned circuit, the theoretical signal is linear with the resonance susceptibility in first order, while in the parallel-tuned circuit the signal is inversely proportional to the susceptibility, i.e. the circuit is inherently non-linear.

These facts were recognized in 1967 by Petricek [1], and various implementations of the series-tuned Q-meter were used for the polarization measurement. In the earliest ones, the RF signal, amplified after the resonant circuit, was converted by a simple diode detector that is sensitive only to the signal amplitude [2, 3]. A substantial improvement in the linearity and noise performance was achieved when Court and coworkers introduced the double balanced mixer that enables phase-sensitive detection of the real part of the signal, and therefore yields a much improved separation of the absorption and dispersion parts of the susceptibility [4, 5]. The team developed and designed the circuit, called Liverpool Q-meter; this was produced semi-industrially and continues to be used successfully by most polarized target teams in the world [6].

6.2.1 The Series-Tuned Q-Meter Circuit Theory

The schematic diagram of the series Q-meter circuit is shown in Figure 6.1. The coupling admittance of the hybrid resonator is

$$Y = \frac{1}{R_i} + \frac{1}{R_o}, \qquad (6.15)$$

where R_o is the oscillator feed resistance and R_i is the amplifier input impedance, both of which are assumed to be purely real parameters. The real and imaginary parts of the voltage

u_i in the input of the amplifier can be written in terms of the real and imaginary parts of the resonator impedance Z as

$$\text{Re}\{u_i\} = u_o \frac{\text{Re}\{Z\} + Y\left[\text{Re}^2\{Z\} + \text{Im}^2\{Z\}\right]}{R_o\left[(1+Y\text{Re}\{Z\})^2 + Y^2\text{Im}^2\{Z\}\right]} \tag{6.16}$$

and

$$\text{Im}\{u_i\} = u_o \frac{\text{Im}\{Z\}}{R_o\left[(1+Y\text{Re}\{Z\})^2 + Y^2\text{Im}^2\{Z\}\right]}. \tag{6.17}$$

The hybrid resonator impedance Z is

$$Z = R + \frac{1}{i\omega C} + Z_c \frac{Z_L + Z_c \tanh \gamma \ell}{Z_c + Z_L \tanh \gamma \ell}, \tag{6.18}$$

where the coil impedance Z_L can expressed in terms of the RF susceptibility and effective filling factor of the spin polarized material η (see Section 6.2.5) by

$$Z_L = R_L + i\omega L\{1 + \eta[\chi'(\omega) - i\chi''(\omega)]\}, \tag{6.19}$$

and the propagation constant and characteristic impedance of the coaxial line of electrical length ℓ are given by

$$\gamma = \sqrt{(R_c + i\omega L_c)(G_c + i\omega C_c)} \cong i\omega\sqrt{L_c C_c}\left(1 + \frac{1}{2iQ_c}\right) \tag{6.20}$$

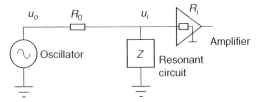

a) Q-meter circuit

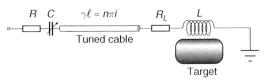

b) Resonant circuit

Figure 6.1 Series Q-meter circuit – (a) block diagram, not showing the matching elements of a coaxial line between the oscillator and the circuit blocks; – (b) elements of the hybrid resonator circuit, with a tuned transmission line connecting the series coil and tuning capacitor

and

$$Z_c = \sqrt{\frac{(R_c + i\omega L_c)}{(G_c + i\omega C_c)}} \cong Z_0 \left(1 + \frac{1}{2iQ_c}\right). \tag{6.21}$$

Here the subscript c labels the distributed parameters of the coaxial transmission line, and we have furthermore defined (as usual)

$$Z_0 = \sqrt{\frac{L_c}{C_c}} \text{ and } Q_c = \frac{\omega L_c}{R_c}. \tag{6.22}$$

These definitions and the approximations of Eqs. 6.20 and 6.21 were first time used in Ref. [7] and it is important to note that the precise understanding of the tuning and shape distortions of wide-sweep NMR circuits should use these or even more accurate expressions, rather than the usual parameters involving a real characteristic impedance and a simple (or zero) attenuation constant.

6.2.2 Series Q-Meter Signal Expansion

Expressions 6.16 and 6.17, in principle, give accurately the output signal of the apparatus used for NMR measurements, once the gain of the amplifying and detecting circuitry is known. These or slightly less accurate expressions have been used for numeric simulation of the output signal, as a response to numeric input spectra of the RF susceptibility. Such simulations are useful for many purposes such as evaluating the distortions and non-linearity of a known circuit. However, these simulations are tedious for the optimization of the circuit itself and are unhelpful for the understanding of the circuit tuning procedure. To gain insight in these, Eq. 6.16 was expanded in power series of the susceptibility, with coefficients themselves expanded as power series of the frequency offset from the Larmor frequency [7]. With appropriate tuning the real part of the RF voltage u_t is

$$\begin{aligned}\operatorname{Re}\{u_t\}\frac{R_o}{u_o} &= A_0(x) + \\ &+ A_1(x)\eta\omega L\chi''(x) + A_1'(x)\eta\omega L\chi'(x) + \\ &+ A_2(x)(\eta\omega L)^2\left[-\chi''^2(x) + \chi'^2(x)\right] + A_2'(x)(\eta\omega L)^2\chi'(x)\chi''(x) + \\ &+ \cdots\end{aligned} \tag{6.23}$$

where x is the relative frequency offset from the Larmor precession

$$x = \frac{\omega - \omega_0}{\omega_0}. \tag{6.24}$$

The term $A_0(x)$ is proportional to the experimental Q-curve, which can be measured when $\chi(\omega) = 0$; this is obtained by shifting the main field off resonance. The aim of the circuit design is to make all higher order coefficients A_n, A_n' small except $A_1(x)$, which should be a large and flat function of x. These features provide selectivity for the absorption lineshape, yielding low distortions and good linearity. The coefficients A_n, A_n' depend only on the circuit parameters, which was discussed in Ref. [8]. They can be approximated for narrow frequency sweep, fairly high Q_c and low Y by

$$A_0(x) = R' + \rho\omega L x + Y\left\{-(R' + \rho\omega L x)^2 + (\omega_0 L)^2 (\delta_t + \delta + 2x)^2\right\} + \cdots \tag{6.25}$$

$$A_1(x) = 1 + \frac{n\pi}{4Q_c^2} + \rho\left(\frac{1}{2Q_c} - \frac{Q}{4Q_c^2}\right) + x\left(\frac{n\pi}{2Q_c^2} - \rho Q\right) +$$
$$-2Y\left\{(R' + \rho\omega L x) + \omega_0 L(\delta_t + \delta + 2x)\left[\frac{\rho}{4Q_c^2}(2Q+1) + \frac{(n\pi)^2}{Q_c}x\right]\right\} + \cdots \tag{6.26}$$

$$A_1'(x) = \left[1 - 2Y(R' + \rho\omega L x)\right]\left[\frac{\rho Q}{4Q_c^2}(2Q+1) + \frac{(n\pi)^2}{Q_c}x\right] + 2\omega_0 L(\delta_t + \delta + 2x) + \cdots \tag{6.27}$$

$$A_2(x) = -\frac{n\pi}{2Z_0 Q_c} + Y\left[1 + \frac{n\pi\omega_0 L}{Z_0}(\delta_t + \delta + 2x)x\right] + \cdots \tag{6.28}$$

$$A_2'(x) = -\frac{2n\pi}{Z_0}x + \cdots, \tag{6.29}$$

where $Q = \omega L / R_L$ is the Q-factor of the coil which has the equivalent series resistance R_L, and δ is the relative offset of the cable resonance frequency from the Larmor precession. For brevity we have used the following parameters:

- effective damping resistance R' of Z at ω_0

$$R' = R + R_L + \frac{n\pi Z_0}{2Q_c}\left\{1 + \left(\frac{R_L}{Z_0}\right)^2(Q^2 - 1)\right\}; \tag{6.30}$$

- frequency sensitivity of effective damping

$$\rho = \frac{n\pi R_L}{Z_0}; \tag{6.31}$$

- tune shift due to damping in the coaxial line

$$\delta \cong \frac{n\pi R_L^2}{2Q_c Z_0^2}(Q^2 - 1); \tag{6.32}$$

- and capacitor tune shift

$$\delta_t = 1 - \frac{1}{\omega_0^2 LC}. \tag{6.33}$$

The above equations were derived using the following approximation for $\tanh \gamma \ell$:

$$\tanh \gamma \ell \cong n\pi \left(\frac{1+x}{2Q_c} + ix \right), \tag{6.34}$$

which is good for $|x| \leq 0.1$ when $n = 1$ and $Q_c \approx 30$ (as defined by Eq. 6.22).

6.2.3. Ab Initio NMR Signal Size

One of the problems in NMR polarization measurement of the deuteron and other spin-1 systems is related with the calibration of the integrated absorption signal at a known polarization, usually at 1 K temperature where the spin-lattice relaxation is rapid ($\approx$1 minute) and where the TE and uniformity are very easy to achieve, by immersing the target in pure superfluid ^{4}He. The size of the signal under these conditions is very small and requires signal averaging over extended periods of time ($\approx$30 min) in order to show up above the noise. This would be relatively simple to do, if the circuit drift during averaging could be eliminated. There is, in fact, an optimum averaging time, beyond which the Q-curve drift results in errors greater than that due to the noise. Only repeated measurements and statistical analysis then will enable one to gain in the statistical accuracy, at the cost of some loss in systematics, because details of the signal such as the line center cannot be resolved with precision. In the following, we shall calculate the size of the NMR signal from first principles using the circuit model parameters and compare it with the amplifier and oscillator noise. Deuterons are used as an example, but the treatment is valid for all NMR signals of comparable size.

In the case of deuteron NMR signal, the susceptibility is so small in all experimental conditions that the contribution of the second and higher order terms in Eq. 6.23 can be ignored. Furthermore, the frequency sweep may be limited to $\pm 1.5\%$ where the frequency dependence of the coefficients A_n, A_n' may be small if the circuit is well designed and tuned. We may then write the output signal in the form (after subtracting the Q-curve)

$$S(\omega) \equiv G\left[\text{Re}\{u_t(\chi)\} - \text{Re}\{u_t(0)\}\right] \cong \frac{Gu_o}{R_o} \eta \omega_0 (1+x) L \left[A_1(x)\chi''(x) + A_1'(x)\chi'(x)\right], \tag{6.35}$$

where u_o is RF source voltage and G is the system voltage gain between the inputs of the preamplifier and the analog-to-digital converter (ADC).

By integrating Eq. 6.35 we can write the relationship between the signal integral, the polarization and the circuit parameters in the form appropriate for wide sweep:

$$\int_{\omega_{\min}}^{\omega_{\max}} S(\omega) \frac{\omega_0^2}{\omega^2} d\omega \cong \frac{Gu_o}{R_o} \eta \omega_0^2 L \int_{x_{\min}}^{x_{\max}} \frac{dx}{1+x} \left[A_1(x)\chi''(x) + A_1'(x)\chi'(x)\right]. \tag{6.36}$$

If $A_1(\omega)$ is a flat function and $A_1'(\omega)$ is a symmetric function about ω_0, and if the absorption part of the susceptibility is also symmetric with cut-offs in both wings, then we can write using Eq. 6.14

$$\int_{\omega_{min}}^{\omega_{max}} S(\omega)\frac{\omega_0^2}{\omega^2}d\omega \cong \eta\omega_0 L \frac{S_0 A_1(0)}{2A_0(0)} \pi\mu_0\hbar\gamma_I^2 n_I IP(I), \tag{6.37}$$

where the gain, oscillator voltage and feed resistance values were expressed in the terms of the signal voltage S_0 at the minimum of the Q-curve, with no NMR signal present:

$$S_0 = A_0(0)\frac{Gu_0}{R_0}. \tag{6.38}$$

The relationship 6.37 gives polarization in the terms of variables that are accurately and directly measurable and/or calculable, with the exception of the effective filling factor η. If the filling factor is known from other measurements, then no calibration is required for the measurement of polarization. The approximate ratio of the coefficients $A_1(0)/A_0(0)$ from Eqs. 6.25 and 6.26

$$\frac{A_1(0)}{A_0(0)} \cong \frac{(1-2YR')(1-\rho Q/Q_c^2)}{R'(1-YR')} \tag{6.39}$$

is obtained using the circuit parameters, the values of which can be verified with precision by fitting the Q-curve with the theoretical expression of Eq. 6.16. The spin density is obtained from the density of the material and from the chemical formulae of its components. We shall discuss below several applications of Eq. 6.37.

6.2.4 Signal-to-Noise Ratio

In the following we shall assume narrow NMR absorption lines, which are the usual case in high field, such as protons and deuterons at 2.5 T. Within the frequency range of interest, we may then put $(\omega_0/\omega)^2 \approx 1$, which allows us to define the absolute effective signal strength S_{eff} from Eq. 6.37 once we know the effective width of the NMR signal $\Delta\omega_{eff}$, defined as

$$\int_0^{+\infty} S(\omega)d\omega \cong S_{eff}\Delta\omega_{eff}. \tag{6.40}$$

As an example, let us estimate the deuteron NMR signal height in glassy deuterated butanol-water, with $\Delta\omega_{eff} = 2\pi \cdot 280$ kHz, at 2.5 T field. With circuit parameters $R' = 15\ \Omega$, $L = 450$ nH, $n = 1$, $Q = 10$, $Q_c = 80$, and $Y \cong 1/(50\ \Omega)$ and filling factor $\eta = 0.2$ usual with embedded probe coils, we find from Eqs. 6.37–6.40

$$\frac{S_{eff}(P=0.5)}{S_0} = 2.7 \cdot 10^{-3} \tag{6.41}$$

and

$$\frac{S_{\mathit{eff}}\left(P=5\cdot 10^{-4}\right)}{S_0}=2.7\cdot 10^{-6}, \tag{6.42}$$

the first case corresponding to maximum deuteron polarization and the second to the calibration signal measured in TE around 1 K temperature. With about 0.3 mA current into the resonant circuit, we have $S_0 = 4.5$ mV; this corresponds to 12 µV and 12 nV changes in the enhanced and calibration signal amplitudes at the input of the preamplifier, respectively. These have to be compared with the amplifier noise floor of 40 nV (with 30 kHz single-sideband bandwidth) and the oscillator noise of 100 nV (using the RF source signal-to-noise ratio of 93 dB). The theoretical TE signal-to-noise ratio is therefore

$$\frac{S_{\mathit{eff}}^{TE}}{V_n} = \frac{2.7 \times 10^{-6} 4.5 \text{ mV}}{\sqrt{1.16} \times 100 \text{ nV}} \cong \frac{1}{8.9}, \tag{6.43}$$

which indicates that the TE signal is always completely invisible without signal averaging. The fully polarized signal, however, is about hundred times higher than the root-mean-square (RMS) noise, but even this is only about one-third of the depth of the Q-curve (with 500 kHz sweep width).

Above we have calculated the signal-to-noise ratio from the first principles, applicable for a raw experimental spectrum obtained by one frequency scan. A large reduction of noise is obtained by the appropriate processing of many such signals. The processing consists of averaging a number of spectra, subtracting the Q-curve obtained by averaging over same number of spectra without NMR signal, correcting for the Q-curve drift and field effect by subtracting a function (usually a polynomial of order 2) obtained by fitting the sides of the spectrum to an expected residual Q-curve, and integrating the resulting spectrum. This procedure is repeated a number of times to reach the desired statistical accuracy and to examine systematic effects in the equipment used for the measurements.

In the CERN deuteron NMR equipment [9], the deuteron NMR signals are digitized and averaged at N_p points of the frequency scan with about 500 kHz width. The scan is made by stepping from the minimum to the maximum frequency and then stepping back to the minimum frequency; a scan thus has two measurements of the spectrum. Averaging N_s such scans reduces the noise by $[2N_s]^{-1/2}$; subtraction of a Q-curve obtained with equal number of scans increases the noise by $2^{1/2}$. N_e points on each end of the scan fall out of the absorption spectrum and are used to determine the residual Q-curve under the signal. Integration of the signal then improves the signal-to-noise ratio by a factor f

$$f = \frac{\sqrt{N_s}}{\sqrt{\frac{1}{N_p - 2N_e} + \frac{1}{2N_e}}}. \tag{6.44}$$

With $N_s = 10^4$, $N_p = 400$ and $N_e = 70$, the above equation gives $f = 954$; applying this to Eq. 6.44 yields a statistical accuracy of about 1% for determining the integrated absorption spectrum from one set of averaged signals and Q-curves. This was confirmed by determining the RMS noise in the experimental signal outside the edges of the DMR absorption lineshape [10].

By repeating the measurement N_n times, a further theoretical improvement by a factor of $[N_n]^{1/2}$ will be obtained in the statistical accuracy of the integrated calibration signal. With $N_n = 100$ the theoretical statistical accuracy is 0.1%. This can be achieved in about 2 days of data taking. The result is interesting because it has been previously thought that precise TE calibration for deuterons is impossible without substantial saturation. This result also calls for improved control of systematics if a comparable systematic accuracy is desired.

6.2.5 Saturation

In the following we shall assume a series-tuned Q-meter circuit in which a double resonant circuit consists of the probe coil in series with a capacitor, with tuned coaxial line connecting these elements, as was described in Section 6.2.1. By appropriate choice of the damping and coupling parameters, a linear measurement of the RF transverse susceptibility $\chi(\omega)$ results. The choice of the circuit parameters can also be made so that the real part of the output signal is an almost undistorted absorption (imaginary) part of the susceptibility and contains only a small contribution from the dispersion (real) part of the susceptibility.

In performing the NMR measurement, a small but finite loss of polarization occurs due to the saturation of the polarization near the probe wire. Here we shall focus on the errors in the polarization measurement due to this saturation. To be able to obtain practical results, we first derive the approximate equations for the sampling efficiency as a function of radial distance from the NMR probe coil wire. As a by-product we shall obtain a fundamental formula for the filling factor η of the probe coil, which is required for absolute calculations of the signal size.

The term 'saturation' describes a parameter that depends on the RF field strength and on the sequence of the measurement. This parameter is needed in the equations for the error in polarization measurement during DNP. It is particularly useful in the evaluation of the decay of the NMR signal after the DNP is stopped and the target is put in the 'frozen spin' state.

Numeric results are finally obtained for a simplified example where the NMR probe is a wire in a circular cavity with radius b, both coaxial with the main field, and the target material fills the cavity from radius a to b.

Self-Inductance and Continuous-Wave NMR Signal

The sinusoidal AC current in a conductor of a coil can be decomposed in rotating and counterrotating components as

$$i_{coil} = I\left(e^{j\omega t} + e^{-j\omega t}\right) = 2I \cos \omega t. \tag{6.45}$$

6.2 NMR Signal Measurement Using a Series-Tuned Q-Meter

The self-inductance[2] L of the coil carrying this current is defined as

$$L = \frac{1}{\langle i_{coil}^2 \rangle} \left\langle \int_v \vec{B} \cdot \vec{H} \, dv \right\rangle, \tag{6.46}$$

where the averaging is over time and the integral of the scalar product of the complex vectors of the magnetic field and induction is over the volume filled by the field. Note that we have decomposed the oscillating current to rotating and counterrotating components, which helps in writing the resulting field in rotating components, for example, at radius r in the proximity of a round wire:

$$H_1(t) = \frac{I}{2\pi r}\left(e^{j\omega t} + e^{-j\omega t}\right). \tag{6.47}$$

These definitions are used in the complex circuit theory which is the basis of the ab initio calculation of the absolute signal size of Section 6.2.3. Assuming constant frequency ω, the oscillating current gives rise to a voltage u when passing through the lumped-element impedance Z:

$$u_{coil} = i_{coil} Z = i_{coil}\left(R_L + j\omega L\right) = i_{coil}\left[R_L + j\omega L_0\left(1 + \eta\chi(\omega)\right)\right], \tag{6.48}$$

where $\chi(\omega)$ is the complex transverse susceptibility due to the resonance of the nuclear spins, and η is the effective filling factor of the probe coil. It is this voltage which shows up in the experimental NMR signal after amplification, phase-sensitive detection and the removal of the Q-curve. In the following we shall not focus on the problems related with the pure circuit theoretical considerations, which were treated in the previous section but wish to elaborate only on the geometrical sampling function of the probe coil, on the spatially inhomogeneous saturation of the resonance signal and on the effective filling factor of the probe coil.

The following simplifying assumptions are made:

(1) High frequency so that the skin depth is negligibly small; this avoids problems related with the complex susceptibility of the conductor material(s).
(2) Longitudinal susceptibility is zero because the longitudinal relaxation time of the nuclear spins is very long in comparison with the period of the RF field, and the longitudinal susceptibility due to first-order forbidden transitions at $2\omega_0$ is negligibly small at ω_0.
(3) Transverse susceptibility due to the nuclear spins is independent of time because the saturation is very small and the speed of passage through the resonance is very slow in comparison with the dipolar width, which is the main cause of homogeneous line broadening.

Sampling Function of the Probe Coil

We may now proceed to write the homogeneous static field in z-direction

$$\vec{H}_0 = H_0 \vec{z}^0, \tag{6.49}$$

---
[2] In the following we use the common abbreviation 'inductance' for the correct term 'self-inductance'.

the RF magnetic field of the probe coil

$$\vec{H}_1 = \left(\vec{H}_\perp + H_\parallel \vec{z}^0\right)\left(e^{j\omega t} + e^{-j\omega t}\right), \qquad (6.50)$$

and the RF magnetic induction:

$$\vec{B}_1 = \mu_0\left[\left(\vec{H}_\perp + H_\parallel \vec{z}^0\right)\left(e^{j\omega t} + e^{-j\omega t}\right) + \vec{H}_\perp \chi(\omega) e^{j\omega t}\right]. \qquad (6.51)$$

Here the vector fields and the complex susceptibility are position dependent, although we have not written this explicitly. Note also that only one of the rotating components of the field gives rise to transverse magnetization and contributes to the resonant susceptibility, because the component counterrotating with the spins is far from resonance.

Assuming that the wavelength at frequency ω is much larger than the size of the probe coil, we can readily write the inductance:

$$L(\omega) = \frac{\mu_0}{2I^2}\int_v \left[2\left(H_\perp^2 + H_\parallel^2\right) + \chi(\omega)H_\perp^2\right]dv = L_0 + \frac{\mu_0}{2I^2}\int_v \chi(\omega)H_\perp^2 dv \qquad (6.52)$$

where

$$L_0 = \frac{\mu_0}{I^2}\int_v H_1^2 dv \qquad (6.53)$$

is the inductance of the empty coil. We note that the susceptibility contributes linearly under the integral 6.52, but it is weighed by the square of the RF field component perpendicular to the static main field.

Equation 6.52 allows us to estimate numerically possible errors in the measurement of volume-averaged polarization if the polarization and hence the susceptibility are not uniform in the sampled volume, and if the perpendicular component of the RF field is known.

Filling Factor

Equation 6.52 can also be used to calculate the effective filling factor, which is defined by Eq. 6.48, and which is due to the fact that the target beads do not fill all the volume where the field is confined. This is done by writing the susceptibility in the form

$$\chi(\vec{r},\omega) = \Psi(\vec{r})\chi(\omega), \qquad (6.54)$$

where the spatial distribution function of the polarized material is separated from the frequency-dependent susceptibility, which is assumed independent of position. Inserting this in Eq. 6.52 and using Eq. 6.48 yields

$$\eta = \frac{\int_v \Psi(\vec{r}) H_\perp^2 dv}{2\int_v H_1^2 dv}. \qquad (6.55)$$

Here Ψ is 1 inside the target beads and 0 elsewhere.

Saturation

Saturation has a clear and simple meaning in magnetic resonance; it refers to the rate equations which describe the time evolution of the longitudinal susceptibility (or polarization) under time-varying transverse fields. For simplicity, let us assume a system of spins with $I = 1/2$ so that we may write the populations of spins with their z-component parallel and antiparallel to the field as N_+ and N_- so that the total number of spins and the population difference are

$$N = N_+ + N_-,$$
$$n = N_+ - N_-. \quad (6.56)$$

Under the action of the RF field, the populations undergo changes at rates

$$\frac{dN_+}{dt} = -Wn,$$
$$\frac{dN_-}{dt} = +Wn, \quad (6.57)$$

which are equal but opposite because of the conservation of the number of spins, and furthermore, are proportional to the matrix elements $|\langle +|V|-\rangle|^2 = |\langle -|V|+\rangle|^2$ of the interaction $V(t)$ due to the perturbing RF field.

Other perturbations also cause changes in the nuclear spin populations; among these are the spin-lattice interaction which is strongly temperature dependent, and the interaction of the nuclear spin system with that of the electronic spins. The former actually mainly proceeds via the latter. In both cases an equilibrium population ratio

$$\frac{N_-^0}{N_+^0} \equiv \frac{N - n_0}{N + n_0} = e^{-\hbar\omega_0/kT_S} \quad (6.58)$$

is asymptotically reached with a time constant which results from the rate equations. Here T_S is the spin temperature of the paramagnetic electron system; this temperature may be very low under favorable DNP conditions, and it may be positive or negative depending on the frequency of the microwave field. If there is no microwave field, the equilibrium spin temperature is equal to the lattice temperature T_L.

Spin-lattice relaxation in the absence of microwaves proceeds with a time constant T_1, which now yields another rate equation:

$$\frac{dn}{dt} = -2Wn + \frac{n_0 - n}{T_1}, \quad (6.59)$$

which has the solution (in equilibrium)

$$n = \frac{n_0}{1 + 2WT_1}. \quad (6.60)$$

We recognize here the saturation time constant

$$\tau_{RF} = \frac{1}{2W}, \qquad (6.61)$$

which is calculable, for example, by using Eq. 1.58. We also note that in the limit of weak saturation $2WT_1 \ll 1$ and $n \approx n_0$, whereas in the limit of strong saturation $n \approx 0$ according to our (unfortunately simplistic) Eq. 6.60. The case of NMR polarization measurement must fall in the limit of weak saturation in which case Eq. 6.60 holds well. The case of strong saturation requires a quantum statistical treatment, similar to DNP.

In our case, the CW NMR signal is measured during a slow passage through the resonance line, which is repeated a large number of times. The most general one can be found in [11] and it can be put in the form

$$P(k) = P(0)e^{-k\varepsilon} \qquad (6.62)$$

where k counts the passages and

$$\varepsilon = \frac{\pi \gamma^2 B_1^2}{|\dot{\omega}|}. \qquad (6.63)$$

If passages are made with intervals of Δt, we find the effective saturation time constant

$$\tau_{RF}^{\text{eff}} = \Delta t \frac{|\dot{\omega}|}{\pi \gamma^2 B_1^2}. \qquad (6.64)$$

The saturation thus varies inversely proportional to the squared transverse RF field strength, which indicates that care must be exercised in order not to partially kill the polarization near the wire of the probe coil if it is embedded in the target material. Some errors have been made in this respect in the past, due to the fact that a thin bare wire has been placed directly in contact with the target material.

We shall now estimate the polarization $P^*(t)$ from the NMR signal due to the slightly saturating NMR measurement. Its decay is non-exponential and is readily calculated from the above formulae. Under the assumption that at time $t = 0$ the polarization is uniform with a value P_0, and that there is no time evolution due to DNP or spin-lattice relaxation, we find

$$P^*(t) = P_0 \frac{\int_v e^{-t/\tau_{RF}^{\text{eff}}(r)} H_\perp^2(r) dv}{\int_v H_\perp^2(r) dv}. \qquad (6.65)$$

The volume-averaged effective instantaneous saturation time $P^*(t)/[dP^*(t)/dt]$ is in this case not a constant but increases as a function of time in a way which depends on the geometry of the probe coil.

Error in Polarization Measurement Due to Saturation

Equations 6.59 and 6.60 can be also used for estimating the saturation due to NMR during DNP, by simply replacing T_1 by the polarizing time constant T_{pol}:

$$\frac{dP}{dt} = -2WP + \frac{P_\infty - P}{T_{pol}};\qquad(6.66)$$

$$P = \frac{P_\infty}{1 + T_{pol}/\tau_{RF}^{eff}}.\qquad(6.67)$$

The error made in the estimation of the steady equilibrium DNP value P_∞ from the slightly saturated NMR signal can now be calculated by

$$\frac{P^*}{P_\infty} = \frac{\int_v \frac{H_\perp^2(r)}{1 + T_{pol}/\tau_{RF}^{eff}(r)} dv}{\int_v H_\perp^2(r) dv},\qquad(6.68)$$

where the explicit radial position dependence has been written for clarity.

Example for Error Due to Saturation of Deuteron NMR during DNP

Let us assume a cylindrical cavity with radius $b = 20$ mm, with wire of radius a in its center coaxial with the main steady magnetic field and the z-axis. The RF current runs along the wire and returns along the cavity, forming a kind of coaxial transmission line, with electrical length much smaller than the wavelength. The sweep interval is 40 ms and the amplitude of the current in the coil is 0.3 mA (approximate values for the Liverpool Q-meters used with the deuterated target of Spin Muon Collaboration (SMC) [10]. Let us denote the wire length by ℓ. For numeric estimates we take $b = 20$ mm and $a = 0.5, 1, 1.5$ and 2 mm.

Let us first write expressions for the inductance with the sampling function:

$$L = L_0 + \chi(\omega)\mu_0 \ell \int_a^b \frac{\Psi(\vec{r})}{4\pi r} dr,\qquad(6.69)$$

where

$$L_0 = \frac{\mu_0 \ell}{2\pi} \ln\frac{b}{a}.\qquad(6.70)$$

The probe 'coil' (a coaxial resonator in our present case) filling factor is

$$\eta = \frac{\int_a^b \frac{\Psi(\vec{r})}{r} dr}{2\int_a^b \frac{1}{r} dr};\qquad(6.71)$$

this is the 1/r rule for the signal sampling function near a current-carrying wire.

The effective saturation time constant at distance r from the center of the probe coil wire is

$$\tau_{RF}^{\text{eff}} = \Delta t \frac{|\dot{\omega}|}{\pi\gamma^2 B_1^2} = 45.7\,\text{h} \times \left(\frac{r}{\text{mm}}\right)^2 \tag{6.72}$$

so that the nearest beads at 2 mm radius see a saturation time constant of about 183 h. The measured average saturation time constant was 250 h; the above calculation did not take into account the real coil geometry and ignored that, between the blocks of a few hundred double sweeps, some time was spent for transferring the NMR data and analysing the signals, before a new block of sweeps was executed.

The error in polarization determined from NMR signal is then obtained from Eq. 6.68 in steady state of DNP:

$$\frac{P^*}{P_\infty} = \frac{\log \dfrac{\dfrac{b^2}{a^2} + \dfrac{T_{\text{pol}}}{\tau_{RF}^{\text{eff}}(a)}}{1 + \dfrac{T_{\text{pol}}}{\tau_{RF}^{\text{eff}}(a)}}}{2\log\dfrac{b}{a}}. \tag{6.73}$$

This ratio decreases quickly for low a. Figure 6.2 gives values for the error with parameters relevant to the 1992 run of NA47. The graph shows the percentage error in the measured polarization as a function of the radial distance from the center of the NMR wire to the polarized material, for DNP time constants varying from 0.1 to 30 h. We conclude that polarization time constants in excess of 1 h lead to $\geq 2\%$ errors in polarization measurement if the radial distance to the material is less than 2 mm.

Example for Deuteron NMR Signal Decay Due to Saturation in Frozen Spin State

Here we assume that DNP is stopped and the target cooled so as to freeze an initially homogeneous polarization and estimate the evolution of the deuteron NMR signal decay.

The time evolution of Eq. 6.65 results in the integral of the volume-averaged instantaneous NMR signal and its saturation time

$$P^*(t) = P_0 \frac{\int_a^b e^{-t/\tau_{RF}^{\text{eff}}(r)} \dfrac{dr}{r}}{\int_a^b \dfrac{dr}{r}}, \tag{6.74}$$

which has no closed-form analytical solution. Developing the integrand in the nominator into series yields the signal decay due to saturation

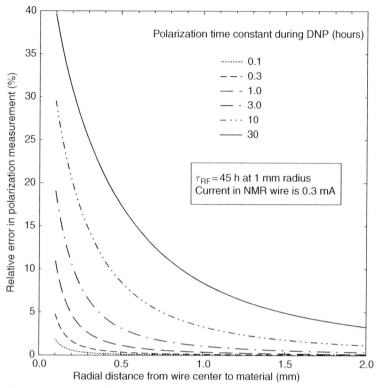

Figure 6.2 Error in deuteron polarization measurement in conditions during NA48 run of 1992, in steady state of DNP, as a function of the minimum distance from the coil wire to the nearest beads of the target. The set of curves correspond to different time constants of DNP build-up

$$P^*(t) = P_0 \left\{ 1 - \frac{1}{2\log\frac{b}{a}} \left[\frac{t}{\tau_a}\left(1 - \frac{a^2}{b^2}\right) - \frac{1}{2 \cdot 2!}\left(\frac{t}{\tau_a}\right)^2\left(1 - \frac{a^4}{b^4}\right) + \frac{1}{3 \cdot 3!}\left(\frac{t}{\tau_a}\right)^3\left(1 - \frac{a^6}{b^6}\right) + \cdots \right] \right\}.$$

(6.75)

This yields the initial instantaneous saturation time at $t = 0$

$$\tau^*_{RF} = \tau_a \frac{2\log\frac{b}{a}}{\left(1 - \frac{a^2}{b^2}\right)}.$$

(6.76)

Here $\tau_a = \tau_{RF}^{\text{eff}}(a)$ is the local saturation time constant at the inner radius of the target. With inner radii of 2, 1.5, 1.0, 0.5, 0.25 and 0.10 mm we find initial saturation time constants of 850, 535, 274, 84, 25 and 4.8 h. These may be compared with the measured values around

250 h for coil wire isolation radius of 1.75 mm. The difference may be due to the fact that the current in the coil was underestimated.

We also note that after a very long time (when there is little polarization left), the saturation time approaches the longest effective time constant at radius b. When starting from dynamic equilibrium with the polarization distribution given by Eq. 6.67 and initial signal by Eq. 6.68, the initial saturation time depends on the polarization time constant at the time of stopping DNP. With both time constants in the same order of magnitude, the initial saturation time is of the order

$$\tau_{RF}^* \cong \tau_{RF}^{\text{eff}}(a)\frac{b}{a} = 45.7 \text{ h} \times \frac{ab}{\text{mm}^2}, \tag{6.77}$$

which yields the saturation time of 1,370 h for $a = 1.75$ mm. This time is much longer than the polarization time constant and therefore the initial decay time after DNP is likely to come closer to the value with initial homogeneous polarization, around 600 h (with 0.3 mA current).

Figure 6.3 shows the decay of the integrated NMR signal after turning on the NMR measurement, starting with homogeneous frozen polarization at time $t = 0$. The risk with material at a small radius is well illustrated by the lower curves.

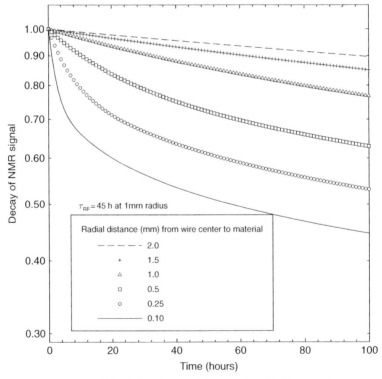

Figure 6.3 Decay of deuteron NMR signal due to RF saturation, in frozen spin state with initially homogeneous polarization

We conclude that when the polarizing time constant is in excess of 1h, target beads nearer than 2mm radius lead to potential errors in the deuteron polarization measurement during DNP. The problem is particularly serious in the case of proton polarization measurement, if 0.3 mA current is used in the probe coil. Consequently, the design of the NMR probe coil and current in the coil of the Liverpool Q-meter was reconsidered.

Regarding deuteron polarization measurement, target material at radius smaller than 2mm is not advisable. The design of the probe coil of the 1994 run of NA47 was improved so that the RF field was very low everywhere, even on the surface of the flat strip conductor. As a result of the low saturation, very high deuteron polarization was reached in the deuterated butanol target, and the error due to saturation was also lower than in the 1992 run of NA47.

The hypothesis that the NMR saturation causes error in polarization measurement can be easily tested by stopping the RF feed to the Q-meter during DNP when a steady state has been reached with constant microwave power. After continuing DNP at constant microwave power for a few hours, the NMR signal is remeasured. If the signal has grown higher, this can be interpreted as a recovery of the polarization in regions close to the NMR coil wire.

Consequently, it is recommendable to use the probe coil current of 0.1 mA for deuteron polarization measurement and of 30 µA for proton polarization measurement. The alternative to keep 0.3 mA current in the probe coil and stop NMR measurement by lowering the duty cycle, from the signal-to-noise point of view, is less efficient, because the same saturation improvement means lowering by 10 or 100 times the measurement duty cycle.

6.3 Optimization of the Series Q-Meter

6.3.1 Design Criteria

With the analytical expressions of Eqs. 6.23–6.33 the optimization of the series Q-meter circuit can be performed for each particular case. These equations give the distortion and non-linearity of the transfer function of the complex RF susceptibility, allowing their optimization. The signal-to-noise ratio can also be maximized using these equations. Furthermore, criteria for the stability for each component can be easily obtained, with requirements imposed by the TE calibration signal size which is calculable.

Equations 6.23–6.33 also show how the circuit should be tuned for best performance. This will be discussed Section 6.3.2. The criteria of the tuning follow from the requirements of the signal symmetry and absence of dispersion contribution, and from the shape of the Q-curve which sets requirements for the dynamic range.

In the circuit design two major parameter specifications emerge: maximum signal size and required scan width. For narrow signals and high susceptibility, such as that of highly polarized protons, the circuit design problem involves mainly the control of the linearity of the response function. For very small signals, the signal-to-noise ratio is to be maximized, while ensuring a reasonable flatness of the functions A_n and A_n'.

The case of very wide signals can also be dealt with Eqs. 6.23–6.33, although more precise description of the coaxial line resonator will be required then. The practical limit of sweep width is determined by the nearest quarter wavelength resonances of the coaxial line below and above the half wavelength resonance near the Larmor frequency; for $n = 1$ the sweep width is thus limited to less than ±50% relative to the Larmor frequency.

The most important component of the circuit is the probe coil, the design of which is discussed in Section 6.3.4.

6.3.2 Q-Meter Tuning

Equation 6.34 for the coaxial transmission line is valid if the cable length is adjusted by minimizing the impedance

$$Z_\ell = Z_c \tanh \gamma \ell \qquad (6.78)$$

of a shorted line at frequency ω_0. This procedure results in a length of

$$\ell \cong \frac{n\pi}{\omega_0 \sqrt{L_c C_c}} \left(1 + \frac{1}{8Q_c}\right) = \frac{n\pi c}{\omega_0 \sqrt{\varepsilon_r}} \left(1 + \frac{1}{8Q_c}\right), \qquad (6.79)$$

which can be shown [8] to give good symmetry properties for the transfer function of the resonant circuit.

If the cable tune is not correct or if the cable length is used for obtaining particular characteristics of the tuned circuit, we may replace x in Eq. 6.34 by $x - x_c$ where x_c is the relative offset of the cable resonance from the Larmor precession:

$$x_c = \frac{\omega_c - \omega_0}{\omega_0}. \qquad (6.80)$$

In Ref. [7] the assumption was made that the transmission line is tuned to resonate at the frequency ω_c, which has a relative deviation of

$$x_c = \frac{Q}{2Q_c}. \qquad (6.81)$$

This tuning that was derived for a special case of proton NMR circuit at 106.5 MHz results in the cancellation of the coefficient of the first-order dispersion term at the center of the NMR line, which, in turn, ensures that the first-order contribution of the dispersion part of the RF susceptibility in the experimental Q-meter signal becomes rather symmetric about the center frequency, if the absorption part is a symmetric function. This holds only for low values of Q/Q_c. Low-frequency and wide-sweep systems require a more precise expression for estimating the required cable mistune, and Eq. 6.37 cannot then be used.

The tune shift δ due to damping in the coaxial line influences the flatness of all parameters A_n and $A_{n'}$. The tune shift can be reduced by designing the coaxial line so that a high effective quality factor results. This is strongly influenced by the cryogenic part of the line, where high-resistivity materials are used for reducing the heat input to the target refrigerator. Outer conductor material made of alloys such as BeCu, CuNi and brass are excellent, because of their low heat conductivity. If the center conductor is silvered, the quality factor is mainly determined by the skin depth of the outer conductor, and Q_c becomes nearly temperature independent because these materials have a small temperature coefficient of the resistivity.

In copper-jacketed semi-rigid coaxial lines with high Q_c, the center conductor surface resistivity determines the quality factor. Because this is a function of temperature, thermal drift of the line results in a tune shift of the circuit. Thus, the temperature of such a line must be stabilized for ultimate stability of the Q-curve and transfer function (see Section 6.5.4). Moreover, the stabilization of the temperature of the tuned cable also reduces thermal drift of its electrical length, which is due to phase transitions occurring in the crystallized fractions of the PTFE dielectric at 19°C and 30°C [12].

Small diameter of the center conductor and low ε_r result in high Z_0, which is beneficial for low circuit distortions. Low ε_r is also desirable because lower n can be achieved in high-frequency applications. Although presently 50 Ω solid PTFE-isolated lines are used almost exclusively, it would be interesting to develop foam-isolated semi-rigid 75 Ω lines for NMR polarization measurement applications.

Finally, the part of the coaxial line which runs in the magnetic field must be made of non-magnetic materials so that the Q-curve will not change when the magnetic field is shifted for the measurement of the baseline curve. The BeCu lines with silvered center conductors have turned out to be excellent in this respect.

The value of the tuning capacitor is adjusted first to obtain a fairly symmetric Q-curve. This value is usually corrected with highly polarized spins by making the NMR absorption signal as symmetric as possible. Equation 6.26 explicitly shows how the capacitor tune shift δ influences the symmetry of $A_1(x)$, while mainly influencing only the magnitude of $A_1'(x)$. The coefficients A_2 are also changed by the capacitor tune, but all the functions cannot be made symmetric with the same value of the capacitor.

6.3.3 Circuit Design

In discussing the design of the series Q-meter circuit we shall use Eqs. 6.23–6.34. Equation 6.25 shows explicitly the frequency dependence of the Q-curve. We note that the Q-curve can be made symmetric by a suitable capacitor tune shift δ, although this is perhaps not the main aim, as was discussed above. The depth of the Q-curve is mainly determined by the inductance L of the probe coil, which should be made as small as is practical. The depth is also strongly influenced by the coupling admittance Y, which should also be made as low as possible, while maintaining the imaginary component in the coupling negligible. In high-frequency systems, this is difficult and a compromise is often necessary.

The flatness of the function $A_1(x)$ is improved by high Q_c, low Y, low L and low Q. The latter cannot be made arbitrarily small by a choice of a high-resistivity material, because these are often magnetic. The probe coil design will be discussed in greater detail later in Section 6.3.4.

The same parameters reduce and flatten the coefficient of the dispersion term in the same way; there is thus no conflict in pushing these parameters to their practical limits. The same is true with the second-order coefficients of Eqs. 6.28 and 6.29.

Specific problems in the circuit design at 106.5 MHz for protons and at 16.35 MHz for deuterons will be addressed below.

Design of the Proton Series Q-Meter

Based on Eqs. 6.25–6.29, the following rules apply for the series Q-meter optimized for the measurement of proton polarization:

(1) The cable should have a loss-factor as low as possible, to yield a high effective Q_c.
(2) The cable should be as short as possible, preferably $n \leq 3$.
(3) The coil should have a low inductance, preferably $\omega L < Z_0$.
(4) The feed resistance and the amplifier input impedance should be as high as possible, while maintaining both real. This requires placing a resistor of about 70 Ω in series with the input of the preamplifier, and using low-inductance resistors in series to build the feed resistance with a value in the range 200 Ω to 600 Ω.
(5) The Q-factor of the coil should be low (≈ 3), and its series resistance should therefore be high; it is preferable to place the additional damping resistor $R \approx 15$ Ω between the coil and the cable rather than between the cable and the tuning capacitor.
(6) One must have maximum $\eta\chi''(\omega)\omega L < 0.3 \, R'$ to preserve good linearity of the integrated absorption signal with respect to the polarization, and to avoid superradiant oscillations at large negative polarization. With embedded coils this requires special precautions for obtaining a low enough effective filling factor (to be discussed below).
(7) One must satisfy maximum $\eta|\chi(\omega)|\omega LY \ll 1$ to avoid non-linear distortions of the lineshape; this condition is difficult to achieve in sizable polarized proton targets, and satisfactory results can be obtained with the less stringent condition 6. The less strict condition is often sufficient because the non-linear signal distortions tend to integrate to zero.
(8) To avoid linear distortions of the lineshape, one must have $2Qx \ll 1$, $2Y\omega Lx \ll 1$, $2n\pi(R_L/Z_0)Y\omega Lx \ll 1$, and $Q^2 n\pi YZ_0 x \ll 1$. One or more of these conditions limit the maximum practical sweep width in high-frequency systems.

Design of the Deuteron Series Q-Meter

The deuteron has the spin $I = 1$ and a sizable quadrupole moment, which broadens the high field NMR spectrum to about 280 kHz in butanol and other glassy hydrocarbons, with a characteristic shape featuring two resolved peaks and broad minimum in between, and relatively flat pedestals outside. The peaks are associated with the two magnetic transitions;

the intensity ratio of these transitions, in principle, gives the spin temperature and therefore polarization, if the thermal distribution of level populations is valid, and if the polarization is homogeneous. Because of its large width and because of the small magnetic moment of the deuteron, the accurate measurement of the DMR absorption spectrum shape, however, is difficult.

At 2.5 T field the deuterons in glassy butanol thus have a total line breadth of about 2×10^{-2} relative to the center frequency of 16.35 MHz; this can be contrasted with protons in the undeuterated butanol, where dipolar interactions give a FWHM of about 4×10^{-4} at 106.5 MHz.

As a consequence, the frequency dependence of the coefficient of the term $A_1(x)$ cannot be ignored, and the real part of the experimental signal becomes distorted. On the other hand, the RF susceptibility of highly polarized deuterons remains very small, so that the terms which are of second or higher order in the RF susceptibility will remain negligible. We may therefore focus the discussion only on Eqs. 6.26 and 6.27.

The linear distortion in a practical circuit can be so large that it is not recommendable to use the experimental NMR line peak asymmetry as a way of determining the polarization, because the heights and shapes of the two superimposed signals do not accurately reflect the transition intensities. The distortion also makes it unreliable to fit the Q-curve drift during recording of the dynamic nuclear polarization, because the admixture of the distorted dispersion signal extends far beyond the edges of the absorption signal.

Provided that the distortions are small or can be sufficiently well corrected, however, the asymmetry of the DMR signal in a large target gives a unique means of estimating the spatial variation of the polarization in the target volume, if the average polarization is known accurately on the basis of the integrated signal calibrated in TE at 1 K, for example. The inhomogeneity of the polarization in a large target leads to a systematically higher asymmetry than that determined from the measured average polarization; this difference, although not highly sensitive to the polarization variation, gives a reasonable estimate of the RMS variation of polarization in the volume sampled by the probe coil.

As an example, if the average polarization $P^*(D) = 0.4$ can be determined to 3% relative accuracy and asymmetry within 5%, the variation of polarization

$$P(x) = P^* \pm \delta P(x) \tag{6.82}$$

is limited to $|\delta P| \leq 0.15$ in the volume sampled by the probe This limit can be made substantially lower with better accuracy in the polarization measurement, distortion control and asymmetry determination.

As the deuteron NMR signal is very small, the Q-meter design should aim at a good sensitivity, which implies high filling factor, high inductance L and a relatively high overall Q of the circuit. A low admittance Y is desirable for optimum signal-to-noise performance. The control of the linear distortion, on the other hand, requires that Qc be as high as possible and L be low, $\rho Qx \ll 1$, $2Y\omega Lx \ll 1$, and $2\rho Y\omega Lx \ll 1$. These clearly set the limit for a maximum Q and indicate that the coupling admittance Y should be as small as is practically possible from the noise performance point of view. Practical design values are $L = 0.4$ μH,

$Y = 0.01\ \Omega^{-1}$ and $R' = 15\ \Omega$, with damping resistor placed between the coil and the coaxial line.

6.3.4 Probe Coil Design

Above it was found repeatedly that low L and Q are desirable for the probe coil. The inductance is controlled by the length and diameter of the wire used for the coil; it is therefore preferable to use a short and thick wire of high resistivity material. Practical coils are made of thin-walled CuNi tubes of 1–4mm diameter and 10–40cm length, bent to a suitable shape around or inside the target.

For proton probe coils the filling factor should be minimized, which can be achieved by placing the coil outside the target, or by surrounding the wire by a PTFE tube which excludes the material from the area of the highest RF field near the wire. Another way of reducing the effective filling factor is to orient the coil (usually one loop) so that the uniform part of the RF field is parallel to the main field; this is made by keeping the wire always perpendicular to the main field. This design concentrates the sampling to the vicinity of the wire, which can be sometimes interesting if localized measurements are desired. A dipole loop made of a tube with length of 15cm and diameter of 4mm gives an inductance of about 100 nH. Such a loop, made of CuNi, gives a high Q-factor, which can be reduced by placing a suitable series resistor between the coil and the coaxial line.

In the case of deuteron probe coils, one wishes to maximize the effective filling factor by using a bare large-diameter embedded wire, predominantly aligned along the main field. A 50cm long tube of 2mm diameter gives an inductance of about 450 nH, which can be easily reduced by selecting a shorter length and a larger diameter. The filling factor can be varied between 0.2 and 0.3 by varying the diameter of a possible PTFE tube around the wire.

An alternative for lowering the Q of the coil is to make the probe conductor out of a thin film of high-resistivity non-magnetic material deposited on a suitable flexible substrate. If a series resistor is used at low temperature, it is very important that its resistance value has low temperature and field coefficients.

6.3.5 Linearity of the Integrated NMR Signal

Linearity of the relationship between the integral of the absorption signal and the spin polarization is better than the distortions in the lineshape, because many of the distorted terms either tend to be linear or to be antisymmetric and therefore integrate to zero. The linearity is harder to achieve in NMR probe systems for protons and other spins ½ with a narrow line, with a high magnetic moment and with a high spin density.

One must have maximum $|\eta\chi''(\omega)\omega L| < 0.3\ R'$ to preserve good linearity of the integrated absorption signal with respect to the polarization, and also to avoid superradiant oscillations at large negative polarization. With embedded coils this requires special precautions for obtaining a low enough effective filling factor (to be discussed below).

One must satisfy maximum $\eta|\chi(\omega)|\omega LY \ll 1$ to avoid non-linear distortions of the lineshape; this condition is difficult to achieve in sizable polarized proton targets, and satisfactory results can be obtained with the less stringent condition above. The less strict condition is often sufficient because the non-linear signal distortions tend to integrate to zero, thus producing little error in the measurement of polarization.

6.3.6 Superradiance

Superradiance occurs when the real part of the impedance of the tuned probe coil gets negative because of the large negative real part of the transverse spin susceptibility. This manifests itself as a limit in the negative polarization that one may achieve by DNP, or as a sudden loss of polarization when the superradiance conditions are met accidentally. The sudden loss may also be associated with a partly reversed polarization, because the probe coil current may 'ring' sufficiently long to cause adiabatic demagnetization in those parts of the target that are close to the probe wire.

Superradiance never occurs at positive polarization, and it is not a major problem for any other spin system than that of protons. At negative polarization it happens when the height of the NMR signal is equal to the height of the Q-curve of the resonant circuit. Before reaching this point, the proton NMR signal becomes abnormally narrow because of the term $-\eta\chi''(\omega)\omega L$ that becomes comparable with R'.

One must have maximum $|\eta\chi''(\omega)\omega L| < 0.3\,R'$ to preserve good linearity of the integrated absorption signal with respect to the polarization (as was discussed in Section 6.3.5), and to be sure to avoid superradiant oscillations at large negative polarization. With embedded coils this requires special precautions for obtaining a low enough effective filling factor.

The conditions for superradiance may also occur when disconnecting the coaxial line between the Q-meter box and the probe coil at high negative polarization, or when ramping the magnetic field so that one of the spin species in the target gets in resonance with a probe coil system. The latter can be avoided by making the magnetic field suitably less homogeneous during the field ramp.

6.4 Improved NMR Circuits

6.4.1 Series Q-Meter Improvements

The receiver selectivity can be greatly enhanced by changing from the present homodyne receiver technique [6] into a heterodyne or superheterodyne technology. These enable good control of the noise sidebands by the appropriate design of filters and frequency management. As the final amplifier stages operate at constant frequency, the filter design can be focused on off-band rejection ratio and time response rather than on flatness in the frequency domain.

The main benefit of heterodyne receiver technique lies in the better elimination of off-band and low-frequency parasitic signals. This, however, can be also accomplished by improved control of the EMIs in all parts of the circuitry, and therefore the use of a highly selective receiver is not necessary. This is due to the fact that the leading source of noise is the oscillator, whose output noise power spectrum is concentrated close to the carrier frequency and cannot therefore be eliminated if a reasonable frequency scan speed is desired.

The damping resistor of the Q-meter circuit is best placed between the coil and the coaxial line, if minimum circuit distortions are desired. This removes the Johnson noise of the resistor and improves the linearity and flatness of the circuit [7].

More fundamental improvements involve changes in the circuit which necessitate different circuit theoretical treatment; some of these are briefly discussed below.

6.4.2 Capacitively Coupled Series Q-meter

In high-frequency systems a major problem is to obtain a low real value of the coupling admittance Y. By examining the circuit of Figure 6.1 we note that the absorption part of the RF susceptibility can also be detected if purely capacitive coupling is used. The expression for the signal is then

$$\text{Re}\{u_i\} = u_o \omega C_o \frac{\text{Im}\{Z\} - \omega C \left[\text{Re}^2\{Z\} + \text{Im}^2\{Z\}\right]}{\left(1 - \omega C \text{Im}\{Z\}\right)^2 + \omega^2 C^2 \text{Re}^2\{Z\}} \tag{6.83}$$

and

$$\text{Im}\{u_i\} = u_o \omega C_o \frac{\text{Re}\{Z\}}{\left(1 - \omega C \text{Im}\{Z\}\right)^2 + \omega^2 C^2 \text{Re}^2\{Z\}}, \tag{6.84}$$

where $C = C_o + C_i$ with C_o and C_i replacing the oscillator feed and amplifier input resistors R_o and R_i, respectively. The absorption thus appears predominantly in the imaginary part of the output signal, which necessitates a 90° hybrid in the reference arm of the RF circuit.

With the low values of $|Y|$ available using small capacitors, an excellent linearity entails, and low distortions could become possible in wide-sweep systems. A potential problem is the stability of the capacitors, which certainly requires temperature stabilization of the circuitry. Furthermore, the preamplifier must be designed so that its input impedance is as purely capacitive as possible.

The coupling capacitors also eliminate the Johnson (thermal) noise of the coupling resistors. The preamplifier should be specially optimized for best noise performance in the resulting circuit.

6.4.3 Crossed-Coil NMR Circuit

One way of reducing the oscillator noise is to use the crossed-coil principle for NMR. Although there exists no adequate circuit-theoretical model for such a system, it is easy to show that the NMR signal is no more distorted by the crossed-coil circuit than by the series

Q-meter circuit. However, the benefit of measuring the oscillator strength simultaneously is lost. Furthermore, as the orthogonality of the coils enables rarely better than 30 dB rejection of the fundamental oscillator signal in the pickup coil, the wide-band noise rejection may be much less good than might seem at first. However, as the two coils can be designed and optimized individually, important system benefits might entail from the new degrees of freedom in the optimization process.

In discussing the crossed-coil NMR circuit we shall call the primary coil connected to the oscillator (RF source) the transmitter coil, and the one connected to the preamplifier the receiver coil. These coils are mounted so that their axes are perpendicular and the capacitive coupling of the wires of the coils is also minimized. Fine adjustment at room temperature helps in reducing the coupling between the transmitter and receiver coils.

One of the obvious benefits is that the transmitter coil can be designed so that the RF field is rather uniform in the volume probed by the receiver coil. This requires a transmitter coil that is substantially larger than the receiver coil, a feature which also facilitates in reducing the (capacitive) coupling between the coils.

Because H_1 is almost constant in the volume probed by the receiver coil, the signal size will be larger compared with the series-tuned Q-meter, with identical saturation produced by an embedded or tightly enclosing coil.

If the transmitter coil is outside the target, its filling factor can be made small. This reduces the influence of the RF susceptibility on the transmitter coil current and results in improved linearity of the circuit. In the first approximation, the RF field H_1 in the target can then be assumed independent of the susceptibility at resonance, which facilitates the circuit-theoretical analysis. This is not, however, a strict requirement for the best measurement of small signals.

The crossed-coil NMR circuit requires one additional coaxial line for feeding the transmitter coil. Because all receiver probes can share a common transmitter coil, multicoil systems will not require doubling the number of coaxial lines.

The tuning of the crossed-coil circuit is more difficult than that of the series Q-meter, because there are now two or more hybrid resonant circuits with a loose mutual coupling. The series or parallel capacitor of the transmitter circuit is tuned first using a vector voltmeter or a Q-meter amplifier connected to the transmitter resonant impedance Z_t. The series or parallel capacitor of the receiver circuit is tuned either using the residual RF signal coupled via the coil pair, or the NMR signal itself. Alternatively, the receiver circuit can be tuned using a small signal injected to the receiver resonant impedance Z_r. A further alternative consists of matched rather than tuned receiver circuit; this might provide the best wideband performance, with some cost in noise performance.

A large transmitter coil requires a larger amount of RF power for getting the oscillating transverse field H_1 equal to that of the embedded probe of a series Q-meter.

If the transmitter coil covers the entire target volume, it might be interesting to use it also for the manipulation of the spin systems in the target. The reversal of polarization by adiabatic passage might thus be performed using the transmitter part of the crossed-coil circuit, with much increased RF power. The cooling of the transmitter coil then needs to be separated from the refrigeration circuit of the target material.

Because the Q-curve is eliminated in first order, a wider dynamic range becomes available and a higher RF gain can be used. Furthermore, this circuit may be designed so that it is well adopted to the measurement of very wide NMR signals, such as that of ^{14}N spins in solid ammonia.

6.4.4 Measurement of Complex RF Susceptibility Using Quadrature Mixer

Quadrature mixer detector will allow to measure simultaneously the real and imaginary parts of the RF signal Gu_i, enabling the reconstruction of the real and imaginary parts of the RF susceptibility without theoretical modelling and fitting of the NMR signal, from [13]:

$$\chi(\omega) = \frac{R_o F^2(\omega)}{\eta Gi\omega L} \times \frac{u_i(\omega,\chi) - u_i(\omega,0)}{1 - \frac{R_o F(\omega)}{Z_0} \cosh\gamma\ell \left\{ YZ_0 + \left[1 + Y\left(R + \frac{1}{i\omega C}\right)\right]\tanh\gamma\ell \right\} \left[u_i(\omega,\chi) - u_i(\omega,0)\right]}, \quad (6.85)$$

where

$$F(\omega) = \frac{1+YZ}{Z_0}\left[Z_0 + (R_L + i\omega L)\tanh\gamma\ell\right]\cosh\gamma\ell \quad (6.86)$$

is a complex function involving only the circuit parameters.

This procedure avoids any complications resulting from circuit distortions and non-linearity, and thus enables one to focus on the noise performance in the Q-meter. The tuning of the input circuit will also be greatly facilitated by quadrature detection, because the system operates effectively as a vector network analyzer.

Reconstruction of the NMR susceptibility from the Q-meter output signal, however, requires the precise knowledge of the circuit parameters. These are best obtained by a fit of the complex Q-curve to the theoretical description of Eqs. 6.16 and 6.17 [13].

6.4.5 Series-Tuned Q-meter Using an RF Hybrid Bridge

The 180° RF hybrid bridges resemble the Wheatstone bridge in that when the bridge is balanced in two of its arms, the input signal is isolated from the output port. This is shown schematically in Figure 6.4: when the signal of the RF source is applied to port A and two identical series-tuned resonators are connected to ports C and D, the port B becomes isolated from port A. If one of the Q-meters has its inductance (coil) in contact with the target material and the other is tuned so that signal in port B is zero, then the oscillator and its noise, in the absence of the signal due to the RF susceptibility, are decoupled from the

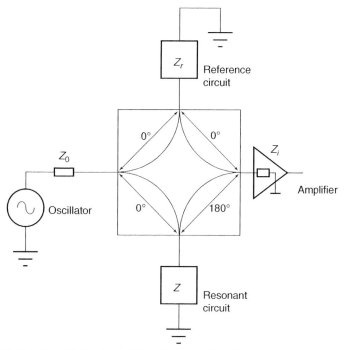

Figure 6.4 Hybrid bridge NMR circuit. The 180° RF hybrid is a wideband four-port device that resembles a Wheatstone bridge in its principle. When the bridge is well balanced, the RF source noise and parasites have a strongly reduced coupling to the preamplifier

preamplifier and mixer detector. The circuit is therefore suitable when the noise of the RF oscillator dominates other sources of noise. Another application is in the measurement of very broad NMR lines, where the depth of the Q-curve prevents the use of adequate gain in the preamplifier.

The alternative to 180° hybrid is its variant called the impedance bridge. This uses a balanced transformer to combine the signals reflected from the reference and measurement arms.

The circuit theoretical analysis of the 180° hybrid bridge with two Q-meters depends on the way the bridge is constructed, and its analysis is beyond the scope of the present book.

6.5 Calibration of NMR Signal for the Measurement of Polarization

6.5.1 Calibration Based on NMR Signal Measurement

Calibration in Thermal Equilibrium with the Lattice

At a temperature in the vicinity of 1 K, the nuclear spin-lattice relaxation time in polarized target materials is in the range of minutes, so that thermal equilibrium (TE) with the lattice and the surrounding helium bath is reached quickly. The temperature close to 1 K is reached and stabilized easily if the cooling system is filled with pure ^{4}He. At this temperature ^{4}He is in superfluid state and has a high thermal conductivity, which ensures a rapid equilibrium and uniform spatial distribution of the temperature of the target lattice.

The TE calibration requires the knowledge of the helium bath temperature. In principle, the vapor pressure of ^{4}He can be used as a practical temperature scale down to 1.25 K, but its accuracy becomes poor at 1 K because of the low pressure and, above all, because of the creep of the superfluid film along the cryostat structures towards the higher temperature regions of the cryostat.

The temperature of the helium bath can be determined more accurately from the vapor pressure of ^{3}He, which can be accomplished by installing a small ^{3}He-filled cell in the helium bath close to the target volume. The ITS-90 vapor pressure scale [14] used previously is now improved around and below 1 K so that its uncertainty does not exceed 0.6 mK down to 0.65 K [15].

Calibration of Small NMR Signal Based on Another Spin Species

Assuming that the NMR signal is small and that the ratio

$$\frac{A'_1(x)}{A_1(x)} \qquad (6.87)$$

is a symmetric function about $x = 0$, Eq. 6.37 can be rewritten in the form

$$P(I) \cong \frac{2}{\pi \mu_0 \hbar \gamma_I^2 n_I I} \frac{A_0(\omega_0)}{\eta \omega_0 L} \int_{\omega_{min}}^{\omega_{max}} \frac{S(\omega) \, \omega_0^2}{A_1(\omega) \, \omega^2} d\omega, \qquad (6.88)$$

which takes into account the frequency response of the Q-meter circuit. The neglect of the correction due to the dispersion signal can be easily justified for relatively narrow signals even if the ratio 6.87 has a small deviation from exact symmetry, provided that A'_1 is made small by circuit design and tuning.

The polarization $P(N)$ of the spin species N can then be obtained from its NMR signal by comparing with the signal of another nucleus H with known polarization $P(H)$, measured at the same center frequency and with the same circuit:

$$P(N) \cong P(H) \frac{\gamma_H^2 n_H I_H}{\gamma_N^2 n_N I_N} \frac{\int_{\omega_{N,min}}^{\omega_{N,max}} \frac{S_N(\omega) \, \omega_0^2}{A_1(\omega) \, \omega^2} d\omega}{\int_{\omega_{H,min}}^{\omega_{H,max}} \frac{S_H(\omega) \, \omega_0^2}{A_1(\omega) \, \omega^2} d\omega}. \qquad (6.89)$$

This provides a convenient and accurate calibration for the measurement of the polarization of rare spin species and with such a wide NMR lines that TE calibration becomes impossible. The method requires that the spin density ratio be known from chemistry or from other measurements, for example, those made in liquid state so that all NMR lines are narrow.

6.5 Calibration of NMR Signal for the Measurement of Polarization

A special application of the method is in the measurement and monitoring of the ^{14}N polarization in NH$_3$ and ND$_3$ targets at 2.5 T field. The Larmor frequency of ^{14}N is 7.694 MHz and the spectrum features a peak separation of 2.37 MHz and total width of 4.74 MHz, requiring a minimum of 5.3 MHz frequency scan from 5 to 10.3 MHz. Although the tuned coaxial line of the series Q-meter becomes a major problem with such a wide sweep, the use of a quadrature mixer and extraction of the absorption part of the RF susceptibility would enable one to overcome the problem related with the dispersion part in Eq. 6.89. Inserting the numeric values for the ratios of the gyromagnetic factors (2.79268/0.40347), of the spin densities (3/1) and of the spins (1/2), yields

$$P\left(^{14}N\right) \cong 71.86 P\left(^{1}H\right) \frac{\int_{\omega_{N,\min}}^{\omega_{N,\max}} \chi''_N(\omega) \frac{d\omega}{\omega}}{\int_{\omega_{H,\min}}^{\omega_{H,\max}} \chi''_H(\omega) \frac{d\omega}{\omega}}, \qquad (6.90)$$

where the absorption parts of the RF susceptibilities are obtained from the complex experimental signals using Eq. 6.85.

The measurement of ^{14}N NMR signal requires the measurement of the Q-curve at the nominal field value, which is best performed at zero polarization before DNP. It is clear that the stability of the Q-curve is a major concern here.

Equation (6.89) can also be used for finding the ratio of spin densities if the polarizations are known from other measurements or arguments such as that based on equal spin temperatures. The special case of TE polarizations at equal lattice temperatures gives

$$\frac{n_1}{n_2} \cong \frac{\gamma_2^2 I_2(I_2+1)}{\gamma_1^2 I_1(I_1+1)} \frac{\int_{\omega_{1,\min}}^{\omega_{1,\max}} S_1(\omega) \frac{\omega_0^2}{\omega^2} d\omega}{\int_{\omega_{2,\min}}^{\omega_{2,\max}} S_2(\omega) \frac{\omega_0^2}{\omega^2} d\omega}, \qquad (6.91)$$

where the signal gain was assumed flat in the narrow frequency scan range where TE signal measurement is possible, and where the integrated signals are the TE signals measured at constant temperature and same center frequency.

In a deuterated target, usually the degree of deuteration and the chemical sites and composition are known to a relatively high accuracy. The exact amount of residual protons can then be obtained using Eqs. 6.89 or 6.91.

6.5.2 Improvement of the International Temperature Scale around 1 K

The provisional low temperature scale PLTS-2000 below 1 K is now based on the melting pressure of ^{3}He. Because the melting pressure thermometer is too complicated to materialize in a polarized target, the practical temperature scale of ITS-90 was commonly used to determine accurately the calibration temperature during the acquisition of the TE NMR signal [14]. This ITS-90 scale is based on the vapor pressure of pure ^{3}He liquid between 3.2 and 0.65 K; the scale was recently improved by PTB[3] to the new scale with the acronym PTB-2006 [15]. The low-temperature scales have been determined and extrapolated below 1 K using the Curie law dependence of the susceptibility of the paramagnetic salt CMN. While the melting pressure scale has the known uncertainty (with regard to the thermodynamic temperature) of about 0.5 mK down to 500 mK, the uncertainty of the new vapor pressure scale is 0.6 mK, which suggests finding an independent method to verify and reduce the uncertainty of the vapor pressure scale.

The TE calibration accuracy depends mainly on the accuracy of the determination of the target temperature during NMR signal measurement. If the calibration of the polarization measurement could be obtained more accurately from another method, the measurement of the integrated NMR signal around 1 K temperature would provide determination of the temperature with similar accuracy, which we shall evaluate below. This other method is based on the knowledge that the common spin temperature of nuclear spins can be accurately measured when it is so low so that the deviation of proton polarization from 100% is very small.

As an example, we may take partly deuterated propanediol $C_3D_6(OH)_2$ with reacted Cr(V) compounds which yields very high DNP and excellent agreement with the equal spin temperature hypothesis between all nuclear spin species [16]. A statistical accuracy of 0.1% was above shown to be possible for the deuteron TE calibration; this is roughly equal to the precision $\delta T \approx 1$ mK of the temperature scale ITS-90 [14] around $T = 1$ K. Assuming that in a dedicated apparatus the systematic accuracy can be even better, deuteron polarization after DNP can hence be measured to the precision in the range of 0.1%.

A value of $P(D) = 0.5$ (and even higher) can be reached in a dilution refrigerator at 2.5 T homogeneous field using microwave frequency modulation. This corresponds to a spin temperature around 1 mK, which can also be now determined to about 0.1% accuracy. Because the proton polarization is nearly complete and is known from the deuteron spin temperature, its precision is roughly given by

$$\delta P_p \cong \frac{\omega_p}{\omega_d} \frac{2 \log R}{R^{\omega_p/\omega_d}} \frac{\delta T_S}{T_S}, \qquad (6.92)$$

where R is

$$R = \exp\frac{\hbar\omega_d}{k_B T_S} = \frac{P_d + \sqrt{4 - 3P_d^2}}{2(1 - P_d)}. \qquad (6.93)$$

The calibration of the proton polarization measurement is thus improved by a numeric factor over that of the deuteron calibration accuracy and over the ITS-90 scale. Using the proton NMR signal now as a thermometer, the temperature scale can be improved over ITS-90 by a similar factor, which is 8 or 20 for deuteron polarizations of 0.4 or 0.5, respectively.

---

[3] Physikalisch-technische Bundesanstalt, Berlin.

The improved temperature scale can be now used for the improvement of the calibration of the deuteron polarization measurement, and the same factor of improvement in the temperature scale will result after repeating the above procedure. The practical limitations for the precision which can be obtained will arise from the statistical accuracy to which the integrated NMR signals can be determined at 1 K, from the homogeneity of the spin temperature and from the validity of the assumption that the two spin systems are in good TE with unique temperature after DNP.

Estimates based on the ab initio calculation of the NMR signal-to-noise ratio indicate that the integrated proton TE signal can be measured to a relative accuracy of 10^{-5} in partly deuterated propanediol. A relative accuracy 10^{-4} of the thermodynamic temperature scale at 1 K would therefore seem possible. Many systematic errors can be controlled by performing the transfer of the calibration at both positive and negative ultimate spin temperatures.

The nuclear spin temperature measured using the above method is the thermodynamic temperature, and its fundamental limit of uncertainty is determined by that of the NMR frequency, because the Planck and Boltzmann constants are defined to have exact values as of 2019.

6.5.3 Electromagnetic Interference (EMI) Control

In Section 6.2.4 we discussed the signal-to-noise ratio in the NMR circuit and showed how the signals completely buried in noise can be accurately measured by signal averaging. The main sources of noise are:

- thermal noise of the preamplifier and other front-end circuitry;
- amplitude and phase noise of the RF source oscillator;
- EMIs with spurious signals in the environment.

The first two can be optimized by careful design and by the choice of low-noise components. The oscillator noise of direct digital synthesizers, both in amplitude and in phase, is very good in comparison with the more traditional phase-locked loop oscillators. The noise of the DAC produced at large frequency offsets in the RF signal is not harmful for the NMR circuitry, and it can be reduced by the circuit design and analog filtering.

The leading item then remains the control of EMIs, a problem that has become a science in its own right since the 1970s; this is called electromagnetic compatibility (EMC). The acronym is a coincidence with that of European Muon Collaboration, but confusion is hardly possible.

EMC is concerned with the electromagnetic energy that is

- generated unintentionally,
- propagated undesirably,
- received harmfully.

These may cause unwanted effects such as EMI or even physical damage in equipment. One example from high-energy physics equipment in the past is the optical spark chamber that caused damage of sensitive circuitry connected to the same electrical power net in

CERN experiment S137. The goal of EMC is the correct operation of different equipment in a common electromagnetic environment.

EMC pursues the three above classes of issue first by reducing the unwanted emissions and by taking countermeasures, many of which are legally required in modern equipment. Secondly, the RF interference is reduced by the design of signal paths, by shielding the sensitive parts of the circuitry and by absorbing unwanted emissions before they reach the 'victim'. Thirdly, immunity is improved by 'hardening' the equipment. This involves careful circuit design that also takes into account the harmonic generation of signals from the unwanted ones far from the frequency spectrum of interest. The hardening often involves the introduction of broadband filters in the power supply lines, for example.

The engineering techniques called 'grounding and shielding' apply to all three issues. However, the discipline of EMC nowadays includes circuit design techniques at all levels: enclosures, power supplies, cables and connectors, filters and isolating devices, printed circuit boards, hybrids and semiconductor microcircuits.

Here are some measures that led to substantial improvements in NMR polarization measurement accuracy:

- The flexible coaxial cables and BNC connectors were replaced by semi-rigid coaxial cables and SMA connectors in all RF signal paths between the RF source and the Q-meter boxes. The BNC connectors were found to be microphonic and they also leaked RF signals from the environment. The sections of the tuned cable inside the cryostat were also replaced by semi-rigid coaxial lines; in these cables the conductors are now made of Be-Cu alloy that has a reasonably low thermal conductivity in comparison with pure Cu used in CuJack cables. The inner conductor was silvered for reducing attenuation.
- RF filters were introduced in all mains power feeds of the power supplies of all equipment connected to the vacuum chamber of the polarized target.
- The power supplies of the digital and analog parts were separated from each other electrically in all instrumentation of the NMR equipment; this was particularly important for the RF source.
- It was found that RF filtering of the instrumentation lines for the thermometry, heaters, level gauges, etc. should not be done at the connectors to the vacuum chamber, because these lines inside the cryostat couple with the RF signals of the NMR probes. This became apparent as a tune change of the DMR probes in the SMC PT after placing RF filters in the instrumentation lines. Instead, the cryogenic instrumentation cables outside the cryostat were then double shielded, and the instrument enclosures were provided with RF filtered mains feeds.
- In several digital signal paths, galvanic lines were replaced by fibre-optic links.

6.5.4 Control of NMR Circuit Drift

Once the sources of noise and spurious signals are minimized, signal averaging was shown to improve the accuracy of the calibration process based on the measurement of the TE signals at an accurately known stable temperature, commonly around 1 K. Because the averaging in a multi-coil system of a large target may take of order 100h, the drift of the circuitry becomes an important concern. Such drifts ultimately limit the improvement of the statistical accuracy that can be reached by signal averaging [10, 17, 18].

The parts of particular concern are the resonant circuits inside and outside the cryostat, the Q-meter boxes containing many temperature-sensitive components, and the RF source. Inside the cryostat the stabilization can be achieved, during the measurement of very small signals and during TE calibration, by controlling all flow rates and by waiting for thermal transients to decay after initial cooldown. Copper and other metals, plastics and also the polarized target materials exhibit slowly relaxing stored heat that takes several days to decay after cooldown. It is particularly important that the tuned coaxial lines inside the cryostat are thermalized onto stable heat sinks.

Outside the cryostat the semi-rigid coaxial lines must be thermally isolated and stabilized at a temperature above the 19 °C phase transition temperature of the PTFE isolation [17]. The transition changes the propagation constant so much that the electrical length is modified by 0.1% when increasing the cable temperature from 13 °C to 19 °C [12].

The Liverpool Q-meter circuit enclosures are made of massive copper and they are designed so that they can be cooled by a stabilized supply of water. The RF source and other possible signal handling electronics are best placed in EMI-shielded racks that are also cooled by stabilized water.

6.5.5 Summary for Series Q-Meter Circuits

We conclude that although the series Q-meter technique has reached the status of mature technology, it can still be improved and developed for specific applications. Some new applications, however, may require more substantially improved circuits. The accurate measurement of the polarization of ^{14}N in ammonia might be one of these.

Apart from the development of improved circuits, it is important to avoid saturation of the spins in the target material close to the probe coil wire. Because the RF field at 2 mm radius from the wire of the classic Liverpool Q-meter causes a measurable error at 0.3 mA probe coil current, it is suggested here that the current should be reduced to 0.1 or 0.03 mA, unless the deuteron polarization is measured with a strongly reduced duty cycle.

For proton target NMR probes, the non-linearity is the main cause of error. In the classical series Q-meter, this can be mitigated by the appropriate choice of resonator circuit parameters and, most importantly, by reducing the filling factor of the probe coil.

References

[1] V. Petricek, M. Odehnal, Analyse du Q-metre utilise pour des mesures de polarisation nucleaire elevée, *Nucl. Instr. and Meth.* 52 (1967) 197–201.
[2] V. Petricek, A linearized Q-meter circuit for measurement of high proton polarization in a target, *Nucl. Instr. and Meth.* 58 (1968) 111–116.
[3] F. Udo, Some new features in a nuclear magnetic resonance detection system for measuring polarization of highly polarized substances, in: G. Shapiro (ed.) *Proc.*

2nd Int. Conf. on Polarized Targets, LBL, University of California, Berkeley, 1971, 397–401.

[4] E. Boyes, G. R. Court, B. J. Craven, R. Gamet, P. J. Hayman, An on-line computer polarization measuring system using R. F. phase lock techniques, in: G. Shapiro (ed.) *Proc. 2nd Int. Conf. on Polarized Targets*, LBL, University of California, Berkeley, 1971, 407–410.

[5] D. Gifford, A Q-meter with RF phase sensitive detector, in: G. R. Court, et al. (eds.) *Proc. 2nd Workshop on Polarized Target Materials*, SRC, Rutherford Laboratory, Chilton, Didcot, 1980, 85–90.

[6] G. R. Court, D. W. Gifford, P. Harrison, W. G. Heyes, M. A. Houlden, A high precision Q-meter for the measurement of proton polarization in polarized targets, *Nucl. Instr. and Methods* A324 (1993) 433–440.

[7] T. O. Niinikoski, Mathematical treatment of the series Q-meter, in: G. R. Court, et al. (eds.) *Proc. of the 2nd Workshop on Polarized Target Materials*, SRC, Rutherford Laboratory, Chilton, Didcot, Oxon, UK, 1980, 80–85.

[8] T. O. Niinikoski, Topics in NMR polarization measurement, in: H. Dutz, W. Meyer (Eds.) *Proc. 7th Int. Workshop on Polarized Target Materials and Techniques*, Elsevier, Amsterdam, 1995, 62–73.

[9] T. O. Niinikoski, A. Rijllart, An MC68000 microprocessor CAMAC system for NMR measurement of polarization, *Nucl. Instrum. Methods* 199 (1982) 485–489.

[10] Spin Muon Collaboration (SMC), B. Adeva, S. Ahmad, et al., Measurement of the deuteron polarization in a large target, *Nucl. Instr. and Meth. in Phys. Res*. A349 (1994) 334–344.

[11] A. Abragam, M. Goldman, *Nuclear Magnetism: Order and Disorder*, Clarendon Press, Oxford, 1982.

[12] S. K. Dhawan, Understanding effect of teflon room temperature phase transition on cax cable delay in order to improve the measurement of TE signals of deuterated polarized targets, *IEEE Trans. Nucl. Sci*. 39 (1992) 1331–1335.

[13] Y. F. Kisselev, C. M. Dulya, T. O. Niinikoski, Measurement of complex RF susceptibility using a series Q-meter, *Nucl. Instr. and Meth. in Phys. Res*. A354 (1994) 249–261.

[14] H. Preston-Thomas, The international temperature scale of 1990 (ITS-90), *Metrologia* 27 (1990) 3–10.

[15] J. Engert, B. Fellmuth, K. Jousten, A new ^{3}He vapour-pressure based temperature scale from 0.65 K to 3.2 K consistent with the PLTS-2000, *Metrologia* 44 (2007) 40–53.

[16] T. O. Niinikoski, Polarized targets at CERN, in: M. L. Marshak (ed.) *Int. Symp. on High Energy Physics with Polarized Beams and Targets*, American Institute of Physics, Argonne, 1976, 458–484.

[17] D. Crabb, S. Dhawan, N. Hayashi, A. Rijllart, Noise and stability improvements in the DMR system for SMC, in: T. Hasegawa, et al. (eds.) *Proc. 10th Int. Symp. on High-Energy Spin Physics*, Universal Academy Press, Inc., Tokyo, 1993, 375–379.

[18] Spin Muon Collaboration (SMC), B. Adeva, E. Arik, et al., Large enhancement of deuteron polarization with frequency modulated microwaves, *Nucl. Instrum. and Methods* A372 (1996) 339–343.

7
Polarized Target Materials

7.1 Criteria for Material Optimization

An ideal polarized target is made of a pure isotope, the nucleus of which is of interest in the reaction under study. This is feasible in the case of optical pumping of noble gases and other gases which can be dissociated into atoms at low pressure. In solid form few pure elemental substances of interest can be dynamically polarized – the only exception is solid deuterium, but so far only rather low polarization has been obtained in it. We shall therefore discuss here the choice of the solid compounds and materials which best suit various types of applications.

The leading application of DNP up till now has been the scattering experiments in high-energy and nuclear physics. Other applications include measurements of slow neutron cross sections, molecular physics using slow neutrons, nuclear magnetism and other solid-state physics experiments and spin filters. The use of polarized solids in magnetic confinement fusion and in magnetic resonance imaging has also been discussed. The material choice evidently depends strongly not only on the application but also on the goal of the experiment or process which is considered. We shall begin by material optimization for scattering experiments that use modern counting methods.

More recently DNP has been used for the signal enhancement in NMR studies of complex chemical and biochemical molecules. In this context DNP and other enhancement techniques are called by the term 'hyperpolarization'.

In this chapter we shall focus mainly on the polarized target applications.

7.1.1 Scattering Experiments

Assuming that systematic errors can be controlled or are independent of the target, the optimization of the target material consists of maximizing the statistical accuracy to which the desired polarization asymmetry can be determined during the experiment. The asymmetry is determined in each kinematical bin from the number of counts with target polarization along or opposite to the magnetic field (field itself being oriented with respect to the beam in a way which depends on the reaction under study):

$$N_\pm = \Phi_\pm a n_t \left(\sigma_0 + f P_\pm P_b \Delta\sigma\right). \tag{7.1}$$

Here $\Phi_\pm$ are the integrated beam fluxes through the target with polarizations $P_\pm$ and $N_\pm$ are the corresponding number of counts, a is the acceptance of the detector in the kinematical bin, σ_0 is the unpolarized cross section, $\Delta\sigma$ is the cross-section difference with opposite orientations of the target nuclear spins, n_t is the target thickness (number of nucleons or nuclei per cm^2) and f is the target dilution factor:

$$f = \frac{n_p \sigma_p}{n_t \sigma_0} = \frac{n_p \sigma_p}{n_p \sigma_p + \sum_i n_i \sigma_i}, \tag{7.2}$$

where the indexes p refer to the polarizable nucleons (or nuclei) and i to unpolarizable background nucleons (or nuclei). Some reactions have an asymmetry only when the beam is also polarized; otherwise, the beam polarization P_b can be omitted.

Supposing that the beam intensity is limited by factors such as the accelerator or the detector, the radiation resistance of the target may not enter in the material optimization. The target asymmetry

$$A = \frac{\Delta\sigma}{\sigma_0} \tag{7.3}$$

is obtained directly from the difference of counts in Eq. 7.1 by normalizing to equal beam fluxes so that $\Phi_\pm = \Phi$ and writing

$$\frac{N_+ - N_-}{N_+ + N_-} = f P_b A \frac{P_+ - P_-}{2 + f P_b A (P_+ + P_-)}, \tag{7.4}$$

which can be simplified by taking the polarizations to be approximately equal but opposite so that $P_+ + P_- = 0$; this yields

$$\frac{N_+ - N_-}{N_+ + N_-} \cong f P_b A \frac{P_+ - P_-}{2} = f P_b A \bar{P}, \tag{7.5}$$

which defines the average absolute value of the target polarization. The experimental target asymmetry is now

$$A \cong \frac{1}{f P_b \bar{P}} \frac{N_+ - N_-}{N_+ + N_-} \tag{7.6}$$

and it has a statistical error due to the finite number of counts

$$\delta A \cong \frac{1}{f P_b \bar{P}} \frac{\sqrt{N_+ + N_-}}{N_+ + N_-} = \frac{1}{f P_b \bar{P}} \frac{1}{\sqrt{2 \Phi a n_t \sigma_0}} = \frac{1}{\mathcal{M}_t} \frac{1}{P_b \sqrt{2 \Phi a \sigma_0}}. \tag{7.7}$$

Here we have separated in the last form all factors related with the polarized target under the figure of merit of the target

7.1 Criteria for Material Optimization

$$\mathcal{M}_t = f \, \bar{P} \sqrt{n_t}, \tag{7.8}$$

which can be determined for each target material and length. Sometimes the target figure of merit is defined as the square of Eq. 7.8.

If the target length is determined by the space available or by the detector requirements, rather than by beam attenuation or multiple scattering of the beam or the secondary particles, the choice of the material can be further simplified by writing the target nucleon thickness in the terms of its average density, length and the nucleon mass

$$n_t = \frac{\rho_t V_t}{m_n} \frac{1}{A_t} = \rho_t \frac{L_t}{m_n}, \tag{7.9}$$

where the liquid helium coolant filling the voids between the target beads must also be taken into account. The figure of merit now reads

$$\mathcal{M}_t = f \, \bar{P} \sqrt{\rho_t} \sqrt{\frac{L_t}{m_n}}. \tag{7.10}$$

and the material-dependent part can be determined for a substance once its filling factor and average polarization are known.

The polarization may evolve during the data taking, because its frequent reversal is often required for reducing systematic errors due to the slow drift of the beam or the detector acceptance, and possibly because of the radiation damage of the target material. The average polarization in Eqs. 7.5–7.10 is then obtained from the square roots of the averages of the squared polarizations, which can be determined when the time evolution of the polarization during DNP, and the dose dependence of the reduction of the polarization are known.

In a high-intensity beam the polarization of the target may be reduced by the direct heating of the material by the beam, and by the radiation damage, which gradually accumulates during the experiment. In this case the figure of merit of the experiment also follows from the minimization of the statistical uncertainty of the target asymmetry of Eq. 7.7, which requires the maximization of

$$\sqrt{\Phi} \mathcal{M}_t = t_{\exp} \sqrt{I} \sqrt{f^2 \langle P^2 \rangle}. \tag{7.11}$$

Here $t_{\exp}$ is the effective duration of the data taking excluding time needed for target annealing or change, I is the beam intensity and the time-average of the polarization needs the knowledge of the polarization build-up during reversal and the reduction of polarization due to the accumulated dose and due to the material heating which depends on the intensity I. It is clear that these parameters can only be obtained by direct measurement, and that also the cooling system will strongly influence the maximization of expression 7.11. These factors will be discussed in Section 7.5.

If multiple scattering limits the length of the target, the best material is one which has the highest material-dependent figure of merit and has a low relative number of heavier nuclei so that the length can be increased. The criteria related with multiple scattering unfortunately cannot be written in simple analytic form and the judgement between materials of roughly equal and high figure of merit must be based on their relative heavy-element contents. The parameter relevant for multiple scattering is the radiation length X_0, which is defined as the mean distance over which a high-energy electron loses all but $1/e$ of its energy by bremsstrahlung, and which is the appropriate scale length for describing high-energy electromagnetic cascades. This parameter has been calculated and tabulated in Ref. [1] and a practical introduction to its use is given in Ref. [2]. The radiation length for a chemical compound or mixture of compounds is given by

$$\frac{1}{X_0} = \sum_i \frac{w_i}{X_i}, \tag{7.12}$$

where w_i and X_i are the weight fraction and radiation length of element or substance i, and

$$X_0 = \frac{716.4 A}{Z(Z+1)\ln(287/Z)} \text{ g/cm}^{-2}, \tag{7.13}$$

for an element with atomic number A and charge Z.

If some of the heavy nuclei become also polarized, their contribution to the scattering asymmetry must be estimated. This requires the estimation or measurement of their polarization. The errors related with these procedures are usually taken into account in the systematic error analysis, because they are usually dominated by the incomplete knowledge of the nuclear structure of the heavy nuclei.

In high-energy physics, polarized proton and neutron targets are of main interest. As free neutrons cannot be confined in solids, polarized neutron targets must contain deuterium or other light nuclei such as ^{3}He or ^{6}Li where the nuclear structure is well understood. A simplified way of guiding the research of polarizable materials is based on the fact that in materials where DNP is successful, the proton polarization is almost complete and deuteron polarization near 0.5. The initial choice of materials can therefore be based only on the hydrogen content, if scattering off all nucleons is indistinguishable with similar cross sections. A convenient approximation for the dilution factor then is obtained by assuming the cross sections of all nucleons in Eq. 7.2 equal, which yields

$$f_p = \frac{n_p}{n_p + \sum_i n_i}, \tag{7.14}$$

where n_p is the number of polarizable nucleons and n_i is the number of nucleons in unpolarized background nuclei. This dilution factor, however, cannot be used for extracting the experimental asymmetry from the data, because the cross-section ratios for scattering off free and bound nucleons can deviate markedly from unity, and the ratio furthermore depends on the kinematic bin.

If the material consists of a pure chemical substance, the dilution factor can be further simplified by assuming all nucleon masses equal so that

7.1 Criteria for Material Optimization

$$f_p = \frac{n_p}{M}, \quad (7.15)$$

where M is the molecular weight of the substance. This allows to compare various chemical substances with each other, as shown in Table 7.1.

Table 7.1 Simplified dilution factors and radiation lengths for hydrogen-rich materials. For carbon, boron, nitrogen and oxygen nuclei natural isotopic abundance is assumed for simplicity, whereas hydrogen and lithium are assumed to be pure isotopes.

Compound	Chemical formula	M (g/mol)	f_p	X_0 (g/cm^2)
Lithium-6 hydride	^{6}LiH	7.02	0.1424	54.87
Lithium-7 hydride	^{7}LiH	8.02	0.2493	62.69
Ammonia	NH_3	17.03	0.1762	40.74
Water	H_2O	18.02	0.1110	23.05
Alkanes				
Methane	CH_4	16.04	0.2494	46.21
Ethane	C_2H_6	30.07	0.1995	45.47
Propane	C_3H_8	44.10	0.1814	45.21
Butane	C_4H_{10}	58.12	0.1721	45.07
Pentane	C_5H_{12}	72.15	0.1663	44.99
Simple heavy hydrocarbons	$\approx(CH_2)_n$	$\approx n \cdot 14.01$	0.1427	44.59
Alcohols				
Methanol	CH_3OH	32.04	0.1248	39.34
Ethanol	C_2H_5OH	46.07	0.1302	40.82
Propanol	C_3H_7OH	60.10	0.1331	41.66
Butanol	C_4H_9OH	74.12	0.1349	42.19
Pentanol	$C_5H_{11}OH$	88.15	0.1361	42.56
Diols				
Ethanediol	$CH_2(OH)CH_2OH$	62.07	0.0967	38.90
1,2-Propanediol	$CH_3CH(OH)CH_2OH$	76.10	0.1051	39.84
1,2-Butanediol	$C_2H_5CH(OH)CH_2OH$	90.12	0.1110	40.52
1,2-Pentanediol	$C_3H_7CH(OH)CH_2OH$	104.15	0.1152	41.03
Other hydrocarbons				
Glycerol	$CH_2(OH)CH(OH)CH_2OH$	92.09	0.0869	38.74
Glucose	$C_6H_{12}O_6$	180.16	0.0666	38.42
Sorbitol	$HOCH_2(CHOH)_4CH_2OH$	182.17	0.0769	38.58
Amines				
Methyl amine	CH_3NH_2	31.06	0.1610	42.42
Dimethyl amine	$(CH_3)_2NH$	45.08	0.1553	43.08
Ethyl amine	$C_2H_5NH_2$	45.08	0.1553	43.08

Table 7.1 (cont.)

Compound	Chemical formula	M (g/mol)	f_p	X_0 (g/cm^2)
Boron compounds				
Ammonium borohydride	NH$_4$BH$_4$	32.88	0.2433	46.73
Diborane	B$_2$H$_6$	27.67	0.2168	54.76
Borane ammonia	BH$_3$NH$_3$	30.87	0.1944	46.03
Borobutane	B$_4$H$_{10}$	53.32	0.1875	54.54
Lithium borohydride	LiBH$_4$	21.78	0.1837	57.22

In Table 7.1 two materials have outstanding dilution factor: ammonium borohydride and methane. Almost complete polarization by DNP has been achieved in irradiated ^{7}LiH and ^{6}LiH, which also have the highest radiation lengths of all the materials listed. Complications with the nuclear structure of lithium nuclei, however, were initially the main limiting reason for the use of these materials in polarized proton targets in high-energy physics experiments. More recently irradiated ^{6}LiD has been used as a polarized deuteron target.

Ammonium borohydride [3] is an exotic material which is stable below 0 °C and where only a small polarization has been obtained when mixed with ammonia. Methane has the melting point slightly below LN$_2$ temperature, which precludes easy doping with paramagnetic molecules. Heat transfer out of irradiated solid methane during DNP limits its potential use to small targets or samples; furthermore, only low paramagnetic center concentrations have been obtained.

Alkanes with straight carbon chain have limited or no capability of dissolving paramagnetic molecules. They are included because some of their paraffin isomers are glass formers where free radicals might be diffused in or created by irradiation. Up till now their use is precluded by the low polarizations obtained in irradiated samples. This leaves the reference position to ammonia, where almost complete DNP has been routinely obtained in large irradiated targets.

The alcohols and diols have been included in Table 7.1 because high polarization can be obtained in nearly all of them, particularly in butanol, pentanol, ethanediol and propanediol, by introducing in their glassy matrix various Cr(V) complexes or free radicals by different techniques, which will be discussed below. These materials, as well as glycerol, glycol and sorbitol, have presented potential new uses in biology and MRI and they are therefore included in the list.

Amines have better dilution factors compared with alcohols. They are also glass formers and are therefore interesting additives to other materials.

The boron compounds have a high dilution factor and radiation length. Some of them can be dissolved in amines: unfortunately, high polarizations have been obtained only in mixtures which have a dilution factor inferior to ammonia [4].

The lighter and more hydrogen-rich materials also tend to have a larger radiation length, which strengthens their position as potential target materials in scattering experiments.

In the case of the deuterated materials of Table 7.2, two materials stand out because of their dilution factors and radiation lengths: irradiated ^{6}LiD and irradiated ND$_3$. Also,

Table 7.2 Simplified dilution factors and radiation lengths for the deuterated compounds. For carbon, boron, nitrogen and oxygen nuclei natural isotopic abundance is assumed, whereas deuterium and lithium are assumed to be pure isotopes. Furthermore, as ^{6}Li and D nuclei reach practically same polarizations carried also by the quasi-free deuteron of ^{6}Li, it is assumed that there are two polarized deuterium nuclei per ^{6}LiD unit cell.

Deuterated compound	Chemical formula	M (g/mol)	f_d	X_0 (g/cm^2)
Lithium deuteride	^{6}LiD	8.03	0.4982	62.75
Ammonium borodeuteride	ND_4BD_4	40.93	0.3909	58.19
Ammonia	ND_3	20.05	0.2993	47.97
Water	D_2O	20.03	0.1997	25.63
Alkanes				
Methane	CD_4	20.06	0.3987	57.82
Ethane	C_2D_6	36.11	0.3323	54.61
Propane	C_3D_8	52.15	0.3068	53.47
Butane	C_4D_{10}	68.18	0.2933	52.88
Pentane	C_5D_{12}	84.22	0.2850	52.52
Simple heavy hydrocarbons	$(CD_2)_n$	$\approx n \cdot 16.02$	0.2496	51.01
Alcohols				
Methanol	CD_3OD	36.06	0.2218	44.29
Ethanol	C_2D_5OD	52.11	0.2303	46.18
Propanol	C_3D_7OD	68.15	0.2348	47.24
Butanol	C_4D_9OD	84.18	0.2376	47.92
Pentanol	$C_5D_{11}OD$	100.22	0.2395	48.40
Diols				
Ethanediol	$CD_2(OD)CD_2OD$	68.11	0.1762	42.68
1,2-Propanediol	$CD_3CD(OD)CD_2OD$	84.15	0.1901	44.06
1,2-Butanediol	$C_2D_5CD(OD)CD_2OD$	100.18	0.1996	45.05
1,2-Pentanediol	$C_3D_7CD(OD)CD_2OD$	116.22	0.2065	45.79
Other hydrocarbons				
Glycerol	$CD_2(OD)CD(OD)CD_2OD$	100.14	0.1598	42.13
Glucose	$C_6D_{12}O_6$	192.23	0.1248	41.00
Sorbitol	$DOCD_2(CDOD)_4CD_2OD$	196.26	0.1427	41.57

Cr(V)-doped butanol has been widely and successfully used in scattering experiments. Biological and other experiments based on contrast variation use deuterated ethanediol, propanediol, butanol and glycerol, where the dilution factor is not important.

In elastic and quasi-elastic scattering, the kinematics of the events can be completely determined. This enables one to separate events originating from protons of hydrogen atoms from those due to heavy nuclei, and even from events due to nucleons bound to heavy nuclei, because of their Fermi momentum which broadens the distribution of scattering angles and therefore enables to separate the narrow elastic peak due to protons from the underlying broad background due to heavy nuclei. The same is true for nucleons in deuterium and ^{6}Li nucleus, because their Fermi momentum is significantly smaller than that of heavier nuclei. In such cases the dilution factor does not enter in the statistical figure of merit of Eq. 7.10, but the contributions of the heavy nuclei must be analyzed from the point of view of systematic errors arising from multiple scattering, which limits the usable target thickness to a fraction of the scattering length.

Most of the hydrogen-rich target materials are liquids or gases at room temperature, which facilitates the introduction of the paramagnetic centers, but makes the later handling, loading into the refrigerator and storing under LN_2 more complicated. Despite of great efforts to develop materials which can be stored and handled at room temperature, no practical solution has been found yet. The handling and loading techniques at LN_2 temperature, on the other hand, have been substantially improved and therefore the need of solid materials has become restricted only to a few specific applications.

In nuclear experiments the scattering off polarized nuclei of interest can often be distinguished from background. In this case the high polarization of these nuclei will guide the material choice. In inclusive nuclear experiments with electron and gamma beams, however, this may not be the case, and then the dilution factor of Eq. 7.2 has to be included in the figure of merit.

7.1.2 Other Types of Applications

In beam transmission type of experiments and in spin filtering, unpolarized background nuclei are usually not too harmful and therefore the dilution factor plays a minor role in the choice of target material. The main requirement is then that only the nucleus of interest has non-zero spin. Sometimes this is difficult to achieve, and one tries to find a substance where the background nuclei with spin have a low density or a much lower magnetic moment, which leads to a low polarization in comparison with the nuclei of interest.

A special beam transmission application is spin filtering of a beam using a polarized target, such as fast neutron polarization using polarized protons. Here the large difference of cross sections between the two spin states of the beam and target particles is important for effective filtering, and obviously a high target polarization is beneficial as well. The dilution factor plays a minor role unless the background nuclei have a large cross section.

Fusion with polarized fuel [5] may have the advantages that a lower ignition temperature can be achieved due to larger fusion cross section, and that the emission of neutrons may be controlled by the direction of the polarization vector which enables to design efficient absorbers and to protect structural materials. The materials should only contain polarizable light nuclei such as deuterons and tritons, and possibly selected isotopes of lithium, beryllium or boron.

The coherent scattering of polarized slow neutrons off spin polarized macromolecules allows one to study their size and shape using the method which is called spin contrast variation. Selective deuteration is used for adjusting the contrast in macromolecules under study, and in the matrix material for minimizing its contrast [6]. The matrix material must be a partly deuterated glass, which can dissolve in liquid state both the molecule under study and the paramagnetic molecules needed for DNP. It is an advantage in many cases if the paramagnetic molecule can also be deuterated. The dilution factor in this case is of minor importance, whereas strongly absorbing nuclei must be avoided in the ingredients. The material has to be a good glass former where no microcrystals are grown when solidifying the matrix.

The use of polarized contrast media in magnetic resonance imaging (MRI) has been demonstrated by inhaled optically pumped ^{3}He, which has a slow enough relaxation time so that human lungs could be imaged. It may be possible in the future to develop injectable spin-polarized liquids which might offer new possibilities to study metabolic reactions and arteries with improved contrast and resolution. Such fluids must contain large amounts of polarized nuclei, be free of toxic substances at the moment of injection and have slow enough relaxation time in order to permit the clinical studies of interest. The dilution factor plays a minor role here.

The design of materials for the studies of nuclear magnetism itself belongs to the realm of solid-state physics. These studies presently are limited to crystalline materials, and it is clear that isotopic purity, low paramagnetic center concentration and low contamination by background nuclei with spin is an advantage. The dilution factor here plays no role, unless the magnetic phase transitions are observed using polarized neutron scattering, which requires that neutron absorption be sufficiently low in the sample.

7.2 Chemically Doped Glassy Materials

The first breakthrough of DNP in organic materials was in 1966 when Borghini et al. [7] succeeded in obtaining 35% polarizations at 1 K temperature and 2.5 T field in ethanol-water, ethanol-methanol and ethanol-propanol mixtures doped with about 3% by weight of the free radical porphyrexide (PX). A few years later (1969) they reached 40% polarization [8] in 1-butanol with 5% water doped with the same radical under the same conditions. Glättli et al. [9] obtained 50% polarization in the same year in ethylene glycol (= ethane-1,2-diol) doped by reacting with potassium dichromate which yields paramagnetic Cr(V) complexes soluble in the liquid and solid phases, at 1 K and 2.5 T. Still in the same year Hill et al. [10] reached 67% polarization in butanol-water-PX at 0.5 K and 2.5 T, and Masaike et al. [11] obtained 80% in ethylene glycol-Cr(V).

In 1971 Gorn and Robrish [12] reported 50% polarization at 1 K and 2.5 T in 1,2-propanediol reacted with potassium dichromate. One year later, De Boer [13] obtained 92% polarization in a similarly prepared material at 0.5 K and 2.5 T. In the following year, de Boer and Niinikoski [14] polarized 1,2-propanediol-Cr(V) to values very close to 100% in a dilution refrigerator.

In the course of the above development, a large number of solvents and paramagnetic compounds were tried with results mostly only somewhat inferior to those quoted above, but sometimes clearly unsuccessful. Glättli [15] reviewed the methods of preparation by dissolving a stable radical and by reacting with Cr compound and concluded: 'unfortunately the description of preparation is limited to the "know-how" rather than the "know why"'. Despite this, butanol and the two diols were extensively and successfully used in a wide range of experiments.

The development of hydrogen-rich organic materials for polarized targets continued to be based on trial-and-error methods until 1979 when it was recognized [16, 17] that the success or failure of DNP was controlled by the glass-forming ability of the solvent in which paramagnetic centers thus became uniformly distributed upon rapid vitrification. This was clearly evidenced by viscometric studies [16] of 1-butanol-water, 1-pentanol-water and 1-pentanol-pinacol solutions and results of DNP studies when stable Cr(V) complexes were dissolved in these solutions [18]. Very soon after this microcalorimetric studies by Hill [19] of these and several other compounds revealed clear features of glass transition and devitrification. These findings also immediately explained why well-polarizing materials lose their characteristics when annealed close or above the devitrification temperature.

In the following sections we shall first briefly outline the physics of the glassy matrix materials, secondly how suitable paramagnetic molecules are introduced and thirdly discuss the fabrication and handling of beads or granules of the material.

7.2.1 Properties of Glassy Matrix Materials

The term 'glass' has been used to describe a state of matter which has properties intermediate between a liquid and a solid, with mechanical characteristics close to crystalline materials but microscopic properties resembling those of liquid state. The glassy state is not stable and not even metastable, and it is not the lowest free-energy state, so that equilibrium thermodynamics cannot be always applied. The liquid-glass transition, in particular, is a non-equilibrium phenomenon in which the time scale of relaxation phenomena becomes comparable with experimental times [20]. The glass formation therefore depends on the quenching rate, i.e. the rate at which the liquid is cooled to such low temperature below the melting point that the liquid disorder becomes frozen-in and nucleation of crystallization becomes hindered or extremely slow.

The transition from supercooled liquid to glass takes place at the temperature T_g, which is called glass transition temperature. It is phenomenologically defined as a temperature where the viscosity reaches 10^{13} P (P = poise = g/(cm·s)), or as a point where the specific heat has an anomalously steep increase when the temperature is increased. Microscopically the glass transition is explained by invoking a rapid change of the degrees of freedom in the supercooled liquid, in various theoretical schemes which are beyond our present scope.

The speed required for the supercooling of the liquid depends on the viscosity and its temperature dependence at the crystal melting point of the substance. Good glass formers are characterized by a high viscosity, which steeply increases when the temperature decreases in the vicinity of and below the melting point. Glass-forming liquid solutions stay clear and transparent near the melting point, thus indicating that no phase separation and no crystal growth takes place. The microscopic origin of the specific heat anomaly at the temperature T_g is thought to be the loss of configurational degrees of freedom, which cannot be understood from first principles but can be related and correlated with several other physical parameters such as the heat capacity, viscosity, diffusion constant and volume expansion. The correlations can be understood in the terms of phenomenological models such as the free-volume model [20], the entropy theory [21] and models explaining transport properties based on the temperature dependencies of the specific heat and volume expansion near the liquid-glass transition temperature.

The viscosity in glass formers (including the organic ones) which is 10^{13} P at T_g follows the law [20]

$$\eta = \eta_0 \exp\left(\frac{b}{T - T_\eta}\right), \qquad (7.16)$$

where η_0 and b are constants and T_η is constant when the viscosity is below 10^4 P and then T_η decreases to zero so that in the high-viscosity region the Arrhenius law is obeyed. The free-volume percolation model [20] successfully explains this functional behavior, which was first established empirically by Vogel [22] and Fulcher [23].

The entropy in supercooled liquid and crystalline phases can be determined from the measured specific heats. The experimental difference between the liquid and crystal entropies, called excess entropy, extrapolates to zero at a temperature T_s, which has been found to be about 20 K to 50 K below T_g for most substances. The entropy theory of the glass

Table 7.3 The crystal melting point T_m, the glass transition temperature T_g, T_η and T_s for some hydrogen-rich hydrocarbons which are easy glass formers. The asterisk for water indicates that the value for glass transition temperature is obtained by extrapolating measurements in ethanediol-water mixtures to zero concentration.

Substance	Formula	T_m (K)	T_g (K)	T_η (K)	T_s (K)	Ref.
2-Methylbutane (isopentane)	$(CH_3)_2CH(CH_2)_2$	113.3	68.2			[24]
2-Methylpentane	$C_3H_7CH(CH_3)_2$	119.5	79.5	59	58	[20]
2,2-Dimethylpropane (neopentane)	$(CH_3)_4C$	256.5	?			[24]
3-Methylhexane (d)	$C_3H_7CH(CH_3)C_2H_5$	156.3	88			[24]
Methanol	CH_3OH	175.2	103	60	63	[20]
Ethanol	CH_3CH_2OH	155.7	98.55		62.39	[25]
1-Propanol	$CH_3(CH_2)_2OH$	146.6	102.35		70.20	[25]
1-Butanol	$CH_3(CH_2)_3OH$	183.6	112.85		79.39	[25]
Ethylene glycol (ethane-1,2-diol)	$OH(CH_2)_2OH$	255.6	152	107	112	[20]
Propylene glycol (propane-1,2-diol)	$OH(CH_2)_3OH$		160		112	[21]
Glycerol	$C_3H_5(OH)_3$	291.2	180	138	134	[20]
Sorbitol	$C_6H_{14}O_6$	383	266	236	236	[20]
Glucose (dextrose)	$C_6H_{12}O_6$	419	305		242	[21]
Water	H_2O	273.15	136*			[26]

transition [21] leads to an expression for viscosity almost identical to Eq. 7.16, with T_η in the exponential replaced by T_s. This gives confidence for the two models to describe many features of the glassy materials.

Table 7.3 lists the crystal melting point T_m, the glass transition temperature T_g, T_η and T_s for some pure organic glasses of interest in polarized targets. It can be noted that T_η and T_s are quite similar in those substances for which they are known.

As T_g depends on the heating rate during the measurement in a microcalorimeter and also on the thermal history of the sample, it has been argued that in the limit of extremely slow heating the temperatures T_g, T_η and T_s could all merge towards a common intermediate value, which could be defined as a phase transition temperature in the sense of equilibrium thermodynamics. The discrepancy, however, is too large and it has been suggested that the phase transition could take place at an intermediate temperature between T_g and T_s at a point where the excess entropy has not reached zero but some critical value.

The glass transition temperature in binary mixtures of glass formers with transition temperatures T_{g1} and T_{g2} can be obtained from [25]

$$T_g(x) = \frac{xT_{g1} + (1-x)T_{g2}C}{x + (1-x)C}, \qquad (7.17)$$

where x is the mole fraction of the component 1 and C is a constant, which can be determined from the entropies for the components of the mixture and is not an adjustable parameter. The entropy can be determined from experimental data on the specific heat. Data on alcohol mixtures [25, 27] very well obey this relation.

Glass transition studies [28] in mixtures between normal alcohols and various Lewis bases such as triethylamine, diethyl ether, acetone and toluene show that these mixtures exhibit a somewhat similar behavior.

7.2.2 Viscosity of Binary Mixtures of Alcohols, Diols and Water

Viscosity in the vicinity of the melting point for pure substances is usually in the range of 10^{-2} P.[1] In this range it can be determined with a rotating viscometer; measurements [16] in an apparatus modified for reduced temperature operation were performed for several substances and mixtures of interest in polarized targets. The sample cup was equipped with a thermometer in the fluid and with a cooling tube mounted outside, through which nitrogen vapor was pumped from an LN_2 bath in a wide glass dewar. The cup was also equipped with a heater, which allowed to control the cooling speed of the sample in the course of the measurements. The speed was usually a few K/min.

Figures 7.1–7.3 from Ref. [16] show the temperature dependence of the viscosity of 1-butanol and 1-pentanol mixtures with water, and of 1-pentanol mixtures with pinacol. In one 1-butanol sample with 5% water, 1.5 g of polyethylene glycol 4000 (PEG4000) was added in 21.2 g of solution. The pentanol-pinacol solutions are clearly good glass formers and exhibit better devitrification characteristics. When using the stable Cr(V)

---

[1] Here we are using the cgs unit poise (1 P = 1 g/(cm·s)) for the viscosity; the SI unit of viscosity is Pa·s, with 1 Pa·s = 10 P).

7.2 Chemically Doped Glassy Materials

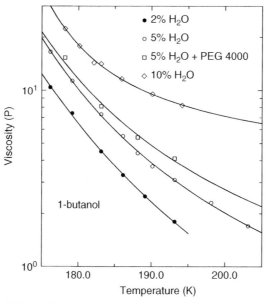

Figure 7.1 Viscosity of 1-butanol with various water concentrations around the melting point of pure 1-butanol (183.6K)

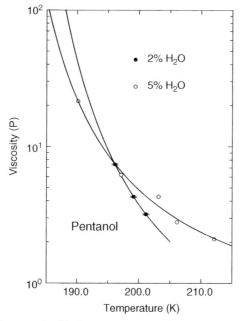

Figure 7.2 Viscosity of pentanol with 2 water concentrations around the melting point of pure pentanol (194K)

complexes, however, solubility in 1-pentanol remains rather low in absence of water. The 1-pentanol-pinacol mixtures might be interesting materials with the diol-Cr(V) complexes which can now be prepared with high concentration [29].

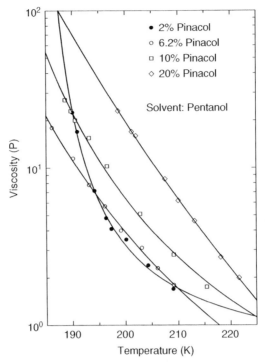

Figure 7.3 Viscosity of pentanol with various concentrations of pinacol

A slow cooling speed during measurements in a rotating viscometer caused alcohol-water mixtures to turn milky indicating the nucleation of crystal growth. The crystals are likely to be water. Pentanol-pinacol solutions did not exhibit such behavior.

7.2.3 Microcalorimetric Studies of Glass Transition and Devitrification

The anomaly of the specific heat in the vicinity of the glass transition can be seen in a microcalorimetric scan of the sample temperature. Other anomalies are also revealed, such as devitrification and melting of the sample. A typical temperature scan from Ref. [30] is shown in Figure 7.4 for 1-butanol sample quenched in LN_2. The device consists of two identical cups supported by constantan wires which are part of a differential thermocouple. When the cups are slowly warmed together with their enclosure, their temperature lags behind that of the surrounding gas by an amount which depends on the heat capacity of the cup and the sample; exothermic/endothermic reactions reduce/add to this lag and show up as peaks up/down. The temperature difference between the cups is thus sensitive to the specific heat of the sample.

The minor step feature G in Figure 7.4 is due to the specific heat anomaly of the liquid-glass transition. The position of this anomaly depends on the thermal history of the sample and on the warming rate, so that slower warming gives a lower transition temperature.

7.2 Chemically Doped Glassy Materials 297

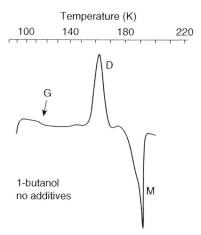

Figure 7.4 Differential microcalorimeter temperature scan of a single bead of pure 1-butanol

The midpoint of the step is usually taken as T_g. Above this temperature the material devitrifies, which is seen as a strong exothermic peak at D, which is due to the heat of crystallization of the whole sample or of a fraction thereof. Further warming results ultimately in the melting of the crystallized material at point M. Sometimes the melting of water frost contamination may be seen close to 273 K temperature (the scan of Figure 7.4 does not extend to this point).

The sample is loaded into the cup in the form of liquid; the subsequent quenching speed is not necessarily the same as is obtained when droplets of the substance are rapidly frozen by letting them solidify on the meniscus of an LN_2 bath. An apparatus suitable for such target material beads [30] consists of thermocouple wire loops symmetrically mounted in a small cylindrical housing and connected to give the differential temperature of the loops, as in the above apparatus. A sample bead is mounted in one of the loops and the other is left empty. The sensitivity and reproducibility of such a device is perhaps not as good as that of a differential calorimeter using closed cups, but it is sufficient to see the features G, D and M when they are pronounced.

In stable glasses the devitrification does not take place at a reasonable sample heating rate, and consequently also the melting anomaly is absent. The lack of D and M features of the differential microcalorimeter scan is thus susceptible to reveal materials which have a potential as a good matrix for accepting a uniform distribution of paramagnetic molecules. For this it is essential that the paramagnetic substance is soluble to the matrix in liquid form, and that the crystallization of the paramagnetic substance does not take place upon rapid cooling of the matrix. Normally the solubility is very limited at reduced temperatures between the melting and devitrification, and the material is far from thermodynamic equilibrium during the quenching. A high viscosity of the substance leads to a slow diffusion of impurity molecules and therefore to a slow (or no) nucleation of crystal growth.

The nucleation and crystal growth, however, also depend on other factors such as trace impurities which may have either promoting or demoting influence on nucleation. It is therefore important to have a diagnostic tool which may reveal the adverse effects of trace impurities in a batch of material to be processed.

Figure 7.5 shows a series of microcalorimeter scans for 1-butanol with various water concentrations. The glass transition is at $T_g = 117$ K for pure butanol and it increases to about 120 K by the addition of water. Pure butanol shows an onset for devitrification at about 160 K, which is not always reproducible, probably due to introduction of impurities in handling or bead preparation. The melting anomaly starts reproducibly at (182 ± 1) K for all scans and agrees well with the handbook value of 183 K. The addition of 0.5% H_2O causes the devitrification to start at 173 K, whereas at 1% the devitrification and melting anomalies are almost absent. At 2% concentration nothing can be distinguished with the sensitivity of the simple device, and it can be concluded that butanol-water mixtures are particularly good glass formers. This and the solubility are undoubtedly among the main reasons why DNP works so well with many paramagnetic molecules in butanol-water matrix.

Microcalorimetric scans with various concentrations of water in 1-pentanol reveal the glass transition at 133 K for all water concentrations including zero. Devitrification starts at 160 K at concentrations of 5% and below, but then becomes sharper and starts already at 145 K for 7% concentration. Melting starts always at about 194 K in agreement with the

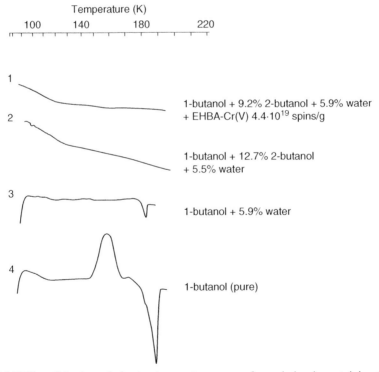

Figure 7.5 Differential microcalorimeter temperature scans of sample beads containing 1-butanol, 2-butanol, water and EHBA-Cr(V)

7.2 Chemically Doped Glassy Materials

handbook value and is well resolved from the devitrification at 6% and higher concentrations, unlike for butanol-water.

1,2-Propanediol and propanediol-Cr(V) show a glass transition at 180 K and no devitrification, whereas 1,2-butanediol has a glass transition at about 200 K and shows no devitrification nor melting.

As an attempt to identify hydrogen-rich materials with improved stability against devitrification between the glass transition and melting point, and better capability of dissolving the stable Cr(V) complexes, a series of microcalorimetric investigations [30] was performed on solutions of butanol isomers, water, pinacol and propanediol. Unlike 1-butanol, pure 2-butanol and isobutanol (2-methyl propanol) show no devitrification and melting; 2-butanol has T_g = 125 K, whereas isobutanol has such a small and gradual anomaly in specific heat that the apparatus was not able to resolve it. The viscosity of the pure solvent, however, is lower than that of materials with water added, because the bead softens and drops through the thermocouple wire loop around the melting point. Solutions containing 1-butanol and 2-butanol exhibit devitrification and melting up to 10% 2-butanol in solution, but at higher concentration no devitrification takes place. Solutions of 1-butanol with isobutanol behave in the same way.

Solutions of 1-butanol with 2.5% to 6.5% water by weight do not devitrify but show a reduced melting anomaly at 183 K. Additions of 9.2% and 12.7% of 2-butanol result in a similar scan but with no melting anomaly. The solution has a saturation point of 8% water at room temperature.

With EHBA-Cr(V) complex in solutions containing 1-butanol with 5% to 16.2% 2-butanol and 4.7% to 59% water, no devitrification nor melting is detected, as shown by the scan 4 of Figure 7.5. The viscosity of the solutions is higher than that of the solutions with no complex compound. Re-cooling the bead from 180 to 77 K resulted in no change in the microcalorimetric scan, unlike in beads prepared using pentanol-water and EHBA-Cr(V) [30].

Pinacol (2,3-dimethyl-2,3-butanediol) and 1,2-propanediol solutions with 1- and 2-butanol exhibit no devitrification nor melting anomalies.

Deuterated 1-butanediol with 5% heavy water, without and with 6.4×10^{19} spins/g EHBA-Cr(V), has a glass transition at the same temperature as normal 1-butanediol. These solutions show no devitrification, but a small melting anomaly is found at 185 K in all of them.

Devitrified and re-cooled materials often exhibit microcalorimetric scan features different from the materials quenched from room temperature. Devitrified beads cooled back to 77 K may appear opaque, which indicates that microcrystals have grown in the high-viscosity liquid and are frozen in the glass. The opacity is particularly pronounced when the target beads are colored by the paramagnetic compound. Glassy beads are translucid and clear with deep color, whereas devitrified beads show a lighter greyish color and are perfectly opaque. The results of DNP in such materials are always dramatically deteriorated and the spin-lattice relaxation time is reduced.

Glassy materials cooled rapidly to helium temperatures or below feature a delayed heat release, which is understood as a relaxation phenomenon of two-level systems present in the glass. Measurements in pure 1-pentanol and 2-pentanol [31] cooled slowly (1–3 K/min) to 77 K indicated that 1-pentanol is crystalline, whereas 2-pentanol is glassy. The specific heat of both vary roughly proportional to the square of temperature between 12 and 2 K, but 2-pentanol has a substantially higher specific heat of

$$c_{2-\text{PeOH}} = 0.05 \left(\frac{T}{K}\right)^2 \frac{\text{mJ}}{\text{gK}}. \tag{7.18}$$

In a sample of 70 g, the heat release is about 10 nW to 100 nW at 3.1 K to 12 K anneal temperature during 20 min after quenching and varies subsequently proportional to $1/(t/\text{min})^a$ where $a = 0.69$. A similar time dependence has been found in epoxy resin [32] with $a = 0.76$, whereas glasses other than organic tend to have $a = 1$.

Other organic glasses are likely to have a delayed heat release similar to the measured value of 2-pentanol [31]. Such heat leaks have been observed in large targets cooled with dilution refrigerator rapidly (< 1 h) to below 50 mK, where nW heat leaks become comparable with the available cooling power.

Table 7.4 summarizes the experimental values for T_g and T_η in the binary solutions of Figures 7.1–7.3. The glass transition temperatures have been obtained from microcalorimetric studies which were described above, whereas T_η is obtained from fits of Eq. 7.16 to the data of the figures. The viscosities of the pure substances are several orders of magnitude below those of the solutions above the melting point and are rather constant down to the melting point where a sudden increase takes place due to solidification.

Table 7.4 Glass transition temperatures, T_η of Eq. 7.13, devitrification temperatures and melting temperatures for binary solutions of water in 1-butanol and 1-pentanol, and of pinacol in 1-pentanol. Numbers in parentheses indicate that the microcalorimetric anomalies are not observed in all samples, or that they are irreproducible. 'None' means that no anomaly is observed, and no entry in table means that no measured data was known of.

Solute conc. (% by wt.)	T_g (K)	T_η (K)	T_D (K)	T_M (K)	Ref.
Water in 1-butanol					
0	117		150	182	[19]
2	120	118	None	None	[16]
5	120	137	None	None	[16]
10	120	163	None	None	[16]
15	(121)		(164)	(177; 254)	[19]
Water in 1-pentanol					
0	133		160	194	
2	133	174	160	194	[16]
5	133	172	160	194	[16]
7	133		150	194	

Table 7.4 (cont.)

Solute conc. (% by wt.)	T_g (K)	T_η (K)	T_D (K)	T_M (K)	Ref.
Pinacol in 1-pentanol					
2		181			[16]
6.2		116			[16]
10		146			[16]
20		21			[16]

7.2.4 Miscellaneous Materials of Interest

Ammonium borohydride NH_4BH_4 is soluble in ammonia and compatible with the stable Cr(V) complexes. Krumpolc [3] describes his method of preparation which is slightly modified from the one previously published, starting from sodium borohydride ($NaBH_4$, 2.0 g), anhydrous ammonia and finely powdered and vacuum-dried ammonium fluoride (NH_4F, 2.0 g). The powders were placed with a magnetic stirrer in a glass vessel with separate reactor and receiver flasks mounted so that reacted liquid can be poured from the reactor to the receiver through a filter. The apparatus was purged with dry nitrogen and the reactor arm was immersed in ethanol/dry ice bath (–80 °C), and about 8 ml of ammonia was condensed. The reaction vessel was then warmed to –45 °C in a beaker with ethanol/dry ice and the mixture was vigorously stirred for 6 h, as ammonium fluoride is only slightly soluble in ammonia. At the end (6–8) ml more ammonia was condensed in the reactor and the whole apparatus was immersed to the large bath at –80 °C. Ammonium borohydride was separated from the insoluble sodium fluoride by passing the solution through a filter separating the two arms of the vessel, slowly applying vacuum to the receiver flask. Ammonia was then partially or totally evaporated at –60 °C under vacuum and 1.4–1.6 g of white crystalline ammonium borohydride was collected from the receiver flask; the corresponding yield is 80–92%. The product is stored under dry nitrogen at –80 °C or in liquid nitrogen.

Ammonium borohydride can spontaneously ignite upon contact with traces of water or in prolonged contact with humid air [4].

Other boron compounds of possible interest are diborane B_2H_6 (called also boroethane), dihydrotetraborane B_4H_{10} (called also borobutane), borane ammonia ($NH_3:BH_3$) and lithium borohydride $LiBH_4$. The two first are gases at room temperature; the second is poisonous.

Toluene is a good glass former [19] and is known to have a low yield of radiolytic paramagnetic centers [33, 34]. The dilution factor, however, is so low that toluene has never been used as a polarized target material and therefore the real radiation resistance is not known. Proton polarization of about 62% has been reached at 0.5 K and 2.5 T using di-tertiary-butylnitroxide (DTBN) as a dopant [33]. Using 1,2-bis-diphenylene-1-phenylallyl (BDPA) as a dopant 41% proton polarization was reached in partly deuterated toluol at 0.7 K sample temperature and 2.5 T field [35].

Using BDPA as a dopant at a low concentration of 0.6×10^{19} spins/cm^3 in partly deuterated m-xylene-d6, proton polarization of 25% and deuteron polarization of 10% were obtained, but at different microwave frequencies [35]. The experiment demonstrated, rather than high polarizations, the distinct mechanisms of DNP which is possible at low concentration, but also demonstrated the 'ill' effects of the low concentration of the paramagnetic centers to DNP.

7.2.5 Preparation of Cr(V) Complexes with Diols

Ethanediol (called also ethylene glycol) reacts with some hexavalent chromium compounds producing a relatively stable metallo-organic complex called here in short ED-Cr(V). Among the dichromates used originally by Garif'yanov et al. [36] were $K_2Cr_2O_7$, $Na_2Cr_2O_7$ and $(NH_4)_2Cr_2O_7$. The reaction of potassium dichromate was studied by Glättli [15] by ESR techniques. The procedure consists of mixing finely ground dichromate with ethanediol (more than the solubility of 20 mmol for 1 mol of ED) and reacting the mixture at 70 °C for 15 min under magnetic stirring. This gives a concentration of 5×10^{19} Cr(V) spins/ml. The reaction speed plotted against inverse temperature shows Arrhenius relation with an activation energy of 20 kcal/mol. It was noted that the reaction water causes the Cr(V) complex to react yielding Cr(III), which has a broad resonance line, and which does not contribute to DNP, but is very effective in relaxing the nuclear spins. Removal of water by evacuation helped reaching higher Cr(V) yield and a better Cr(V)/Cr(III) ratio of about 5. The proton polarization of 85% in a 7 g sample was reached at 2.5 T using a ^{3}He refrigerator for cooling. This was clearly a breakthrough which stimulated interest in Cr(V)-diol complexes.

Bontchev et al. [37] found that the reactions described above were sensitive to light in addition to the presence of water.

Hill [38] describes a slightly refined process where 40 ml of ethanediol (spectroscopic grade) is reacted with 3.75 mg of potassium dichromate (reagent grade) in a 150 ml flask. The flask is equipped with a water-cooled reflux condenser, which prevents significant loss of the diol but lets water to be removed by a vacuum pump. Ethanediol is pre-heated to 66 °C and stirred under vacuum to remove air. The flask is then backfilled with inert gas (helium) and potassium dichromate is added. The mixture is stirred for 40 min at 66 °C, after which the flask is backfilled with inert gas and cooled rapidly to room temperature.

A similar reaction takes place in 1,2-propanediol, glycerol and 2,3-butanediol. De Boer [13] describes the preparation of 1,2-propanediol-Cr(V) complex (called PD-Cr(V) for brevity) as follows: 250 ml of vacuum-distilled PD and 20% by weight of vacuum-dried $K_2Cr_2O_7$ were mixed and magnetically stirred at 55 °C for 80 min under a few torr pressure obtained by pumping via a water-cooled distillation column. The flask was illuminated by a 250 W incandescent lamp. It is believed that the formation of Cr(III) due to light was prevented because high speed of magnetic stirring was used, which forced the undissolved dichromate to form a layer of solid particles along the glass wall; the outside of the flask was covered by black tape in areas which were uncovered by solid residues. The formation of Cr(V) was thought to be more due to the light than to the thermal activation at the reduced reaction temperature of 55 °C. Trying even lower reaction temperatures and longer reaction times resulted in higher Cr(III) concentration, probably due to less efficient distillation of the reaction water. The described recipe gives about 3×10^{19} C(V) spins/cm^3 with a very low concentration of Cr(III).

The concentration of Cr(V) was then increased to the desired value by quickly (10–20 min) evaporating the diol at 1 Torr pressure in the absence of light. The optimum concentration for DNP at 2.5 T field in a ^{3}He refrigerator was found to be around 16×10^{19} Cr(V) spins/cm^3, with a rather flat maximum of about 83%. The maximum polarization increased to 92% when increasing the pumping speed of ^{3}He from 170 m^3/h to 250 m^3/h. The initial polarization build-up was 2 min to $0.7 \cdot P_{max}$; this fast time could be expected from the large electron spin concentration. The spin lattice relaxation time, however, remained 170 min at 0.5 K, suggesting that the concentration of Cr(III) was indeed quite low.

Krumpolc, Hill and Stuhrmann [29] further refined the procedure for preparing the diol-Cr(V) complexes. They observed that the photosensitivity for Cr(III) formation was actually higher than previously reported [37], and therefore performed the reaction in total darkness. By using a nearly stoichiometric ratio between the diols and sodium dichromate, a considerable increase in the amount of Cr(V) complex was achieved. Furthermore, because the decomposition of Cr(V) into Cr(III) is accelerated by the presence of water formed in the course of the oxidation, water was continuously removed by anhydrous magnesium sulphate which is a common dehydrating agent. The absence of the absorption line of Cr(III) at 600 nm wavelength showed that the decomposition of Cr(V) was indeed stopped, and this was confirmed by column chromatography using silica gel.

The diol-Cr(V) complexes have not yet been isolated in pure form, but the above simple procedure allows one to obtain high concentrations of the complexes without Cr(III) for dissolving in suitable solvents forming a hydrogen-rich glass matrix.

In the original work of Garif'yanov [36] Cr(V) complexes were observed in dimethylformamide (DMFA), acetone and acetonitrile in addition to diols and glycerol. Svoboda [39] has prepared Cr(V) complexes in dimethylformamide HCON(CH$_3$)$_2$ solution using pyrocatechol as a ligand. The EPR linewidth is 2.4 mT in 0.89 T field and 4.2 K temperature, comparing favorably with that of the diol complexes which have linewidths of 3.2 mT to 3.8 mT under the same conditions. The material, however, reaches rather low DNP at this field and 1.4 K temperature. Svoboda reports also that the same technique can be used to produce Cr(V) complexes in alcohols and other solvents provided that it is able to dissolve the dichromate or other chromium salt. Among other reducing agents than pyrocatechol, alizarin or 2,3-naftalenediol can be used.

Bunyatova [40] has succeeded in observing a photochemical reaction of ammonium dichromate in pentanol, which results in a paramagnetic complex with the g-factor of 1.973, a value typical of Cr(V). Laser light with 441 nm wavelength was obtained from a helium-cadmium laser. The intensity-width relationship of the hyperfine components as well as the g-anisotropy estimated from the EPR line in solid at 77 K confirms that the leading paramagnetic substance is a Cr(V) complex belonging to the same family as the diol-Cr(V) complexes.

The preparation of Cr(V) complexes by reaction with deuterated diols was first believed to require that hydrogen atoms should remain at least in the hydroxyl groups, and therefore deuterated targets were prepared by mixing perdeuterated propanediol (DPD-8) with concentrated solution of partly deuterated propanediol-Cr(V) obtained by reacting potassium dichromate with $C_3D_6(OH)_2$ (abbreviated DPD-6) [41]. Bunyatova has shown [40] that the reduction of potassium dichromate proceeds also in perdeuterated propanediol, and that the resulting perdeuterated complex has a line width due to g-anisotropy slightly narrower than that in normal (undeuterated) propanediol.

7.2.6 Preparation of Cr(V) Complexes in Stable Pure Form

In 1976 Krumpolc and Rocek [42] discovered that a Cr(V) complex formed in water solution by reacting 2-hydroxy-2-methylbutyric acid with chromic acid. The Cr(V) absorption line at 750 nm indicated that a complex with exceptional stability at room temperature was formed. The reaction kinetics and structure of the complex were studied [43] and it was obtained in crystalline form when the reaction was made with CrO_3 [44]. Further refinements to the reaction led to much simpler procedure [45] with a yield of 60% to 90% based on the sodium dichromate, which was used as a starting material. This procedure takes about two days and can be performed without sophisticated equipment.

Krumpolc and Rocek describe the process for EHBA-Cr(V)[2] as follows [45]: To a solution of 19.8 g (0.150 mol) of 2-ethyl-2-hydroxybutyric acid[3] in a 300 ml Erlenmayer flask is added 6.5 g (0.025 mol) of anhydrous sodium dichromate (the dihydrate dried in vacuo at 100 °C for about 30 min to 40 min and finely pulverized), and the heterogeneous mixture is magnetically stirred until the dichromate is completely dissolved (about 10 min). The formation of dark red brown solution of the Cr(V) complex can be observed instantaneously. The flask is fitted with a glass stopper and immersed in a water bath controlled at 25.0 °C temperature for a period of 23 h to 24 h. The solution is then poured into 375 ml of hexane, upon which the Cr(V) complex precipitates out in the form of dark red violet solid. The crude product is collected, dried in vacuo at room temperature for about 30 min to remove water and volatile materials (solvents, diethyl ketone), dissolved in 125 ml of acetone and reprecipitated in 375 ml of hexane. The crystalline product is washed with 20 ml of hexane and dried in vacuo at room temperature to constant weight (about 30 min) giving 14.8 g to 16.5 g (0.040 mol to 0.045 mol) of sodium bis(2-ethyl-2-hydroxybutyrato)oxochromate(V) monohydrate. The yield is between 80% and 90% based on sodium dichromate. The analysis for the product with the chemical formula $Na[((C_2H_5)_2COCO_2)_2CrO]\cdot H_2O$ is given in Table 7.5.

Table 7.5: Analysis of EHBA-Cr(V) complex monohydrate [46].

Substance	Calculated	Found
C	39.0	39.4
H	6.01	6.00
Cr	14.1	14.1
Na	6.23	6.6
H_2O	4.88	5.3

[2] Abbreviation for sodium bis(2-ethyl-2-hydroxybutyrato)oxochromate monohydrate.
[3] Aldrich.

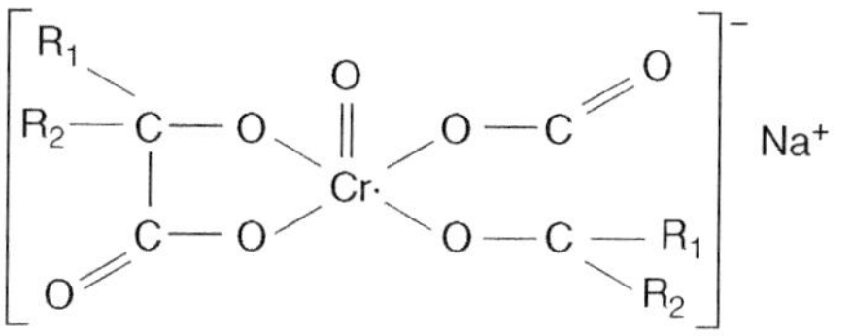

$$\left[\begin{array}{c} R_1 \\ R_2-C-O \\ | \\ C-O \\ O \end{array} \begin{array}{c} O \\ \| \\ Cr \\ \end{array} \begin{array}{c} O \\ O-C \\ O-C-R_1 \\ R_2 \end{array} \right]^{-} Na^+$$

Figure 7.6 Chemical formula of the stable Cr(V) complexes of tertiary hydroxy acids synthetized by Krumpolc and Rocek [45]

The presence of one molecule of water in the crystal was established by deuterium exchange technique and was determined by NMR.

The product EHBA-Cr(V) is a dark red violet crystalline solid. It is very stable at room temperature: after exposure to air and light for several weeks no visible decomposition was observed. It does not show a melting point but slowly decomposes at 170 °C. It is readily soluble in polar solvents (water, acetone pyridine, dimethylformamide, dimethyl sulphoxide, acetic acid, liquid ammonia), but insoluble in hydrocarbons, carbon tetrachloride, chloroform and diethyl ether. The compound is relatively stable to hydrolysis: only about 20% of 0.01 M solution of the complex is decomposed over a period of 24 h at 25 °C. The stability can be substantially enhanced by the addition of a small amount of 2-hydroxy-2-methylbutyric acid. Upon acidification the complex undergoes fast disproportionation to chromium(VI) and chromium(III); the addition of sodium hydroxide results in instantaneous disproportionation. The infrared and visible spectra feature several lines given in Ref. [45].

The complex is paramagnetic with d1 electron configuration giving the spin 1/2; the hyperfine spectrum and g-factor were discussed in Chapter 3.

The same process was extended by Krumpolc and Rocek [45] to other tertiary hydroxy acids including 2-hydroxy-2-methylpropionic acid (HMPA), 2-butyl-2-hydroxyhexanoic acid (BHHA), 1-hydroxycyclopentanecarboxylic acid (HCpCA), 1-hydroxycyclohexanecarboxylic acid (HChCA) and 2-hydroxy-2-phenylpropionic acid (HPPA). The molecular structure is shown in Figure 7.6, where the radicals R1 and R2 are given in Table 7.6 together with data on the solubility and stability of the complexes.

Two of the stable Cr(V) complexes, HMPA-Cr(V) and EHBA-Cr(V), have also been prepared in perdeuterated form [29]. The latter, called by the acronym EDBA-Cr(V), has now gained wide use in deuterated polarized targets for high-energy scattering experiments and in molecular biology experiments using spin contrast variation. Deuteration seems to promote the stability of the complex even further. Unfortunately, the starting material perdeuterated EHBA is not available commercially and its synthesis [29] requires chemistry skills and a well-equipped laboratory.

The effect of deuteration has also been studied in a broader class of similar Cr(V) complexes by the same team [47].

Bunyatova reported that HMBA-Cr(V) is unstable when dissolved in diols and in ethanol [40].

Table 7.6 Composition, solubility and stability of stable Cr(V) complexes with tertiary hydroxy acids, from Ref. [45].

Complex	R_1	R_2	Solubility[a] Water (g/l)	Solubility[a] Acetone (g/l)	Decomp[b] (%) In H_2O	Decomp[b] (%) In 0.1 M of HA	Thermal decomp. (°C)
HMBA-Cr(V)	CH_3	C_2H_5	290	103	58	13	180
HMPA-Cr(V)	CH_3	CH_3	270	4.1	81	64	180
EHBA-Cr(V)	C_2H_5	C_2H_5	190	203	28	2	180
BHHA-Cr(V)	C_4H_9	C_4H_9	16	87	27	–[c]	180
HCpCA-Cr(V)	$(CH_2)_4$	$(CH_2)_4$	204	730	100	53	170
HCxCA-Cr(V)	$(CH_2)_5$	$(CH_2)_5$	340	710	100	13	170
HPPA-Cr(V)	C_6H_5	C_6H_5	170	810	100[d]	100[d]	140

[a] At 25 °C.
[b] About 0.01 M solutions of Cr(V) complexes in water and in aqueous solutions of corresponding hydroxy acids (HA), followed over a period of 24 h at 25 °C.
[c] 2-Butyl-2-hydroxyhexanoic acid is insoluble in water.
[d] Decomposed in about 1 h.

7.2.7 Preparation of Glassy Beads

Polarized target material must be relatively finely divided to provide good heat transfer from the solid to the dilute $^3He/^4He$ solution of a dilution refrigerator, or to the two-phase flow of ^4He or ^3He in the case of evaporation refrigerators. In the first case the heat transfer is predominantly limited by the thermal boundary resistance (Kapitza resistance); in the second it is determined by the two-phase boiling heat transfer coefficient in series with the Kapitza resistance. In both cases there is an optimum size of the target grains, because very small grains size will cause obstruction to the convectional heat transfer in dilute solution and will limit the flow in the case of two-phase boiling heat transfer. This optimum size appears to be about 2 mm in both cases, although there have been no systematic studies so far. Clearly the optimum grain size depends also on the filling factor and grain shape, but in polarized targets one wishes to have the closest possible packing of the material, and the grain shapes are limited by practical fabrication techniques. The heat transfer will be discussed in greater detail in Chapter 8; we shall discuss here first how the target material is prepared in liquid form, and then how regular grains can be prepared by quenching the liquid in droplets into liquid nitrogen.

The preparation of chemically doped target materials involves dissolving the matrix material ingredients into a homogeneous solution, dissolving in it the free radical or metallo-organic complex and solidifying the material by rapid cooling. In the case that the paramagnetic compound does not support the temperature at which the matrix material is liquid, two other methods might be tried: The paramagnetic compound is diffused in the solid matrix material, or the matrix material is dissolved to another solvent, which is vacuum evaporated after dissolving the paramagnetic compound.

Dissolved air and in particular oxygen in the matrix material has a strong influence in the proton spin lattice relaxation time, for example, in toluene and in OTP [6].

The solubilities of the various substances have to be tried out in most cases, because it may take more time to find data in the literature (unless a chemistry department is within easy reach). The ingredients are deoxygenated by bubbling dry nitrogen and, if they are not anhydrous, they are dried before weighing the right proportions. The handling of deuterated solvents is best made in a glove box with dry nitrogen atmosphere; otherwise, they become contaminated by the protons of the ordinary water from the humidity of air.

After dissolving all ingredients, the liquid is once more deoxygenated by briefly bubbling dry nitrogen, usually directly in the vessel from which the liquid is dropped onto LN_2 or injected in a mould at 77 K temperature.

If one of the ingredients is a gas or is highly volatile at room temperature, the handling must be done at a lower temperature. Hill [4] has described a simple apparatus which can be easily modified for including the feeds for condensable gases, for example.

A simple method of preparing round glassy beads from a liquid solution consists of placing it in a syringe with a thin hypodermic needle. The needle size and shape control the drop size, and the dropping rate can be controlled manually such that only a few beads float on the meniscus of the LN_2 bath during the process. The beads made in this way have a diameter between 1.5 mm and 2.0 mm and the larger ones tend to be slightly flattened. Sometimes the bead begins to spin while floating; this leads to flattened shapes. It can also be noted sometimes that the top surface of the bead is concave, and it is not rare to find a bead which has a gas-filled bubble totally enclosed by the glassy material. Such beads usually disintegrate when warmed rapidly from liquid helium temperature, because superfluid helium may fill the void gradually during target operation, but cannot escape when warmed above 2.17 K.

The droplet size can be also controlled by applying a high voltage between the needle and an annular electrode mounted about five millimeters below the tip of the needle; in this case the droplets also become so much charged that they repel each other, thus avoiding collisions while floating on LN_2. The charge may, however, make them stick to the wall of the dewar before solidifying. A discharging gun is practical for neutralizing the charges in the various dielectric materials of the apparatus and the beads. Discharging facilitates also the handling of beads fabricated without high voltage.

The beads are collected into storage bottles using an attached funnel, the rim of which is a few millimeters above the liquid surface.

The manual method works well for quantities of a few mL. Quantities of normal (2.3 kg) and deuterated (1.8 kg) butanol-5% water beads, doped with stable EHBA-Cr(V) and

EDBA-Cr(V) complexes, respectively, were prepared [48, 49] for the large polarized target of the Spin Muon Collaboration experiment NA47 at CERN. We shall describe here the preparation of the deuterated beads [49], which was slightly more elaborate. The apparatus, mounted inside a glove box, consisted of a rotating foam-isolated large vessel subdivided into 24 compartments by 12 radial walls and one circumferential wall. The target material solution was poured into two syringes with a hypodermic needle; the syringes were temperature and pressure controlled. The pressure was adjusted so as to get a drop rate of 1/s from each, and the LN_2 vessel was rotated at a speed controlled so that at most one bead was floating in each compartment. Helium gas was blown onto the surface of the bath to depress the partial pressure of nitrogen, thus preventing boiling inside the bath and making the beads sink sooner (about 4 s to 5 s) as they were cooled faster below the critical temperature of film boiling. In this way the quantity of about half million beads with a diameter of 1.8 mm were produced in batches of 60 mL of solution, each of which took about 90 min to process.

Casting in a mould at 77 K temperature is the technique which was first used for organic materials. The mould was often the microwave cavity itself, precooled close to 77 K before quickly filling with liquid, and then quenched into the LN_2 bath. This works for small samples, and the heat transfer is likely to be rather limited already for targets of less than 1 cm^3 volume.

The mould casting technique is needed in applications which require a thin target, and in small-angle neutron scattering experiments which are sensitive to the background caused by the small-angle magnetic scattering (specular mirror reflections) by the surfaces of the round beads. In the first case thin flat pieces with parallel surfaces are needed, and in the second case pieces about $3 \times 17 \times 17$ mm^3 are made [50]. In both cases a demountable copper mould is prepared with right size and shape. The mould is cooled in an LN_2 bath but not filled with LN_2. The cavity is rapidly filled with the target liquid by injecting a measured dose from a syringe. The mould is then dismounted under LN_2. The blocks thus made are not always as transparent as the beads, but results of DNP seem to confirm that the cooling speed is sufficient for most materials, at least for good glass formers such as ethanediol, propanediol, glycerol-water and butanol-water.

7.2.8 Recent Research and Future Developments with Glassy Materials

DNP in Scintillating Plastics

In the nitroxyl radicals TEMPO, DTBN and oxo-TEMPO (see Chapter 3) the unpaired electron is localized predominantly in the N–O bond and is surrounded and shielded by four methyl groups [51]. These molecules can therefore combine and react with many other molecules without first losing the free electron. This opens many possibilities to dissolve or diffuse the paramagnetic substances thus generated into substances that contain the desired nuclei to be dynamically polarized. Among these are solids such as polyethylene (PE), polymethyl methacrylate (PMMA) and polystyrene; the two latter can be used also as scintillating detectors. This gives perspectives to many experiments that require a polarized target combined with particle detection very close to the polarized target nucleus [51, 52]. The applications are notably in reactions where the kinetic energy is so low that the recoil particle is stopped very close to the vertex of the event.

Magic Angle Spinning (MAS) NMR Enhanced by DNP

The use of DNP for the sensitivity enhancement of NMR has led to the systematic development of new radicals, bi-radicals and tri-radicals that can be dissolved, together with the molecules under study, in solid form inside a MAS NMR probe [53]. The sample preparation involves dissolving the paramagnetic polarizing agent (PA) at RT in a glass-forming solution, often glycerol-water, together with the molecules under study (analyte). The solvent matrix may be deuterated or partially deuterated. Alternative sample preparation techniques aim at reducing or eliminating the glassy matrix which dilutes the analyte molecules or structures to be studied; matrix-free approaches involve chemical binding of the paramagnetic molecule to the analyte structure, by introducing a 'gluing' intermediate molecule or by suitably modifying the paramagnetic or analyte molecule [53].

In the MAS DNP the sample is rotated by a small turbine driven by nitrogen or helium that is also cooling the sample. Operating around LN_2 temperatures (down to 30 K with helium) and at very high magnetic fields up to 9 T and even above, the sample inside a rotor within a simple resonator cavity is fed from an EIO or gyrotron microwave source.

The MAS DNP based on solid effect uses narrow-line radicals of BDPA-type and of trityl-type. BDPA is soluble in polystyrene and it could, in principle, be used for the DNP of scintillation detectors. Its solubility in water is poor, but its derivatives sulphonated BDPA and sulphonamide BDPA are water soluble and they are used for DNP MAS spectroscopy [54]. Among the trityl-type radicals the OX063 and CT-03 were originally developed for EPR oximetry MRI based on Overhauser effect [55] at RT and in a low field. They are water soluble and are also successfully applied for the DNP at a high field ≥ 5 T.

In MAS DNP the nitroxide-based radicals and biradicals have been found to have EPR line widths and relaxation times suitable for cross-effect DNP. Here the biradicals in particular are highly efficient [53] and over hundred new radicals of the TEMPO family have been developed for these applications.

Dissolution DNP NMR Spectroscopy

The trityl family of triphenylmethyl-based radicals has been shown to work well also at 5 T magnetic field, which has been applied for dissolution DNP [56]. Dissolution NMR uses DNP of glassy solid samples at low temperature (1 K) and in a very high magnetic field. The polarized sample is then rapidly transferred and warmed up in order to melt it before placing into the probe of an NMR spectrometer [57, 58]. The nuclear spin relaxation time remains sufficiently long to permit fairly long acquisition time of signals that may be enhanced by a factor of 10^4 over the equilibrium signal at RT.

7.3 Irradiated Materials

There are two main motivations for preparing polarized target materials by generating radiolytic paramagnetic centers by irradiation:

(1) The method is likely to work with pure and hydrogen-rich materials that cannot dissolve paramagnetic dopants in liquid state.
(2) The target thus produced is likely to be radiation resistant, or at least can be regenerated by annealing at an elevated temperature in situ.

The generation of paramagnetic radicals by irradiation of solids was discovered by Zavoiskii in 1945 and this marked a breakthrough in radiation chemistry because detailed studies of the products with extended lifetime could be undertaken using EPR in organic materials. These have been reviewed in Ref. [59].

Two types of materials are susceptible to yield useful irradiated polarized targets: simple cubic crystalline solids and glassy solids consisting of simple molecules. The cubic structure is of interest because this may lead to a small anisotropy of the g-tensor and of the hyperfine tensor so that the resultant EPR linewidth in the solid is suitable for DNP. This is particularly important in the case of polycrystalline materials, because otherwise only a fraction of the crystallites would absorb microwave power and polarize. Ammonia and lithium hydrides belong to this category of materials, and they are used successfully in scattering experiments. Organic glasses belong to the second category, and recently UV-irradiated butanol containing $\leq 1\%$ phenol was polarized to 1.4% at 1.5 K and 1.2 T; it was shown that the butyl free radicals generated in situ disappeared rapidly upon warming up of the sample [60]. This is likely to open up the UV photolysis for dissolution DNP enhanced NMR spectroscopy.

Irradiated single crystals do not need to cubic, although LiF and CaF_2 which are best studied have a cubic structure.

All types of radiation can be used for producing radicals, although the yield of different types depends on the particle and energy used. Electrons and energetic photons lose their energy in an electromagnetic cascade which ionizes the material and causes some displacements of atoms. Massive charged particles also give rise to electromagnetic cascades but are more effective in displacing atoms. Fast neutrons are particularly effective in displacing atoms and only give rise to electromagnetic cascades and ionization via charged secondary particles; their effectiveness in creating free charge is 10 to 100 times lower than that of protons. One may thus control the types and relative yields of paramagnetic centers by the choice of the particle and its kinetic energy.

In the following subsections, we shall discuss the preparation, irradiation, handling, DNP and other important properties of ammonia, lithium hydrides, other crystalline materials and organic materials. Materials with triplet-state paramagnetic molecules excited from the diamagnetic ground state by optical transitions are discussed also here, before concluding with safety aspects related specifically with the chemical instability of irradiated materials.

7.3.1 Preparation of Ammonia Beads

Ammonia is a gas at room temperature and normal pressure and has normal boiling and melting temperatures of −33.35 °C and −77.7 °C. Two methods have been used for producing beads of solid ammonia before irradiation: letting small droplets of liquid ammonia solidify on the meniscus of LN_2 and slowly freezing liquid ammonia in a glass before crushing and selecting the size of the fragments.

The first method [61, 62] is similar to what is used for preparing solid beads of organic liquids doped with dissolved paramagnetic molecules. Ammonia gas from a pressurized cylinder is led to a condenser which terminates with a hypodermic needle. The condenser is cooled by flow of methanol from a bath with 'dry ice' (solid carbon dioxide, sublimation point −78.5 °C) or by flow of a coolant from a bath of a refrigerator. Suitable coolants in the latter case include methanol, ethanol, fluorinated hydrocarbons or silicone oil. By adjusting the condensation pressure slightly above atmospheric pressure, droplets of 1.5 mm to 2.0 mm diameter detach from the tip of the needle with intervals of several seconds to a LN_2 bath. The beads float on the meniscus of LN_2 for several seconds until they solidify and cool below the film boiling temperature, at which point they may sink into the liquid. The dropping rate must be adjusted so that only a few beads float at any time, in order to minimize the collision of a liquid droplet with a solid bead which results in the fusing of both together.

The resulting white polycrystalline beads are not perfectly round and some of them may shatter during manipulation. Shattered pieces can be separated by a suitable size sieve, after the possible large fused beads are separated by a larger size sieve. These operations are performed under LN_2.

The density of the polycrystalline NH_3 at liquid nitrogen temperature is 0.836 g/cm³, just above that of LN_2 which is 0.8081 g/cm³. The polycrystalline beads have an effective density slightly lower than the single crystal density quoted. Convectional currents of the liquid N_2 easily carry the ammonia beads; this does not facilitate their collection in a recipient. A further difficulty is that the method of dropping results in electrically charged beads which repel each other and stick to the walls of the dewar and containers. The charge can be neutralized sufficiently, however, by using a source of negative ions close to the apparatus. Commercial dust removers and discharging devices work well; their effectiveness is easily tested by pointing the device to the bath where charged beads stick to the wall.

The above method of preparing ammonia is suitable for the production of large quantities of beads and, because little is lost in the process, it is also suitable for isotope labelled ND_3, $^{15}NH_3$ and $^{15}ND_3$.

The slow freezing method results in a transparent solid where the crystal size depends on the speed of solidification. Small samples can be prepared by the method of Cameron [63] in a test tube, immersed in a methanol dry ice bath, where ammonia gas from a cylinder is first condensed and then solidified after shutting off the flow of gas. The test tube is then transferred to a liquid nitrogen bath and broken. After separating the block of solid ammonia from debris of glass, the sample is broken into smaller pieces by crushing through a metal screen with drilled holes and captured in a slightly finer mesh screen. The advantages of this method are a slightly higher density and solidity of the pieces produced, and a higher thermal conductivity. Because of the irregular shape of the pieces, the surface-to-volume ratio is also higher than for spherical beads; this improves the heat transfer to the helium both in the case of evaporative cooling and in the case of cooling by a dilution refrigerator.

The slow freezing method has also been used for producing large quantities of solid ammonia target material [64] by liquefying 150 cm^3 batches in an ethanol bath at −85 °C, cooled by the flow of nitrogen gas passing through an LN$_2$ bath. When shutting off the ammonia gas inlet, the liquid begins to solidify on the glass tube wall and its pressure drops. After reaching the triple point pressure of about 60 mbar, the tube is pressurized to 500 mbar by argon gas. The resulting clear and transparent block of solid ammonia is then shock cooled to LN$_2$ temperature by filling the tube directly with LN$_2$, which causes the material to crack. Crushing and sifting to desired size is made under LN$_2$, and the material is stored in LN$_2$ until irradiation. The process yields 125 cm^3 of 2 mm to 3 mm fragments from the 150 cm^3 initial block and takes about 6 h for each batch. The material has a density of 0.853 g/cm^3 at LN$_2$ temperature; this can be compared with that derived from the lattice parameter $a = 5.084 \pm 0.001$ Å, which yields the density 0.8619 g/cm^3. The material is therefore likely to have a relatively large fraction of vacancies or voids, which are smaller than the wavelength of visible light.

7.3.2 Irradiation of Ammonia

Irradiation of solid ammonia at 77 K yields only ·NH$_2$ radicals which result from the following reactions taking place fairly easily [34]:

$$NH_3 \rightarrow \cdot NH_3^+ + e^-$$
$$\cdot NH_3^+ + NH_3 \rightarrow \cdot NH_2 + NH_4^+$$
$$NH_4^+ + e^- \rightarrow NH_3 + \cdot H$$
$$NH_3 + \cdot H \rightarrow \cdot NH_2 + H_2.$$

Irradiation at 4 K or below may yield also ·H, NH$_3^+$ and pairs of radicals which are stable up to a limit of concentration. The EPR lines of the NH$_2$ radical with various nitrogen and hydrogen isotopes were discussed in Chapter 3.

The first DNP results [61] were obtained in a sample irradiated under LN$_2$ by 580 MeV protons to an accumulated flux of $0.95 \cdot 10^{15}$ protons/cm^2, which is equivalent to a deposited dose of 40 Mrad in the material. The initially white beads had after irradiation a pale violet color, which slightly faded under a prolonged storage of 6 months in LN$_2$. The paramagnetic spin density was estimated to be about 5×10^{18} cm^{-3} from the measured proton relaxation and polarization time constants, under the assumption that electron spin lattice time is similar to that in PD-Cr(V). The EPR line was centered at $g_{av} = 2.0090$ (which is slightly higher than presently known, probably due to the calibration of the cavity wavemeter which was used for frequency measurement) and had FWHM linewidth of about

5 mT, estimated from a bolometric spectrum at 0.15 K temperature and 2.5 T field. This sample yielded maximum polarizations of +90.5% and –93.6%, which immediately triggered vivid interest among users of polarized targets in high-energy physics experiments.

The polarization build-up time increased 30% after 6 months storage, indicating a similar drop in the paramagnetic center concentration [65]. Polarization tests at 5 T in a dilution refrigerator indicated slow growth of polarization. The EPR spectrum at 5 T had an FWHM linewidth of 6 mT, suggesting that the main cause of broadening was hyperfine interactions with the nuclei of the $\cdot NH_2$ radical.

The results were soon confirmed with samples irradiated at 77 K by neutrons in a reactor [66], at 1 K in 22.6 GeV electron beam [67], at 90 K in 70 MeV electron beam [68], at 90 K in 20 MeV electron beam [69, 70] and at 90 K using a ^{60}Co source [63].

The irradiations at 90 K were performed in liquid argon because explosions of LN_2 had occurred in some of the previous irradiations. This and other safety aspects of irradiations will be discussed in Section 7.5.

Irradiation in situ at 1 K [67] resulted in the disintegration of the target beads, prepared by dropping into LN_2, into a white powder after an accumulated electron beam exposure of about 1.6×10^{15} e$^-$. It was soon observed that beads prepared by the slow freezing method were more resistant against disintegration, and that irradiation at temperatures above 20 K does not cause the material to break up [71] while higher and faster polarization is obtained.

Härtel et al. [72] irradiated NH_3 at 90 K using 20 MeV electrons with accumulated electron flux of 10^{17} cm^{-2}. DNP at 0.5 K and 2.5 T yielded +66% and –64% polarizations, with a build-up time of 9 min to 70% of the maximum value. A first test in a beam also confirmed the hypothesis that the material is about three times more resistant to radiation damage than butanol doped with porphyrexide; furthermore, annealing at 120 K recovered the polarization completely.

The optimum radical concentration in NH_3 was studied by the Bonn group [70] by performing DNP at 1 K and 2.5 T. Three samples were irradiated in the 20 MeV electron LINAC beam current of 2×10^{14} e$^-$/s for 3 h, 6 h and 10 h and polarization build-up and relaxation times were measured. Comparison with PD-Cr(V) with known paramagnetic center concentration yielded the radical densities of 4.8×10^{19} cm^{-3}, 6.1×10^{19} cm^{-3} and 6.9×10^{19} cm^{-3} indicating that the density does not grow linearly with the dose at 90 K, in contrast with the finding that proton relaxation rate increases linearly with the dose received *in situ* at 1 K [73, 74].

The decay of the radicals in ammonia stored at 77 K also appears to be non-exponential [70]; an NH_3 sample had a spin density of 12×10^{19} cm^{-3} 1 day after irradiation, which decayed to 9.6×10^{19} cm^{-3} in 5 weeks and to 4.2×10^{19} cm^{-3} in 1.5 years. The decay of radicals in ND_3 appears to be slower, judging again from the polarization growth time constant which increased by a factor of only 2 after 2.5 years of storage in LN_2 [70].

Irradiation with 250 MeV electrons [75] was carefully optimized for operation at 0.5 K and 2.5 T, on the one hand, and for 1 K and 5 T on the other. In both cases the optimum appears to be 10^{17} e$^-$ with a beam spot of about 1.5 cm diameter. Dose rate estimates were not made, but uniformity of the irradiation was improved by returning the sample holder regularly. It was noticed that irradiation at a high current of 2 µA resulted in lower radical concentration than one at 1 µA, and 1.5 µA produced intermediate results; this was explained by the heating and annealing due to the higher currents [76].

The highly irradiated ammonia samples yielded 96% proton polarization at 5 T and 1 K [77] which, combined with the significantly improved radiation resistance, meant an improvement of about 10 in the squared figure of merit P^2I of polarized targets operated in high-intensity beams, when compared with operation at 0.5 K and 2.5 T. This is due to the higher tolerance to the beam heating at 1 K temperature, and will be discussed in Section 7.5.

First tests [72] of DNP in ND_3 irradiated with electron fluence of 10^{17} cm^{-2} yielded 11% deuteron polarization at 0.5 K and 2.5 T; the same sample at 1 K and 2.5 T polarized only to 1%. This interesting temperature dependence suggested using a dilution refrigerator, where +32% and −29% deuteron polarizations were obtained in the same field [69]. Furthermore, it was discovered that additional irradiation in situ at low temperature during the experiment improved the material so that +40% and −44% polarizations were reached, and with a faster build-up. Radiation damage could be annealed, and further irradiation improved the material up to a photon fluence of 6×10^{14} cm^{-2}.

Post-irradiation of ND_3 was later found [78] to increase the deuteron polarization at least by a factor of 2 at 0.2 K and 2.5 T, with substantially decreased build-up time, so that values of 49% can be achieved. The same effect was observed at 1 K and 5 T even more dramatically: the deuteron polarization increased by a factor of at least 3 from 12% to 13% after pre-irradiation to over 40% after post-irradiation during operation in a high-energy electron beam [79].

Isotopic substitution of ^{14}N by ^{15}N does not substantially affect the DNP results of ammonia or deuterated ammonia [70, 79], although the hyperfine structure of the EPR line, discussed in Chapter 3, changes somewhat. The mechanisms of DNP in ammonia with different isotopic compositions, however, have been given several controversial interpretations, which were discussed in Chapter 4. It is likely that the concentration of the paramagnetic centers and the radiation damage obtained in situ have a strong influence on the electron spin-lattice relaxation as well as cross-relaxation rates between the different spin species, probably due to lattice stress generated by interstitial radiolytic centers that are diamagnetic. The stress controls the electron spin dipolar relaxation and flip-flop rate, which, in turn, control the thermal contact with the nuclear spins and therefore the thermalization of the nuclear spin species towards a common temperature.

7.3.3 Irradiated Lithium Hydrides

Lithium hydride has face-centered cubic lattice structure very similar to sodium chloride. It has a melting point of 680 °C and decomposes in contact with water. It has very slight solubility in alcohols. Its handling is not particularly difficult if humidity is well controlled. The density of lithium hydride with natural content of 7.42% of ^6Li is 0.82 g/cm^3.

Table 7.7 Square roots of the second moments for the EPR linewidth in lithium hydrides [81].

Compound	^{7}LiH	^{6}LiH	^{7}LiD	^{6}LiD
EPR linewidth (mT)	3.06	1.43	2.85	1.01

In 1966 Borghini [80] suggested that ^{6}LiD might be a material where a high DNP of deuterons could be achieved, based on the relatively small EPR linewidth obtained in irradiated material. He listed the then known linewidths [81] for all isotopic compounds of lithium hydrides, as shown by Table 7.7:

It appears clearly that the linewidth is associated with the hyperfine coupling of the paramagnetic electron with the lithium nucleus. This is due to a free electron trapped in a vacant site of a hydrogen ion, surrounded by lithium ions the nuclei of which have hyperfine interaction with the electron; such a paramagnetic impurity is called the F-center. Irradiation at low temperatures below 20 K produces these and the X_2-centers [82], both of which have spin 1/2. The F-center has an isotropic g-factor close to that of the free electron, but a rather complex hyperfine structure due to the overlap of the wave function with nearest lithium nuclei and also with the hydrogen nuclei of the second shell. The spin-lattice relaxation time is obviously very long because of little coupling with the lattice owing to the absence of a g-shift and related anisotropy.

The X_2-centers [82] (called also H-center) are interstitial H_2- or D_2-ions and have relatively simple EPR spectra with some g-tensor anisotropy and hyperfine structure. The ion is also covalently bonded to the two H$^-$ ions at the ends of its bond axis.

Pairs of separated F- and H-centers are effectively produced by UV photons; typically, 5 eV to 10 eV is required for this process. High-energy electrons and protons also produce such pairs due to the electromagnetic cascade, which is the main cause of energy deposit in the material. Energetic particles also produce damage to the cation sub-lattice which leads to complicated recombination and annealing effects due to overlapping cascades. Metallic clusters of lithium atoms can be seen in the EPR spectrum at high doses of energetic particles.

Roinel and Bouffard [83] succeeded in 1977 in polarizing irradiated ^{7}LiH at 0.6 K and 6.5 T to such high values that Abragam et al. could observe first time the nuclear antiferromagnetic phase by the characteristic Bragg reflection in slow neutron scattering [84]. Polarizations $P(H) = 95\%$ and $P(^7Li) = 80\%$ were reached after 2 d to 3 d of DNP, in a single crystal of $5 \times 5 \times 0.5$ mm^3 size, irradiated at 77 K by an electron fluence of 10^{18} cm^{-2} of 3 MeV electrons. The paramagnetic electron concentration was estimated at 2×10^{19} cm^{-3} and they were identified as ^{7}Li F-centers.

Subsequently Bouffard and coworkers [85] irradiated with 3 MeV electron fluence of 2.25×10^{17} cm^{-2} sintered polycrystalline ^{6}LiD platelets of $5 \times 5 \times 0.5$ mm^3 size using liquid argon as coolant, in view of improving the material performance in polarized targets. The EPR line was measured using the NEDOR technique, and the DNP tests were performed in 2.5 T, 5 T and 6.5 T fields using a dilution refrigerator operating at 0.2 K temperature. The free electron concentration of the F-centers was found to be 1.3×10^{19} cm^{-3} and the electron spin-lattice relaxation time was 1 s at 5 T and 6.5 T, and 2.5 s at 2.5 T; these are the initial NEDOR pulse recovery times, the final recovery being slower probably due to a distribution of different types of relaxing centers in addition to the pure F-centers. The nuclear spin-lattice relaxation results point in the same direction.

The NEDOR linewidths (FWHM) of ^{6}LiD were 3.8 mT at 2.5 T, 5 T and 6.5 T, while in a sample of pure ^{7}LiH it was 9 mT, as can be expected because the hyperfine splitting due to ^{6}Li must be about 2/5 times that due to ^{7}Li as nearest neighbors of the F-center.

Roinel [86] continued to refine the sample preparation and DNP of ^{7}LiF, ^{7}LiH and ^{6}LiD. It was observed that there was a difference in irradiating LiH and LiD samples submerged in liquid argon or just above the surface. In the first case the samples had red color, while in the second one they came out blue. The blue samples had relatively much faster polarization build-up, and the red ones were so slow that maximum polarization was not determined. The difference was believed to originate from the sample temperature during irradiation due to the different cooling conditions; samples cooled in the vapor phase clearly have a much lower heat transfer coefficient and therefore a higher temperature.

It was also noted that F-centers in LiF are stable at room temperature, whereas in LiH and LiD they were unstable at temperatures above 130 K. A fluence of 10^{17}–10^{18} cm^{-2} of 3 MeV electrons yielded a paramagnetic spin density of $(0.5$–$2.5) \cdot 10^{19}$ cm^{-3}, and it was noted that the DNP time constant scaled linearly with the electron spin-lattice relaxation time and inversely proportional to the electron spin density. The DNP results are summarized in Table 7.8 from where the potential of lithium hydride as a polarized target material appears very clearly.

In lithium hydride the electron spin-lattice relaxation time varied between 1 s and 20 s and was found not to be an intrinsic property of the material. For lithium deuteride electron spin density 2×10^{19} cm^{-3} the electron spin-lattice relaxation time was 1.2 s and the nuclear

Table 7.8 Results [86] of electron spin-lattice relaxation, DNP build-up time and maximum DNP for irradiated lithium fluoride and hydride samples of size $5 \times 5 \times 0.5$ mm^3. Only positive maximum polarization was determined. The values are typical for different samples, except for lithium hydride for which the best values are quoted. The 0.7 K sample temperature during DNP was obtained in a ^{3}He evaporation refrigerator, whereas 0.2 K was obtained in a dilution refrigerator. The ^{6}LiD samples were polycrystalline, while all others were single crystals.

Substance	T_{1e} (s)	T_{DNP} (h)	P_{max}(Li) (%)	P_{max}(H/F) (%)	B (T)	T (K)
^{7}LiF	0.1	4	60	80	5.5	0.7
^{7}LiH	1	40	94	99	6.5	0.2
^{6}LiD	1	40	70	70	6.5	0.2

spin-lattice relaxation time at 4.2 K was 120 min, both values unusually large in polarized target materials. Clearly the long nuclear spin-lattice relaxation time shows that the paramagnetic centers were its main cause, a condition that leads to high nuclear polarization. Although the polarizing time in lithium deuteride was 40 h to 70%, a polarization of 68% was reached after 20 h of DNP.

Attempts to produce larger quantities of well-polarizing lithium hydrides by irradiation at higher temperature and with high-energy electrons [87, 88] met with problems such as metallic lithium, which showed up in the EPR signals, until a special argon cryostat was fabricated which allowed to perform irradiations at 180 K temperature. With a total flux of 2×10^{17} cm^{-2}, Durand et al. [89] succeeded in obtaining ^{6}Li and D polarizations of 37% and 43%, respectively, after 8 h of DNP at 2.5 T field in a dilution refrigerator, in a sizable sample. A 1.6 cm^3 sample of irradiated ^{6}LiD produced in this way was used by van den Brandt et al. [90] in a first experiment of medium-energy pion scattering with excellent results of +54% and −49% polarizations for both spin species. The polarization of the residual ^{7}Li nuclear spins was also determined, with maximum values of +89% and −91%, in thermal equilibrium with the low-moment species within experimental accuracy. Two other experiments [91] used a ^{7}LiH target of the same size, irradiated to electron fluence of $\approx 2 \times 10^{17}$ cm^{-2} at 180 K with ^{4}He gas cooling [92]; the polarization of ^{7}Li of +50% and −30% was reached using phase-locked Impatt diode microwave source adjusted in sequence to proton, ^{7}Li and ^{6}Li solid-effect frequencies. Using the source in free-running mode, carefully temperature stabilized, the polarization growth was faster and $P(^7\text{Li})$ of +49% and −38% was reached. The gain was particularly important in negative polarization speed and value.

The setup consisted of a dilution refrigerator and a 2.5 T magnet of a homogeneity not better than 3×10^{-4} over the sample volume. The polarizing time was 2 d to 3 d, but 2/3 of the final value was reached in 16 h.

Similar values were reported by Jarmer and Penttilä [93] for ^{7}LiH samples irradiated to integrated fluences of $(0.5$–$2.1)\cdot 10^{17}$ cm^{-2} with 30 MeV electrons at temperatures between 180 and 200 K. In all cases polarization growth was rather slow and compatible with earlier results in similar conditions.

Goertz et al. [94] studied the EPR spectra and DNP in samples irradiated using 20 MeV electrons to $(1$–$4)\cdot 10^{17}$ cm^{-2} fluence at 90 K to 200 K temperatures (pre-irradiation) and post-irradiated in situ during DNP at 1 K temperature. Some samples were also annealed at room temperature after pre-irradiation. Samples irradiated at the best temperature of 180 K with electron fluence of 10^{17} cm^{-2} show the following features in their EPR line: ^{7}LiH has 5.6 mT FWHM line consisting of a main line at $g = 2.00$ and a smaller satellite at $g = 2.05$, probably due to impurities. ^{6}LiD and ^{6}LiH show FWHM of 3.0 mT and 3.3 mT, respectively, with a superposition of a broader main line due to F-centers and a narrower (0.5 mT) line due to metallic clusters, both centered at $g = 2.00$. These numbers are in fair agreement with the second moments given in Table 7.7 and agree with the interpretation that they are due to F-centers.

Goertz et al. [94] also observed that samples irradiated at a temperature of 160 K could be polarized twice faster and yielded a higher polarization if they were annealed at room temperature for 5 min. The EPR line also became more intense and narrower, both by a factor of about two. This was interpreted as a change in the atomic structure of the F-centers. Studies of the saturation of the EPR spectrum indicated that it must originate from several different types of F-centers with superposed lines but different relaxation times. Annealing at room temperature would then cause the F-centers to migrate and to form clusters, which is favorable for DNP. Irradiation at high enough temperature enables this to happen already during the process, but if temperature is increased above 190 K, recombination of the F-centers may reduce their yield. This explains plausibly the rather narrow optimum of the irradiation temperature, and also the gain in the speed of DNP, which is most likely due to the faster spin-lattice relaxation time of electron spins in clusters with exchange narrowing.

The EPR spectra of ^{6}LiD F-centers at 2.5 T field was measured by Heckmann et al. using their EPR spectrometer with a tuned Fabry–Perot resonator (see Chapter 3); their result for the FWHM linewidth is 1.8 mT [95], which agrees with that measured at 0.3 T and shows clearly the hyperfine structure with $2NI + 1 = 13$ equally spaced lines with the right intensity ratios. This is also in better agreement with the width of 1.5 mT reported in Ref. [85].

Post-irradiation in situ was performed in 1.2 GeV electron beam at 1 K and 2.5 T. A ^{7}LiH sample was pre-irradiated in standard conditions and yielded 11.5% proton polarization with 50 min build-up time to 63% of the ultimate polarization value [94]. At a fluence of 10^{14} e$^-$/cm^2 the maximum polarization increased to 14.5% and the build-up time was reduced to 8 min. These remained the same until the run was stopped at an accumulated flux of 5×10^{15} cm^{-2}. Deuteron polarization in ^{6}LiD showed a similar reduction in DNP build-up, but the maximum polarization remained constant at about 12% under the same conditions.

Clearly the low-temperature irradiation speeds up the electron spin-lattice relaxation, which is reflected directly in the speed of DNP and is theoretically easily understood by the models of Chapter 4. Due to the symmetry of the surroundings of the F-center and consequently the absence of a g-anisotropy and shift, the spin-lattice coupling remains very small. Lattice stress, due to damage which does not anneal at 1 K, may lift the symmetry entailing improved coupling with the lattice phonons and faster spin-lattice relaxation. This happens because the lattice stress directly affects the lattice potential terms of Eq. 3.72 and therefore the electron spin-lattice relaxation time due to the direct process.

Goertz et al. [94] succeeded in obtaining in ^{6}LiD deuteron polarizations exceeding 50% at 0.2 K and 5 T, and in ^{6}LiH at 0.2 K and 2.5 T proton polarizations exceeding 40%, both with reasonable build-up times of 4.4 and 1.4 h, respectively. These two materials have been used in many high-energy physics experiments by this writing.

The lithium hydrides are also useful in high-intensity beams when operated at 1 K and 5 T, where −32% deuteron polarization was obtained in ^{6}LiD [96]. This target, prepared by crushing and sieving the sintered polycrystalline starting material, was operated in SLAC E155 experiment in high-intensity electron beam. The targets were pre-irradiated at 183 K using 30 MeV electrons with a total fluence of 3.7×10^{17} cm^{-2} on the sample container of 20 mm diameter. Such a target was exposed to the 50 GeV/c beam with 80 nA current during E155. The maximum polarization increases and the DNP speed improves up to an electron fluence of 5.0×10^{15} cm^{-2}, similar to the results with deuterated ammonia in the course of the experiment E143 using the same target set-up, but with the usable fluence about five times higher for ^{6}LiD [96]. During the beam exposure the optimum frequency for negative polarization was 140.310 GHz in the beginning, constantly increasing to 140.320 GHz at the end, suggesting that the in situ irradiation at 1 K caused permanent material damage that increases the lattice stress, which, in turn, increases the g-shift and therefore shortens the spin-lattice relaxation time, similar to ammonia targets.

The disadvantage of the ^{6}LiD in comparison with ND_3 is its slower DNP speed, probably due to the small g-shift of the F-center used for DNP. The modulation of the microwave frequency was not tried during E155; it would be interesting to know if it could help with the EPR line the main broadening of which is due to the hyperfine interactions of the F-center with the surrounding six nearest Li nuclear spins.

^{6}LiD was also used by the COMPASS collaboration in their large target set-up with two or three target cells operated in 2.5 T field and cooled by a dilution refrigerator [97–100]. Using DNP with frequency modulation, maximum deuteron polarizations of +57% and −53% were reached during the run of 2003, as averaged over the target cell volumes. The two- to three-cell upgrade was made in 2006 [101].

The ^{6}LiD target material of COMPASS was prepared by reacting enriched ^{6}Li with pure deuterium gas in a furnace operated between 700 K and 1100 K temperature. The sintered polycrystalline material was cut into pieces passing through a sieve of 2 mm, with a density of 0.84 g/cm^3. These were irradiated by 20 MeV electrons of the Bonn synchrotron injector linac with total electron fluence of 2×10^{17} cm^{-2} at 190 K temperature [97]. The EPR spectra of the produced F-centers were recorded using the V-band Fabry–Perot interferometer showing the 13-line hyperfine spectra that yield the density of 2×10^{19} g^{-1} for the centers. The low density, combined with the narrow EPR line, result in a slow growth of DNP. However, by using microwave power and frequency optimization combined with microwave frequency modulation, the high final polarizations were reached in all target cells in less than two weeks [97]. Clearly this makes the use of field rotation techniques vitally important in order to suppress false asymmetries in the experiment.

7.3.4 Other Crystalline Inorganic Materials

Early work on DNP with inorganic crystals was reported on LiF containing F-centers. Subsequently irradiated ^{6}LiD, CaF_2 and $Ca(OH)_2$ were studied; this work was briefly reviewed by Henderson in [82]. In these simple cubic ionic crystals, ionizing radiation below 20 K produces only F-centers and X_2^--centers (also called H-center); the former is an anion vacancy trapping a free electron, whereas the latter is an interstitial molecular ion occupying a single anion site. In LiF the ion is F_2^-, for example. They are produced in F-center/H-center pairs at rather low radiation energy, for example, by UV radiation. The wave function of the trapped electron extends out over the neighboring nuclei, generating hyperfine structure. The X_2^--ion is also covalently bonded to the other two X$^-$-ions at the ends of the bond axis.

The F-center ESR spectrum is quite complex, even though $S = ½$ and the g-value is isotropic and very close to the value of free electron. The complexity arises from the diffuseness of the electron wave function which spreads over many shells of nuclei. In ^{7}LiH, for example, the first shell contains $N = 6$ equivalent Li$^+$ ions, and since ^{7}Li has $I = 3/2$, there are $2NI + 1 = 19$ lines with intensity variation (at high temperature) 1:18: ... :18:1 determined by the number of ways compounding the total nuclear spin. The sequence is that of the binomial coefficients obtained from Pascal's triangle, which was described in Section 3.2.2 and in Ref. [102].

The second shell interaction with 12 H nuclei with $I = ½$ gives a 13-line splitting of each of the lithium hyperfine lines. The net result is an inhomogeneously broadened line, with some resolved structure, which is strongly anisotropic. The spin-lattice relaxation time around 1 K is in the range of 10–100 ms.

The X_2^--centers have relatively simple ESR spectra, which show g-tensor anisotropy and hyperfine structure. There are discrete hyperfine splitting lines due to the dominant interaction from the two nuclei of the X_2^- radical. Each of these lines is however inhomogeneously broadened by weaker interactions with the neighboring ions. The spin-lattice relaxation time is shorter than that of the F-centers: it is in the range of 0.1 ms to 1 ms. Thus, the saturation phenomena may be very different.

The F-center/H-center pairs can be produced by low-energy radiolysis, for example, by UV radiation; only 5 eV to 10 eV may be required for a pair. X-rays are also efficient for producing such pairs. Electrons and protons also produce these pairs, because each electromagnetic cascade involves numerous X-rays. Energetic protons and neutrons may damage the lattice more substantially and yield more complex paramagnetic defects.

Beyond the lithium hydrides, high DNP has been reported in calcium fluoride CaF_2 and lithium fluoride LiF containing F- and H-centers, and in calcium hydroxide $Ca(OH)_2$ containing molecular O_2^--centers ($S = ½$) [82].

The results of DNP in irradiated lithium hydrides were reviewed in Section 7.3.3. The DNP of hyperfine nuclei was discussed in Section 4.3.

7.3.5 Organic Glassy Materials

The first test with irradiated butanol were made in University of Liverpool[4] with the purpose of better understanding the radiation damage mechanisms due to radiolytic paramagnetic centers. An undoped butanol sample was placed in a 4 GeV bremsstrahlung beam, and DNP was made at 1 K and 2.5 T while the material was irradiated in situ. It was found that the polarization increases linearly with dose, reaching values of +7% and −6% after a dose equivalent of the dose, which would reduce the polarization to 0.75 times its initial maximum value in a material normally doped with porphyrexide. The proton relaxation time was 1,500 s before irradiation and 550 s at the end. The material was then annealed in the same way as the doped material. The proton relaxation time was found to have returned to its initial value while the DNP did not appreciably change.

A further small irradiation was possible, and it was found that polarization continued to increase with dose until the test had to be stopped; the polarization reached at that time was 8% and appeared to continue rising with dose.

A plot of polarization against microwave frequency after the anneal showed a shape and width similar to that for an undamaged sample with porphyrexide but shifted 0.2% lower

---

[4] B. Craven, Thesis, University of Liverpool (1973).

in frequency, suggesting that the non-annealable centers have a g-value of approximately 2.001. When the plot was added to that of an undamaged doped sample, the result was in excellent agreement with that obtained for damaged normally doped samples, both with respect to shape and absolute size.

Butanol samples were subsequently irradiated with different methods, but the results were hardly encouraging [67, 103] because 12% proton polarization at 0.5 K and 2.5 T was the best result, and the evidence for continued growth of the polarization with dose was less clear. It was concluded, however, that it is about seven times more efficient to irradiate at 4 K than at 77 K.

Normal and deuterated 1-butanol beads cooled by liquid argon at 87 K were irradiated with 20 MeV electrons of the Bonn injection Linac to an electron fluence of 10^{15} cm^{-2}. It was observed that the yield of paramagnetic centers was about twice higher when compared with inorganic materials [104]. Two samples of deuterated 1-butanol reached 1×10^{19} g^{-1} and 2×10^{19} g^{-1} electron spin densities. DNP at 1 K and 2.5 T confirmed earlier findings, but subsequent DNP tests in 2.5 and 5 T fields, cooled by a dilution refrigerator at about 0.2 K temperature, yielded −55.1% and −70.8% maximum deuteron polarizations, respectively [104]. The higher polarization was reached in the sample with higher spin density, as can be expected.

Borghini [80] reviewed ESR tests with irradiated polyethylene samples displaying three types of radicals, with different annealing temperatures, all above room temperature. At that time no irradiation tests were known at lower temperatures, and the sample characteristics may not have been studied well enough.

7.4 Materials with Optically Excited Triplet-State Paramagnetic Molecules

Ultraviolet irradiation can excite some molecules in triplet states which exhibit paramagnetism. Borghini [80] noted that naphthalene ($C_{10}H_8$) dissolved in 1,2,4,5-tetramethyl benzene ((CH_3)$_4C_6H_2$ or durene) has in the triplet state a spin Hamiltonian [105], which might be suitable for DNP. He also noted that if nuclear spin relaxation is mainly caused by the excited paramagnetic centers, relatively low concentrations of the excited molecules can be used because they appear and disappear randomly everywhere in the material. Deuteration of the naphthalene molecule would reduce the 1.2 mT linewidth further; this also extends the lifetime of the excited molecule by a factor of 10. van Kesteren, Wenkebach and Schmidt [106] have obtained 42% proton polarization in fluorene $C_{13}H_{10}$ single crystals doped with deuterated phenanthrene $C_{14}D_{10}$, which was photoexcited in triplet state using ultraviolet light from a 100 W mercury arc lamp, filtered using nickel and cobalt sulphate solution and Schott UG-5 glass filter. The experiments were performed at 1.4 K temperature and 2.7 T field; the resonance frequencies of the excited paramagnetic triplet coincide at 75 GHz when the crystal is suitably oriented. Single crystals were grown from the doped melt of fluorene. The high polarization is reached in about 3 h when using magnetic field modulation.

Optically induced paramagnetic triplet state in phenantrene-d10 molecules [107] was used for DNP with 75 GHz microwaves at 2.7 T field. Using this method, deuterons in single crystal fluorene-d10 were polarized to 3.7×10^{-3} and then aligned to -3.8×10^{-3} using adiabatic fast passage [107].

7.5 Safety Aspects of Irradiated Materials

In the course of irradiations of target materials, several incidents have taken place. It is believed that these are due to liquid nitrogen which was used for cooling the material, or was introduced in the argon cryostat when moving samples from storage dewars to the irradiation cryostat.

In the first case, an ammonia sample was irradiated in the CERN Synchrocyclotron behind an absorber block of a 580 MeV proton beam in an open glass dewar filled initially with LN_2. The dewar exploded when about two-third of the liquid had boiled off due to the beam heat load. The dose deposited to LN_2 was much larger than that received by the sample because of the unfocused beam. A previous irradiation with a focused primary beam had been carried out without incidents. It was suggested by some specialists that the explosion was due to oxygen-nitrogen compounds formed in the radiation field, and that it could be avoided by using liquid nitrogen of very high purity.

The next irradiation was carried out using 99.999 pure nitrogen in a closed dewar with extreme precautions to avoid oxygen contamination. The irradiation was completed without incidents, but when the ammonia sample cup was extracted from the dewar, the cup filled with ammonia and LN_2 exploded immediately after entering in contact with air. This led to the conclusion that LN_2 alone developed unstable molecular forms during irradiation, and to the ban of LN_2 in the subsequent irradiations.

After irradiation of ammonia beads submerged in liquid Ar by 20 MeV electrons from the Bonn synchrotron injector, flames have been observed when removing the irradiated material batch from the cryostat. No explosions were reported in such cases. It is thought that this is due to residual N_2 on the beads introduced upon the transfer of the target batch from LN_2 storage vessel into the irradiation cryostat.

Beyond the chemical risks due to irradiation, it is also important to observe all radioprotection precautions if the irradiated material, coolant and irradiation equipment contain heavy elements that may be activated by their interaction with the beam particles.

7.6 Crystalline Materials with Substitutional Paramagnetic Ions

The first successful polarized target material was lanthanum magnesium double nitrate $La_2Mg_3(NO_3)_{12} \cdot 24H_2O$ (LMN) doped with Nd^{3+} ions [108]. Earlier work [109, 110] at lower fields clearly indicated that at high field and low temperature the 'solid effect' method should produce nearly complete proton polarization, and indeed Schmugge and Jeffries [108] succeeded in obtaining 72% proton polarization in a sizable crystal in 1.95 T field and 1.5 K temperature.

7.6 Crystalline Materials with Substitutional Paramagnetic Ions

The LMN crystal was grown [108] from saturated aqueous solution in a desiccator at 0 °C temperature, using $La(NO_3)_3 \cdot 6H_2O$ and $Mg(NO_3)_2 \cdot 6H_2O$ in stoichiometric amounts, and Nd_2O_3 to yield Nd^{3+} ions 1% relative to La^{3+} ions. The crystals grow in flat hexagonal plates. The high polarization was obtained using Nd enriched to 98.5% in even-even Nd isotopes.

Borghini [80] lists diamagnetic salts in which similar paramagnetic ions can be introduced by the crystal growth method. We reproduce the list in Table 7.9.

Borghini [80] also lists g-values of some suitable ions, reproduced here in Table 7.10. He points out that all but the LMN are difficult to prepare because only small crystals can be obtained or because of widely different solubilities. On the other hand, only special applications may require such crystals, because of their poor hydrogen content.

One such application is in 'spin refrigerator' polarized targets, which will be discussed in Chapter 10. A particularly suitable material is yttrium ethylsulphate doped with Yb^{3+} ions [111].

Nuclear physics experiments with polarized nuclei heavier than the hydrogens have, in principle, a wide choice of crystalline materials available. Among first DNP materials was ruby Al_2O_3 with Cr^{3+} ions replacing some of the Al^{3+} ions [112, 113]. The EPR line has four resolved components in high field due to the spin 3/2 of the Cr^{3+} ion; consequently the understanding and analysis of the results is complicated. Polarization enhancements of ^{27}Al nuclei by a factor of 31 were reached at 1.7 K and 0.9 T field [114]. A simpler system of Ce^{3+} ions with spin 1/2 appears in $CaWO_4$ [115], as well as Er^{3+} ions in BaF_3 [116]. The Er^{3+} ion with spin 1/2 in BaF_3 was found mostly (90%) with trigonal symmetry with four main lines at $g_\parallel = 5.94$ and $g_\perp = 7.13$, but also 10% in cubic symmetry. Weak hyperfine lines due to 23% of ^{167}Er were also observed.

^{139}La target with 20% polarization has been prepared using $LaAlO_3$ crystals doped with Nd^{3+} [117, 118].

Table 7.9 Hydrogen-rich diamagnetic salts in which magnetic rare-earth ions may replace non-magnetic ions. The last column gives the simplified free-proton dilution factor of Eq. 7.11.

Substance	Chemical formula	f_p
Methylammonium aluminium sulphate	$(CH_3NH_3)Al(SO_4)_2 \cdot 12H_2O$	0.064
Ammonium aluminium sulphate	$(NH_4)Al(SO_4)_2 \cdot 12H_2O$	0.062
Yttrium ethylsulphate	$Y(C_2H_5SO_4)_3 \cdot 9H_2O$	0.056
Lanthanum ethylsulphate	$La(C_2H_5SO_4)_3 \cdot 9H_2O$	0.051
Yttrium acetate	$Y(CH_3COO)_3 \cdot 4H_2O$	0.050
Potassium aluminium sulphate	$KAl(SO_4)_2 \cdot 12H_2O$	0.048
Cesium aluminium sulphate	$CsAl(SO_4)_2 \cdot 12H_2O$	0.042
Lanthanum magnesium nitrate	$La_2Mg_3(NO_3)_{12} \cdot 24H_2O$	0.031

Table 7.10 The g-factors of some magnetic rare-earth ions in selected diamagnetic crystals. The g-tensor is axially symmetric in trigonal crystals and can therefore be characterized by g-values with field parallel and perpendicular to the crystal symmetry axis, whereas in triclinic crystals the three diagonal values can be different.

Diamagnetic substance; crystal symmetry	Ion	$g_\parallel$	$g_\perp$
Lanthanum ethylsulphate; trigonal	Nd^{3+}	3.53	2.07
	Er^{3+}	1.47	8.8
	Dy^{3+}	10	0
Yttrium ethylsulphate; trigonal	Ce^{3+}	3.81	0.20
	Nd^{3+}	3.66	1.98
	Er^{3+}	1.50	8.77
	Yb^{3+}	3.328	0.003
Lanthanum magnesium nitrate; trigonal	Nd^{3+}	0.36	2.70
	Dy^{3+}	4.28	8.92
	Er^{3+}	4.21	7.99

	Ion	g_x	g_y	g_z
Yttrium acetate; triclinic,	Dy^{3+}	13.60	3	4
	Yb^{3+}	4.57		

Abragam et al. [119] have polarized ^{19}F nuclei in CaF_2 single crystals doped with U^{3+} and Tm^{2+} ions to 90%. Such a material has been used in the studies of nuclear magnetism at ultralow temperatures, but could also be useful for nuclear physics experiments.

7.7 Radiation Resistance of Polarized Targets

The main effects of in situ irradiation of a polarized target upon the DNP arise from the creation of paramagnetic free radicals, and from the build-up of lattice stress, which modifies the g-factor of paramagnetic centers and shortens their spin-lattice relaxation time. The created paramagnetic centers usually deteriorate DNP, whereas the shortened relaxation time sometimes improves DNP.

Other effects of radiation include the creation vacancies and radiolytic impurities which are not paramagnetic. Among these, H_2 is the most common, with a yield of several molecules for each 100 eV of deposited energy in alcohols and diols. As 75% of the H_2 molecules are ortho-hydrogen with $J = 1$ and their conversion to the ground state is slow, substantial leakage relaxation may result because of the rapid spin-lattice relaxation due to the coupling of the rotational and spin degrees of freedom. Also CH_4, CO_2 and CO are formed with yields around 1 molecule per 100 eV. Such impurities cause important stress in the lattice, which improves the coupling of the paramagnetic centers with the lattice phonons, and which may ultimately lead to the mechanical destruction of the lattice or matrix.

7.7 Radiation Resistance of Polarized Targets

The decay of maximum polarization during DNP in a beam is assumed to follow

$$P(\Phi) = P_0 e^{-\Phi/\Phi_A}, \qquad (7.19)$$

where $P(\Phi)$ is the polarization after irradiation with particle flux Φ, P_0 is the maximum polarization before the beam exposure starts and Φ_A is the characteristic flux for the polarization to decrease by $1/e$. The subscript A refers to the fact that most of the damage created can be repaired by annealing the material at a suitable higher temperature. The irradiation is usually carried out only to the particle flux of some 10^{14} cm^{-2} before the anneal; tests with higher fluxes on butanol, for example, indicate that the decay of polarization slows down after about 4×10^{14}/cm^2 in the case of protons [120]. Within errors Φ_A is constant up to an accumulated fluence of 2×10^{14} cm^{-2} and is independent of the beam intensity up to 3×10^9 cm$^{-2} \cdot$s^{-1}. At higher beam intensities the beam heating begins to influence the polarization and requires turning off the beam for the study of the influence of the radiation damage alone.

Repeated annealing of the target between beam exposures recovers the polarization immediately after annealing to a value, which is lower than P_0 and is parametrized by

$$P(\Phi) = P_A + (P_0 - P_A) e^{-\Phi/\Phi_{NA}}, \qquad (7.20)$$

where P_A is the asymptotic value of the polarization after annealing at very large flux Φ, and Φ_{NA} is the characteristic flux for the non-annealing damage to reach $1/e$ of its asymptotic value. The characteristic fluxes Φ_A and Φ_{NA} depend on the particle and its energy, on the material itself and on the sign of the polarization.

The studies of radiation damage are often difficult to interpret because the beam profile may be such that the dose is not homogeneous on the target. Rastered small beams are better in this respect than larger beams which usually expose the central part of the target much more heavily than the sides. If the NMR coil is wrapped around the target, it measures the polarization predominantly from the outer layer, which is less exposed. Therefore, the quoted polarization values must be taken with some precautions, although mostly the conclusions are not significantly changed qualitatively. The spin-lattice relaxation will exhibit a clear non-exponential decay if the material is not homogeneously exposed; this can sometimes be used for estimating the dose homogeneity and the error made in the measurement of the polarization.

Chemically Doped Hydrocarbons

A systematic study by Fernow [120] with ethanediol-Cr(V) and propanediol-Cr(V) in a 10 GeV proton beam is summarized in Table 7.11. The ethanediol sample reached 81% polarization, and propanediol yielded 76% polarization before damage.

As can be seen from the data, positive polarization is less sensitive for radiation damage in all diol samples, but it suffers a larger asymptotic damage at high fluxes. Ethanediol seems to degrade slightly more slowly than propanediol, but the effect might be masked by

Table 7.11 Characteristic proton fluxes Φ_A and Φ_{NA} for the annealing and non-annealing radiation damage, and the asymptotic polarization degradation for large fluxes, from Ref. [120]. The target was operated at 0.5 K and 2.5 T, cooled by a ^{3}He refrigerator. ED = ethanediol with 8×10^{19} spins/cm^3 Cr(V); PD = propanediol with 11×10^{19} spins/cm^3 Cr(V); BuOH-PX = 1-butanol with 5% water and 1% of porphyrexide. Other data are also shown for comparison: [1] from Ref. [121], temperature 1 K; [2] from Ref. [122], proton energy 24 GeV.

Material	$\Phi_A(+)$ (10^{14}/cm^2)	$\Phi_A(-)$ (10^{14}/cm^2)	$\Phi_{NA}(+)$ (10^{14}/cm^2)	$\Phi_{NA}(-)$ (10^{14}/cm^2)	P^+_A/P^-_0	P^-_A/P^-_0
ED-Cr(V)	2.33±0.20	1.75±0.15	0.76±0.16	0.80±0.15	0.69±0.03	0.89±0.02
ED-Cr(V)[1]	1.64±0.10	1.82±0.17				
PD-Cr(V)	1.24±0.27	0.94±0.10	0.80±0.25	1.00±0.30	0.73±0.04	0.90±0.04
PD-Cr(V)[2]	2.00±0.25	1.60±0.20				
BuOH-PX	2.2±0.5	4.5±1.0				

different concentrations of the paramagnetic Cr(V) centers. The annealing was performed by heating the target to 160 K to 180 K temperature in 8 min to 13 min and then rapidly cooling back to the operating temperature. The diol targets are exceptional compared with other glassy hydrocarbons in that the material does not devitrify when heated up to this high temperature.

The annealing is known to deplete the trapped electrons completely, and most other paramagnetic centers are also believed to be reduced to a rather low concentration. The non-annealable damage would appear to be due to such radical species that are stable upon annealing; this is supported by the measurement of the optimum microwave frequency in an ethanediol sample before and after radiation damage, and immediately after annealing at 143 K temperature [120]. The maximum positive polarization was found at the same frequency for all three cases within experimental error, but the maximum negative polarization after annealing was obtained at a frequency about 25 MHz higher than the optimum frequency before irradiation. The optimum negative frequency of the radiation damaged sample was about 75 MHz higher than that before damage.

Fernow also studied the radiation damage of various isomers of butanol with water and porphyrexide [120]. The characteristic flux Φ_A is larger than that for the diols, and the negative polarization resists better radiation than positive, in contrast with the diols. The damage remaining after annealing was not investigated. Annealing of butanol must be made at a temperature below 140 K in order to avoid the crystallization of the paramagnetic impurity, and this may leave more of the radiolytic paramagnetic centers in the materials after annealing.

The radiolytic paramagnetic centers in butanol can also be 'bleached' by light which removes most of the visible coloring [123]. It has been observed that this can be made by shining light via a light guide or waveguide. Some of the centers, however, may be stable against this and require thermal annealing.

The nuclear spin-lattice relaxation time is shortened by radiation damage; in the samples studied in [120] this amounted to about 25% in ethanediol and 15% in propanediol before annealing. This can be partly recovered; in butanol, for example, the relaxation time is some 5% longer after an anneal at 120 K [123]. The working life of the target, however, cannot be infinitely extended because of the non-annealing damage, which requires changing the material after about 10 anneals.

Irradiated Ammonia

Irradiated ammonia is the preferred material for polarized targets in high-intensity beams, because the radiation damage followed by anneal results in the same radical $NH_2\cdot$, which is formed during the primary irradiation. It has also turned out that the characteristic flux Φ_A is very much higher for ammonia than for chemically doped hydrocarbons, and that annealing after irradiation not only restores the original polarization, but in many cases improves it.

The first systematic damage studies were made at 2.5 T with protons by Crabb et al. [68] and with electrons by Althoff et al. [124]. In the first case a ^{3}He evaporation refrigerator was used, but ^{4}He was added because it was found that much higher heat loads could be tolerated and therefore higher microwave power could be used. In the electron beam studies, the samples were cooled with a ^{4}He evaporation refrigerator.

The resistance against proton beam damage turned out to be some two to three times better for ammonia than for butanol-porphyrexide [68], and annealing at 90 K temperature recovered totally the polarization up to the total number of 2.5×10^{16} protons through the target of 2.9 cm diameter. The accuracy of the damage constant was limited by the rather small exposure, poor knowledge of the beam profile and lack of optimization of the microwave frequency. Later experience at 1 K and 5 T shows that the polarization can be recovered by anneal up to a total flux of about 10^{17} protons through the target; beyond this value the polarization continues to decay despite annealing [76].

The electron beam studies revealed that the loss of maximum polarization is non-exponential both in butanol and in ammonia. The most interesting findings, however, were that the ammonia polarization did not converge towards zero but towards a reasonably high value at very high doses [124]. The characteristic fluxes are given in Table 7.12, where it appears clearly that the initial decay agrees with the measurements using protons, but the subsequent decay is very much slower. In these measurements the microwave frequency was optimized during the run.

The polarization, microwave frequency, polarization time constant and relaxation time were determined for positive polarization at several doses; these are shown in Table 7.13. It is evident that the electron spin resonance line broadens substantially due to the radicals created by the electron beam irradiation, and that the electron spin-lattice relaxation time becomes at the same time shorter. The non-exponential relaxation of the proton polarization is probably due to spin diffusion, which becomes visible at these short time constants

Table 7.12 Characteristic electron flux for the decay of maximum proton polarization to $1/e$ during electron beam irradiation at 1 K and 2.5 T [124]; comparison with ammonia and butanol.

Material	Total flux (10^{14} cm^{-2})	Φ_A (10^{14} cm^{-2})
NH$_3$	0–2	10
	2–9	41
	9–32	300
BuOH-PX	0–2	3.8
	2–6	6.6

Table 7.13 Maximum proton polarization P_{max}, optimum microwave frequency f_{opt}, polarization growth time to $0.7 \cdot P_{max}$ and proton spin-lattice relaxation time T_1 initially (i) and after decay to $1/e$ of the initial polarization (f), for irradiated NH$_3$ [124].

Dose (10^{14} e/cm^2)	P_{max} (%)	f_{opt} (GHz)	$T_{0.7}$ (min)	T_1(i) (min)	T_1(f) (min)
0	36	70.004	5.8	18.5	39.0
7	28	69.975	–	9.8	22.0
20	25	69.945	2.5	7.5	16.5
33	25	69.935	1.5	6.3	13.3

at 1 K temperature and at the electron spin concentration, which was initially of the order of 10^{19} cm^{-3} based on the spin-lattice relaxation time.

The best annealing temperature of NH$_3$ turned out to be 75 K to 80 K [124]. In a sample irradiated in situ without pre-irradiation, annealing at 75 K to 80 K yielded proton polarizations around 45%, clearly in excess of the values obtained with the samples with pre-irradiation only at high temperature (90 K).

The radiation damage characteristics of deuterated ammonia ND$_3$ are qualitatively similar to that of NH$_3$ at 1 K, with the main difference that the ESR line broadening is more dramatic because the width only after pre-irradiation is much smaller in comparison with NH$_3$ [124]. The behavior at 2.5 T and 0.2 K, however, revealed another interesting feature. The sample which yielded initial polarizations of +31% and –29% after pre-irradiation polarized up to +40% and –44% after in situ photon irradiation to about 15×10^{14} cm^{-2} [69]. Subsequent irradiation showed a resistance against damage about 10 times higher than that of deuterated butanol in the same conditions. The polarizing time to $0.7 \cdot P_{max}$ was 40 min for ND$_3$, which was slower than the value ≈ 15 min for butanol-d$_{10}$ in the same conditions.

The characteristic flux for the loss of polarization in NH$_3$ at 1 K and 5 T is even better than at 0.5 K and 2.5 T. Proton beam irradiation yielded the initial value $\Phi_A = 40 \times 10^{14}$ cm^{-2} up to 10^{15} cm^{-2} total flux [76]; the subsequent decay is slower with $\Phi_A = 130 \times 10^{14}$ cm^{-2}. These are in agreement with the earlier results [73] and with the electron beam damage results $\Phi^A = 70 \times 10^{14}$ cm^{-2} initially and $\Phi^A = 130 \times 10^{14}$ cm^{-2} after a dose of about 10^{15} cm^{-2} [79]. These characteristic fluxes are same for both polarizations in NH$_3$, whereas in ND$_3$ in the same conditions the decay is much slower and asymmetric, with $\Phi_A(+) = 140 \times 10^{14}$ cm^{-2} and $\Phi_A(-) = 250 \times 10^{14}$ cm^{-2} [79]. Similar to 2.5 T and 0.2 K, the deuteron polarization in ND$_3$ increases with in situ dose after each anneal, whereas the proton polarization in NH$_3$ returns to the same level as without irradiation in situ.

The difference between the optimum frequencies for positive and negative proton polarization in NH$_3$ at 5 T and 1 K is 250 MHz initially and increases to 375 MHz when radiation damage accumulates [76]. This increase is the same at 2.5 T field, which allows to conclude that the dominant ESR broadening mechanism is hyperfine interactions. Some of the additional broadening after in situ irradiation, however, must be due to increased g-anisotropy, because the relaxation times appear to be significantly shorter. Furthermore, the frequency separation at 5 T suggests that the main mechanism for DNP is not the solid effect nor any of its variants.

The beam heating at 1 K and 5 T causes the proton polarization to fall from 87% to 85% at a proton beam current of 7×10^{10} s^{-1}, which is about 10 times higher beam flux than is possible when operating the same target at 2.5 T and 0.5 K [76]. The difference stems from the much higher heat transfer coefficient and burnout heat flux from the target to the boiling liquid ^{4}He, in comparison with the boiling liquid ^{3}He. With rastered high-energy electron beam, ammonia targets have been operated at 5×10^{11} s^{-1} with a loss of polarization of 15% relative to the value obtained at the beam current of 2×10^{11} s^{-1} [79].

Irradiated Lithium Hydrides

Lithium hydrides have a face-centered cubic structure and therefore a potential for DNP in polycrystalline or powder form. As with ammonia, however, the cubic structure leads to small g-factor anisotropy and therefore slow electron spin lattice relaxation time, which slows down DNP. Additional irradiation at low temperature is likely to create a high stress in the lattice, which increases the g-anisotropy and the spin lattice coupling. It can therefore be expected that the behavior of high-temperature irradiated lithium hydrides is qualitatively similar to ammonia, i.e. that additional irradiation in situ will speed up DNP and shorten all relaxation times, and possibly also increase the maximum DNP.

Systematic investigations of irradiated lithium hydrides were made at Bonn with samples pre-irradiated at 180 K with electron fluence of 10^{17} cm^{-2} using the Bonn 20 MeV pre-accelerator LINAC.

An irradiated ^{7}LiH sample was polarized to the maximum value 11.5% at 1 K and 2.5 T and was exposed to a 1.2 GeV electron beam. The build-up time for protons was initially 50 min and was reduced to 8 min at a dose of 10^{14} e/cm^2, and the maximum polarization increased to 14.5% [94]. These characteristics remained unchanged up to the maximum electron fluence of 5×10^{15} cm^{-2} at which the test was stopped.

Deuterons in a sample of ^{6}LiD were polarized under similar conditions initially to 12% and the sample was exposed to the same beam 1.2 GeV electron beam. The polarization growth time was shortened from 50 min to 8 min after about 2×10^{14} cm^{-2}, but the maximum polarization was slightly reduced to 10.5%. The polarization then increased back to 12% by the dose of 10^{16} cm^{-2} [94].

The radiation resistance of pre-irradiated ^{6}LiD in the SLAC E155 experiment was described already in Section 7.3.3. The set-up operated in a 5 T field at 1 K temperature which yields higher and faster polarization in comparison with the tests at Bonn. The radiation resistance was better by a substantial factor of five in comparison with deuterated ammonia targets in similar beam conditions [96].

7.8 Storage and Handling of Targets

The polarized target materials are most conveniently stored in liquid nitrogen using vacuum-insulated dewars designed for biological applications. These come equipped with numbered holders which allow a practical organization of the samples. Dewars made with a low-conduction glass-epoxy composite neck and vacuum superisolation have a liquid N_2 holding time in excess of one month.

To avoid accumulation of water and oxygen in the storage dewar, it is a good practice to always use the original stopper of the access port of the dewar. This also reduces the heat leak and therefore liquid boiloff. For the storage of very expensive targets such as those made of isotope labelled or enriched materials, it may be worth equipping the dewar with an alarm for increased boiloff due to vacuum failure and for liquid dry-off. As the LN_2 level gauges are not reliable enough for this task, the best way appears to be to mount two platinum wire thermometers to the stopper, one so that it measures the vapor temperature in the lower part of the neck, and the other so that it measures the temperature of the vapor exit of the dewar. Dry-off is detected as stopped or reduced evaporation, the first manifestation of which is the warmup of the neck which is no longer cooled by the boiloff vapor. Vacuum failure leads to a vigorous boiling, which is immediately detected as an abnormal cooling of the boiloff gas outlet. The samples must be saved within an hour following the alarm.

The sample storage containers can be made of Pyrex glass or a suitable plastic such as soft polyethylene. Small samples can be stored in glass test tubes closed with a cork stopper or with an aluminium foil attached with a copper wire around the tube. Larger samples are conveniently stored in sample bottles of various sizes equipped with a ground glass stopper. Such a stopper, however, has the unpleasant tendency of sticking, but this can be avoided by placing an aluminium foil strip between the ground surfaces. The stopper must then be secured by copper wires attached around the neck; these wires are also convenient for handling and moving the bottles between the handling bath and the storage dewar. Labelling of the bottles or test tubes is conveniently done by using a permanent felt-pen made for writing overhead transparencies. If labelling is forgotten before cooling the bottle, one may also write on an aluminium foil attached with a copper wire around the bottle.

It is a good practice to test the ink before use, because some of them tend to flake off at 77K, particularly from a polyethylene or metal support which is flexible.

The above way of storing is convenient for targets of a few hundred mL. Storage and handling of larger amounts of material is more conveniently done in nylon socks [64] held in the metallic containers which come with the biological dewars. This facilitates the transfer of the material under liquid nitrogen to the target cell in the cryostat because the sock is deformable and transparent to LN_2. The sock may also provide some shielding against the frost which tends to accumulate in LN_2 during handling.

The transfer of target beads to and from a glass or plastic container under liquid nitrogen is difficult if the beads are charged, because they adhere to the walls of the container. This can be prevented to a good extent by using a piezoelectric discharge gun designed for dust discharging.

The handling bath can be a wide glass or stainless steel dewar. It is a good practice to prepare for all tools needed in the sample manipulation holders which keep the tool tips under LN_2 and the handle at room temperature. A low-cost handling bath is obtained from polyurethane or polystyrene foam; boxes used for packaging of goods for transport are often suitable. The thermal losses to such a bath are quite high but the resulting vigorous boiloff has the good advantage of preventing frost from accumulating from the ambient and exhaled air.

Direct contamination of target bead surfaces by water frost should be carefully avoided, because the removal of the frost afterwards is practically impossible. The water frost is seen in the proton NMR signal but cannot be distinguished from the signal arising from the target material protons; this affects the calibration of the NMR apparatus used for polarization measurement, because the protons in the frost are not (or little) polarized during DNP. It is therefore necessary to keep the beads always submerged in LN_2 unless totally dried atmosphere surrounds them. One may visually observe the speed at which frost contaminates the beads, because they lose their translucid color in the same way as a window glass when frosted. The amount of water contamination also depends on the humidity of the air, so that on a dry winter day little frost is collected, whereas on a humid summer day one cannot get a view into the liquid nitrogen bath because it is covered by a stagnating cloud of mist and frost. One might then consider performing critical operations in a dehumidified room, and if this is not possible, for example, in large experimental halls, raising a dehumidified tent around the target installation for the duration of the target loading.

References

[1] Y.-S. Tsai, Pair production and bremsstrahlung of charged leptons, *Rev. Mod. Phys.* **46** (1974) 815–851.

[2] Particle Data Group, M. Tanabashi, K. Hagiwara, et al., Review of particle physics, *Phys. Rev. D* **98** (2018).

[3] M. Krumpolc, Ammonium Borohydride – A novel, hydrogen-rich material for polarized targets, in: G.M. Bunce (ed.) *High Energy Spin Physics – 1982*, American Institute of Physics, New York, 1983, 502–504.

[4] D. Hill, J. Hill, M. Krumpolc, Polarization in chemically doped hydrogen-rich glasses, in: W. Meyer (ed.) *Proc. 4th Int. Workshop on Polarized Target Materials and Techniques*, Physikalisches Institut, Universität Bonn, Bonn, 1984, 84–93.

[5] R. M. Kulsrud, H. P. Furth, E. J. Valeo, M. Goldhaber, Fusion reactor plasmas with polarized nuclei, *Phys. Rev. Letters* **49** (1983) 1248–1251.

[6] M. G. D. van der Grinten, H. Glättli, C. Fermon, M. Eisenkremer, M. Pinot, Dynamic proton polarization on polymers in solution: creating contrast in neutron scattering, *Nucl. Instr. and Meth. in Phys. Res.* **A356** (1995) 422–431.

[7] M. Borghini, S. Mango, O. Runolfsson, J. Vermeulen, Sizeable proton polarizations in frozen alcohol mixtures, *Polarized Targets and Ion Sources*, La Direction de la Physique, CEN Saclay, Saclay, France, 1966, 387–391.

[8] S. Mango, Ö. Runolfsson, M. Borghini, A butanol polarized proton target, *Nucl. Instr. and Meth.* **72** (1969) 45–50.

[9] H. Glättli, M. Odehnal, J. Ezratty, A. Malinovski, A. Abragam, Polarisation dynamique des protons dans le glycol ethylique, *Phys. Letters* **29A** (1969) 250–251.

[10] D. A. Hill, J. B. Ketterson, R. C. Miller, et al., Dynamic proton polarization in butanol water below 1K, *Phys. Rev. Letters* **23** (1969) 460–462.

[11] A. Masaike, H. Glättli, J. Ezratty, A. Malinovski, High proton polarization at 0.5 °K, *Phys. Letters* **30A** (1969) 63–64.

[12] W. Gorn, P. Robrish, Polarization in assorted hydrocarbons, in: G. Shapiro (ed.) *Proc. 2nd Int. Conf. on Polarized Targets*, LBL, University of California, Berkeley, Berkeley, 1971, 305.

[13] W. de Boer, High proton polarization in 1,2-propanediol at ^{3}He temperatures, *Nucl. Instr. and Meth.* **107** (1973) 99–104.

[14] W. de Boer, T. O. Niinikoski, Dynamic proton polarization in propanediol below 0.5K, *Nucl. Instrum. and Meth.* **114** (1974) 495–498.

[15] H. Glättli, Organic materials for polarized proton targets, in: G. Shapiro (ed.) *Proc. 2nd Int. Conf. on Polarized Targets*, LBL, University of California, Berkeley, Berkeley, 1971, 281–287.

[16] T. O. Niinikoski, Viscometric study of binary mixtures of butanol and pentanol with water and pinacol, in: G. R. Court, et al. (eds.) *Proc. of the 2nd Workshop on Polarized Target Materials*, SRC, Rutherford Laboratory, 1980, 69–71.

[17] M. Symons, Nature and preparation of glasses, in: G. R. Court, et al. (eds.) *Proc. of the 2nd Workshop on Polarized Target Materials*, SRC, Rutherford Laboratory, 1980, p. 68.

[18] T. O. Niinikoski, Dynamic nuclear polarization with the new complexes, in: G. R. Court, et al. (eds.) *Proc. of the 2nd Workshop on Polarized Target Materials*, SRC, Rutherford Laboratory, 1980, 60–65.

[19] D. A. Hill, J. J. Hill, *An Investigation of Polarized Proton Target Materials by Differential Calorimetry – Preliminary Results*, Argonne National Laboratory Report ANL-HEP-PR-81-05, 1980.

[20] G. S. Grest, M. H. Cohen, Liquids, glasses and the glass transition: a free-volume approach, in: S. A. Rice, E. Prigogine (eds.) *Advances in Chemical Physics*, Wiley, New York, 1981, 455–525.

[21] G. Adam, J. H. Gibbs, On the temperature dependence of cooperative relaxation properties in glass-forming liquids, *J. Chem. Phys.* **43** (1965) 139–146.

[22] H. Vogel, The law of the relation between the viscosity of liquids and the temperature, *Phys. Z.* **22** (1921) 645–646.

[23] G. S. Fulcher, Analysis of recent measurements of the viscosity of glasses, *J. Am. Ceram. Soc.* **8** (1925) 339–355.

[24] C. A. Agnell, J. M. Sare, E. J. Sare, Glass transition temperatures for simple molecular liquids and their binary solutions, *J. Phys. Chem.* **82** (1978) 2622–2629.

[25] J. M. Gordon, G. B. Rouse, J. H. Gibbs, W. M. Risen, The composition dependence of glass transition properties, *J. Chem. Phys.* **66** (1977) 4971–4976.

[26] D. H. Rasmussen, A. P. MacKenzie, The glass transition in amorphous water, *J. Phys. Chem.* **75** (1971) 967–973.

[27] A. V. Lesikar, Comment on 'The composition dependence of glass transition properties', *J. Chem. Phys.* **68** (1978) 3323–3325.

[28] A. V. Lesikar, On the glass transition in mixtures between the normal alcohols and various Lewis bases, *J. Chem. Phys.* **66** (1977) 4263–4276.

[29] M. Krumpolc, D. Hill, H. B. Stuhrmann, Progress in the chemistry of chromium(V) doping agents used in polarized target materials, in: E. Steffens, et al. (eds.) *Proc. 6th Workshop on Polarized Solid Targets*, Springer Verlag, Heidelberg, 1991, 340–343.

[30] S. Takala, T. O. Niinikoski, Measurements of glass properties and density of hydrocarbon mixtures of interest in polarized targets, in: E. Steffens, et al. (eds.) *Proc. 6th Workshop on Polarized Solid Targets*, Springer Verlag, Heidelberg, 1991, 347–352.

[31] S. Sahling, Low temperature thermal properties of pentanol-2 – a perspective polarized target material, in: E. Steffens, et al. (eds.) *Proc. 6th Workshop on Polarized Solid Targets*, Springer Verlag, Heidelberg, 1991, pp. 356–357.

[32] S. Sahling, A. Sahling, M. Kolác, Low temperature long-time relaxation in glasses, *Solid State Comm.* **65** (1988) 1031–1033.

[33] R. C. Fernow, Dynamic polarization in radiation resistant materials, *Nucl. Instr. and Meth.* **159** (1979) 557–560.

[34] M. Symons, Radiation induced paramagnetic centres in organic and inorganic materials, in: G. R. Court, et al. (eds.) *Proc. of the 2nd Workshop on Polarized Target Materials*, SRC, Rutherford Laboratory, 1980, 25–28.

[35] M. Borghini, W. de Boer, K. Morimoto, Nuclear dynamic polarization by resolved solid-state effect and thermal mixing with an electron spin-spin interaction reservoir, *Phys. Lett.* **48A** (1974) 244–246.

[36] N. S. Garif'yanov, B. M. Kozyrev, V. N. Fedotov, Width of the EPR line of liquid solutions of ethylene glycol complexes for even and odd chromium isotopes, *Sov. Phys. Dokl.* **13** (1968) 107–110.

[37] P. R. Bontchev, A. Malinovski, M. Mitewa, K. Kabassanov, Intermediate chromium(V) complex species and their role in the process of chromium(VI) reduction by ethylene glycol, *Inorg. Chim. Acta* **6** (1972) 499–503.

[38] D. Hill, Preparation of ethanediol-Cr(V), in: G. R. Court, et al. (eds.) *Proc. Second Workshop on Polarized Target Materials*, SRC, Rutherford Laboratory, 1980, 72–73.

[39] J. Svoboda, Dynamic polarization and relaxation of protons in Cr(V) complexes in dimethylformamide, *Czech. J. Phys.* **B28** (1978) 473–475.

[40] E. I. Bunyatova, Investigation of organic substances for development of targets with polarized hydrogen and deuterium nuclei, in: E. Steffens, et al. (eds.) *Proc. 6th Workshop on Polarized Solid Targets*, Springer Verlag, Heidelberg, 1991, 333–339.

[41] T. O. Niinikoski, Polarized targets at CERN, in: M. L. Marshak (ed.) *Int. Symp. on High Energy Physics with Polarized Beams and Targets*, American Institute of Physics, Argonne, 1976, 458–484.

[42] M. Krumpolc, J. Rocek, Stable chromium(V) compounds, *J. Am. Chem. Soc.* **98** (1976) 872–873.

[43] M. Krumpolc, J. Rocek, Three-electron oxidations. 12. Chromium(V) formation in the chromic acid oxidation of 2-hydroxy-2-methylbutyric acid, *J. Am. Chem. Soc.* **99** (1977) 137–143.

[44] M. Krumpolc, B. G. DeBoer, J. Rocek, A stable Cr(V) compound. Synthesis, properties, and crystal structure of potassium bis(2-hydroxy-2-methylbutyrato)-oxochromate(V) monohydrate, *J. Am. Chem. Soc.* **100** (1978) 145–153.

[45] M. Krumpolc, J. Rocek, Synthesis of stable chromium(V) complexes of tertiary hydroxy acids, *Journal of the American Chemical Society* **101** (1979) 3206–3209.

[46] D. Hill, R. C. Miller, M. Krumpolc, J. Rocek, A new CrV doping agent for polarized targets, *Nuclear Instruments and Methods* **150** (1978) 331–332.

[47] S. N. Mahapatro, M. Krumpolc, J. Rocek, Three-electron oxidations. 17. The chromium(VI) and chromium(V) steps in the chromic acid cooxidation of 2-hydroxy-2-methylbutyric acid and 2-propanol, *Journal of the American Chemical Society* **102** (1980) 3799–3806.

[48] Spin Muon Collaboration (SMC), D. Adams, B. Adeva, et al., The polarized double-cell target of the SMC, *Nucl. Instr. and Meth. in Phys. Res.* **A437** (1999) 23–67.

[49] S. Bültmann, G. Baum, P. Hautle, et al., Properties of the deuterated target material used by the SMC, in: H. Dutz, W. Meyer (eds.) *7th Int. Workshop on Polarized Target Materials and Techniques*, Elsevier, Amsterdam, 1994, 102–105.

[50] H. B. Stuhrmann, N. Burkhardt, G. Dietrich, et al., Proton and deuteron targets in biological structure research, in: H. Dutz, W. Meyer (eds.) *7th Int. Workshop on Polarized Target Materials and Techniques*, Elsevier, Amsterdam, 1994, 124–132.

[51] E. I. Bunyatova, Free radicals and polarized targets, *Nuclear Instruments and Methods in Physics Research* **A 526** (2004) 22–27.

[52] B. van den Brandt, P. Hautle, J. A. Konter, E. Bunyatova, Progress in scintillating polarized targets for spin physics, in: M. Anginolfi, et al. (eds.) *Second International Symposium on the Gerasimov-Drell Hearn Sum Rule and the Spin Structure of the Nucleon*, World Scientific, Singapore, 2002, 183–187.

[53] A. S. Lilly Thankamony, J. J. Wittmann, M. Kaushik, B. Corzilius, Dynamic nuclear polarization for sensitivity enhancement in modern solid-state NMR, *Progress in Nuclear Magnetic Resonance Spectroscopy* **102–103** (2017) 120–195.

[54] O. Haze, B. Corzilius, A. A. Smith, R. G. Griffin, T. M. Swager, Water-soluble narrow-line radicals for dynamic nuclear polarization, *Journal of the American Chemical Society* **134** (2012) 14287–14290.

[55] J. H. Ardenkjær-Larsen, I. Laursen, I. Leunbach, et al., EPR and DNP properties of certain novel single electron contrast agents intended for oximetric imaging, *J. Magn. Res.* **133** (1998) 1–12.

[56] S. Jannin, A. Comment, F. Kurdzesau, et al., A 140 GHz prepolarizer for dissolution dynamic nuclear polarization, *J. Chem. Phys.* **128** (2008) 241102.

[57] J. H. Ardenkjaer-Larsen, On the present and future of dissolution-DNP, *J. Magn. Res.* **264** (2016) 3–12.

[58] J. Ardenkjaer-Larsen, S. Bowen, J. Raagaard Petersen, et al., Cryogen-free dissolution dynamic nuclear polarization polarizer operating at 3.35 T, 6.70 T, and 10.1 T, *Magnetic Resonance in Medicine* **81** (2018) 2184–2194.

[59] S. Y. Pshetzhetskii, A. G. Kotov, V. K. Milinchuk, V. A. Roginskii, V. I. Tupikov, *EPR of Free Radicals in Radiation Chemistry*, John Wiley & Sons, New York, 1974.

[60] T. Kumada, Y. Noda, T. Hashimoto, S. Koizumi, Dynamic nuclear polarization study of UV-irradiated butanol for hyperpolarized liquid NMR, *J. Magn. Res.* **201** (2009) 115–120.

[61] T. O. Niinikoski, J.-M. Rieubland, Dynamic nuclear polarization in irradiated ammonia below 0.5 K, *Phys. Lett.* **72A** (1979) 141–144.

[62] S. Brown, G. Court, G. Hayes, et al., The production of large quantities of irradiated ammonia for use as a polarized target material, in: W. Meyer (ed.) *Proc. 4th Int. Workshop on Polarized Target Materials and Techniques*, Physikalisches Institut, Universität Bonn, Bonn, 1984, pp. 66–78.

[63] P. R. Cameron, Preparation and irradiation of ammonia target beads, in: W. Meyer (ed.) *Proc. 4th Int. Workshop on Polarized Target Materials and Techniques*, Physikalisches Institut, Universität Bonn, Bonn, 1984, 79–80.

[64] Spin Muon Collaboration (SMC), B. Adeva, E. Arik, et al., Measurement of proton and nitrogen polarization in ammonia and a test of equal spin temperatures, *Nucl. Instr. and Meth. in Phys. Res.* **A 419** (1998) 60–82.

[65] T. O. Niinikoski, Proton irradiated ammonia, in: G. R. Court, et al. (eds.) *Proc. of the 2nd Workshop on Polarized Target Materials*, SRC, Rutherford Laboratory, 1980, 36–37.

[66] G. Court, Reactor irradiated ammonia, in: G. R. Court, et al. (eds.) *Second Workshop on Polarized Target Materials*, SRC, Rutherford Laboratory, Cosener's House, Abingdon, Chilton, Didcot, UK, 1979, 38–39.

[67] A. D. Krisch, Electron irradiated ammonia and Butanol, in: G. R. Court, et al. (eds.) *Proc. Second Workshop on Polarized Target Materials*, SRC, Rutherford Laboratory, 1980, 39–43.

[68] D. G. Crabb, P. R. Cameron, A. M. T. Lin, R. S. Raymond, Operational characteristics of radiation doped ammonia in a high intensity proton beam, in: W. Meyer (ed.) *Proc. 4th Int. Workshop on Polarized Target Materials and Techniques*, Physikalisches Institut, Universität Bonn, Bonn, 1984, 7–12.

[69] R. Dostert, W. Havenith, O. Kaul, et al., First use of ND_3 in a high energy photon beam, in: W. Meyer (ed.) *Proc. 4th Int. Workshop on Polarized Target Materials and Techniques*, Physikalisches Institut, Universität Bonn, Bonn, 1984, 13–22.

[70] R. Dostert, W. Havenith, O. Kaul, et al., Dynamic nuclear polarization studies in irradiated ammonia at 1 K, in: W. Meyer (ed.) *Proc. 4th Int. Workshop on Polarized Target Materials and Techniques*, Physikalisches Institut, Universität Bonn, Bonn, 1984, pp. 33–52.

[71] M. L. Seely, M. R. Bergström, S. K. Dhawan, et al., Dynamic nuclear polarization of irradiated targets, in: C. Joseph, J. Soffer (eds.) *Proc. 1980 Int. Symp. on High-Energy Physics with Polarized Beams and Polarized Targets*, Birkhäuser Verlag, Basel, Boston and Stuttgart, 1981, 453–453.

[72] U. Härtel, O. Kaul, W. Meyer, K. Rennings, E. Schilling, Experience with NH_3 as target material for polarized proton targets at the bonn 2.5 GeV electron synchrotron, in: C. Joseph, J. Soffer (eds.) *Proc. 1980 Int. Symp. on High-Energy Physics with Polarized Beams and Polarized Targets*, Birkhäuser Verlag, Basel, Boston and Stuttgart, 1981, 447–450.

[73] M. L. Seely, A. Amittay, M. R. Bergström, et al., Dynamic nuclear polarization of irradiated targets, *Nuclear Instruments and Methods in Physics Research* **201** (1982) 303–308.

[74] M. L. Seely, A. Amittay, M. R. Bergstrom, et al., Dynamic nuclear polarization of irradiated targets, in: G. M. Bunce (ed.) *High Energy Spin Physics – 1982*, American Institute of Physics, New York, 1983, 526–533.

[75] R. S. Raymond, P. R. Cameron, D. G. Crabb, T. Roser, Dynamic polarization in a ^{3}He/^{4}He evaporation cryostat, in: S. Jaccard, S. Mango (eds.) *International Workshop on Polarized Sources and Targets*, Birkhäuser, Montana, Switzerland, 1986, 777–780.

[76] D. G. Crabb, Polarization studies with radiation doped ammonia at 0.5 T and 1 K, in: E. Steffens, et al. (eds.) *Proc. 6th Workshop on Polarized Solid Targets*, Springer Verlag, Heidelberg, 1991, 289–300.

[77] D. G. Crabb, C. B. Higley, A. D. Krisch, et al., Observation of a 96% proton polarization in irradiated ammonia, *Phys. Rev. Letters* **64** (1990) 2627–2629.

[78] B. Boden, V. Burkert, G. Knop, et al., Elastic electron deuteron scattering on a tensor polarized solid ND$_3$ target, *Zeitschrift für Physik C Particles and Fields* **49** (1991) 175–185.

[79] D. G. Crabb, D. B. Day, The Virginia/Basel/SLAC polarized target: operation and performance during experiment E143 at SLAC, in: H. Dutz, W. Meyer (eds.) *7th Int. Workshop on Polarized Target Materials and Techniques*, Elsevier, Amsterdam, 1994, 11–19.

[80] M. Borghini, *Choice of Substances for Polarized Proton Targets*, CERN Yellow Report CERN 66–3, 1966.

[81] F. E. Pretzel, G. V. Gritton, C. C. Rushing, et al., Properties of lithium hydride-IV: F-center formation at low temperatures, *Journal of Physics and Chemistry of Solids* **23** (1962) 325–337.

[82] B. Henderson, Inorganic materials, in: G. R. Court, et al. (eds.) *Proc. Second Workshop on Polarized Target Materials*, SRC, Rutherford Laboratory, 1980, 29–32.

[83] Y. Roinel, V. Bouffard, Polarisation dynamique nucléaire dans l'hydrure de lithium, *J. Phys. France* **38** (1977) 817–824.

[84] A. Abragam, Polarized targets in high energy and elsewhere, in: G. H. Thomas (ed.) *High Energy Physics with Polarized Beams and Polarized Targets*, American Institute of Physics, New York, 1979, 1–14.

[85] V. Bouffard, Y. Roinel, P. Roubeau, A. Abragam, Dynamic nuclear polarization in ^{6}LiD, *J. Phys. France* **41** (1980) 1447–1451.

[86] Y. Roinel, Electron irradiated lithium fluoride (^{7}LiF) lithium hydride (^{7}LiH) and lithium deuteride (^{6}LiD), in: G. R. Court, et al. (eds.) *Proc. Second Workshop on Polarized Target Materials*, SRC, Rutherford Laboratory, 1980, 43–46.

[87] P. Chaumette, J. Deregel, G. Durand, et al., Attempt to polarize a large sample of ^{7}LiH irradiated with high energy electrons, in: W. Meyer (ed.) *Proc. 4th Int. Workshop on Polarized Target Materials and Techniques*, Physikalisches Institut, Universität Bonn, Bonn, 1984, 81–83.

[88] P. Chaumette, J. Deregel, G. Durand, et al., Progress report on polarization of ^{6}LiD and ^{7}LiH irradiated with high energy electrons, in: S. Jaccard, S. Mango (eds.) *International Workshop on Polarized Sources and Targets*, Birkhäuser, Montana, Switzerland, 1986, 767–771.

[89] P. Chaumette, J. Deregel, G. Durand, J. Fabre, L. van Rossum, Progress report on polarization of irradiated ^{6}LiD and ^{7}LiH, in: K. J. Heller (ed.) *Proc. Int. Symp. on High Energy Spin Physics*, AIP, Minneapolis, 1988, 1275–1280.

[90] B. van den Brandt, J. A. Konter, S. Mango, M. Weßler, Results from the PSI ^{6}LiD target, in: E. Steffens, et al. (eds.) *Proc. 6th Workshop on Polarized Solid Targets*, Springer Verlag, Heidelberg, 1991, 320–324.

[91] B. van den Brandt, J. A. Konter, A. I. Kovalev, S. Mango, M. Wessler, Operation of a polarized ^{7}LiH target, in: T. Hasegawa, et al. (eds.) *Proc. 10th Int. Symp. on High-Energy Spin Physics*, Universal Academy Press, Inc., Tokyo, 1993, 369–370.

[92] G. Durand, C. Gaudron, J. Ball, M. Combet, J. Sans, Progress report on polarizable lithium hydrides, in: T. Hasegawa, et al. (eds.) *Proc. 10th Int. Symp. on High-Energy Spin Physics*, Universal Academy Press, Inc., Tokyo, 1993, 355–361.

[93] J. J. Jarmer, S. Penttilä, Polarization of irradiated lithium hydride, in: T. Hasegawa, et al. (eds.) *Proc. 10th Int. Symp. on High-Energy Spin Physics*, Universal Academy Press, Inc., Tokyo, 1993, 363–367.

[94] S. Goertz, C. Bradtke, H. Dutz, et al., Investigations in high temperature irradiated ^{6,7}LiH and ^{6}LiD, its dynamic nuclear polarization and radiation resistance, in: H. Dutz, W. Meyer (eds.) *7th Int. Workshop on Polarized Target Materials and Techniques*, Elsevier, Amsterdam, 1994, 20–28.

[95] J. Heckmann, S. Goertz, W. Meyer, E. Radtke, G. Reicherz, EPR spectroscopy at DNP conditions, *Nucl. Instrum. Methods Phys. Res.* **A 526** (2004) 110–116.

[96] S. Bültmann, D. G. Crabb, D. B. Day, et al., A study of lithium deuteride as a material for a polarized target, *Nucl. Instr. Meth.* **A 425** (1999) 23–36.

[97] J. Ball, G. Baum, P. Berglund, et al., First results of the large COMPASS ^{6}LiD polarized target, *Nuclear Instruments and Methods in Physics Research Section A: Accelerators, Spectrometers, Detectors and Associated Equipment* **498** (2003) 101–111.

[98] K. Kondo, J. Ball, G. Baum, et al., Polarization measurement in the COMPASS polarized target, *Nuclear Instruments and Methods in Physics Research* **A 526** (2004) 70–75.

[99] N. Doshita, J. Ball, G. Baum, et al., The COMPASS polarized target, *Czechoslovak Journal of Physics* **55** (2005) A367–A374.

[100] C. Adolph, M. Aghasyan, R. Akhunzyanov, et al., Final COMPASS results on the deuteron spin-dependent structure function g_1^d and the Bjorken sum rule, *Physics Letters B* **769** (2017) 34–41.

[101] P. Abbon, E. Albrecht, V. Y. Alexakhin, et al., The COMPASS experiment at CERN, *Nuclear Instruments and Methods in Physics Research Section A: Accelerators, Spectrometers, Detectors and Associated Equipment* **577** (2007) 455–518.

[102] C. P. Slichter, *Principles of Magnetic Resonance*, 3rd ed., Springer-Verlag, Berlin, 1990.

[103] D. Crabb, Irradiated butanol, in: G. R. Court, et al. (eds.) *Proc. Second Workshop on Polarized Target Materials*, SRC, Rutherford Laboratory, 1980, 33–36.

[104] S. T. Goertz, J. Harmsen, J. Heckmann, et al., Highest polarizations in deuterated compounds, *Nuclear Instruments and Methods in Physics Research* **A 526** (2004) 43–52.

[105] C. A. Hutchison, B. W. Mangum, Paramagnetic resonance absorption in naphthalene in its phosphorescent state, *J. Chem. Phys.* **34** (1961) 908–922.

[106] H. W. van Kesteren, W. T. Wenckebach, J. Schmidt, Production of high, long-lasting, dynamic proton polarization by way of photoexcited triplet states, *Phys. Rev. Lett.* **55** (1985) 1642–1644.

[107] P. F. A. Verheij, W. T. Wenckebach, J. Schmidt, Microwave induced optical deuteron polarization at 75 GHz: a quantitative analysis, *Applied Magnetic Resonance* **5** (1993) 187–205.

[108] T. J. Schmugge, C. D. Jeffries, High dynamic polarization of protons, *Phys. Rev.* **138** (1965) A1785–A1801.

[109] O. S. Leifson, C. D. Jeffries, Dynamic polarization of nuclei by electron-nuclear dipolar coupling in crystals, *Phys. Rev.* **122** (1961) 1781–1795.

[110] T. J. Schmugge, C. D. Jeffries, Sizeable dynamic proton polarizations, *Phys. Rev. Lett.* **9** (1962) 268–270.

[111] J. Sowinski, L. D. Knutson, A spin refrigerator polarized target for nuclear physics experiments, *Nuclear Instruments and Methods in Physics Research Section A: Accelerators, Spectrometers, Detectors and Associated Equipment* **355** (1995) 242–252.

[112] V. A. Atsarkin, A. E. Mefed, M. I. Rodak, Connection between dynamic polarization of nuclei and electron spin-spin reservoir temperature, *Sov. Phys. JETP* **6** (1967) 359–362.

[113] V. A. Atsarkin, A. E. Mefed, M. I. Rodak, Connection of electron spin-spin interactions with polarization and nuclear spin relaxation in ruby, *Sov. Phys. JETP* **28** (1969) 877–885.

[114] V. A. Atsarkin, A. E. Mefeod, M. I. Rodak, Electron cross relaxation and nuclear polarization in ruby, *Phys Lett.* **27A** (1968) 57–58.

[115] V. A. Atsarkin, Verification of the spin-spin temperature concept in experiments on saturation of electron paramagnetic resonance, *Soviet Phys.- JETP* **31** (1970) 1012–1018.

[116] A. Atsarkin, Experimental investigation of the manifestations of the spin-spin interaction reservoir in a system of EPR lines connected with cross relaxation, *Sov. Phys. JETP* **32** (1971) 421–425.

[117] A. Masaike, Toward precise tests of time reversal invariance using very low energy neutrons, *First Int. Symp. on Symmetries in Subatomic Physics*, Taipei, 1994.

[118] T. Adachi, K. Asahi, M. Doi, et al., Test of parity violation and time reversal invariance in slow neutron absorption rections, *Nucl. Phys.* **A577** (1994) 433 c–442 c.

[119] A. Abragam, M. Chapellier, M. Goldman, J.F. Jacquinot, V. H. Chau, A clean example of large dynamic polarization of F^{19} in CaF_2, in: G. Shapiro (ed.) *Proc. 2nd Int. Conf. on Polarized Targets*, LBL, University of California, Berkeley, Berkeley, 1971, 247–256.

[120] R. C. Fernow, Radiation damage in polarized target materials, *Nucl. Instr. and Meth.* **148** (1978) 311–316.

[121] H. Petri, G. Abshire, Polarization decay of an ethylene glycol polarized proton target in an intense proton beam, *Nucl. Instr. and Meth.* **119** (1974) 205–207.

[122] D. Crabb, Measurement of the polarization parameter in pp elastic scattering at 24 GeV/c, in: M. L. Marshak (ed.) *High Energy Physics with Polarized Beams and Targets*, American Institute of Physics, New York, 1976, 120–125.

[123] G. R. Court, D. G. Crabb, R. C. Fernow, et al., Report of the workshop on polarized target materials, in: G. H. Thomas (ed.) *High Energy Physics with Polarized Beams and Polarized Targets*, American Institute of Physics, New York, Argonne, 1979, 15–40.

[124] K. H. Althoff, V. Burkert, U. Hartfiel, et al., Radiation resistances of ammonia at 1 kelvin and 2.5 tesla, in: W. Meyer (ed.) *Proc. 4th Int. Workshop on Polarized Target Materials and Techniques*, Physikalisches Institut, Universität Bonn, Bonn, 1984, 23–32.

8
Refrigeration

Polarized targets need continuous cooling of a relatively large heat load during dynamic nuclear polarization (DNP) at temperatures around or below 1 K. This can be achieved by continuous-flow refrigerators based on the evaporation of liquid ^{4}He or ^{3}He, or on the dilution of ^{3}He by ^{4}He. The refrigerator components have unusual requirements due to the large helium mass flow rates and the demand of long uninterrupted runs of operation.

We shall begin by describing the heat transfer mechanisms from the solid target material to the coolant fluid in Section 8.1 and then evaluate the various cooling cycles in detail in Sections 8.2 and 8.3. The heat loads, ranging from some W/cm^3 to some tens of µW/cm^3, and the choice of the cooling method are outlined in Section 8.4. We shall then discuss the design of other cryogenic parts of the apparatus, including the precooling heat exchangers, in Section 8.5, thermometry and other instrumentation in Section 8.6, and the pump and gas purification systems in Section 8.7.

8.1 Heat Transfer between Target Material and Helium

In order to promote heat transfer between the coolant and the solid material, the target is usually made of small spherical beads or irregularly shaped crushed and sieved pieces, as was discussed in Chapter 7. The mean diameter of the beads or pieces is normally just below 2 mm, which is conveniently defined as an equivalent surface-to-volume diameter d_e. This is obtained by considering the real geometric surface σ_t of the target beads confined in the volume of the target holder cartridge, with the volume filling factor η:

$$\frac{\sigma_t}{\eta V} = \frac{4\pi (d_e/2)^2}{(4/3)\pi (d_e/2)^3} = \frac{6}{d_e}, \qquad (8.1)$$

where V is the total volume filled with the target beads.

A definition of d_e identical to Eq. 8.1 will be used for evaluating the effective area of sintered heat exchange surfaces. The surface thus defined is equal to the hydrodynamic or hydraulic surface area and can be determined experimentally from the measured flow resistivity of the sintered sponge at room temperature. The hydrodynamic surface area thus measured is also the most reliable one for the surface boundary conduction, limited by the Kapitza resistance.

Moreover, the equivalent surface-to-volume diameter d_e is useful for determining the flow resistance of the coolant within the target volume, needed for the estimation of convectional heat transfer from the target beads to the surrounding helium bath.

8.1.1 Kapitza Resistance

Acoustic Mismatch Models

The heat transfer from solid material to liquid helium is limited at low temperatures by the Kapitza thermal boundary resistance, which is phenomenologically understood as an acoustic phonon mismatch or acoustic impedance between the two media. Phonons of relatively long wavelength[1] may be reflected from the interface because their velocity is different on the two sides of the boundary. The model of Khalatnikov [1] explains the steep temperature dependence of the heat transfer between solid materials and liquid helium but gives always too low magnitude for the surface conductance. The model yields the emitted phonon energy flux from one side of the interface

$$\frac{W(T)}{A} = \frac{4\pi^5 \rho c_1 (kT)^4}{15 \rho_L (hc_t)^3} F\left(\frac{c_l}{c_t}\right), \qquad (8.2)$$

where ρ is the density of the liquid, c_1 is its acoustic velocity (that of first sound in ^{4}He), ρ_L is the density of the solid lattice, c_l and c_t are the velocities of the longitudinal and transverse phonon modes in the solid and F is a definite integral involving the velocity ratio; this integral has the value around 2 for all solids, including glasses, dielectrics, metals and so on [1].

Assuming that the flux of Eq. 8.2 is emitted from solid with temperature T_L to liquid helium with temperature T_{He}, and that the same formula applies to flux emitted from liquid back to the solid, the amount of heat transferred is

$$\frac{\dot{Q}}{A} = \frac{W(T_L)}{A} - \frac{W(T_{He})}{A} = \frac{4\pi^5 \rho c_1}{15 \rho_L (hc_t)^3} F\left(\frac{c_l}{c_t}\right)\left((kT_L)^4 - (kT_{He})^4\right)$$

$$= \alpha\left(T_L^4 - T_{He}^4\right); \qquad (8.2')$$

$$\alpha = \frac{4\pi^5 \rho c_1 k^4}{15 \rho_L (hc_t)^3} F\left(\frac{c_l}{c_t}\right).$$

Thus, the theoretical calculation of the heat transfer by phonons gives the right dependence on the temperatures of the lattice T_L and of liquid helium T_{He} for the heat flux and describes correctly the fact that the heat transfer is reversible, but the experimental Kapitza conductance constant α is about one order of magnitude higher than the theoretical values obtained

---

[1] The mean wavelength, which dominates the phonon specific heat and heat transport at 1 K in solids, is around 0.24 μm (see Eq. 2.177). In liquid helium such phonons propagate about 20 times slower, thus having much shorter wavelength.

for most materials [2]. The discrepancy has many possible explanations, among which the following can give important contributions:

(a) The physical surface may be up to three times larger than the geometric surface, owing to microscopic details after machining or mechanical polishing.
(b) The solid surface has a different density than the bulk solid, and therefore different acoustic characteristics.
(c) The first atomic layers of helium behave more like solid rather than liquid.
(d) In metals the electron-phonon interaction near the surface may be modified.
(e) The surface (Rayleigh) waves provide additional degrees of freedom and contribute to the heat transmission.

The irreproducibility of the results with different samples of the same material, and even with the same sample, suggests that the state of the surface and its impurities (adsorbed gases, oxides, sulphides, etc.) provide additional mechanisms for the heat transfer by phonons and by other excitations.

Kapitza Conductance at Metal-Liquid Interfaces

When comparing the theoretical values for solids with different elastic characteristics, Challis [3] noticed that while the theoretical expression for the phonon mismatch depends on the Debye temperature as Θ^{-3}, the experimental values for Kapitza conductance fit quite well a much less steep dependence Θ^{-1}; this seems to hold for metals as well as for dielectric materials. As the polarized target materials have a rather low density and they have Debye temperatures in the range of 100 K to 200 K, they can be expected to have higher Kapitza conductance than Cu, for instance. Snyder [4] has also reviewed the experimental results in view of the Debye temperatures and molar masses of several metals and dielectrics.

In comparing the Kapitza conductances of the interfaces of Cu with pure ^{3}He and with pure ^{4}He, one notices that the theoretical phonon mismatch models need to take into account that below 0.2 K the main excitations of the liquid state are quasiparticles rather than phonons. This provides additional mechanisms for the heat transfer, thus increasing the theoretical conductance. The experiments give the opposite results:

$$\alpha = 12 \frac{W}{m^2 K^4} \quad \text{for Cu in pure } ^3\text{He at 0.1 K;}$$

$$\alpha = 28 \frac{W}{m^2 K^4} \quad \text{for Cu in pure } ^4\text{He at 0.1 K.}$$

Thus, the model based on the phonon flux as a limiting factor in the solid material is the leading theoretical basis for more refined and detailed models, but it fails to give the right magnitude for the experimental conductance.

For small temperature differences Eq. 8.2 can be written as

$$\dot{Q} = 4\sigma_t \alpha T_L^3 \Delta T = \frac{\sigma_t \Delta T}{R_K}, \tag{8.3}$$

where R_K is the Kapitza resistance, related to the Kapitza conductance by

$$\alpha = \frac{1}{4R_K T_L^3} \tag{8.4}$$

In dilution refrigerators the relevant Kapitza conductances between the metallic surfaces of the heat exchangers is reviewed in Appendix A.5.4.

Kapitza Conductance between Non-metallic Substances and Liquid

Dielectric materials are much less studied than pure metals; here we shall briefly discuss these and the rare results available for real target materials.

In early experimental work on Kapitza conductance it was noted already that dielectric substances behave in the same way as metals, including the quantitative discrepancy with the phonon mismatch models and the poor reproducibility of the results [2]. Furthermore, Glättli [5] discovered already in 1968 that paramagnetic cerium ethylsulphate (CeES) featured a high Kapitza conductance of about $\alpha = 100$ W/[m² (T/K)$^{3.4}$] but with a less steep temperature dependence; the experiments were carried out in ⁴He below the lambda point down to 1.4 K. Measurements on three paramagnetic salts $FeNH_4$, $Mn(NH_4)_2$ and CrK alum were then carried out by Vilches and Wheatley at temperatures down to 17 mK in contact with ⁴He [6]; they found the normal temperature dependence and value similar to that of Glättli:

$$\alpha = 50 \frac{W}{m^2 K^4} \quad \text{3 paramagnetic salts in }^4\text{He}$$

Wheatley and his coworkers measured an anomalously high boundary conductivity between the paramagnetic salt cerium magnesium nitrate (CMN) and pure ³He [7]. However, when about one monolayer of ⁴He was added to the sample cell, the thermal conductivity drastically diminished. The high conductivity was interpreted as direct magnetic coupling between the paramagnetic spins of CMN and the nuclear magnetic moments of pure liquid ³He. The measurements were carried out at low external field values and were of great technical interest for the cooling of ³He using adiabatic demagnetization.

In polarized targets the ³He is never pure enough for the magnetic coupling to help substantially in the heat transfer from the target material to the coolant. Furthermore, in a dilution refrigerator there is anyway much ⁴He in both phases, and in particular all solid surfaces are covered with about five monolayers of practically pure ⁴He.

Boyes et al. [8] have measured the Kapitza resistance between butanol beads and liquid ⁴He around 1 K by applying microwave heating off-resonance while measuring the proton spin-lattice relaxation time from which the lattice temperature was determined. The vapor pressure was used for determining the liquid temperature, and the power dissipation in the beads was estimated from the boiloff rate. In these estimates correction was made for power dissipation on the walls of the cavity by making measurements also with empty cavity. The average temperature rise, assumed to be small relative to the temperature of the beads with radius r, is

8.1 Heat Transfer between Target Material and Helium

$$\Delta T = \frac{\dot{Q}}{V}\left(\frac{R_K r}{3} + \frac{r^2}{9\kappa}\right), \tag{8.5}$$

where κ is the thermal conductivity of the bead material. By repeating the measurement with beads of radii varying from 0.5 to 3 mm, they could determine the linear and quadratic parts of Eq. 8.5 and thus extract both the Kapitza resistance and the thermal conductivity at 1.08 K, with values [8]

$$\left.\begin{array}{l} R_K = 59 \pm 28 \; \dfrac{\text{cm}^2\text{K}}{\text{W}}; \\[6pt] \kappa = (2.1 \pm 1.1)\cdot 10^{-3}\dfrac{\text{W}}{\text{cmK}} \end{array}\right\} \text{ at } T = 1.08\text{ K.} \tag{8.6}$$

The Kapitza conductance constant from the above is

$$\alpha = 34 \pm 16 \; \frac{\text{W}}{\text{m}^2\text{K}^4} \; (^4\text{He} - \text{butanol}); \tag{8.7}$$

this is probably the best systematic measurement of the Kapitza conductance for polarized target materials at 1 K.

The Kapitza conductance was similarly determined between propanediol encapsulated in thin FEP foil and dilute solution in the mixing chamber of a dilution refrigerator below 0.5 K, with a result [9]

$$\alpha = 25 \pm 5 \; \frac{\text{W}}{\text{m}^2\text{K}^4} \; \left(\text{dilute solution} - \text{FEP film}\right). \tag{8.8a}$$

However, in this case the power dissipated in the cavity walls was estimated as 50% of the total power, rather than being determined from measurement; the related error was not included in the error estimation of Eq. 8.8a. This resistance is formed of two boundary resistances in series, but because the boundary conductance between two rather similar solid materials can be one order of magnitude higher than that between helium and any of the two materials, the measured value is characteristic of dilute ^{3}He and FEP.

The boundary resistance between ^{3}He and Epibond 100A epoxy, which also resembles glassy target materials with regard to its acoustic phonon spectrum, has been measured [10] to be 3×10^{-3} K^4 m^2/W below 0.1 K and yields the conductance

$$\alpha = 83 \; \frac{\text{W}}{\text{m}^2\text{K}^4} \; (^3\text{He} - \text{Epoxy}). \tag{8.8b}$$

This is inconsistent with the results on pure Cu and other metals that have a general tendency for the conductance to be lower for pure ^{3}He than for ^{4}He and dilute solutions. There is also a tendency that the conductance in ^{3}He decreases between 1 K and 0.1 K, as can be observed in the compilation of Lounasmaa [11].

The Kapitza conductance between different metals and plastics or glues [11] varies between 200 W/(K^4 m^2) and 1000 W/(K^4 m^2) and is rather constant around 0.1 K temperature. A thin isolation layer on a copper wire thus does not reduce the boundary conductance between the wire and helium, and there is evidence that it might in fact increase the heat transfer by providing a sort of acoustic matching layer.

The same might be true for adsorbed gas layers, such as water and nitrogen that usually cover the beads of the polarized target materials after preparation, handling and loading into the cryostat.

Dynamic polarization that results as a function of microwave power can also be used for the determination of the Kapitza resistance. In such cases it must be assumed that most of the power is absorbed in the target material, and that the ultimate polarization is calculable from the lattice temperature and the EPR lineshape, which is known quite well for propanediol-Cr(V) and other Cr(V) compounds. Deuterated butanol in dilute solution below 300 mK temperature has a somewhat lower Kapitza conductance than given by Eq. 8.7:

$$\alpha = 24 \pm 10 \ \frac{W}{m^2 K^4} \left(\text{dilute solution} - \text{d-butanol} \right), \tag{8.8c}$$

basing the determination on the best experimental results [12], and on the calculated relation between the spin and lattice temperatures, as was discussed in Chapter 4.

The Kapitza resistance between sintered metals and concentrated and dilute solutions of ^{3}He in ^{4}He is numerically quite close to the above values of Eqs. 8.7–8.8 c; these will be discussed in better detail in Appendix A.5.4, together with the contributions of the thermal conductance of the metallic powder and of the liquid filling the pores of the powder.

8.1.2 Boiling Heat Transfer

The discussion below on boiling focuses on polarized target applications, in all of which the boiling takes place at low pressures, and the liquid temperature is in the vicinity of 1 K for pure ^{4}He, close to 0.5 K for pure ^{3}He, and in the range 0.7 K to 1 K for dilute solutions in the still of a dilution refrigerator.

The heat transfer within the boiling coolant must also be considered, particularly because the thermal conductivity of pure ^{3}He and of dilute solutions is very low. In contrast, in pure ^{4}He superfluid the conductivity is mainly dependent on the geometry because heat is carried by ballistic phonons which are scattered by boundaries only. For low heat fluxes therefore the heat transfer in ^{4}He evaporation refrigerators is given by the Kapitza conductivity,[2] whereas at high heat fluxes thermal gradients arise and nucleate boiling on the bead surfaces begins to dominate. In ^{3}He evaporators the nucleate boiling heat transfer dominates at all heat fluxes during DNP, whereas in dilution refrigerators it is the convection of the dilute solution of the mixing chamber which carries the heat out of the interior of the target.

---

[2] In this case evaporation takes place only on the surface of the liquid pool, where the heat load is transported by the acoustic phonons.

8.1 Heat Transfer between Target Material and Helium

The nucleate boiling heat transfer coefficient h_{NB}, defined as

$$\frac{\dot{Q}}{\sigma_t} = h_{NB} \Delta T, \tag{8.9}$$

can be obtained from the Forster–Zuber equation for pool boiling

$$h_{NB} = 0.00122 \Delta T_{sat}^{0.24} \Delta p_{sat}^{0.75} \frac{C_{pL}^{0.45} \rho_L^{0.49} \kappa_L^{0.79}}{\sigma^{0.5} L_0^{0.24} \eta_L^{0.29} \rho_G^{0.24}}, \tag{8.9a}$$

where ΔT_{sat} is the difference between the saturation temperature of the fluid and the temperature of the heated surface, and Δp_{sat} the difference between the vapor pressure of the fluid and its saturation pressure at the wall temperature; the subscripts L and G refer to the properties of the liquid and gas phases. This equation gives a non-linear dependence between the heat flux and temperature difference, and it holds quite well at pressures above atmospheric, but increasingly large deviations are seen at low boiling pressures for which no good universal correlations exist. There are no systematic measurements of boiling heat transfer in helium at low pressures, and we shall therefore review here the rare indirect information which can be deduced from polarized targets.

The experimental heat transfer coefficient from propanediol to boiling ^{3}He can be deduced similarly from the measured liquid temperatures and the lattice temperatures determined from the DNP; using the data of the Refs. [9, 13] we find the boiling heat transfer coefficient between propanediol and ^{3}He (assuming linear relationship between heat flux and temperature difference)

$$h_{NB} = 3 \pm 0.3 \ \frac{W}{m^2 K} \quad (^3\text{He} - \text{propanediol at 0.5 K}). \tag{8.10}$$

Another measurement yields boiling heat transfer between the heated wall of the still of a dilution refrigerator [14] and the boiling dilute solution [15]:

$$h_{NB} = 50 \ \frac{W}{m^2 K} \quad \left(\text{dilute solution} - \text{Cu at 1 K}\right). \tag{8.11}$$

This much larger value is probably due to the larger temperature differences used in these measurements, compared with those between ^{3}He and propanediol.

Rather good data exist for the boiling heat transfer to ^{4}He. At low pressures the heat flux is related to the temperature difference and vapor pressure by [16]

$$\frac{\dot{Q}}{\sigma_t} = 31100 \left(\frac{p}{\text{Torr}}\right)^{1/2} \left(\frac{\Delta T}{K}\right)^{2.035} \frac{W}{m^2}, \tag{8.12}$$

whereas around 4.2 K the heat flux behaves linearly for $\Delta T < 0.2$ K

$$\frac{\dot{Q}}{\sigma_t} = 5000 \left(\frac{\Delta T}{K}\right) \frac{W}{m^2}, \tag{8.13}$$

and then increases faster. The heat flux at 0.1 Torr and $\Delta T = 0.1$ K is about 100 W/m² from Eq. 8.12; this corresponds to the usual operating conditions of polarized targets at 1 K temperature.

In the evaporation refrigerators the heat transfer is understood in the terms of pool nucleate boiling heat transfer, the resistance of which is in series with the Kapitza surface boundary resistance describing the heat transfer from the solid to the liquid phase. At very high heat fluxes the turbulent two-phase flow correlations should be used for determining the heat transfer between the surface of the solid and the coolant fluid. The dimensionless correlations are determined only for boiling at or above atmospheric pressure, and they cannot be extrapolated to low boiling pressures. It should be noted, however, that at high flow speeds the forced convection of the fluid begins to contribute to the heat transfer, and the effective heat transfer coefficient may be substantially higher than the pool boiling heat transfer coefficient and very much higher than the liquid-only single-phase heat transfer coefficient. Under the continuous-flow conditions it is unlikely that a transition from nucleate to film boiling occurs, but there is some evidence that at high heat fluxes a dryout of the target beads may take place. This happens when the heat flux is so high that the replenishment of the liquid film on the beads from the vapor-carried liquid mist is insufficient compared with the required rate of evaporation.

8.1.3 Convection of Dilute Solution

In dilution refrigerators the target is immersed in the dilute solution of ³He in ⁴He in the mixing chamber of the device, and the heat transfer is mainly limited by the Kapitza surface boundary resistance. The convectional heat transfer of the dilute solution is rather good and limits to a lesser extent the heat extraction out of the target. The convection eddy with volume flow rate $\dot{V}$ transfers the heat

$$\dot{Q} = \dot{V} \frac{C_p \Delta T}{V_m} \tag{8.14}$$

out of the target volume, where C_p is the molar specific heat and V_m the molar volume of the dilute solution, and ΔT is the difference between the average temperature of the fluid exiting the volume and that of the phase boundary where the cooling takes place. The molar quantities must be taken both per mole of total solution or per mole of ³He; we use here the latter.

The volume flow rate can be estimated from the hydrostatic pressure difference due to the density difference between the heated fluid inside the target and the cold fluid in the return path (assuming that dilution takes place also below the target):

$$\dot{V} = \frac{\Delta \rho_d g h}{Z \mu_d} = \frac{g h}{Z \mu_d} \frac{d\rho_d}{dT}\bigg|_{\mu_4} \Delta T, \tag{8.15}$$

where h is the height of the target, μ_d is the viscosity of the dilute solution, Z is the flow impedance of the target and the derivative is taken at constant osmotic pressure, which is the same as the constant chemical potential of the superfluid ^{4}He at such low temperature that the fountain pressure is insignificant. The derivative is obtained from the temperature derivative of the ^{3}He concentration X at constant osmotic pressure

$$\left.\frac{d\rho_d}{dT}\right|_{\mu_4} = \left.\frac{d\rho_d}{dX}\frac{dX}{dT}\right|_{\mu_4} \cong -0.58\rho_4 \left.\frac{dX}{dT}\right|_{\mu_4}, \tag{8.16}$$

where ρ_4 is the density of ^{4}He and the numeric factor comes from the atomic mass difference and molar volume difference between ^{3}He and ^{4}He:

$$\frac{d\rho_d}{dX} = \frac{m_4}{V_4}\frac{d}{dX}\frac{1+X(m_3-m_4)/m_4}{1+X(V_3-V_4)/V_4} \cong -0.58\rho_4. \tag{8.17}$$

Here small terms proportional to X have been dropped because $X < 0.2$ in the usual operating conditions below 0.5 K. The negative value means that the hot fluid is heavier than the cold fluid, and the heated fluid thus moves downwards rather than upwards, which would be the case for convection in the gaseous phase.

The flow impedance is obtained by considering the target as a porous material with hydrodynamic surface area equal to the geometric area of the beads of Eq. 8.1, and by applying the equation of Kozeny (see Eq. 8.78) which relates this to the flow resistivity z of the material

$$Z = \frac{h}{A}z = \frac{h}{A}5.0\left(\frac{\sigma_t}{V}\right)^2 \frac{\eta^2}{(1-\eta)^3} = \frac{h}{A}5.0\left(\frac{6\eta}{d_e}\right)^2 \frac{\eta^2}{(1-\eta)}. \tag{8.18}$$

Here A is the cross-sectional area of the target perpendicular to the convection eddy. Assuming for simplicity that the volume of the target is a vertical cylinder with height h and base with area A, the power per unit volume can now be related to the temperature difference by inserting the terms of Eqs. 8.15–8.18 to Eq. 8.14

$$\frac{\dot{Q}}{V} = \frac{\dot{Q}}{hA} \cong \frac{g}{5.0h\mu_d}\left(\frac{d_e}{6\eta}\right)^2 \frac{(1-\eta)^3}{\eta^2}\left(0.58\rho_4\right)\left.\frac{dX}{dT}\right|_{\mu_4}\frac{C_p}{V_m}(\Delta T)^2, \tag{8.19}$$

which yields the temperature difference at the temperature of 0.2 K of the phase boundary

$$\Delta T \cong 4.6\,\mathrm{mK}\left(\frac{\dot{Q}}{\mathrm{mW}}\right)^{\frac{1}{2}}\left(\frac{\mathrm{cm}^3}{hA}\right)^{\frac{1}{2}} \tag{8.20}$$

using the values $C_p = 16.47$ J/(mol·K), $V_m = V_{m,total}/X \approx 30/X$ cm^3/mol, $h = 5$ cm, $\eta = 0.6$ and $d_e = 0.2$ cm. As the normal power for DNP at 0.2 K temperature of the coolant is a fraction of 1 mW/cm^3, the coolant temperature within the target volume is at most 10 mK higher than that of the phase boundary. Convection is therefore a very efficient way of transporting the heat out of the volume of the porous media, if the input of the concentrated ^{3}He is placed under the target so as to cool the fluid in the lower part of the mixing chamber which will cause the cold and light dilute solution to rise up, thus closing the convection eddy. Moreover, in long targets the convection ensures good horizontal transport of heat from those areas where there may be lower input flow of the concentrated solution.

If the entry of the concentrated ^{3}He is placed in the upper part of the mixing chamber, the fluid becomes convectionally stable and heat transport from the target to the phase boundary will be only by thermal conduction, which is very poor. Although a thermometer at the phase boundary would show a relatively low temperature, the fluid in the target would be much hotter and therefore DNP would be unexpectedly low. The same result would be obtained if the convection is prevented by enclosing the target beads into a cell with too few or too small perforations. The best results are obtained by making the cell out of thin nylon mesh [12] and by distributing the ^{3}He inlet under it along the length of the target. It is also best to place a thermometer under the target close to the exit of the dilute stream, in order to detect possible problems with the heat transport by convection.

The temperature dependence of the convective heat transfer, Eq. 8.19, is rather weak and follows from that of the specific heat, viscosity (about 40×10^{-6} g/(cm·s) at 0.2 K) and the temperature derivative of the concentration at constant osmotic pressure. The latter is given in Table 8.1, where the values are calculated from Table 17 of Ref. [17]. We note that both of these parameters make the convective heat transfer better at higher temperatures.

The temperature difference solved from Eq. 8.19 is inversely proportional to the bead size d_e. On the other hand, the heat flux due to the Kapitza conductance, Eq. 8.2, is also proportional to the inverse of the bead size. The dependence of the target temperature T_t on the bead size can be therefore solved from

$$\frac{\dot{Q}}{V} = \frac{6\eta}{d_e}\alpha\left[T_t^4 - \left(T_m + \Delta T\right)^4\right] \qquad (8.20a)$$

if the mixing chamber temperature is known from the cooling power curve. This would, in principle, allow to optimize the bead size so as to get the lowest possible bead temperature for a given power input. The range of bead sizes so obtained is below what is used

Table 8.1 ^{3}He concentration X and osmotic temperature gradient of the concentration $dX/dT|_{\mu_4}$ at the solubility curve of dilute ^{3}He in ^{4}He.

T_m (K)	X	$-dX/dT$ (1/K)
0.05	0.06530	0.2600
0.10	0.07042	0.3675
0.20	0.09144	0.3832
0.30	0.12363	0.4634
0.40	0.16650	0.5868
0.50	0.22010	0.8608

commonly, about 1.5 mm to 2.0 mm, which is dictated by the fabrication techniques and by ease of handling. There is, however, room for some improvement if a special application would call for a target which operates at temperatures lower than is usual.

Because in the dilution refrigerator the Kapitza resistance is in series with a much smaller resistance due to convective heat transfer, whereas in evaporation heat transfer the Kapitza resistance is dominated by the much larger boiling heat resistance, the temperature of the target is always significantly lower in a dilution refrigerator for the same heat flux. This will be discussed after working out the cooling powers of the various cooling cycles used in the polarized targets.

8.1.4 Heat Transport by Subcooled ^{4}He and by Superfluid ^{4}He Film

The coherent scattering of slow polarized neutrons has been applied for the biological structure research using cold neutrons of 8.5 Å wavelength [18, 19]. Such neutrons have a very large absorption cross section by ^{3}He nuclei, and therefore the coolant must be eliminated in the beam path. This can be obtained by using ^{4}He as a coolant, but because it is desirable to get a fairly large deuteron polarization and to be able to run in frozen spin mode that enables selective saturation of the protons or the deuterons in the target, the only alternative is to use a dilution refrigerator that cools the target via a subcooled ^{4}He bath. The heat transport by superfluid ^{4}He is very good above 1.4 K but gets low below 0.6 K, where boundary scattering of phonons totally dominates heat transport. On the other hand, the boundary scattering is specular, and one may draw profit of this in the geometrical design of the subcooled fluid containment. Practical formulas for the heat transport in ^{4}He are given in Appendix A.5.3.

In the above experiments the heat sink of the subcooled ^{4}He bath was a dilution refrigerator and the target was a square plate of 3 mm thickness having 17×17 mm^2 cross section.

The application of subcooled ^{4}He bath in high-intensity beams was studied by the Michigan group; in their case the heat sink was a ^{3}He evaporation refrigerator [20]. The preliminary test results did not encourage pursuing further; the team failed to consider the heat transport within the restricted geometry of the target that consisted crushed solid ammonia fragments of 1.5 mm mean size.

One of the potential advantages of immersing a target in a subcooled ^{4}He bath is that the density of the coolant is independent of the beam heating and microwave power for DNP. This reduces the possible false asymmetry due to correlations between the target thickness and the target and beam polarization.

In very thin targets it may be desirable to reduce the amount of coolant to minimum by using the evaporation of superfluid film. The film has the thickness of 20 nm to 30 nm and it replenishes at the volume flow speed of

$$\dot{V} = 7.5 \cdot 10^{-9} \frac{\mathrm{m}^3}{\mathrm{s}} \times \frac{P}{\mathrm{m}},$$

where P is the smallest perimeter available for the film flow. This yields the molar flow of

$$\dot{n}_4 = \frac{\dot{V}}{V_4} = 2.8 \cdot 10^{-6} \frac{\text{mol}}{\text{s}} \times \frac{P}{\text{cm}}$$

and leads to an approximation for the maximum cooling power

$$\dot{Q}_{max} \cong \dot{n}_4 L_4^0 = 190 \cdot 10^{-6} \text{W} \times \frac{P}{\text{cm}}.$$

As DNP is possible at a microwave power of a fraction of mW/cm^3, cooling of a thin target by evaporation of a superfluid film appears quite possible around 1 K temperature.

8.2 Evaporation Refrigerators

8.2.1 Cooling Cycle

Figure 8.1 shows the schematic diagram of a continuous-flow ^{4}He refrigerator. Liquid helium is drawn from a transport dewar through a low-loss vacuum-isolated transfer line to a phase separator, from which liquid at 4.2 K is brought via a heat exchanger to a Joule–Thomson (J-T) expansion device which is an adjustable needle valve. The orifice of the valve is typically 0.5 mm in diameter. The two-phase fluid with low quality factor is blown directly into the cavity filled with target beads. The cold vapor cools the exchanger upstream of the J-T valve.

In the ^{3}He refrigerator of Figure 8.2 the continuous-flow ^{4}He refrigerator cools the condenser, from which ^{3}He liquid is led to a subcooling exchanger and a J-T valve. The

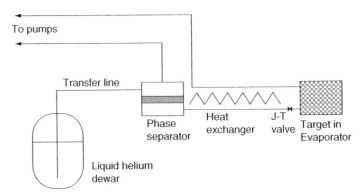

Figure 8.1 Schematic diagram of a continuous-flow open-cycle ^{4}He refrigerator. Liquid helium is siphoned from a transport or storage dewar along a vacuum-isolated transfer line into a liquid-gas phase separator that works by two-phase flow laminarization using sintered sponge materials. The liquid is then subcooled in a heat exchanger before expansion through a needle valve into the target cavity. The expanded fluid undergoes flow boiling while traversing and cooling the target beads, and the boiloff vapor passes through the subcooling heat exchanger and possible other heat exchangers that absorb heat losses in support structures and shields. The details of the exchangers for the latter are not shown

8.2 Evaporation Refrigerators

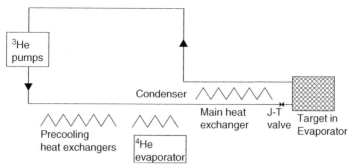

Figure 8.2 Schematic diagram of a continuous-flow closed-cycle ^{3}He refrigerator, in which ^{3}He from a hermetic pump and purifier system is condensed after passing through precooling heat exchangers; the condenser is cooled by the evaporator of an open-cycle ^{4}He refrigerator shown in Figure 8.1. The main heat exchanger then subcools the ^{3}He using the cold vapor flow out of the target cavity, before expansion through a needle valve into the cavity. The expanded two-phase fluid undergoes flow boiling while traversing and cooling the target beads. The low-pressure ^{3}He vapor then returns to the hermetic pump system for recompression and purification

two-phase ^{3}He flows through the target beads in the microwave cavity where it is evaporated, and the cold low-pressure vapor returns through the subcooling exchanger. The valve has an orifice of about 0.5 mm which allows molar flow rates in the range up to 20 mmol/s to be obtained.

The enthalpy-pressure diagram is often used for evaluating the J-T cycle. It helps to visualize the ^{3}He that is cooled isobarically in a series of exchangers and in the condenser down to the J-T valve; the isobaric cooling is represented by a horizontal line. At the valve the expansion is isenthalpic, represented by a vertical line which enters from the subcooled liquid region into the two-phase region; this yields the quality factor of the expanded fluid. Boiling in the two-phase region absorbs heat from the target isothermally, and the enthalpy of the cold vapor can be seen to be sufficient for subcooling the liquid because its specific heat is larger than that of the liquid. The cold low-pressure vapor cools the refrigerator structures, the waveguide and the coaxial lines before returning the pump and purifier system. The compressed purified gas re-enters the refrigerator via precooling heat exchangers which use the cold vapor of ^{4}He. The ^{3}He cycle is thus closed, which preserves the rather expensive fluid.

8.2.2 Cooling Power of Evaporation Refrigerators

The maximum cooling power of an evaporation refrigerator based on Joule–Thomson (J-T) cooling cycle is

$$\dot{Q}_{max} = \dot{n}_{max} L^0 \tag{8.21}$$

where $\dot{n}_{max}$ is the maximum mass (or molar) speed of the pump, and L^0 is the latent heat of evaporation. This ignores possible small effects due to expansion into the two-phase region and flash evaporation. The upper limit of helium mass flow is quite low for diffusion pumps (below 1 mmol/s for oil diffusion pumps and a few mmol/s for oil vapor boosters) and therefore polarized targets are usually operated using mechanical Root's blowers which have an almost constant volume speed in the pressure range 10^{-2} mbar to 10 mbar. The Root's blowers have a maximum mass flow rate which depends on the power of the electrical motor, on the capacity of the next stage pumps and on the cooling of the pump and the gas. These pumps are available in hermetic versions with speeds from 100 m^3/h to over 10^4 m^3/h, with maximum mass flows in the range of 500 mmol/s for the gas-cooled versions. The versions with a shaft seal are capable of even larger mass flow rates. The maximum cooling power can thus reach some 40 W with ^{4}He around 1.3 K, and 15 W with ^{3}He around 0.7 K. These numbers are similar to the output of high-power microwave tubes.

In evaporation refrigerators the cooling power becomes limited at lower temperatures by the drop of the vapor pressure p which follows in the vicinity of a reference temperature T_0 the exponential law

$$\frac{p}{p_0} = \exp\left[-\frac{L^0}{N_A k}\left(\frac{1}{T} - \frac{1}{T_0}\right)\right], \tag{8.22}$$

where p_0 is the vapor pressure at T_0. Assuming a constant volume speed $\dot{V}_p$ of the pump and a pipe connecting to the target with a negligible pressure drop so that the mass flow can be related to the volume speed by the ideal gas law

$$\dot{n} = \frac{p\dot{V}_p}{RT_a}, \tag{8.23}$$

where $R = N_A k$ is the gas constant and T_a is the ambient temperature, the cooling power in the evaporator becomes

$$\dot{Q} = L^0 \frac{p\dot{V}_p}{RT_a} = L^0 \frac{p_0 \dot{V}_p}{RT_a} \exp\left[-\frac{L^0}{k}\left(\frac{1}{T} - \frac{1}{T_0}\right)\right]. \tag{8.24}$$

At low pressures the frictional pressure drop in the pipe between the refrigerator and the pump, however, cannot be neglected. The pipe is usually dimensioned so as to obtain laminar flow and has a constant flow impedance per unit length which is for a circular pipe of length L and diameter D

$$\frac{Z}{L} = \frac{128}{\pi D^4}. \tag{8.25}$$

This causes a pressure gradient

$$\frac{dp(z)}{dz} = \frac{Z}{L}\mu \dot{V}(z) = \frac{Z}{L}\mu \frac{\dot{n}'RT_a}{p(z)}, \tag{8.26}$$

8.2 Evaporation Refrigerators

where μ is the viscosity of the gas. By integrating Eq. 8.26 we obtain the relation between the pressure at the evaporator p and the pressure at the pump inlet p_p

$$p^2 = p_p^2 + 2Z\mu \frac{\dot{n}RT_a}{L}. \tag{8.27}$$

The mass flow can now be solved by recalling that the pressure at the pump inlet is

$$p_p = \frac{\dot{n}RT_a}{\dot{V}_p}; \tag{8.28}$$

this yields

$$\frac{\dot{n}'}{\dot{n}} + \sqrt{1 + \left(\frac{\Delta p}{p}\right)^2} - \frac{\Delta p}{p}, \tag{8.29}$$

where

$$\Delta p = Z\mu \dot{V}_p; \tag{8.30}$$

and $\dot{n}$ is given by Eq. 8.23.

The expression for the cooling power becomes now

$$\dot{Q} = L^0 \frac{p\dot{V}_p}{RT_a}\left[\sqrt{1 + \frac{\Delta p^2}{p}} - \frac{\Delta p}{p}\right], \tag{8.31}$$

where the pressure p of the evaporator is related to the fluid temperature by Eq. 8.22.

The real cooling power is further reduced by gas dynamic effects due to the heating of the vapor from the evaporator to the room temperature, by the thermomolecular effect, by flash evaporation and by heat exchanger inefficiency, among others. The superfluid film creep limits downwards the temperature range of ^{4}He refrigerators to about 0.8 K, depending on the perimeter which limits the flow rate due to the creep. These can, however, be minimized by careful design of the geometry of the evaporator and of the pump line, as well as of the heat exchanger.

Figure 8.3 shows the cooling power of a ^{4}He evaporation refrigerator given by Eq. 8.31, with a constant volume speed of 2000 m^3/h and a pump line of 3 m length between the refrigerator and the pump inlet, for 3 different diameters of the pumping line.

8.2.3 ^{3}He-^{4}He Evaporation Refrigerators

In 1983 the U. Michigan group discovered [21] that the replacement of pure ^{3}He in an evaporation refrigerator by a mixture of ^{3}He and ^{4}He enabled their polarized ammonia target to tolerate a three times higher proton beam flux in their elastic scattering experiment. After a detailed analysis of the mechanisms of heat transfer in this complicated

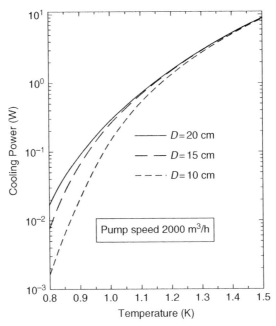

Figure 8.3 Cooling power of a ^{4}He evaporation refrigerator using a pump with constant volume speed of 2000 m^3/h connected to the refrigerator with a pumping line of 3 m length having 3 different diameters. It is assumed that the subcooling heat exchanger has 100% efficiency, so that expansion in the needle valve yields single-phase liquid into the target cavity; this is not always the case and therefore the real cooling power may be 10% to 20% lower than shown here

case of two-phase boiling, it was concluded that the osmotic phenomena together with the higher dryout (or burnout) heat flux were the leading benefactors [22]. Further tests with the intense polarized proton beam of Brookhaven AGS confirmed the analysis and enabled to determine the optimum beam intensity of slightly less than 5×10^{10} protons per pulse [23]; the beam pulse length was 1 s and repetition cycle length 3 s. At this intensity the beam heating dominated locally the microwave heating due to DNP, which lets conclude that, beyond the optimum beam intensity, the dryout of the target beads was located at the peak of the beam flux.

In these tests the molar fraction of ^{3}He in ^{4}He was 0.4 in the total amount of coolant loaded in the closed circuit, and in the recirculated gas the molar fraction was 0.6; these lead to suppose that the evaporation of ^{4}He contributed locally to the cooling of the beads, that the superfluid ^{4}He film contributed to the liquid film replenishment and that osmotic phenomena contributed to the distribution of the liquid phase of the coolant.

Similar tests were also made in the 1.5 GeV electron beam of Bonn synchrotron using a deuterated polarized target in a beam current up to 35 nA ($\approx 22 \times 10^{10}$ s^{-1}) rastered over the 3 cm^2 front face of the target [24]. The beam duty cycle was 5%. A clear improvement was observed in the DNP of deuterons after the replacement of the pure ^{3}He coolant by a mixture with ^{3}He molar fraction of 0.6, although the effect was not as marked as that with the proton beam of AGS. It was concluded that the electron beam heating effect in the helium bath temperature was negligible, suggesting that the loss of DNP in deuterated ammonia was due to the local heating of the solid material. Furthermore, the local loss of polarization may be substantially larger than that measured by the NMR coil that integrates the polarization over the volume of the target.

8.3 Dilution Refrigerators

8.3.1 Cooling Cycle

At low temperatures liquid solutions of ^{3}He and ^{4}He undergo a spontaneous separation into two phases with different concentrations which depend on temperature as shown by the solubility lines in the phase diagram of Appendix A.5.1. The phase which is concentrated in ^{3}He is lighter and floats on top of the heavier dilute phase. The principle of dilution refrigeration is based on the absorption of energy upon passing ^{3}He atoms from the concentrated ^{3}He-^{4}He solution into the dilute ^{3}He-^{4}He solution.

The solubility in the dilute phase at $T = 0$ is about 0.064 [17].[3] The fact that the solubility remains finite makes it possible to obtain continuous cooling in a wide range of temperatures. This is done by connecting the dilute phase of the cooling chamber (mixing chamber) to a distillation device (still) where ^{3}He is evaporated and then recirculated back to the concentrated phase of the mixing chamber. This requires external pumps for continuous recirculation. In another scheme, continuous dilution is obtained by circulating ^{4}He rather than ^{3}He; this is done by using a superleak through which superfluid only can be injected into the dilute phase. In the other end of the superleak heat must be removed using a continuous-flow ^{3}He refrigerator because the ^{4}He is separated from the ^{3}He using the heat-flush effect which requires a heat sink at a temperature well below 1 K.

It has turned out that the first scheme is more practical for obtaining a large cooling power in a wide range of temperatures. Although it requires sophisticated heat exchangers for a high cooling power, this is well compensated by the fact that the additional ^{3}He refrigerator can be eliminated.

Figure 8.4 shows the schematic diagram of the dilution refrigerator operating with the principle of the first scheme. Gaseous ^{3}He enters in the refrigerator into the precooling exchangers which use the returning cold ^{4}He vapor from the continuous-flow ^{4}He refrigerator. The ^{3}He is condensed in the evaporator of this refrigerator in the same way as in the ^{3}He evaporation refrigerator; the condenser is usually a capillary pipe which runs through the ^{4}He evaporator. The condensed fluid is further subcooled in another similar exchanger in the still and then passes through an expansion device without entering into the two-phase (gas and liquid) region. The expansion device can be a small-diameter capillary pipe in small refrigerators, whereas in larger ones it is an adjustable needle valve.

---

[3] Experimental values vary from 0.064 to 0.067. For consistency, we use here the value 0.064 adopted by Radebaugh [17] although 0.066 is currently favored [25].

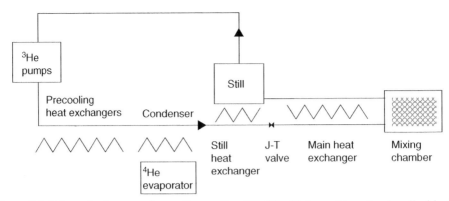

Figure 8.4 Schematic diagram of a continuous-flow ^{3}He-^{4}He dilution refrigerator described in the text. The cycle down to the condenser is the same as that of the ^{3}He evaporation refrigerator, but after condensation the concentrated ^{3}He is further subcooled in the still heat exchanger, before expansion into the main heat exchanger without boiloff. The concentrated fluid is further subcooled in the main heat exchanger, before entering under the target located in the mixing chamber. The heat load of the target is cooled by the convection of the dilute phase that transports the heat to the boundary of the concentrated and dilute phases; this boundary is partly trapped below the target and partly surrounding the droplets of the concentrated phase that underflows the traps. ^{3}He in the dilute phase then diffuses along the dilute fluid column into the still, where ^{3}He gas is distilled out of the dilute solution before returning to the pumps via precooling heat exchangers (not shown for simplicity)

The ^{3}He then passes through the main heat exchanger and enters into the mixing chamber where the target beads are in the dilute solution of ^{3}He in ^{4}He. Cooling there takes place as the ^{3}He passes the phase boundary from the concentrated into the dilute phase. ^{3}He then moves along the column of the dilute phase through the main heat exchanger into the still, where it is distilled out of the solution. The distillation is based simply on the fact that the vapor pressure of ^{3}He in the solution is much higher than that of ^{4}He. The vapor is pumped out, compressed to a pressure determined by the condenser, purified and sent back to the refrigerator.

Because of the large mass flows required for high cooling power, it is tempting to use the enthalpy of the cold vapor of the still for absorbing some of the heat from the incoming ^{3}He, thus lowering the need of liquid ^{4}He. This, however, requires very large heat exchangers, and there is evidence that the pressure drop in them may actually deteriorate the cooling performance in the mixing chamber. The origin of this is not fully understood, but it is clear that because the still then operates at a higher temperature, there will be more ^{4}He vapor in the circulated helium. Furthermore, the hot end of the main heat exchanger will then also operate at a higher temperature with a lower diffusion constant of ^{3}He in the column of dilute solution. This may reduce the amount of ^{3}He which can be circulated, and will be discussed at the end of the next subsection.

It is theoretically possible to operate the dilution refrigerator without a ^{4}He evaporator running at a low pressure. It has turned out, however, that this reduces the range of ^{3}He mass flows for stable operation and needs a very efficient heat exchanger between the low-pressure vapor of the still and the incoming ^{3}He. The pressure drop on the low-pressure side increases the still temperature and therefore also reduces the maximum cooling power available. The ^{4}He evaporator cooling the condenser should operate at a temperature below 2 K for the best performance of the dilution refrigerator in terms of the cooling power and temperature range.

The main heat exchanger is often bypassed by leading a pipe from the condenser directly into the mixing chamber; for clarity this is not shown in Figure 8.4. The bypass is used for cooling down the target from LN_2 temperature to helium temperature and is closed by a needle valve during normal operation.

Other details not shown in Figure 8.4 include the ^{4}He circuit and the exchangers for cooling the thermal screens and heat sinks for waveguides, coaxial cables and instrumentation leads in the refrigerator.

8.3.2 Cooling Power of the Dilution Refrigerator

The cooling power $\dot{Q}_m$ available in the mixing chamber (MC) at a flow rate $\dot{n}_3$ of the ^{3}He is

$$\dot{Q}_m(T_m) = \dot{n}_3 \left[H_\ell(T_m) - H_c(T_o) \right], \tag{8.32}$$

where $H_\ell(T_m)$ is the enthalpy of the saturated dilute solution at the mixing chamber temperature T_m and $H_c(T_o)$ is the enthalpy of the concentrated incoming ^{3}He at the temperature T_o of the outlet stream of the counterflow heat exchanger. In this equation it is assumed that the phase boundary between the dilute and concentrated phases of the ^{3}He is isothermal and its temperature T_m is equal to that of the dilute solution in the MC.

The maximum of this cooling power is found by requiring that its derivative with respect to $\dot{n}_3$ be zero, while keeping constant T_m, the still temperature T_s and the ^{4}He fraction x_4 of the concentrated solution, i.e.

$$\left(\frac{d\dot{Q}_m(T_m)}{d\dot{n}_3} \right)_{T_m, T_s, x_4} = 0, \tag{8.33}$$

The optimum ^{3}He flow $\dot{n}_{opt}$ which maximizes the cooling power is immediately obtained by applying this to Eq. 8.32 and solving for $\dot{n}_3$:

$$\dot{n}_{opt} = \frac{H_\ell(T_m) - H_c(T_o)}{C_c(T_o) \left(\frac{dT_o}{d\dot{n}_3} \right)_{T_m, T_s, x_4}}. \tag{8.34}$$

Here $C_c(T_o)$ is the specific heat of the concentrated stream fluid in the outlet of the heat exchanger, assumed to be at the same temperature as the mixing chamber inlet. This equation can be put to explicit form by finding T_o and its derivative with respect to $\dot{n}_3$; these depend on the performance of the heat exchanger and require the solution of the differential equation describing the heat transfer and flow friction in this device.

In the continuous-flow heat exchanger the incoming concentrated ^{3}He is cooled by heat transfer to the colder outcoming ^{3}He diffusing in the dilute solution towards the still. By denoting the longitudinal coordinate of the heat exchanger along the direction of flow by z, with $z = 0$ in the mixing chamber and $z = L$ in the still, and by $\sigma(z)$ the amount of effective heat exchange surface between 0 and z, the following equation relates the thermal gradient in the concentrated stream with the fluid and exchanger parameters at any position z:

$$\frac{dT_c(z)}{dz} = \underbrace{\frac{\alpha(T_c,T_d)}{\dot{n}_3 C_c(T_c)} \frac{d\sigma(z)}{dz}}_{\text{heat exchange}} - \underbrace{\frac{A_c(z)}{\dot{n}_3 C_c(T_c)} \left[\kappa_c(T_c) \frac{d^2 T_c(z)}{dz^2} + \frac{d\kappa_c(T_c)}{dT_c} \left(\frac{dT_c(z)}{dz} \right)^2 \right]}_{\text{axial heat conduction}}$$

$$- \underbrace{\frac{\eta_c(T_c) V_c^2 \dot{n}_3}{C_c(T_c)} \frac{dZ_c(z)}{dz}}_{\text{frictional heating}} \quad (8.35)$$

Here the subscripts c and d refer to the fluid properties in the concentrated and dilute streams, respectively; $C_c(T_c) = d(H_c(T_c))/dT_c$ is the specific heat of the helium mixture in the concentrated stream, and $\alpha(T_c,T_d)$ is the function of the heat transfer between the two streams, Eq. 8.2, in the low-temperature limit. The molar volume, heat conductivity and viscosity of the concentrated fluid are denoted by V_c, $\kappa_c(T_c)$ and $\eta_c(T_c)$; $A_c(z)$ is the cross-sectional area available for the flow of the concentrated fluid, and $Z_c(z)$ is the impedance of the concentrated flow channel between the points z and 0, the outlet of the concentrated stream.

The Eq. 8.35 is obtained by requiring energy balance in each infinitesimal volume $A_c(z)dz$ of the concentrated stream, and we note immediately that axial conduction and flow friction in the dilute stream and conduction in the mechanical structures are all neglected. These can be justified if T_d is so much lower than T_c that its effect on the heat transfer can be approximated by a small adjustment of the heat transfer constant, and the mechanical design is made so as to give negligible conduction in structural materials. The Eq. 8.36 thus deals with the optimum design of the concentrated stream, whereas additional requirements must be satisfied by the dilute stream and the mechanical structures, in order that these will not limit the performance based on the optimized design of the concentrated stream which is usually more critical.

It can be seen immediately that the heat exchange term in Eq. 8.35 should be large, and the terms describing the axial heat conduction and frictional heating in the concentrated stream should be everywhere as small as possible. If these terms can be made negligibly small compared with the heat transfer term, Eq. 8.35 can be truncated to

$$\frac{dT_c(z)}{dz} = \frac{\alpha(T_c,T_d)}{\dot{n}_3 C_c(T_c)} \frac{d\sigma(z)}{dz}, \quad (8.36)$$

which can be integrated from the outlet temperature $T_c(0) = T_o$ at $z = 0$ to the still temperature $T(L) = T_{still}$ at $z = L$, in order to obtain the relationship between T_o and $\dot{n}_3$. The integration can be made in closed form if the dependence of $\alpha(T_c,T_d)$ on T_d can be solved and

$T_d = f(T_c)$ can be written explicitly. This is possible, because at any position z of the heat exchanger the energy balance requires

$$H_c(T_c) = H_d(T_d) + \frac{\dot{Q}_m}{\dot{n}_3} \tag{8.37}$$

in the absence of axial conduction and viscous heating. When the dilution refrigerator runs at optimum power, it is easy to demonstrate [26, 27] that $T_d/T_c = \rho_T \approx 0.50$ everywhere in the heat exchanger, a condition which allows to easily obtain analytical solutions of Eq. 8.36.

8.3.3 Heat Transfer in Sintered Sponge Surfaces by Kapitza Conductance

The heat transfer parameter is determined by the thermal conductances of the helium fluids and their separating wall materials and, in particular, by the thermal boundary (Kapitza) conductance between the helium fluids and the solid walls. If the wall surface is extended by fine sintered powder, for example, the temperature drop due to the Kapitza resistance may become smaller than those due to other thermal resistances; this tends to be the case for high temperatures $T > 0.1$ K typically. At lower temperatures the Kapitza conductance often is the limiting factor, and the heat transfer between the concentrated and dilute streams obeys then

$$\Delta \dot{Q} = \Delta \sigma_c \beta_c \left(T_c^n - T_{Cu}^n \right) = \Delta \sigma_d \beta_d \left(T_{Cu}^n - T_d^n \right), \tag{8.38}$$

where $\beta_c(d)$ are the Kapitza conductances at the interfaces of the sintered powder and the concentrated (dilute) fluids, and $n \cong 4$ is approximately constant in a wide range of temperatures. The temperature of the separating wall T_{Cu} can be eliminated from Eq. 8.38 to get

$$\Delta \dot{Q} = \Delta \sigma_c \beta_c T_c^n \frac{1 - \rho_T^n}{1 + \rho_\beta} = \Delta \sigma_c S T_c^n. \tag{8.39}$$

This defines the parameter S which is approximately constant when the ratio

$$\rho_\beta = \frac{\Delta \sigma_c \beta_c}{\Delta \sigma_d \beta_d} \tag{8.40}$$

of the Kapitza conductances per unit length of the heat exchanger between the concentrated and dilute streams is made constant by design. The expression Eq. 8.40 is often also made small by adjusting the thickness of the sintered powder in the dilute stream greater than that in the concentrated stream; in our case, however, both are similar and we therefore have $\rho_\beta = \beta_c / \beta_d \cong 0.5$ owing to the lower Kapitza resistance in the dilute stream.

The fraction in the middle of Eq. 8.39 is thus approximately constant in the low temperature limit. At high temperature, however, the thermal conductivities of the sintered powder as well as those of the fluids filling the pores of the powder begin to limit the heat transfer, and this must be taken into account by replacing $\Delta \sigma$ by an effective exchange surface $\Delta \sigma_{eff}$ with a temperature dependence

$$\Delta\sigma_{\mathit{eff}} = \frac{\Delta\sigma}{\left(1+\dfrac{T}{T_\sigma}\right)^2}, \qquad (8.41)$$

where the adjustable parameter $T_\sigma \approx 0.25\,\mathrm{K}$ is obtained from a fit to data on exchangers with 1 mm sinter depth of 325 mesh Cu powder. This parameter can be also derived directly from the thermal conductivity data but with a reduced precision. Other sinter thicknesses and grain sizes will have a different temperature dependence for the effective surface area; this is discussed in Appendix A.5.4, together with the thermal penetration in the liquid filling the pores and in the sintered metal sponges.

We conclude that we have thus eliminated T_d from the expression of $\alpha(T_c, T_d)$ and may write

$$\alpha(T_c, T_d) \cong \alpha_c(T_c) = S \frac{T_c^n}{\left(1+\dfrac{T_c}{T_\sigma}\right)^2}, \qquad (8.42)$$

where

$$S = \beta_c (1-\rho_T^n)/(1+\rho_\beta) \qquad (8.43)$$

is approximately constant.

8.3.4 Maximum Cooling Power and Optimum Flow Rate

The truncated Eq. 8.36 can now be integrated to get

$$\sigma_c = \dot{n}_3 \int_{T_o}^{T_s} \frac{C_c(T_c)}{\alpha_c(T_c)} dT_c, \qquad (8.44)$$

where σ_c is the integrated surface area of the concentrated stream. The temperature dependence of the thermal penetration in the sintered powder is thus included in the surface conductance term $\alpha_c(T_c)$. The derivative required for Eq. 8.33 is now easily obtained by differentiating with respect to $\dot{n}_3$ while keeping all other variables but T_o constant and rearranging the terms to get

$$\frac{\sigma_c}{\dot{n}_3} = \dot{n}_3 \frac{C_c(T_o)}{\alpha_c(T_o)} \left(\frac{dT_o}{d\dot{n}_3}\right)_{T_m, T_s, x_4}. \qquad (8.45)$$

This gives

$$C_c(T_o)\left(\frac{dT_o}{d\dot{n}_3}\right)_{T_m, T_s, x_4} = \frac{\sigma_c \alpha_c(T_o)}{\dot{n}_3^2}, \qquad (8.46)$$

which can be inserted in Eq. 8.33 to yield the optimum flow rate of $^3\mathrm{He}$:

8.3 Dilution Refrigerators

$$\dot{n}_3^{opt} = \frac{\sigma \alpha(T_o)}{H_\ell(T_m) - H_c(T_o)} \quad (8.47)$$

and the maximum cooling power

$$\dot{Q}_m^{max}(T_m) = \sigma \alpha(T_o). \quad (8.48)$$

Eqs. 8.47 and 8.48 can be used for the design of the heat exchanger of a dilution refrigerator with given power specifications, once the outlet temperature T_o of the heat exchanger is known. This can be obtained by combining Eqs. 8.44 and 8.47 to get

$$\alpha(T_o) \int_{T_0}^{T_s} \frac{C_c(T_c)}{\alpha_c(T_c)} dT_c = H_\ell(T_m) - H_c(T_o) \quad (8.49)$$

whose solution yields the ratio T_o/T_m as a function of T_m which is universal for a given temperature dependence of the Kapitza resistance. The numeric solution of this equation [26, 27] shows that this ratio approaches an asymptotic value at low temperatures, where the enthalpies can be approximated by

$$H_\ell(T_m) = aT_m^2; \\ H_c(T_c) = bT_c^2, \quad (8.50)$$

and where the Kapitza conductance is, using the asymptotic form of Eq. 8.42,

$$\alpha_c(T_c) = ST_c^4. \quad (8.51)$$

Here a, b and S are constants. Inserting these into Eq. 8.49 yields, for $x_4 = 0.1$,

$$\frac{T_o}{T_m} \cong \sqrt{\frac{a}{2b}} \cong 1.9, \quad (8.52)$$

which gives the asymptotic expressions for the maximum cooling power and the corresponding optimum ³He flow rate:

$$\dot{Q}_m^{max} = \left(\frac{a}{2b}\right)^2 \sigma_c ST_m^4 \cong 12.5 \, \sigma_c ST_m^4; \quad (8.53)$$

$$\dot{n}_3^{opt} = \frac{a}{2b^2} \sigma_c ST_m^2 \cong 0.27 \sigma_c ST_m^2 \frac{K^2 \text{mol}}{J}. \quad (8.54)$$

At low temperatures the maximum cooling power is therefore proportional to T_m^4 and to the effective total surface area of the exchanger; the maximum power is obtained at an optimum flow which is proportional to T_m^2 and to the effective surface area.

At mixing chamber temperatures above 50 mK Eq. 8.49 must be solved numerically. This can be done most conveniently by using functions fit to the experimental data on

$H_\ell(T)$, $H_c(T)$ and $\alpha_c(T)$. The following relationships have been used in the design of several large refrigerators [12, 15, 27–33]:

$$H_\ell = \frac{aT^2}{1+T/T_a}, \qquad (8.55)$$

$$H_c = \frac{bT^2}{1+T/T_b} \qquad (8.56)$$

and

$$\alpha(T) = \frac{ST^4}{\left(1+T/T_\sigma\right)^2}, \qquad (8.57)$$

with the numerical values for the constants given in Table 8.2 for various ^{4}He concentrations in the concentrated stream.

Figure 8.5 shows the maximum cooling power and optimum flow rate of CERN Frozen Spin Target dilution refrigerator [27]. The above parameters were obtained by adjusting the curves so as to fit the data visually; the adjustment mainly concerns S that was defined by Eqs. 8.39 and 8.43; the value deduced from the fit corresponds to $R_K \sigma T^3 = 85$ cm^{2}K^4 W^{-1}. Several larger dilution refrigerators were consequently designed using these parameters and the design principles of dilute and concentrated flow channel to be described below; Figure 8.6 shows the cooling power of the largest one, that of the SMC Polarized Target [32]. The refrigerator will be described in more detail in Section 8.5.3.

8.3.5 Design of Dilute and Concentrated Stream Channels

Eq. 8.35 allows to design the flow passages for the concentrated and dilute streams so that the effects of the axial heat conduction and viscous heating are minimized when operating the refrigerator at optimum flow. For this we need the first and second derivatives of the temperature in the concentrated stream $T_c(z)$. These are readily obtained in the asymptotic low-temperature limit by combining Eqs. 8.53–8.54 with the truncated Eq. 8.36:

Table 8.2 Parameters fit to the experimental data of Ref. [17] on the enthalpies of the dilute and concentrated streams and to the effective Kapitza conductance of sintered 325 Mesh Cu powder of 40% filling factor and 1 mm depth [25], using the functions of Eqs. (8.55–8.57).

x_4	a (J mol^{-1}K^{-2})	b (J mol^{-1}K^{-2})	T_a (K)	T_b (K)	T_σ (K)	$\frac{1-\rho_T^n}{1+\rho_\beta}$
0.00	92.0	12.0	0.316	0.400	0.250	0.625
0.10	92.0	13.0	0.316	0.500	0.250	0.625
0.27	92.0	14.7	0.316	1.000	0.250	0.625

8.3 Dilution Refrigerators

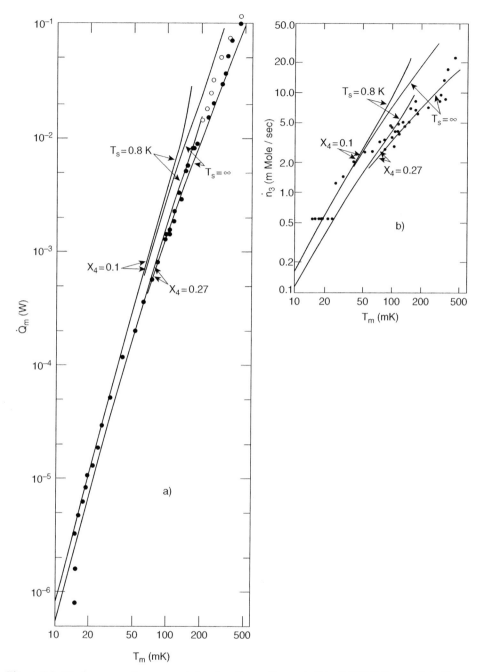

Figure 8.5 Maximum cooling power (a) and optimum flow rate (b) of CERN Frozen Spin Target dilution refrigerator. The solid lines are visual fits to Eq. 8.48 and Eq. 8.47, using the enthalpy and heat transfer functions of Eqs. 8.55 to 8.57. Reprinted from Ref. [26].

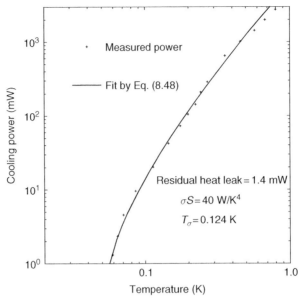

Figure 8.6 Maximum cooling power of SMC polarized target refrigerator that uses design principles similar to that of CERN Frozen Spin target, with the difference that the condenser runs close to 2 K temperature rather than 1.2 K

$$\frac{dT_c}{dz} \cong \frac{b}{a}\frac{T_c^3}{T_m^2}\frac{d\sigma}{\sigma dz}, \tag{8.58}$$

$$\frac{d^2T_c}{dz^2} \cong \frac{3b^2T_c^5}{a^2T_m^4}\left(\frac{d\sigma}{\sigma dz}\right)^2. \tag{8.59}$$

In Eq. 8.59 two small terms were eliminated which contribute exactly zero if the heat exchange surface per unit axial length is piecewise constant. Using these and in addition assuming that heat exchange surface is uniformly distributed along z, we can write the requirement that the second term on the right side of Eq. 8.35 is always much smaller than the first one:

$$\frac{\sigma}{L}ST_c^4 \gg A_c\frac{b^2}{a^2T_m^4L^2}\left(3\kappa_c T_c^5 + \frac{d\kappa_c}{dT_c}T_c^6\right) \tag{8.60}$$

or

$$A_c \ll \frac{4L\max\{\dot{Q}_m\}}{3T_c\kappa_c}\frac{1}{1+\frac{1}{3}\frac{T_c}{\kappa_c}\frac{d\kappa_c}{dT_c}}. \tag{8.61}$$

8.3 Dilution Refrigerators

The axial conduction is likely to be most important at the lowest temperature of operation. Under this condition the logarithmic derivative of the heat conductivity in Eq. 8.61 can be replaced by its maximum of 1, which yields a simple result

$$A_c \ll \frac{L\dot{Q}_m^{min}}{T_c \kappa_c(T_c)}, \tag{8.62}$$

where the maximum cooling power is replaced by the minimum heat leak to the mixing chamber.

In the high-temperature end of the heat exchanger there is usually less surface per unit length; under these conditions the requirement for the cross-sectional area can be written from Eqs. 8.60 to 8.61 as

$$A_c \ll \frac{\dot{Q}_m^{min}}{T_c \kappa_c(T_c)} \frac{\sigma}{d\sigma/dz}. \tag{8.63}$$

If the thermal penetration in the sintered powder is not complete in the upper end of the heat exchanger, the surface area should be the temperature-dependent effective area and two remarks must be made: (1) Eq. 8.63 is no longer exactly valid; (2) and, however, only a short section of the exchanger suffers from the possible disagreement with Eq. 8.63, because the thermal gradient at that point is steepest. In practice it is a good design principle to ensure that Eq. 8.63 is obeyed for the low-temperature ends of all sections of the heat exchanger with a constant axial distribution of the exchange surface.

For the dilute stream one can write equations similar to the above ones. We note that much larger cross sections are allowed and also required because the thermal gradient in the dilute stream is small everywhere except in the immediate vicinity of the still.

The effect of the viscous heating is also minimized by requiring that the last term in Eq. 8.35 be much smaller than the first one on the right side. This yields

$$\frac{dZ_c}{dz} \ll \frac{\alpha(T_c)}{\dot{n}_3^2 \eta V_c^2} \frac{d\sigma}{dz}, \tag{8.64}$$

which is clearly hardest to obey at the low-temperature end of the concentrated stream. There we have $T_c = T_0$; using Eqs. 8.51 and 8.54 we get the requirement

$$\frac{dZ_c}{dz} \ll \frac{b^2}{\sigma S \eta V_c^2} \frac{d\sigma}{\sigma dz}. \tag{8.65}$$

For a circular passage of diameter D of the concentrated stream in the outlet of the exchanger, we find

$$D^4 \gg \frac{128 \eta_c(T_0) V_c^2 S}{\pi b^2} \frac{\sigma^2}{d\sigma/dz} \tag{8.66}$$

and for a slit with width w and height t we get

$$wt^3 \gg \frac{12\eta_c(T_0)V_c^2 S}{b^2} \frac{\sigma^2}{d\sigma/dz}. \tag{8.67}$$

The viscous heating in the dilute stream has been debated in the past and it has been argued that because the osmotic pressure is the driving force for the quasiparticle 'gas', a loss in this pressure must be associated with isenthalpic expansion and therefore cooling or heating of the fluid depending on the slopes of the isotherms in the plot of osmotic pressure versus enthalpy. At low temperatures where viscous heating could be a problem, the quasiparticle fluid has the properties of a degenerate Fermi fluid with specific heat close to $C_{v,\mu4}$ = 107.2 (T/K) J/(mol·K) and isenthalpic expansion therefore results in heating which has been evaluated experimentally by Wheatley [34]. Above 0.2 K the specific heat approaches a constant value $(5/2)R$ and the quasiparticle fluid therefore behaves as an ideal gas; in this case an isenthalpic expansion causes no heating and may even result in cooling, as has been shown by Radebaugh [17]. At high temperatures and flow rates, furthermore, the concentration is high and viscosity low, and therefore the possible heating effects in isenthalpic expansion are reduced.

Flow resistance in the dilute stream causes a loss in the osmotic pressure which can be observed as a drop in the ^{3}He concentration in the vapor pumped out of the still. It has turned out in practice that losses of the osmotic pressure scale with the cross-sectional area of the dilute stream flow channel of a dilution refrigerator. This indicates that the dilute flow is limited by diffusion rather than by viscous effects, and that the drop tends to occur close to the still. In dilution refrigerators with maximum ^{3}He flows ranging from 2 mmol/s to 350 mmol/s, the measured value of the flow limit at high temperature obeys approximately

$$\max(\dot{n}_3) \leq 0.5 \frac{\text{mmol}}{\text{s}} \times \left(\frac{A_d}{\text{mm}^2}\right), \tag{8.68}$$

where A_d is the cross-sectional area in the dilute stream of the refrigerator. This rule, however, is applicable only to refrigerators operating at high flow rate and at relatively high temperature. The flow in the dilute stream at very low temperatures will be discussed at the end of this subsection.

Although there is no microscopic description of the origin of the diffusion, one may imagine a 'mutual friction' which limits the flow speed of ^{3}He quasiparticles in the superfluid ^{4}He. The concentration gradient under flow along the z-axis, and under isothermal conditions,[4] is given by

$$D\frac{\partial X}{\partial z} - Xv_n = 0, \tag{8.69}$$

---

[4] This allows one to ignore the concentration change due to the thermal gradient.

8.3 Dilution Refrigerators

where D is the mass diffusion coefficient of dilute ^{3}He in ^{4}He, and v_n is the drift velocity of the normal fluid in superfluid. The two-fluid model involves relatively little ^{4}He atoms in the normal fluid below 0.6 K if the concentration is above 1%. By expressing the drift velocity in terms of the mass flow rate of the ^{3}He in a channel of cross-sectional area A

$$v_n = \frac{\dot{n}_3 V_{3d}}{A} \cong \frac{\dot{n}_3 V_4}{AX}, \tag{8.70}$$

where the molar volume of the dilute ^{3}He is approximated by V^4/X, the concentration gradient becomes

$$\frac{\partial X}{\partial z} \cong \frac{\dot{n}_3 V_4}{AD}. \tag{8.71}$$

By inserting the critical design flux of ^{3}He from Eq. 8.68 to Eq. 8.71 one gets an idea of the order of magnitude of the diffusion constant under the critical operating conditions of the dilution refrigerators when maximum flow conditions are required. This yields the value $D = 1300$ cm^2/s for an estimated allowable concentration gradient of 10^{-3} cm^{-1}.

The diffusion coefficient can be estimated, on the other hand, from the measured thermal conductivity of the dilute solution by using the two-fluid model which relates the thermal gradient to the normal fluid drift velocity

$$\frac{\dot{Q}}{A} = S_{4L}^0 T \frac{v_n}{V_4^0} = -\kappa_d \frac{\partial T}{\partial z} \tag{8.72}$$

in a pipe with cross-sectional area A. Here the entropy S_{4L}^0 is that of ^{4}He in the dilute solution and is approximated by the entropy of pure ^{4}He. In dynamic equilibrium, on the other hand, the gradients of the concentration and temperature are related by requiring that the chemical potential of ^{4}He be constant, which leads to the requirement that the osmotic pressure be constant, because the contribution of the fountain pressure is small below 1 K for concentrations above 1%. This can be written as

$$RT \frac{\partial X}{\partial z} = -S_{4L}^0 \frac{\partial T}{\partial z}. \tag{8.73}$$

Solving now the drift velocity from Eq. 8.72 and the temperature gradient from Eq. 8.73, and inserting these to Eq. 8.69, yields the diffusion constant

$$D = \kappa_d \frac{V_4^0 RX}{\left(S_4^0\right)^2}. \tag{8.74}$$

This relation should be approximately valid in the temperature region below 1 K with concentrations for which the ^{4}He phonon contribution to the thermal conduction remains smaller than that due to the internal superfluid convection. The diffusion constant is calculated in Table 8.3 from the experimental conductivity data of Ref. [35] and from the

Table 8.3 Entropy of pure ^{4}He, and the thermal conductivity [35] and diffusion constant of two dilute solutions. The diffusion constant is estimated using the Eq. 8.74 and is expected to be of the right order of magnitude at and above 0.4 K temperature, but is known not to diverge at lower temperatures as shown in the table (see text).

T	S_4^0	X = 0.013		X = 0.050	
		κ_d	D	κ_d	D
(K)	(10^{-3} J/Mol/K)	(W/m/K)	(cm^2/s)	(W/m/K)	(cm^2/s)
0.1	2.723×10^{-5}	0.40	1.6×10^7	0.051	7.9×10^6
0.2	2.179×10^{-4}	0.31	194000	0.065	157000
0.3	7.353×10^{-4}	0.23	12600	0.060	12700
0.4	1.743×10^{-3}	0.21	2060	0.060	2260
0.5	3.404×10^{-3}	0.21	540	0.058	573
0.6	5.881×10^{-3}	0.20	172	0.055	182
0.7	9.725×10^{-3}	0.20	63	0.050	61
0.8	0.01724	0.19	19.0	0.048	18.5

entropy compiled in Ref. [17]. The values for the diffusion constant must be considered as upper limits because the internal convection is certainly not the only mechanism of heat transport contributing to the experimental thermal conductivity. Furthermore, the internal convection becomes small below 0.6 K and therefore the divergence of the values in Table 8.3 at low temperatures is an artefact of this method of determining the diffusion constant.

We note that the diffusion constant reaches the critical value of 1300 cm^2/s at about 0.45 K which corresponds quite well with the observation that the maximum flow rate becomes saturated at a mixing chamber temperature of about 0.5 K. Higher flow rates thus require a larger cross-sectional area of the dilute stream, or a shorter heat exchanger. At lower temperatures the optimum flow rate is limited by the heat exchanger surface area, and the diffusion constant becomes rapidly larger so that the concentration in the still can be expected to correspond to nearly the same osmotic pressure as that of the mixing chamber.

Below 0.2 K, however, the two-fluid model cannot be used for relating the diffusion constant and the thermal conductivity, because these are determined increasingly by different mechanisms. Furthermore, at temperatures where the phonon density becomes negligible, other mechanisms than simple diffusion begin to limit the flow speed of ^{3}He. The first mechanism which was suggested involved the viscous flow of the quasiparticle gas, but it was discovered in several laboratories that measurements of viscosity by different methods did not agree among themselves and were not in agreement with other related parameters either, such as the viscous heating and the viscous pressure drop. The dilemma has been experimentally and theoretically studied by de Waele et al. [25, 36, 37], who found that below 0.2 K the flow of ^{3}He generates a ^{4}He vortex tangle which leads to the observed mutual friction between ^{3}He and ^{4}He, and that the vortex tangle is strongly pinned to the walls.

The experimental observations of de Waele et al. below 150 mK can be summarized as follows:

(1) The mechanical pressure drop across a flow channel is consistent with zero at all flow rates and temperatures, in strong variance with the viscous flow model predictions (laminar Poiseuille flow).
(2) The concentration difference between the two ends of a pipe is proportional to its length L and depends on its cross-sectional area A and on the flow rate by

$$x_1 - x_2 = (3.2 \pm 0.3) \cdot 10^{-8} \left(\frac{L}{m}\right) \left(\frac{\dot{n}_3}{A} \frac{m^2 s}{Mol}\right)^{2.8 \pm 0.4}. \tag{8.75}$$

(3) The temperature of the dilute solution in the outlet pipe of a mixing chamber is related to the concentration in the pipe by

$$T^2 - T_m^2 = \beta_3 (x - x_m), \tag{8.76}$$

where β_3 is independent of L, A and the flow rate, and varies slightly from $0.21 K^2$ at $T_m = 12$ mK to $0.19 K^2$ at $T_m = 70$ mK.

By inserting the high-temperature critical flow limit from Eq. 8.68 into 8.75 we get a concentration gradient of 1.2 m^{-1} which is too high by a factor of about 10. It is rather clear that the flow is limited by different mechanisms at temperatures well below and well above 0.2 K, but unfortunately there is no systematic study in the high-temperature regime comparable to that of de Waele et al.

The strong temperature dependence of the diffusion constant in Table 8.2 suggests that if a high cooling power and therefore a high mass flow are required, the volume speed of the ^{3}He pump is important because this lowers the still temperature and therefore the temperature of the hotter end of the heat exchanger. Direct evidence of this was obtained by preceding a pump with a nominal speed of 1000 m^3/h by one with a speed of 3000 m^3/h; this increased the maximum ^{3}He flow and cooling power by 20% to 50% in the temperature range 0.40 K to 0.55 K [27, 30].

8.3.6 Sintered Copper Heat Exchangers: Preparation and Evaluation

Tubular Heat Exchangers

The upper part of the main heat exchanger of the dilution refrigerator should have a short and wide dilute stream so that the osmotic pressure gradient in it remains reasonable. The usual length of this part of the heat exchanger is between 20 cm and 50 cm. The cross-sectional area of this part then determines the maximum mass flow, given by the empirical Eq. 8.68.

The upper part of this main heat exchanger houses the tubular heat exchanger. For flow rates below 1 mmol/s a simple tube-in-tube exchanger is sufficient. At higher mass flows a coiled tube should be mounted into the dilute channel. The heat exchange surface can, in principle, be extended by mounting several parallel tubes, but this has the risk of channelling the flow preferably into the hottest tube. It is therefore preferable to extend the surface area by adjusting the length of the coiled tube.

Kozeny's Law and Measurement of the Surface Area Effective for Heat Exchange

The heat exchange surface area of sintered powders is an important parameter, and therefore several methods have been used for evaluating the active surface of the sintered powder. The use of gas adsorption measures accurately the physical area, but this includes also almost closed voids that are dead ends for heat conduction. We have therefore used the gas flow method that measures the hydrodynamic surface area of the powder [38]; this is insensitive to closed voids and maps predominantly multiply connected flow channels that simulate the flow of conducted heat, both in the liquid filling the pores and in the metallic grains of the sinter.

The flow in such porous media is described by the Kozeny–Carman equation that can be written, for a sintered plug of cross-sectional area A and length L, in the form

$$\Delta p = 5.0 \frac{(\sigma/V)^2}{(1-\beta)^3} \mu \dot{V} \frac{L}{A}, \tag{8.77}$$

where the numeric factor 5.0 is empirical, β is the volume filling factor of the powder, σ is the total hydraulic surface area of the powder plug, $V = LA$ and μ is the viscosity of the fluid and $\dot{V}$ its volume flow rate. Powders are characterized by a distribution of grain sizes and shapes: the grain i has a volume $V_i = \frac{\pi}{6} D_i^3$ and surface area $A_i = \lambda_i \pi D_i^2$, where D_i is the diameter of a sphere of equal volume and λ_i is the form factor. With these we can define the surface-to-volume diameter D

$$D = \frac{\sum_i D_i^3}{\sum_i \lambda_i D_i^2}$$

which allows to relate the hydraulic surface area and the surface-to-volume diameter:

$$\frac{\sigma}{V} = \frac{\sum_i A_i}{(1/\beta)\sum_i V_i} = \frac{6\beta}{D}.$$

By inserting this into Eq. 8.77 we get

$$\Delta p = \frac{180}{D^2} \frac{\beta^2}{(1-\beta)^3} \mu \dot{V} \frac{L}{A}. \tag{8.78}$$

The 325 Mesh Cu powder has a maximum grain size of 44 μm and the size distribution extends down to fine dust. The grains are irregularly shaped with no predominance of form;

the surface-to-volume diameter is $D = 18$ μm [38] before sintering and changes little upon sintering if the powder is not compressed. A more recent study of a 325 Mesh Cu powder, sintered under similar conditions, yielded $D = 20.6$ μm [39]. In the same study a Cu powder of nominally 1 μm grain size yielded $D = 1.9$ μm.

Choice of Sintered Powder

Appendix A.5.4 describes a simple method to evaluate the thermal penetration in sintered metal contacts used in the heat exchangers of dilution refrigerators. Because the temperature range in frozen-spin operation is limited by the heat loads due to the beam, for example, we are here interested mainly in temperatures of the mixing chamber above 20 mK. In dilute solution the thermal penetration depth for a 325 Mesh Cu powder above 30 mK is numerically

$$\lambda_T^2 \cong \frac{D(1-\beta)}{6\beta} \frac{\kappa(T_1)}{\alpha T_1^3} = (67 \, \mu m)^2 \left(\frac{T_1}{K}\right)^{-3}, \tag{8.79}$$

assuming that the heat conductivity has a constant value of 25 mW/(Km) up to about 0.5 K temperature, for liquid concentrations near the solubility line. Thus, a sintered heat exchanger in the dilute solution at 0.5 K temperature has the thermal penetration mainly to the first 0.2 mm layer, which is why polarized target refrigerators usually feature sinter depths between 0.5 mm and 1 mm only. With yet finer powders the penetration depth varies roughly with the square root of surface-to-volume diameter, so that when using $D = 1.9$ μm powder, the thermal penetration depth is about one-third of the above value of 0.2 mm at 0.5 K.

In concentrated solutions the thermal conductivity and thermal penetration depth behave in the same way as in the dilute solution, but we may note that because the concentrated side of the heat exchanger is at a temperature 1.5 to 2 times that of the dilute side, the thermal penetration is even more severely restricting the active layer for heat exchange.

In our temperature range of interest the heat conduction in the metallic part of the sintered sponge may also limit the heat transport, but this is negligible with respect to that of the liquid conduction, unless very thick sintered layers are used.

Continuous Sintered Heat Exchangers

The axial conduction along the flow of the main heat exchanger can be limited by the design of the liquid streams as was described in Section 8.5.3. The metallic parts of the sintered Cu exchangers, however, are practically isothermal, and therefore they approximate poorly the continuous counterflow principle, in particular in operation at the lowest temperatures. A study was made on the numeric error made by ignoring the thermal conduction in the copper heat exchanger elements, and it was concluded that when the main exchanger consists of order 10 isothermal elements, the residual error is in the range of a few percent [40]. Even with two sintered elements only, if their design is such that the axial heat conduction in the fluids can be ignored and if there is a coaxial-tube countercurrent

8.4 Heat Load Evaluation and Choice of the Cooling Method

The operation of a polarized target in a high-intensity beam was discussed in Section 7.1.1 from the point of view of target material optimization. Here we shall focus mainly on the choice of the cooling method.

A simple rule-of-thumb lets estimate the heat deposit by a charged beam by the minimum ionization energy loss per target thickness, which is about $dE/dx = 2$ MeV/(g/cm^2). This gives a volumetric heat dissipation in the target

$$\frac{\dot{Q}}{V_t} = \Phi_b \rho_t \frac{dE_{min}}{dx}, \tag{8.80}$$

where Φ_b is the beam flux, ρ_t is the mean target density and $x = \rho z$. With a mean target density of 0.6 g/cm^3 and beam flux of 10^{10} /(s·cm^2), this yields roughly

$$\frac{\dot{Q}}{V_t} = 1.9 \frac{\text{mW}}{\text{cm}^3}, \tag{8.81}$$

which is of the same order of magnitude as the maximum power for DNP below 0.5 K. Therefore, operation in higher beam fluxes requires the use of evaporative cooling by ^{4}He.

The highest reported beam flux was on the solid polarized target of the SLAC E143 experiment [41], in which an electron beam of intensity 5×10^{11} s^{-1} was rastered over the 5 cm^2 cross section of the target, cooled by a powerful ^{4}He refrigerator operated close to 1 K. The beam heating was witnessed by raising the beam intensity from 2×10^{11} s^{-1} to 5×10^{11} s^{-1} which caused a relative loss of 15% of the proton polarization of the solid ammonia target [41]. The mean flux of the beam was 10^{11} s^{-1}·cm^{-2} which deposits at least 19 mW/cm^3 of heat in the target; this is about as much as the microwave power used for DNP at 1 K. By using Eq. 8.12 we may estimate that the beam flux caused about 0.1 K temperature rise of the target beads, thus explaining quantitatively the loss in the target polarization, which was about 60% to 75% during the high-intensity runs. It should be noted that these numbers include also the polarization losses due to radiation damage, which required annealing of the target material after each cumulated electron flux of about 10^{16} cm^{-2}.

On the other hand, frozen spin operation requires reduced thermal loads tolerable in the range 50 mK coolant temperature; this is relatively easy to obtain from the point of view of the dilution refrigerator. The Kapitza resistance between the target beads and the dilute solution, however, then sets an upper limit of charged beam operation

$$\frac{\dot{Q}}{V_t} = \frac{6\beta}{D} \alpha \left(T_t^4 - T_{helium}^4 \right). \tag{8.82}$$

Assuming a butanol target with 2 mm bead diameter and that the target temperature should not be raised by the beam above 0.1 K, the beam heating should be limited to 4.3 µW/cm³. This, in turn, sets the limit of minimum ionizing beam flux to

$$\Phi_b = \frac{\dot{Q}}{V_t}\left[\rho_t \frac{dE_{min}}{dx}\right]^{-1} = \frac{6\beta}{D}\alpha\left(T_t^4 - T_{helium}^4\right)\left[\rho_t \frac{dE_{min}}{dx}\right]^{-1} = 2.2 \cdot 10^7 \text{ s}^{-1}\text{cm}^{-2}. \quad (8.83)$$

We conclude that the beam heating in frozen spin operation begins to dominate polarization losses at a mean beam flux of about $10^7/(\text{s cm}^2)$, while in continuous DNP mode the dilution refrigerator can cool an additional heat load due to the beam flux of $10^{10}/(\text{s cm}^2)$. At a higher beam flux evaporative cooling by ⁴He and operation in 5 T magnetic field are recommendable.

The polarized muon beam M2 in the North Area of CERN SPS had an intensity of 4×10^7 muons in each 2.4 s long spills with a repetition rate of 14.4 s during the experimental runs of SMC. The flux was rather well confined to the 5 cm diameter of the SMC polarized target cells, with a broad maximum at the center. The maximum flux in the central part of the target was therefore about $10^6/(\text{cm}^2\text{s})$, and this did not cause any observable polarization loss during frozen spin operation.

Evaporative cooling by ³He can tolerate a slightly higher beam flux when compared with dilution refrigeration, at the cost of slightly lower maximum polarization. Still higher beam flux can be tolerated when adding ⁴He to the circuit, as was discussed in Section 8.2.3.

The heat load distribution and its variation with the microwave power and beam current influence the coolant density in the target volume:

- In all evaporative cooling methods the higher heat load entails higher quality factor (vapor fraction) of the coolant fluid, both spatially and in time.
- In dilution refrigerator the higher heat load leads to spatially slightly higher coolant density owing to osmotic pressure, and higher overall heat load leads to slightly lower coolant fluid density because of higher solubility of ³He in the dilute phase at a higher temperature.

These variations of the coolant density may lead to a systematic error in the spin asymmetry measurement, due to a possible correlation of the variation with the target spin orientation. The methods of estimating and avoiding the resulting false asymmetries will be discussed in Chapter 11.

8.5 Design of Precooling Heat Exchangers

The precooling heat exchangers reduce the heat load to the ⁴He reservoir at 4.2 K and to the condenser pot near 1 K. This heat load is obviously largest at the highest ³He flow rates, and therefore the heat exchangers must be dimensioned and designed for this flow rate. This implies calculation of the effectiveness and pressure drop of the heat exchangers.

8.5.1 Effectiveness of the Precooling Heat Exchangers

Before entering the still heat exchanger, the concentrated stream of the dilution refrigerator must be cooled down from RT to below 4K and then condensed in a heat exchanger immersed in a boiling bath of ^{4}He at a temperature below 2K. The design of this series of exchangers is in the domain of classical thermodynamic engineering, and it is based on the application of the *number of thermal units* (N_{TU}) method or *effectiveness* method, both of which use dimensionless numbers, such as the Nusselt number (Nu), Prandtl number (Pr), Reynolds number (Re) and so on, and empirical correlations of the dimensionless fluid properties and flow geometry. This method applies to all flow regimes and covers also situations where there is boiling and condensation on one or both sides of the heat exchanger.

The Reynolds number Re is

$$\mathrm{Re} = \frac{\rho u D_h}{\mu}, \quad (8.84)$$

where ρ is the fluid density, D_h is the hydraulic diameter of the channel (see below), u is the average fluid velocity and μ is the dynamic viscosity of the fluid. The Prandtl number depends only on the properties of the fluid:

$$\mathrm{Pr} = \frac{\mu c_p}{\kappa}, \quad (8.85)$$

where c_p is the isobaric specific heat of the fluid, and κ its thermal conductivity.

In the N_{TU} method and *effectiveness* method we first define the *heat capacity flow rates*

$$\begin{aligned}
\dot{C}_h &= \dot{n}_h C_h; \\
\dot{C}_c &= \dot{n}_c C_c; \\
\dot{C}_{min} &= \min(\dot{C}_h, \dot{C}_c); \\
\dot{C}_{max} &= \max(\dot{C}_h, \dot{C}_c); \\
R &= \frac{\dot{C}_{min}}{\dot{C}_{max}},
\end{aligned} \quad (8.86)$$

where the subscripts *h* and *c* refer to the hot and cold fluids, respectively; min and max refer to the smaller and greater of the two; *n* and *C* are the number of moles and the molar specific heat of the fluid concerned. The method is particularly suitable to cover flows of ideal gases such as helium isotopes and their mixtures, whose specific heat is rather constant in the temperature and pressure range of usual dilution refrigerator precoolers.

In counterflow heat exchangers, which have the advantage of using the cold fluid heat capacity in an optimum way and which we always use in dilution refrigerators, we then define the theoretical maximum rate of heat transfer between the two fluids and the real heat transfer rate $\dot{q}$:

8.5 Design of Precooling Heat Exchangers

$$\dot{q}_{max} = \dot{C}_{min}\left(T_{h,in} - T_{c,out}\right); \qquad (8.87)$$
$$\dot{q} = \dot{C}_h\left(T_{h,in} - T_{h,out}\right) = \dot{C}_c\left(T_{c,out} - T_{c,in}\right),$$

which define the effectiveness

$$\varepsilon = \frac{\dot{q}}{\dot{q}_{max}}. \qquad (8.88)$$

If we know the effectiveness and because $\dot{q}_{max}$ is known, we find the amount of heat transferred and the outlet temperature of the hot fluid. The effectiveness is a function

$$\varepsilon = f\left(N_{TU}, R\right), \qquad (8.89)$$

where N_{TU} is the number of transfer units

$$N_{TU} = \frac{UA}{\dot{C}_{min}} \qquad (8.90)$$

calculated from the heat transfer area A and the overall heat transfer coefficient U. These are obtained from the sum of the heat resistances

$$\frac{1}{UA} = \frac{1}{h_c A_c} + \frac{1}{h_h A_h} + R_{wall}, \qquad (8.91)$$

where R_{wall} is the thermal resistance of the separating pipe wall; for thin-walled coaxial tube exchangers this can be assumed to be negligible.

In a counterflow heat exchanger the effectiveness is

$$\varepsilon = \frac{1 - \exp\left[-N_{TU}(1-R)\right]}{1 - R\exp\left[-N_{TU}(1-R)\right]}, \qquad (8.92)$$

for $R = 1$ the effectiveness is

$$\varepsilon = \frac{N_{TU}}{1 + N_{TU}}. \qquad (8.93)$$

If there is boiling or condensation on one side of the heat exchanger, $R = 0$ whence

$$\varepsilon = 1 - \exp(-N_{TU}). \qquad (8.94)$$

The heat transfer coefficients h, on the other hand, are obtained from the Nusselt number Nu:

$$\text{Nu} = \frac{\text{Convective heat transfer}}{\text{Conductive heat transfer}} = \frac{hD_h}{\kappa} = f(\text{Re}, \text{Pr}), \qquad (8.95)$$

where h is the convective heat transfer coefficient of the flow, D_h is the hydraulic diameter of the flow channel, κ is the heat conductivity of the fluid and f is an empirical universal function that depends on the heat exchanger geometry. The hydraulic diameter is the diameter for a circular pipe, and for the annular space between two coaxial tubes it is $D_h = D_{outer} - D_{inner}$; for a more complex shape it is defined as

$$D_h = \frac{\text{Cross-sectional area of flow channel}}{\text{Wetted perimeter}/4} = \frac{4A_{flow}}{P}. \tag{8.96}$$

For laminar flow, we have simply Nu = 3.66 if the entrance and exit effects of the flow channel are insignificant. For turbulent flow with Re > 10000, the simplest correlation to obtain h is that of Dittus and Bölter:

$$\text{Nu} = \frac{hD_h}{\kappa} = 0.023\left(\frac{GD_h}{\mu}\right)^{0.8}\left(\frac{\mu c_p}{\kappa}\right)^n = 0.023\,\text{Re}^{0.8}\,\text{Pr}^n, \tag{8.97}$$

where G is the mass flow divided by A_{flow}, $n = 0.4$ for the cold fluid and $n = 0.33$ for the hot fluid, and Re is the dimensionless Reynolds number. Given that the Prandtl number for helium is practically constant Pr = 0.67 from 300 K to 4 K, and because the heat conductivity κ and viscosity μ vary as square root of T, the Nusselt number and h vary little in the entire range down to 4 K (see Appendix A.5.3). Unfortunately, the correlation is good only for high flow rates with Re > 10000. An improved agreement for 0.1 < Pr < 1 (helium gas has Pr = 0.67) covers a wider range of flows $3000 < \text{Re} < 10^6$; the correlation is obtained by Taler and Taler [42]

$$\text{Nu} = 0.02155\,\text{Re}^{0.8018}\,\text{Pr}^{0.7095}. \tag{8.98}$$

The heat transfer of heat exchangers designed for the cooling of shields and support structures is obtained also using the above Nusselt numbers.

The boiling heat transfer was discussed in Section 8.1.3 (see Eqs. 8.9–8.13). For the condensation of the recirculated ^{3}He there are no empirical nor theoretical correlations available; one should note that the low viscosity of the vapor and liquid phases lead easily to turbulence in both phases. We have resorted to using the Dittus–Bölter equation above and applying a generous safety factor: for the maximum flow of 30 mMol/s, for example, this leads to using a pipe of 2 mm inner diameter and 4 m length, wound into a coil in the ^{4}He evaporator bath.

8.5.2 Pressure Drop in the Heat Exchangers

The incoming ^{3}He, the shields and the thermal losses of support structures are cooled mainly by flow of ^{4}He in heat exchangers consisting of circular pipe channels. For heat exchange these channels should be as small as is practical, while allowing a flow rate that is sufficient to absorb the heat load. Therefore, it is important to evaluate the pressure drop in such pipework so that the designed maximum flow rate can be maintained. In general, the pressure drop and heat transfer are coupled problems and they often use the same dimensionless parameters.

In this chapter here we shall ignore the gas dynamic effects, i.e. pressure drop due to the acceleration in heated flow or deceleration in cooled flow. These effects are not negligible, but they are relatively small in the flow regimes common in polarized targets. Thus, the formulas below apply mainly to incompressible flow.

The pressure drop per unit length in single phase flow is determined from

$$\frac{p}{\Delta L} = 4 f_d \frac{\rho u^2}{D_h}, \tag{8.99}$$

where u is the mean fluid velocity and f_d is the Fanning friction factor:

$$f_d = \frac{16}{\text{Re}} \text{ for laminar flow Re} < 2000;$$

$$f_d = \frac{0.0791}{\text{Re}^{0.25}} \text{ for turbulent flow } 2000 < \text{Re} < 10000; \tag{8.100}$$

$$f_d = \left[1.737 \ln \left(\text{Re} \, f_d^{0.5} \right) - 0.396 \right]^{-2} \text{ for turbulent flow Re} > 10000.$$

In the last form the friction factor must be solved numerically.

8.5.3 Examples of Polarized Target Refrigerators

The first polarized targets were cooled with small horizontal ^{4}He evaporation refrigerators [43, 44] that were soon modified by adding a ^{3}He circuit [45, 46, 47], and a dilution refrigerator circuit [14]. These enable the cooling of targets up to 5 cm length. Slightly more sizable versions of these were then constructed to cool targets up to 20 cm length that was optimized for physics with hadron beams [15, 46, 47]. Most of these targets used iron core magnets with vertical field which imposed a horizontal geometry for the refrigerator. Here we shall describe in more detail the recent polarized targets using superconducting magnets; these targets were optimized for the studies of the nucleon spin structure using the deep inelastic scattering of intense electron and muon beams. In these targets the main field and polarization were axial with the beam, but also a transverse field and polarization were needed.

Virginia-Basel-SLAC Polarized Target

This target features a powerful ^{4}He evaporation refrigerator built into the vertical cryostat of a 5 T superconducting magnet. The field axis is horizontal and the magnet can be operated with the field axial or transverse to the intense polarized electron beam of SLAC [41]. The bore of the magnet is 20 cm in diameter with 100° angle of access about the solenoid axis; the split is 8 cm with angles of access 34° horizontally and 50° vertically. Figure 8.7 shows the side view of the target schematically.

378 Refrigeration

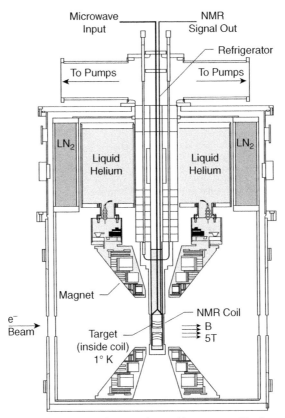

Figure 8.7 Side view of the Virginia-Basel-SLAC ^{4}He refrigerator used in the polarized target of the SLAC E143 experiment. The vacuum of the split-coil magnet and its ^{4}He cryostat is separated from the vacuum of the ^{4}He evaporation refrigerator, which simplifies the target loading and annealing operations. Reprinted from Ref. [41], with permission from Elsevier

The central target insert moves vertically along the central bore of the refrigerator permitting the positioning of four different targets into the beam: irradiated ND_3, irradiated NH_3, empty target and a dummy carbon or aluminium target. The vertical move can be done remotely without interrupting the cooling of the targets.

The refrigerator is pumped by a Root's blower cascade with a volume speed of about 12000 m^3/h, yielding 1.3 W cooling power at 1.06 K temperature. Liquid ^{4}He is transferred into the phase separator of the refrigerator from the LHe vessel of the magnet via a short vacuum-isolated line. The gas out of the separator cools the radiation baffles, waveguides and coaxial NMR lines inside the central bore, and the liquid flow to the nose is controlled by a needle valve.

The microwave guide terminates into a horn transition just above the targets; this and the NMR coaxial lines move vertically inside the insert together with the targets.

The refrigerator and magnet system is based on the 1 K/5 T tests of the Michigan group which made a breakthrough in the radiation-resistant polarized targets for experiments in

8.5 Design of Precooling Heat Exchangers

the intense proton beams of BNL [48]. A similar target is used at Bonn (with a 4 T magnet) [49] and in the Jefferson Lab [50].

SMC Double-Cell Polarized Target

The NA47 experiment of SMC in the polarized muon beam of CERN SPS used a 1.5 m long double-cell polarized target to measure very small asymmetries in the deep inelastic scattering off polarized protons and deuterons [12]. The target needed a high cooling power for the DNP and a low base temperature for frozen spin operation when reversing the target polarizations by rotating the axial field orientation and when operating in the transverse field mode. The target layout, quite similar to that of its predecessor in the NA2 experiment of EMC [28, 29], is shown schematically in Figure 8.8 [32, 33].

The target holder has two main parts: a stainless steel vacuum chamber with 0.1 mm thick beam windows and a lightweight plastic part confining the target material in two cells made of polyester net with 60% transparency for providing good convectional cooling of

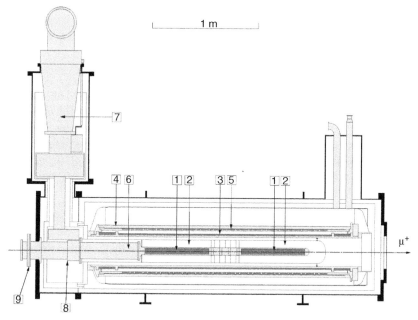

Figure 8.8 Side view of the Spin Muon Collaboration polarized target used in CERN NA47 experiment in 1993–1995. The muon beam enters along the target axis from the left and traverses the two target cells (1). The microwave cavity (2) is electrically split in two halves by isolating structures. The main solenoid coil (3) and its compensation coils (5) are coaxial to the targets. The dipole coil (4) is outside the compensation coils, and the main heat exchanger (6) and lower part of the still are on the axis of the target. The precooler heat exchangers are housed in the vertical tower (7). The target holder has a cold indium seal (8) and a warm seal at RT (9); 10 coaxial lines for NMR polarization measurement are housed in the target holder. These and the microwave guides are not shown for clarity. The magnet current leads and helium lines pass through the service tower of the magnet system on the right

the targets by dilute solution. The polyester and other plastic parts contribute about 1% to the proton NMR calibration signals and this can be corrected with a minor residual uncertainty. The 10 NMR lines traverse the vacuum chamber where heat sinks are provided for the thermalization of the lines and radiation baffles with thin Al windows for the beam.

The mixing chamber is made of 0.6 mm thick glass reinforced epoxy and is located, together with the main heat exchanger, coaxially inside the microwave cavity. The cavity, split in two compartments by the absorbing and reflecting structures of a microwave isolator, is hermetic with indium seals at both ends and is fed by two Ka-band guides with tapered slots for side coupling with the cavity compartments.

The still has a lower compartment coaxial to the main heat exchanger, communicating with the vertical compartment housing the electrical heater of 0.57 m² boiling surface to stay in the nucleate boiling regime at a flow of 500 mmol/s requiring 16 W of heating power. This is based on test data with smaller machines indicating that the critical heat flux to film boiling is not much higher than 3 mW/cm² on the still boiling surface.

The main heat exchanger has a higher-temperature part made of flattened stainless steel tubes and has a total inner surface of 0.1 m². This is in series with the lower-temperature part made of 12 sintered copper units arranged in two parallel streams, with 6 units in each, crossed at several points to avoid the formation of a cold plug due to the increasing viscosity of ^{3}He below 0.5K. The sintered elements consist of flat plates of P-doped Cu sheet of 0.2 mm thickness covered on both sides by 12×4 cm² layers of sintered 325 Mesh Cu powder with 18 μm surface-to-volume diameter; after sintering the plates were bent to the shape of the dilute channel and then electron beam welded together. The sintered layers have 0.5 mm deep grooves and an average thickness of 0.75 mm; this yields 375 g sintered sponge and the exchange surface area of 12 m² in both streams. The inner grooves are skewed to the direction of the flow, thus providing mixing of the flow of the concentrated fluid. The outer grooves are oriented along the direction of the flow in the dilute channel. The dilute stream is machined on a glass fibre epoxy spacer between the inner and outer shells of the main heat exchanger; the channel is 1.7 m long and has an average free cross section of 11.5 cm².

The measured cooling power, plotted in Figure 8.6, was fit by the expression

$$\dot{Q}_m(T_m) = -\dot{Q}_{\text{res}} + \frac{\gamma T_m^4}{\left(1 + \frac{2T_m}{T_\sigma}\right)^2}, \qquad (8.100a)$$

which is based on Eqs. 8.42, 8.48, 8.55 to 8.57; the fit yields $\gamma = 500$ W/K^4, $T_\sigma = 0.248$ K and the residual heat load = 1.4 mW. These indicate that either the effective heat exchange surface or Kapitza conductance of the sintered heat exchanger units are by a factor of 2 to 3 smaller than the values calculated for a $d = 18$ μm powder, but that the thermal penetration in its pores behaves as expected. These make us suspect that at low flow rate the convection makes the upflow parts of the dilute stream isothermal in the exchanger.

The dilution refrigerator is pumped by a cascade of 8 Root's blowers with a nominal volume speed of 13500 m³/h; the pumping line has a length of 20 m and diameter 320 mm. The line begins to limit the conductance at mass flow rates below 200 mmol/s.

The precooling requires from 15 l/h to 40 l/h of LHe between the minimum flow of the recirculated ³He of 30 mmol/s and maximum flow of 350 mmol/s during normal operation below 500 mK. Higher flow can be tolerated at the cost of a higher mixing chamber temperature; stable operation is possible up to 0.8 K temperature. The 27 l evaporator bath of the condenser is pumped with a Root's blower of 2000 m³/h nominal volume speed; the bath temperature is around 2 K. The condenser heat exchanger is made of Cu tubes with total inner surface of 0.35 m².

Prior to arriving into the condenser, ³He is precooled in a parallel-tube heat exchanger where ⁴He gas from the separator is counterstreaming. This heat exchanger is in contact with fin-type heat exchangers cooled by the outgoing low-pressure streams of the evaporator and the still.

8.6 Thermometry and Other Instrumentation

8.6.1 Thermometry

The calibration of the NMR polarization measurement circuit requires accurate temperature measurement around 1 K temperature; this temperature must be stable over extended periods of time, and it must be measured with a high absolute accuracy, as was described in Chapter 6.

To get a rough temperature dependence of the vapor pressure one may integrate the Clausius–Clapeyron equation

$$\frac{dp}{dT} = \frac{p}{kT^2} L^0$$

$$\text{to get} \quad \ln p = -\frac{L^0}{kT} + C. \tag{8.101}$$

Here L^0 is the latent heat of evaporation and C is a constant of integration. The formula is not very accurate, because the latent heat of evaporation depends also on the temperature. Therefore, one needs a precise empirical formula for the relation between the vapor pressure and temperature.

This is accomplished in the absolute temperature scale between 0.65 K and 3.2 K that is defined by the International Temperature Scale of 1990 (ITS-90) as the relation between the vapor pressure in Pa of ³He and the temperature T_{90} in K [51]:

$$\frac{T_{90}}{K} = A_0 + \sum_{i=1}^{9} A_i \left[\frac{\ln \frac{p}{Pa} - B}{C} \right]^i, \tag{8.102}$$

where the coefficients A_i, B and C are given in Appendix A5.2.

There are reported deviations of less than 0.2 mK of the thermodynamic temperature from ITS-90 above 1 K; these are lesser in the new scale PTB-2006 [52]. The vapor pressure of ^{3}He is not sensitive to high magnetic field, and the capacitive pressure gauges are stable and accurate when the gauge cell temperature is controlled in an oven.

The practical temperature scale uses platinum resistor thermometers for the interpolation between the fixed points of ITS-90. Resistance thermometers are the most commonly used devices for the monitoring and control of temperature, and calibrated devices are commercially available also for the range of dilution refrigeration below 1 K. When choosing the resistive temperature transducer (RTD), one should keep in mind its stability, radiation resistance and sensitivity to magnetic field. The nuclear transmutation-doped germanium RDTs are excellent in all of these, but their sensitivity to magnetic field limits their use to the transfer of calibration to other thermometers that are less sensitive to the field. These are ruthenium oxide, carbon composite and carbon glass RTDs. Cryogenic resistance thermometry has been reviewed in several papers recently [53–56].

The resistance thermometers suffer from self-heating at temperatures below 100 mK. This calls for the use of specialized readout instruments that are available commercially[5]; such instruments use self-balancing AC bridge techniques and their electromagnetic interference (EMI) control is compatible with the low signal levels. The AC technique avoids thermocouple effects in the measurement leads, and the AC frequency is subharmonic to 50 Hz or 60 Hz, thus minimizing interference due to the parasitic signals at the mains frequency and its harmonics. The self-balancing AC bridge technique has been extended to a multichannel system that reduces the cost per measurement channel [57].

In order to avoid the propagation of EMI signals to the resistance thermometers inside the cryogenic enclosure, it is tempting to use hermetic feedthroughs of the instrumentation wires equipped with RF filters. This is not, however, recommendable in polarized target instrumentation, because the RF signals of the NMR coils couple with the thermometry wiring in the target. The result is that the NMR coil impedance and tuning are affected by the presence of the magnetic materials in these filters, causing tune shift and tune drift due to temperature effects and due to external field shift during the NMR calibration process. These are particularly harmful for the accuracy of deuteron TE calibration signals, which is why their use must be avoided in polarized targets.

8.6.2 Pressure and Vacuum Measurement

Apart from the capacitive pressure gauges for thermometry, the cryogenic circuits of polarized targets need the measurement of the pressure of helium and vacuum in the various areas:

(1) boiling pressure in the still, in the ^{4}He vapor-liquid separator and in the evaporator;
(2) condensation pressure of ^{3}He;
(3) outlet and/or inlet pressures of the various heat exchangers;

---

[5] RV Elektroniikka PICOWATT.

(4) ^{3}He pump outlet pressure;
(5) vacuum measurement in the outer and inner vacuum enclosures.

All of the above can be covered with commercial gauges of hermetic construction, operating at room temperature. The items (1) to (4) are best measured with metal membrane devices with semiconductor strain gauges sensing the deformation due to inner pressure. In these gauges the ambient pressure may influence the reading, unless the reference pressure is vacuum.

The boiling pressures in the still and/or evaporators are best measured with thermal Pirani gauges based on the thermal losses of a heated wire sensing the molecular conduction due to the gas. A particularly useful concept is that of a Pirani wire operating at constant temperature, because of its fast response time. Unfortunately, the Pirani gauge must be calibrated for each gas separately. The calibration can be easily carried out by comparison with a McLeod gauge.

The vacuum measurement is useful at the time of the evacuation of the enclosures, because it enables the diagnostics of the presence of residual gases and possible large leaks into the vacuum circuits. During stable operation, however, it is best to turn off these gauges that are based on the ionization of the gas atoms or molecules. The ionizers break hydrocarbon molecules of the residual gas which entails generation of neutral hydrogen atoms at a low density. Because the wall recombination of these atoms is low due to the low density and small sticking probability, the atoms survive travel to the low-temperature parts of the insulation vacuum, thus causing a residual heat leak.

8.6.3 Mass Flow Measurement

Calorimetric mass flowmeters measure directly the mass flow of ^{3}He and ^{4}He and their mixtures, because their principle is based on the thermal mapping of a heated capillary pipe carrying a laminar flow of the gas. As a consequence, the sensor is calibrated for all gases having the same molar specific heat, such as the helium isotopes in a wide range of pressures. The range of each flowmeter is selected by the choice of the laminar flow element bypassing the measurement capillary.

Most commercially available mass flowmeters are designed for operation in high-pressure gas lines drawn from pressurized cylinders. Consequently, their pressure drop is too high for the helium circuits of polarized targets, where none of the gases are at a pressure above atmospheric. Fortunately, there is at least one supplier[6] that specializes in mass flowmeters with low pressure drop, with execution that can be made hermetic.

The mass flowmeters are particularly handy for the control of the counterflow heat exchanger described in Section 8.5.1. One mass flowmeter measures the flow of ^{3}He and gives a signal to a flow controller valve that adjusts the ^{4}He counterflow to the same or slightly higher value, measured with a second mass flowmeter.

---

[6] Teledyne Hastings Instruments.

8.6.4 Helium Level Gauges

Polarized target refrigerators work mostly with continuous flow of liquid helium from a transport dewar of liquefier dewar. The transfer line terminates in a liquid-gas separator, which does not need level control. The superconducting magnet may have its own helium circuit, and many of the magnets are cooled by a liquid bath with a level and flow control system adopted to the needs of cooling the coils and the current leads.

The ^{4}He evaporator in dilution refrigerators works in continuous refill mode that needs a liquid level measurement. Unfortunately, the commercially available superconducting wire level gauges don't work in a superfluid ^{4}He bath, so the only simple alternative is the use of heated resistance thermometers that sense the level based on the heat transfer difference between the gas and liquid phases. The level detection is improved by connecting the sensor to a bridge circuit that provides positive feedback to the power dissipated in the sensor when it is not submerged in the liquid phase.

Superconducting wire level gauges, however, can work also in superfluid bath, if the level measurement is based on the quench propagation in the wire covered with superfluid film. A pulsed measurement principle was tested in CERN Cryolab, based on a pulsed current source that caused the normal zone to propagate downwards in the gas phase until the liquid level was reached, leaving the length of the wire submerged in the liquid in the superconducting state. The method works well with 20 μm NbTi filaments extracted from a multifilament wire after removing the stabilizing Cu matrix by etching.

8.6.5 Residual Gas Analysis and Measurement of ^{4}He Content of Recirculated ^{3}He

The analysis of residual gases is useful if the cryogenic circuit of ^{3}He gets partially or completely blocked and needs remedy. A sample of the gases emitted by the blocked circuit during its warm-up is collected and analyzed using a quadrupole mass spectrometer. This will be discussed in better detail in Section 8.7.

During the initial testing phase of a dilution refrigerator, and also if problems arise during routine operation, it is useful to measure the quantity of ^{4}He in the recirculated ^{3}He. The quadrupole mass spectrometer can then be used for sampling the gas in the still pumpout line. This, in principle, is straightforward, but if high accuracy is required, one needs to calibrate the sensitivity of the sampling valve with each isotope, because ^{3}He traverses the valve more quickly owing to its lower atomic mass and therefore higher mean velocity. The calibration is best obtained by preparing a known mixture of the two isotopes, diluting it to a volume connected to the sampling valve and then measuring the two peaks of the spectrometer as a function of time; the extrapolation of the resulting peak area ratio back to time zero yields the sensitivity ratio in continuous measurement.

8.7 Helium Circulation Systems

Here we shall discuss the systematic elimination of impurities in cryogenic closed-cycle hermetic helium circuits involving oil-containing pumps and elastomer seals. Such circuits and pumps are usually made of metallic components joined and sealed with torical o-rings.

8.7 Helium Circulation Systems

The pumps may have canned electrical motors (Root's blowers) or rotating lip seals on shaft (rotary blade pumps). There are now also rotary blade pumps equipped with magnetic coupling through a hermetic wall.

Such circuits and pumps have been run uninterrupted for periods in excess of six months in high-energy physics experiments. Substantial experience with them has been gained in circuits with flow speeds up to 600 mmol/s. The careful analysis of many incidents has led to several systematic measures, with proven effectiveness against cryogenic circuit blockages.

These measures include the preparation and tests of the pumps, the installation of active purifiers and the analysis of the incidents in order to be able to eliminate or cure failing components. These will be briefly discussed in the subsections below.

8.7.1 Preparation of Pumps

Purge

The delivered pump oils always contain a large amount of evaporable impurities, which will be carried out of the pump by the helium stream. These impurities will saturate the purifiers if no measure is taken before initial startup after installation or oil change. The analysis of these impurities shows large amounts of water (around 50% of the fraction collected in a condenser at LN_2 temperature) and a wide spectrum of light and heavy hydrocarbons.

Because it is impractical to purge the pump and the oil separately, we have used the following procedure for cleaning the pump and the subsequent oil mist filter before connection to the sorbent filters and the cryogenic circuit:

- The first stage of the purge is made by pumping dry air for six hours at a speed of about 80% of the nominal pump system rating. The exhaust is to the air through a check valve and sometimes via an LN_2-cooled condenser for observing and monitoring the impurities. The functioning of the oil mist filter and its automatic return valve is verified. The dry air is obtained by passing ambient air through another LN_2-cooled condenser.
- The second stage is made by pumping dry nitrogen gas overnight at about 20% of the nominal pump speed. The exhaust can be to air with a check valve, or the circuit can be closed if an LN_2-cooled condenser and an X13 molecular sieve filter[7] are used in the circuit. Helium can also be used in closed-circuit purge; this can eliminate the third step of the purge.
- The third step consists of evacuating the nitrogen from the pump and degassing the oil by pumping dry helium through it. We use helium with max. 40 ppm impurities, stored in pressure cylinders and cleaned with X13 molecular sieve. The helium purge must last several hours.

If the pump system consists of several pumps (usually there is one or more Root's blowers in front of a rotary blade pump), the whole pump cascade is purged in the same process. This obviously requires that all the pumps are operated so that their oil warms up to normal operating temperature.

---

[7] Sigma-Aldrich Inc.

Tests

Leak through the shaft seal of the rotary blade pump may depend on the operating pressure, and therefore the pump may show no leak under vacuum, while leaking badly at exhaust pressures slightly above atmospheric. Such leaks are usually associated with oil leakage through the lip seal. The possible losses of oil out of the shaft seal are therefore monitored carefully during the purge.

In closed-circuit purging the loss or gain of gas is always monitored by recording the exhaust pressure with an electronic manometer connected to a pen recorder. This allows to detect air leaks down to the range around 0.1 litre·atm/day at which blockages of the refrigerator will occur only after several weeks of running, provided that the adsorbent purifiers will have a normal capacity.

The leak testing of helium pumps by a helium leak detector is difficult if the circuit is filled with helium, and in this mode only the pump inlet circuit can be tested while the pump is running. The leak detector is then connected to the inlet of the first pump in the cascade, after purging by air and nitrogen. If the oil in all the pumps has never been exposed to helium or is sufficiently well degassed, leaks of even the outlet of the primary pump can be detected on a level which is well below the critical capacity of the purifiers.

The functioning of the oil mist filter is tested in the first step of the purge by observing the absence of oil mist in the gas outlet of the pump. A simple method of verification consists of placing a clean glass plate for about 10 minutes to the outlet gas stream and inspecting it then under microscope. A quantitative measure can be obtained by weighing precisely a thin aluminium foil before and after exposure to the outlet gas stream.

The operation of the automatic return valve of the oil mist filter can be tested by filling some oil into the chamber and observing the level change after turning the pump on. This should be made after initial installation and after all oil changes in the beginning of the purge operations.

8.7.2 Helium Purifiers

Oil Mist Filters

The oil mist filters are commercially available in hermetic execution, with stainless steel outer casing and elastomer o-ring seals. The mist filters themselves were earlier fabricated using paper stacks, and they are now made of porous solid materials and are commercially available with automatic oil return valves; these work satisfactorily in helium circuits if the filter element is dimensioned correctly for a reasonable pressure drop at the maximum mass flow rate.

Charcoal and Molecular Sieve Filters at RT

The gas coming out of a primary pump or a Root's blower is always contaminated also by oil vapors, even if a careful purge has been performed. These vapors can be eliminated to a large extent by adsorbent filters made of charcoal or chemisorption filters using molecular sieve 13X. The charcoal is made by dehydrating carbonizing coconut shell granules with water steam flow at 800 °C to 1100 °C temperature; such activated granules have a surface of physisorption of about 2000 m^2/g. The activated charcoal is particularly effective for adsorbing oil vapors, but it also removes smaller amounts of water vapor. The molecular sieves are particularly effective for removing water vapor.

Commercial filter cartridges and shells are available in several sizes of shells. They can be used with cartridges which are specified for flow rates up to 20 mMol/s and 100 mMol/s nominal helium mass flow rates. For short times, three times higher mass flow rate can be tolerated. The pressure drop in these filters is rather high, and the flow rating can only be extended by installing several of them in parallel. The seals and the pipe fittings of the commercial filter shells can be easily improved by machining a groove for a torical seal and by brazing metal-seal fittings.

For higher flow rates it is best to fabricate custom-made shells in which one can fill bulk sorbent material, and which one can provide with a heater system for rapid regeneration of the sorbent.

Activated charcoal cartridges are installed after the oil mist filter of the primary pump. As the gas coming out of the mist filter is quite hot at high flow, the charcoal filter is placed always in the far end of the line leading to the gas handling and LN$_2$ trap system.

A better choice of sorbent might be the molecular sieve 13X, which can absorb almost as much oil vapors, and is effective against water as well. Water is known to diffuse through elastomer seals, and also sometimes refrigerators are started under conditions which may lead to serious water contamination of the ^{3}He circuit. Under such circumstances the use of 13X sorbent should be considered before further purification in an LN$_2$-cooled activated charcoal trap.

The main motivations for using an RT sorbent filter are to avoid blocking of the heat exchanger of the LN$_2$-cooled trap and to avoid the gradual saturation of the activated charcoal in it. There is also good evidence for the LN$_2$-cooled trap alone to be ineffective against some types of heavy contaminants, including water. It has been speculated that condensable gases may form microcrystals similar to snow when cooled rapidly in the heat exchanger, or after degassing off from the heat exchanger surface when the flow varies or LN$_2$ level drops. These microcrystals may be carried through the sorbent granules and the sintered bronze end caps by the gas stream and they will then be evaporated again when flowing up in the heat exchanger. Further evidence for this will be provided in Section 8.7.3.

Charcoal Filter at LN$_2$ Temperature

The LN$_2$-cooled activated charcoal trap usually has a volume of about 500 cm^3 and it is equipped with a tube-in-tube countercurrent heat exchanger dimensioned for the required gas flow. The trap is activated by heating it up to about 200 °C temperature and then evacuating with a primary pump while purging through the trap a small flow of pure helium gas. The purge gas is obtained from a pressure cylinder; the initial impurity level in the range of 40 ppm is improved to below 1 ppm by a commercial molecular sieve 13X filter. Sometimes helium from a pressurized dewar is used for purging the trap.

The tube-in-tube heat exchanger of the trap is dimensioned for the maximum flow rate of the refrigerator. A length of about 1 m is found to be adequate for flows up to 30 mMol/s; the largest polarized target setups have an exchanger of 4 m length and can be used for ^{3}He flows up to 350 mMol/s. The inlet to the trap is through the annular space between the tubes and the outlet is in the central tube; this arrangement is believed to give some degree of immunity against the blocking of the exchanger by condensable impurities which may obstruct the inlet channel of the trap after the first charcoal filter becomes saturated during the run.

The filling of the LN_2 dewar of the trap can be made automatic by using a commercial LN_2 level gauge and an ordinary bellows-sealed electromagnetic valve in the filling line.

8.7.3 Operation of Gas Purification Systems and Analysis of Blockage Events

Snow Blockages

Before the need of activated charcoal filters between the oil mist filter and the LN_2-cooled trap became known, it was rather common to observe the gradual blockage of the refrigerator in about one week from the initial startup. The LN_2-cooled trap also usually becomes blocked; depending on the geometry of the heat exchanger, the trap gets blocked soon thereafter or some weeks later.

The analysis of the vapors collected from these types of blockages has revealed only minor amounts of argon, which is thought to indicate the absence of air leakage into the circuit. Other evidence against an air leak comes from the relative difficulty in removing the blockage by warmup and from the observation that the LN_2-cooled trap is not saturated by air.

The water signal has also been always detected in the blockage vapors. The most abundant gases present in the blockage vapors, however, were hydrocarbons of various masses ranging up to 400, the range of the quadrupole mass analyzer. After speculating that these condensable vapors may traverse the LN_2-cooled trap in the form of frozen microcrystals or snow, this finding led to the addition of the RT charcoal filter, as was already discussed above. The improvement immediately enabled one to run for months with no signs of this type of blockage.

The 'snow theory' of blockages due to condensable vapors is also supported by the findings coming from refrigerators equipped with an adjustable J-T-valve. Before a complete blockage occurs, it has been common to observe the rise of the condensation pressure while operating in otherwise stable conditions. The initial condensation pressure can be almost always recovered by closing the valve for a few seconds and immediately opening it to the initial position. The recovery of the condensation pressure is usually spectacular, and we found no other explanation than snow-like impurities which are compressed by the needle and partly pushed through the orifice.

A snow trap in the refrigerator could be effective against these types of blockage. Sintered copper filters were tried for this purpose, but it was found that the sintered 325-Mesh powder had to be heated to impractically high temperatures in order to be able to degas the impurities and recover the normal flow at RT. A better material and geometry for trapping the snow should be possible to find, and this would be highly desirable because the preventive measures described above may sometimes fail, for example, due to an accidental warmup of the LN_2 trap.

Hydrogen Blockages

The cracking of hydrocarbon or silicon oils always leads to the formation of gaseous hydrogen. The statements below are based on experience with systems using Root's blowers followed by a rotary blade pump, but the same problem is known to occur with oil diffusion or booster pumps, where the oil is very hot.

Some of the blockages have been recovered after a heatup to only 30K temperature, and we have noted that this kind of incident usually entails the rotary blade pump failure soon thereafter. The likely cause of these failures is the extended operation of the pump at low back pressures, which usually occurs during frozen spin target runs. The problem appears to be due to the lack of oil circulation in the pump, which is normally forced by the pump back pressure. Other types of pump failure have also resulted in detecting suspect hydrogen blockages prior to the jamming of the pump. After the pumps were equipped with an internal oil circulation device, no evidence was found of hydrogen blockages when using this type of pumps.

It should be noted that the LN_2-cooled trap is capable of absorbing reasonable amounts of hydrogen (≈ 0.1 liters NTP) before becoming saturated, if its heat exchanger is correctly dimensioned.

Air Blockages

The clearest signature of an air leak is the saturation of the LN_2-cooled trap, which must be always tested when unblocking the cryostat. The saturation corresponds to the RT pressures varying from a few to 10 bar, depending on the dead volumes in the trap and in the exchanger.

The pressure resulting from the warmup of the saturated trap can be tested by deliberately letting air flow into the cold trap until about 10 mbar pressure is observed; the untrapped air is then purged away by flowing clean helium through the cold trap during about one hour, and by finally evacuating the trap before closing and warming it up. We should warn that the trap, the heat exchanger and the manometers should be compatible with the resulting pressures; otherwise the air must be allowed to expand to an additional closed volume which is large enough.

Operation of Gas Purifier Systems

With the measures described above the blockage problems have become rarer but have not vanished entirely. Most of the remaining problems are ascribed to the malfunctioning or failure of one the filters or traps described above. The failure of the oil mist filter, for example, results in the quick saturation of the activated charcoal trap at RT; this in turn soon leads to the snow-type blockage of the refrigerator and possibly blockage of the heat exchanger of the LN_2-cooled trap. The problem can only be cured by changing or improving the oil mist filter and then cleaning or activating the other two traps.

The gradual saturation of the activated charcoal trap at RT leads also to the symptoms of blockages. The interval for changing the filter cartridge depends on its capacity and on the running conditions; this interval must be found empirically. If operation periods are extended over six months, it may be worth installing two parallel sets of RT filters to facilitate their change without interrupting the gas circulation.

The activated charcoal traps at LN$_2$ temperature have retained a reasonable adsorption capability despite observation of oil in them while recharging with new charcoal.

Among all incidents leading to the blockage of the refrigerator, the failure of the oil mist filter is probably the worst event to occur. This is most likely to happen during operation at high flow. The presence of oil both in the line leading to the refrigerator and inside the refrigerator itself should be checked as soon as possible after the first symptoms are detected for a blockage of either the LN$_2$-cooled trap or the refrigerator itself.

References

[1] I. M. Khalatnikov, Theoretical model of Kapitza resistance, *Zh. Eksp. Teor. Fiz.* **22** (1952) 687.

[2] G. L. Pollack, Kapitza resistance, *Rev. Mod. Phys.* **41** (1969) 48–81.

[3] L. J. Challis, Experimental evidence for a dependence of the Kapitza conductance on the Debye temperature of a solid, *Phys. Lett.* **26 A** (1968) 105–106.

[4] N. S. Snyder, *Thermal Conductance at the Interface of a Solid and Helium II (Kapitza Conductance)*, National Bureau of Standards 385, 1969.

[5] H. Glättli, Kapitza resistance of cerium ethylsulphate, *Can. J. Phys.* 46 (1968).

[6] O. E. Vilches, J. C. Wheatley, Measurements of the specific heats of three magnetic salts at low temperatures, *Phys. Rev.* **148** (1966) 509.

[7] W. C. Black, A. C. Mota, J. C. Wheatley, J. H. Bishop, P. M. Brewster, Thermal resistance between powdered cerium magnesium nitrate and liquid helium at very low temperatures, *J. Low Temp. Phys.* **4** (1971) 391–400.

[8] E. Boyes, G. R. Court, B. Craven, R. Gamet, P. J. Hayman, Measurements of the effect of absorbed power on the polarization attainable in a Butanol target, in: G. Shapiro (ed.) *Proc. 2nd Int. Conf. on Polarized Targets*, LBL, University of California, Berkeley, Berkeley, 1971, pp. 403–406.

[9] W. de Boer, T. O. Niinikoski, Dynamic proton polarization in propanediol below 0.5 K, *Nucl. Instrum. and Meth.* **114** (1974) 495–498.

[10] A. C. Anderson, J. I. Connolly, J. C. Wheatley, Thermal boundary resistance between solids and helium below 1°K, *Phys. Rev.* **135** (1964) A910–A921.

[11] O. V. Lounasmaa, *Experimental Principles and Methods below 1K*, Academic Press, New York, 1974.

[12] Spin Muon Collaboration (SMC), D. Adams, B. Adeva, et al., The polarized double-cell target of the SMC, *Nucl. Instr. and Meth. in Phys. Res.* **A 437** (1999) 23–67.

[13] W. de Boer, High proton polarization in 1,2-propanediol at ^{3}He temperatures, *Nucl. Instr. and Meth.* **107** (1973) 99–104.

[14] T. O. Niinikoski, A horizontal dilution refrigerator with very high cooling power, *Nucl. Instrum. and Meth.* **97** (1971) 95–101.

[15] T. O. Niinikoski, F. Udo, 'Frozen spin' polarized target, *Nucl. Instr. and Meth.* **134** (1976) 219–233.

[16] G. Hildebrandt, Heat transfer to boiling helium-1 under forced flow in a vertical tube, *ICEC4*, IPC Business Press, Guildford, UK, 1972, pp. 295–300.

[17] R. Radebaugh, *Thermodynamic Properties of ^{3}He-^{4}He Solutions with Applications to the ^{3}He-^{4}He Dilution Refrigerator*, National Bureau of Standards Technical Note 362, 1967.

[18] H. B. Stuhrmann, N. Burkhardt, G. Dietrich, et al., Proton and deuteron targets in biological structure research, in: H. Dutz, W. Meyer (eds.) *7th Int. Workshop*

on Polarized Target Materials and Techniques, Elsevier, Amsterdam, 1994, pp. 124–132.

[19] J. Zhao, W. Meerwinck, T. O. Niinikoski, et al., The polarized target station at GKSS, in: H. Dutz, W. Meyer (eds.) *Proc. 7th Int. Workshop on Polarized Target Materials and Techniques*, Elsevier, Amsterdam, 1995, pp. 133–137.

[20] P. R. Cameron, D. G. Crabb, Construction and preliminary tests of a subcooled ^{4}He polarized proton target in: S. Jaccard, S. Mango (eds.) *International Workshop on Polarized Sources and Targets*, Birkhäuser, Montana, Switzerland, 1986, pp. 781–785.

[21] D. C. Peaslee, J. R. O'Fallon, M. Simonius, et al., Large-P_{perp} spin effects in p+p->p+p, *Phys. Rev. Lett.* **51** (1983) 2359–2361.

[22] P. R. Cameron, A simple model of bead cooling, in: W. Meyer (ed.) *Proc. 4th Int. Workshop on Polarized Target Materials and Techniques*, Physikalisches Institut, Universität Bonn, Bonn, 1984, pp. 136–142.

[23] D. G. Crabb, P. R. Cameron, A. M. T. Lin, R. S. Raymond, Operational characteristics of radiation doped ammonia in a high intensity proton beam, in: W. Meyer (ed.) *Proc. 4th Int. Workshop on Polarized Target Materials and Techniques*, Physikalisches Institut, Universität Bonn, Bonn, 1984, pp. 7–12.

[24] K. H. Althoff, B. Boden, V. Burkert, et al., First experience with a ^{3}He/^{4}He mixture in a ^{3}He- refrigerator in high intensity electron beams, in: W. Meyer (ed.) *Proc. 4th Int. Workshop on Polarized Target Materials and Techniques*, Physikalisches Institut, Universität Bonn, Bonn, 1984, pp. 149–154.

[25] A. T. A. M. de Waele, J. C. M. Keltjens, C. A. M. Castelijns, H. M. Gijsma, Flow properties of ^{3}He moving through ^{4}He-II at temperatures below 150 mK, *Phys. Rev.* **B28** (1983) 5350.

[26] T. O. Niinikoski, Cooling power of dilution refrigerator, in: M. Krusius, M. Vuorio (eds.) *Proc. 14th Int. Conf. on Low Temperature Physics (LT 14)*, North Holland, Amsterdam, 1975, pp. 29–30.

[27] T. O. Niinikoski, Dilution refrigeration: New concepts, in: K. Mendelssohn (ed.) *6th Int. Cryogenic Engineering Conf.*, IPC Science and Technology Press, Guilford, 1976, pp. 102–111.

[28] T. O. Niinikoski, J.-M. Rieubland, Large dilution refrigerators, in: K. Yasukochi, H. Nagano (eds.) *9th Int. Cryogenic Engineering Conference*, Butterworth, Guilford, 1982, pp. 580–585.

[29] T. O. Niinikoski, Dilution refrigerator for a two-litre polarized target, *Nucl. Instr. and Meth.* **192** (1982) 151–156.

[30] T. O. Niinikoski, Polarized targets at CERN, in: M.L. Marshak (ed.) *Int. Symp. on High Energy Physics with Polarized Beams and Targets*, American Institute of Physics, Argonne, 1976, pp. 458–484.

[31] T. O. Niinikoski, Recent developments in polarized targets at CERN, in: G. H. Thomas (ed.) *High Energy Physics with Polarized Beams and Polarized Targets*, American Institute of Physics, New York, 1979, pp. 62–69.

[32] P. Berglund, J. Kyynäräinen, T. O. Niinikoski, A large dilution refrigerator for polarized target experiments, *Cryogenics* **34** (1994) 235–238.

[33] J. Kyynäräinen, The SMC polarized target, in: H. Dutz, W. Meyer (eds.) *Proc. 7th Int. Workshop on Polarized Target Materials and Techniques*, Elsevier, Amsterdam, 1995, pp. 47–52.

[34] J. C. Wheatley, R. E. Rapp, R. T. Johnson, Principles and methods of dilution refrigeration. II, *J. Low Temp. Phys.* **4** (1971) 1–39.

[35] W. R. Abel, J. C. Wheatley, Experimental thermal conductivity of two dilute solutions of He^3 in superfluid He^4, *Phys. Rev. Lett.* **21** (1968) 1231–1234.

[36] G. M. Coops, A. T. A. M. de Waele, H. M. Gijsma, Experimental evidence for mutual friction between ^{3}He and superfluid ^{4}He, *Phys. Rev.* **B25** (1982) 4879.

[37] C. A. M. Castelijns, J. G.M. Kuerten, A. T. A. M. de Waele, H. M. Gijsma, ^{3}He flow in dilute ^{3}He-^{4}He mixtures at temperatures between 10 and 150 mK, *Phys. Rev. B* **32** (1985) 2870–2886.

[38] T. O. Niinikoski, Construction of sintered copper heat exchangers, *Cryogenics* **11** (1971) 232–233.

[39] G. Burghart, Baseline Design of the Cryogenic System for Eureca, Dr. Techn. Thesis, 2010, Atominstitut (E141), Technical University of Vienna

[40] P. Wikus, T. O. Niinikoski, Theoretical models for the cooling power and base temperature of dilution refrigerators, *J. Low Temp. Phys.* **158** (2010) 901–921.

[41] D. G. Crabb, D. B. Day, The Virginia/Basel/SLAC polarized target: operation and performance during E143 experiment at SLAC, *Nucl. Instrum. and Meth. in Phys. Res. A* **356** (1995) 9–19.

[42] D. Taler, J. Taler, Simple heat transfer correlations for turbulent tube flow, *E3S Web of Conferences*, EDP Sciences, 2017, pp. 02008.

[43] P. Roubeau, Horizontal cryostat for polarized proton targets, *Cryogenics* **6** (1966) 207–212.

[44] M. Borghini, P. Roubeau, C. Ryter, Une Cible de Protons Polarises pour la Physique des Hautes Energies I. Cible, *Nucl. Instr. and Meth.* **49** (1967) 248.

[45] P. Roubeau, J. Ezratty, H. Glättli, J. Vermeulen, M. Borghini, Organic polarized proton target, using a continuous flow ^{3}He cryostat, *Nucl. Instrum. and Methods* **82** (1970) 323–324.

[46] P. Roubeau, Progress in polarized target cryogenics, in: G. Shapiro (ed.) *Proc. 2nd Int. Conf. on Polarized Targets*, LBL, University of California, Berkeley, Berkeley, 1971, pp. 47–55.

[47] J. Vermeulen, ^{3}He-cooled polarized targets, in: G. Shapiro (ed.) *Proc. 2nd Int. Conf. on Polarized Targets*, LBL, University of California, Berkeley, Berkeley, 1971, pp. 69–72.

[48] D. G. Crabb, C. B. Higley, A. D. Krisch, et al., Observation of a 96% proton polarization in irradiated ammonia, *Phys. Rev. Letters* **64** (1990) 2627–2629.

[49] A. Thomas, C. Bradtke, H. Dutz, et al., Behaviour of polarized ammonia in an intense electron beam, in: H. Dutz, W. Meyer (eds.) *7th Int. Workshop on Polarized Target Materials and Techniques*, Elsevier, Amsterdam, 1995, pp. 5–8.

[50] J. Pierce, J. Maxwell, T. Badman, et al., Dynamically polarized target for the g^p_2 and G^p_E experiments at Jefferson Lab, *Nucl. Instr and Meth. A* **738** (2014) 54–60.

[51] H. Preston-Thomas, The international temperature scale of 1990 (ITS-90), *Metrologia* **27** (1990) 3–10.

[52] J. Engert, B. Fellmuth, K. Jousten, A new ^{3}He vapor-pressure based temperature scale from 0.65 K to 3.2 K consistent with the PLTS-2000, *Metrologia* **44** (2007) 40–53.

[53] L. G. Rubin, B. L. Brandt, H. H. Sample, Cryogenic thermometry: a review of recent progress, II, *Cryogenics* **22** (1982) 491–503.

[54] H. H. Sample, B. L. Brandt, L. G. Rubin, Low-temperature thermometry in high magnetic fields above 2 K, *Rev. Sci. Instrum.* **53** (1982) 1129.

[55] B. L. Brandt, L. G. Rubin, H. H. Sample, Low-temperature thermometry in high magnetic fields. VI. Industrial grade resistors above 66 K; Rh-Fe and Au-Mn resistors above 40 K, *Rev. Sci. Instrum.* **59** (1988) 642–645.

[56] A. C. Anderson, Instrumentation at temperatures below 1 K, *Rev. Sci. Instrum.* **51** (1980) 1603.

[57] J. Ylöstalo, P. Berglund, T. O. Niinikoski, R. Voutilainen, Cryogenic temperature measurement for large applications, *Cryogenics* **36** (1996) 1033–1038.

9
Microwave and Magnet Techniques

9.1 Microwave Sources

Most polarized targets operate in a magnetic field of 2.5 T or 5 T. The corresponding EPR frequencies of free radicals are around 70 GHz and 140 GHz and the free-space wavelengths are 4 mm and 2 mm, respectively. Some of the systems operate in 3.4 T or 6.5 T fields, which correspond to 95 GHz or 182 GHz frequency bands. The power required for DNP at 70 GHz and 0.5 K is typically about 1 mW/g when reversing the polarization, and about 0.2 mW/g when optimizing for high polarization. For targets operating in high-intensity beams, the typical requirement for DNP is 100 mW/g at 140 GHz and 1 K. Given the power transmission losses between the source and target, the power output of the microwave source therefore puts limits to the sizes of practical targets: about 1 kg at 0.5 K and 50 g at 5 T and 1 K.

High-power microwave sources covering such frequencies use vacuum tubes to generate continuous-wave microwave power. These devices are based on the ballistic motion of electrons in vacuum, under the influence of controlling electric and magnetic fields, and include crossed field tubes (magnetrons), linear beam tubes and fast-wave tubes such as the gyrotron. Among these only the linear beam tubes have been commonly used in polarized targets; in this category the main devices are klystrons, backward wave oscillators (BWO) and extended interaction oscillators (EIO). They work on the basis of bunches of electrons flying ballistically through a drift space inside resonant structures, rather than using a continuous stream of electrons. Gyrotron sources have been recently introduced for DNP in high-field MAS NMR spectroscopy [1–3].

9.1.1 Backward Wave Oscillator (BWO)

The BWO or Carcinotron tube has been used in many polarized targets because of its ability of being electrically tunable in a wide enough range of frequencies (≈ 1 GHz) so that no mechanical tuning or magnetic field shift was necessary to reach the upper and lower edges of the ESR line frequencies. The principle is similar to traveling wave tube (TWT) oscillators, but the resonant structure is a series of cavities rather than a helix, and the amplified wave travels in opposite direction to the electron beam; power is extracted close to the cathode, and the frequency is controlled by the cathode potential that sets the speed of the electron beam. The output power is controlled by the anode potential that sets the cathode current. The frequency control is very fast, which enables the efficient use of BWO in radar jammers. Moreover, the output power spectrum is rather narrow, which has applications in radio astronomy and metrology. More than 10 W of power in the 70 GHz band and up to 2 W in the 140 GHz band have been available. The disadvantage is that the cathode lifetime is rather limited in high-power operation, in comparison with that of the EIO tube.

Similar to the EIO tube, BWO tubes have a strong permanent magnet for focusing the electron beam.

9.1.2 Klystrons

Klystron millimeter wave sources also have an electron beam accelerated between a heated cathode and an anode. The beam is bunched when traversing an input cavity, and the bunched beam loses its energy to the electromagnetic field in the output cavity. A part of the output power is directed back to the input cavity, thus closing the positive feedback loop. The electric tuning range is some MHz at 70 GHz, and the source has to be tuned mechanically to switch between positive and negative polarization frequencies. It has been a common practice to use two klystrons tuned to these two frequencies, because the cavity tuning has a limited lifetime of about 1000 tuning cycles. The output power is a few hundred mW, which is now accessible also with Gunn and IMPATT diode sources having a much broader electric tuning range. Therefore the klystron sources have been replaced by IMPATT and Gunn sources in many small polarized targets, test set-ups and instrumentation applications.

9.1.3 Extended Interaction Oscillator (EIO)

The EIO is a single-cavity tube that is capable of producing a very high continuous wave power, since the collector is separate from the resonant structure, and the device is not limited by the power dissipation capability of the RF elements. The EIO tube has a separate cathode that leads to a lower current density and therefore to a longer cathode life expectancy of several thousand hours at maximum specified power that can be 100 W at 70 GHz and 20 W at 140 GHz; operation at lower power extends the cathode life further. The mechanical tuning range is about 3% and electric tuning up to 0.3%; operation at reduced power also reduces the electric tuning range. Because of the latter, targets using EIO tubes usually have two tubes mechanically tuned to the frequencies of the lower and upper edges of the ESR line. The switching between the two sources can then be made by a mechanical waveguide switch.

Low-power microwave sources use solid-state devices such as tunnel diodes, Gunn diodes and IMPATT diodes. The tunnel diode, known also as Esaki diode, is a low power device that has been replaced by the Gunn diode devices capable of generating 100 mW of millimeter wave power.

9.1.4 Gunn Diode Sources

A Gunn diode is also known as a transferred electron device (TED); these two-terminal devices have heavily n-doped regions for contacts to the lightly n-doped layer in between. Unlike the usual diodes where a p-n structure works as a rectifier, TED is symmetric and therefore cannot rectify. The frequency can be multiplied and amplified in several steps after splitting the power from a phase-locked Gunn oscillator operating at a lower frequency. The power output can be thus increased by combining the power from several Gunn diode sources driven by the common phase-locked source.

These Gunn diodes are nowadays made mostly of GaAs; InP is used in their upper frequency range.

The Gunn diode sources are mechanically tunable in a wide range of frequencies, and they are used for microwave instrumentation where they have advantageously replaced the reflex klystrons. Custom-designed sources can deliver 100 mW power in the V band. The sources can be frequency modulated or phase locked to a lower-frequency stabilized reference source.

9.1.5 IMPATT Diode Sources

The IMPact ionization Avalanche Transit-Time (IMPATT) diode is a high-power semiconductor device used in millimeter wave electronics applications to make oscillators and amplifiers. These operate at frequencies between about 3 GHz and 100 GHz or more. The main advantage is the high-power capability of the device structure, with a diamond substrate as a heat sink. These diodes are used in a variety of applications from low-power radar systems to proximity alarms. A major drawback of IMPATT diodes is their high level of phase noise that results from the statistical nature of the avalanche process, in which a fast electron strikes an atom breaking a covalent bond and liberating several valence electrons that are again accelerated by the electric field so as to cascade the process. The transit time must match with the oscillation frequency to which the resonant structure is tuned. The frequency of an IMPATT oscillator can be tuned by adjusting the resonant structure by a micrometer, and to some extent also by the bias current.

The IMPATT sources can generate up to 0.5 W power at 70 GHz and therefore they can be used for the DNP of small polarized targets or for the tests of small samples in laboratory environment.

One of the advantages of semiconductor sources is that they do not require a strong magnet for focusing the electron beam. Therefore, they can be placed closer to the polarized target magnet. However, these sources must be equipped with an output isolator that has a magnetic part in the junction of a waveguide circulator, and this cannot tolerate a large stray field close to a superconducting magnet.

9.2 Waveguide Circuits

9.2.1 Waveguide Modes of Propagation

The transmission of microwave power requires a low-loss line that is also free of discontinuities, which may reflect a substantial part of the power back to the source or its output isolator. Coaxial lines can, in principle, transmit millimeter waves, but their transmission losses are much higher than those of hollow metallic waveguides.

The electromagnetic waves in a metal-pipe waveguide may be imagined as travelling down the guide in a zig-zag path, being repeatedly reflected between opposite walls of the guide. For the particular case of rectangular and round waveguides, it is possible to base an exact analysis on this view. In general, electromagnetic waveguides are analyzed by solving Maxwell's equations (see Appendix A1.1), or rather their reduced form based on sinusoidally varying fields, the electromagnetic wave equation, with boundary conditions determined by the properties of the materials and their interfaces. These equations have multiple solutions, or modes, which are eigenfunctions of the differential equations. Each mode is characterized by a cut-off frequency below which the mode cannot transmit power in the guide; however, the TEM_{00} mode of a coaxial line transmits down to zero frequency.

Waveguide propagation modes depend on the operating wavelength and polarization (here meaning the orientation of the electric field vector) and on the shape and size of the guide. The transverse modes are classified into different types, with index m giving the number of half-wavelengths in the x-direction and index n giving the number of half-wavelengths in the y-direction (along the smaller wall of a rectangular guide):

- TE_{mn} modes (transverse electric) have no electric field in the direction of propagation.
- TM_{mn} modes (transverse magnetic) have no magnetic field in the direction of propagation.
- TEM_{mn} modes (transverse electromagnetic) have neither electric nor magnetic field in the direction of propagation; these modes occur in coaxial transmission lines.
- Hybrid modes have both electric and magnetic field components in the direction of propagation.

9.2.2 Rectangular Waveguides

We shall discuss here only the TE and TM modes and limit ourselves to standard rectangular waveguides; round waveguides have a different definition of the indexes, which will be discussed in the following subsection. Among the modes the fundamental mode has the lowest cut-off frequency and therefore the transmission along the guide takes place always in single mode up to the next lowest cut-off frequency. Only one polarization is then transmitted and the design of waveguide couplers, junctions, attenuators, phase shifters, terminator loads, circulators and power loads is reliable and predictable in the single-mode band.

9.2 Waveguide Circuits

Table 9.1 Rectangular waveguide bands most commonly used in the microwave sources and circuits in polarized targets. Transmission of power between the recommended frequencies takes place in single mode, which facilitates and simplifies the characterization of the waveguide components. X band and Ka band guides are oversized at 70 GHz and 140 GHz frequencies; their straight sections with tapered transitions to single mode guides transmit power with substantially lower losses at these frequencies.

Waveguide band designation	Recommended low-frequency limit (GHz)	Recommended high-frequency limit (GHz)
X band	8.2	12.4
Ka band	26.5	40.0
V band	40	75
E band	60	90
D band	110	170

Let us assume the cross section of the guide shown in Figure 9.1 having metallic walls with width a and height b, and let us align the coordinate axis z along the direction of propagation of the electromagnetic wave, axis x along the wide wall and axis y along the side wall, as shown by Figure 9.1. Without going into the details of the solution of Maxwell's equations (see Appendix A1.1) in the case of propagation of waves with angular frequency ω, we just note that the electric field in a charge-free region must obey [4] the Laplace equation:

$$\nabla^2 \mathbf{E} + \left(1 + \frac{i\sigma}{\varepsilon\omega}\right)\mu\varepsilon\omega^2 \mathbf{E} = 0, \tag{9.1}$$

which yields the equation of the electric field along y-direction in the case of the TE_{0m} modes:

$$\frac{\partial^2 E_y}{\partial z^2} + \frac{\partial^2 E_y}{\partial x^2} = \left(-\omega^2 \mu_0 \varepsilon + i\omega\mu_0\sigma\right)E_y. \tag{9.2}$$

This simplified equation was obtained by considering propagation at a constant angular frequency ω in the free space limited by high-conductivity walls. The solution of

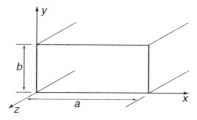

Figure 9.1 Dimensions of a rectangular waveguide, and the coordinate system used in the text

the differential Eq. 9.2 gives $E_y(x, z)$ from which the magnetic field components can be obtained by taking the derivatives

$$\frac{\partial E_y}{\partial x} = -i\omega\mu_0 H_z;$$

$$-\frac{\partial E_y}{\partial z} = -i\omega\mu_0 H_x.$$

(9.3)

If the guide is filled with air or vacuum whose conductivity σ is zero, and if we assume that the wave propagates in the z-direction as

$$E_y = K(x,y)e^{-\gamma z},$$

(9.4)

where $\gamma = \alpha + i\beta$ is the complex propagation constant, we get

$$\frac{\partial^2 E_y}{\partial x^2} = -k^2 E_y$$

(9.5)

where

$$k^2 = \gamma^2 + \omega^2\mu_0\varepsilon$$

(9.6)

and E_y is only a function of x and y. Given the general form of the solution of Eq. 9.5

$$E_y = (A_1 \cos kx + A_2 \sin kx)e^{-\gamma z}$$

(9.7)

and the boundary conditions of rectangular section with high conductivity so that the electric field vanishes at $x = 0$ and $x = a$, we get

$$A_1 = 0,$$
$$A_2 \sin ka = 0,$$

(9.8)

which yields

$$E_y = Ae^{-\gamma z} \sin\frac{m\pi x}{a}$$

(9.9)

and

$$k = \sqrt{\gamma^2 + \omega^2\mu_0\varepsilon} = \frac{m\pi}{a}.$$

(9.10)

So there can be an infinite number of modes at high enough frequencies. The lowest frequency propagation mode is obtained for $m = 1$, which is the case of the fundamental single-mode propagation with the constant

$$\gamma = \sqrt{\left(\frac{\pi}{a}\right)^2 - \omega^2\mu_0\varepsilon}.$$

(9.11)

Equation 9.11 can have a real or imaginary value, among which the real value corresponds to an exponentially attenuated penetration of the microwave field into the guide, and the imaginary value

$$\gamma = \alpha + i\beta = 0 + i\sqrt{\omega^2 \mu_0 \varepsilon - \left(\frac{\pi}{a}\right)^2} \tag{9.12}$$

yields propagation without attenuation. Therefore the low-frequency limit of propagation is at the cut-off angular frequency:

$$\omega_c = \frac{\pi}{a\sqrt{\mu_0 \varepsilon}} = \frac{\pi c}{a}. \tag{9.13}$$

For an air-filled or evacuated line the cut-off frequency is then $f_c = c/(2a)$ and the cut-off wavelength is

$$\lambda_c = 2a. \tag{9.14}$$

The phase velocity $u_p = \omega/\beta$ for the same line is

$$u_p = \frac{c}{\sqrt{1 - \left(\frac{\pi c}{a\omega}\right)^2}} \tag{9.15}$$

and is seen to be frequency dependent and higher than the speed of light. This makes the guide behave like a dispersive medium, and the group velocity of a pulse propagated along the guide is

$$u_g = \frac{c^2}{u_p} = c\sqrt{1 - \left(\frac{\pi c}{a\omega}\right)^2} \tag{9.16}$$

is always smaller than c and becomes zero at the cut-off frequency.

The impedance of the waveguide transmission line is useful for understanding the reflections and matching of the devices connected to it. The guide impedance is usually defined as the ratio of the maximum amplitudes of E_y and H_x; the magnetic field can be obtained from the electric field of Eq. 9.9 by taking the derivatives of Eq. 9.3:

$$Z_g = \frac{\max(E_y)}{\max(H_x)} = \frac{\omega \mu_0}{\beta} = \mu_0 u_p = \frac{\mu_0 c}{\sqrt{1 - \left(\frac{\pi c}{a\omega}\right)^2}} = \frac{Z_f}{\sqrt{1 - \left(\frac{\pi c}{a\omega}\right)^2}}, \tag{9.17}$$

where we can see that the impedance of the rectangular guide in TE_{01} mode is always higher than the impedance of the free space Z_f, and that it depends on frequency, unlike the impedance of a coaxial line in the fundamental mode. Here $Z_f = 377 \, \Omega$ is the impedance of free space:

$$Z_f = \sqrt{\frac{\mu_0}{\varepsilon_0}} = 377\Omega.$$

The attenuation of the waveguide is mainly due to the finite conductivity of its inner walls. In a given mode of propagation, the power losses that can be compared with the transmitted power are given by the Poynting vector. Without following through the details of the derivation that is beyond our scope here, we shall just give the result for the attenuation constant α:

$$\alpha = \frac{2}{\omega\mu_0 \beta ab} \sqrt{\frac{\omega\mu}{2\sigma}} \left[\frac{\pi^2 b}{a^2} + \frac{\pi^2}{2a} + \frac{\beta^2 a}{2} \right], \quad (9.18)$$

where the propagation constant is, from Eq. 9.12,

$$\beta^2 = \left(\frac{\omega}{c}\right)^2 - \left(\frac{\pi}{a}\right)^2, \quad (9.18a)$$

σ is the conductivity of the wall metal, and μ its magnetic permeability. By inserting the permeability $\mu = \mu_0$ and conductivity of copper $\sigma = 0.596 \times 10^7$ 1/(Ωm) (at RT), we find the attenuation of a V band guide of about 0.3 dB/m for 70 GHz microwave frequency.

In Eq. 9.18 the conductivity of the metallic walls appear explicitly, because its finite value makes the electromagnetic field attenuate exponentially with the depth from the metallic surface. Both the electric and the magnetic fields fall to $1/e$ in a distance

$$\delta = \sqrt{\frac{2}{\mu\sigma\omega}}, \quad (9.19)$$

which is known as the skin depth [4]. At 70 GHz in pure Cu, this is 0.25 μm, which is about one order of magnitude higher than the mean free path of the conduction electrons (a few tens of nm). Therefore at the millimeter wave frequencies, the surface conductivity is not yet limited by the anomalous skin depth [5], the effects of which begin to appear only when the frequency approaches the electron collision frequency

$$\frac{1}{\tau} = \frac{n_e e^2}{m_e \sigma}, \quad (9.20)$$

where m_e and n_e are the mass of the free electron and their number density. For copper this is in the far infrared region at RT, but if the temperature is lowered below 1 K, the conductivity can be several orders of magnitude higher and the skin depth may be higher than that given by the low-frequency Eq. 9.19.

Attenuation in a straight oversized guide can be substantially lower than that of a single-mode guide, if the transition from single-mode guide at both ends is done progressively so as to reduce the reflections and mode conversions at both ends. It is usual to place the microwave source and its control circuitry at a distance of order 10 m from the polarized target, and to cover the essential part of the distance by an X-band guide where the attenuation is about three times lower. Discontinuities, bends, twists and other possible transitions are excluded in this line; these must be made in the single mode sections of the transmission line behind the tapered transitions.

9.2.3 Round Waveguides

Because the transmission line inside the refrigerator often needs to traverse hermetically through various walls and because there needs to be at least one waveguide window that separates the air-filled and evacuated sections of the guide, it is practical to make a section of the transmission line out of a round tube having a low thermal conductivity, for example, 70/30 cupronickel, coated inside by silver or copper layer of a few skin depths thickness, usually 0.5 µm to 1 µm.

In round waveguides the first subscript designates the order of the Bessel function that is the solution of Laplace's equation derived from the wave equation in cylindrical coordinates, and the second subscript gives the number of the root of this Bessel function; these stem from the boundary conditions that must be satisfied at the inner surface of the metallic wall of the round guide.

In a round guide the circular symmetry leads to the propagation with no preferred transverse orientation. Any imperfection of the guide can then change the orientation of the transverse electric field of the TE_{11} mode and also to conversion to another possible mode. Therefore, it is common to maximize the transmission of the microwave power in situ by rotating the plane of the polarization so that the transmission through the round guide and coupling to the target cavity are maximized. This is easily accomplished by using a suitable resistance thermometer as a microwave bolometer inside the refrigerator near the target.

The simplest mode of propagation in a round waveguide is TM_{01} in which the magnetic field lines are circles concentric with the axis of the guide, the z-axis of our cylindrical coordinate system, with no component in the z and radial directions. The electric field lines meet the walls perpendicular to it and to the magnetic field lines, with components in the z and radial directions. One uses the term circular polarization for this mode of transmission, because the electric field has components in all transverse directions.

The dominant mode (with lowest cut-off frequency), however, is TE_{11} in which the electric and magnetic field lines resemble those of the TE_{01} mode of a rectangular guide. The electromagnetic wave equation, with boundary conditions determined by the properties of the materials and their interfaces, can be solved in cylindrical system of coordinates. These equations have multiple solutions, or modes, which are eigenfunctions of the differential equations; because of the cylindrical symmetry, the eigenfunctions are Bessel functions, whose zeros are determined by the boundary conditions given by the cylindrical walls.

The details of the mathematics are beyond our present scope, and we shall just summarize here the results for the mode TE_{11} in which the electric field has no component in the z direction, and which has the lowest cut-off frequency.

Maxwell's equations for sinusoidally varying fields yield Laplace Eq. 9.1; in cylindrical coordinate system, this can be written, for a round guide with propagation in the direction of the z-axis in the TE_{11} mode (with no electric field in the z direction):

$$\frac{\partial H_z}{r\partial\theta} - \gamma H_\theta = (\sigma + i\omega\varepsilon)E_r;$$

$$-\gamma H_r - \frac{\partial H_z}{\partial r} = (\sigma + i\omega\varepsilon)E_\theta;$$

$$\frac{1}{r}\left(\frac{\partial(rH_\theta)}{\partial r} - \frac{\partial H_r}{\partial\theta}\right) = 0; \quad (9.21)$$

$$\gamma E_\theta = -i\omega\mu_0 H_r;$$

$$-\gamma E_r = -i\omega\mu_0 H_\theta;$$

$$\frac{1}{r}\left(\frac{\partial(rE_\theta)}{\partial r} - \frac{\partial E_r}{\partial\theta}\right) = -i\omega\mu_0 H_z.$$

Here γ is again the propagation constant that has to be determined. These can be (tediously) further simplified to obtain the differential equation for the radial component of the transverse electric field

$$r\frac{\partial^2(rE_r)}{\partial r^2} + \frac{\partial(rE_r)}{\partial r} + \frac{\partial^2 E_r}{\partial\theta^2} + h^2 r^2 E_r = 0, \quad (9.22)$$

where

$$h^2 = \omega\mu_0\varepsilon + \gamma^2 - i\sigma\omega\mu_0. \quad (9.23)$$

The solution of Eq. 9.22 can be tried by the technique of separated variables, which is done by the replacement

$$rE_r(r,\theta) = R(r)T(\theta) \quad (9.24)$$

that yields

$$\frac{r^2}{R(r)}\frac{\partial^2 R(r)}{\partial r^2} + \frac{r}{R(r)}\frac{\partial R(r)}{\partial r} + \frac{1}{T(\theta)}\frac{\partial^2 T(\theta)}{\partial\theta^2} = -h^2 r^2. \quad (9.25)$$

This can be written as two independent differential equations

$$\frac{\partial^2 T(\theta)}{\partial\theta^2} = -C_1 T(\theta);$$

$$\frac{\partial^2 R(r)}{\partial r^2} + \frac{1}{r}\frac{\partial R(r)}{\partial r} + \left(h^2 - \frac{C_1}{r^2}\right)R(r) = 0. \quad (9.26)$$

9.2 Waveguide Circuits

The solution of the first of these is a sinusoidal dependence on θ; by choosing the electric field orientation so that $T(\theta)$ vanishes for all r when θ is 0 and $n\pi$, and by choosing the simplest mode, we find $C_1 = 1$ and

$$T(\theta) = C_2 \sin\theta. \tag{9.27}$$

The solution of the second Eq. 9.26 is

$$R(r) = FJ_1(hr) + GY_1(hr), \tag{9.28}$$

where J_1 and Y_1 are first-order Bessel functions of the first and second kind, respectively. The function Y_1 diverges for $r = 0$, so that G must be zero for a physical solution. These yield the complete solution for the radial electric field:

$$rE_r(r,\theta) = AJ_1(hr)\sin\theta. \tag{9.29}$$

The azimuthal component $E_\theta(r,\theta)$ is obtained by taking the derivative of the fourth of Eq. 9.21 and integrating

$$\frac{\partial E_\theta(r,\theta)}{\partial \theta} = -\frac{\partial(rE_r(r,\theta))}{\partial r} = AhJ_1'(hr)\sin\theta, \tag{9.30}$$

which yields

$$E_\theta(r,\theta) = AhJ_1'(hr)\cos\theta + B. \tag{9.31}$$

The boundary conditions for $E_\theta(r,\theta)$ are

$$\begin{aligned} E_\theta(r,\theta) &= 0 \text{ for all } r = b \text{ (radius of the guide);} \\ E_\theta(r,\theta) &= 0 \text{ for all values of } r \text{ when } \theta = \frac{\pi}{2} + n\pi. \end{aligned} \tag{9.32}$$

The second of these gives that the constant of integration must be $B = 0$. The first condition yields

$$J_1'(hb) = 0, \tag{9.33}$$

which implies that hb must be a root of the derivative of $J_1(x)$. For the TE_{11} mode this is the first root is $p_1 = 1.84118$, which leads to

$$hb = p_1 = 1.84118 = b(\omega^2\mu_0\varepsilon + \gamma^2 - i\sigma\omega\mu_0). \tag{9.34}$$

Because the conductivity inside the evacuated or air-filled guide is zero, we can solve

$$\gamma^2 = \left(\frac{p_1}{b}\right)^2 - \omega^2\mu_0\varepsilon \equiv (\alpha + i\beta)^2, \tag{9.35}$$

and as we are here interested in unattenuated ($\alpha = 0$) propagation, we find the propagation constant

$$\beta = \sqrt{\omega^2 \mu_0 \varepsilon - \left(\frac{p_1}{b}\right)^2} = \sqrt{\left(\frac{\omega}{c}\right)^2 - \left(\frac{p_1}{b}\right)^2}. \qquad (9.36)$$

The first roots of the low-order Bessel functions can be found in Appendix A.8; higher order functions and their roots can be found in all major tables of functions, for example, Ref. [6].

The electric field components can now be summarized for the TE_{11} mode:

$$E_r(r,\theta) = \frac{A}{r} J_1(hr) \sin\theta;$$
$$E_\theta(r,\theta) = Ah J_1'(hr) \cos\theta; \qquad (9.37)$$
$$E_z = 0.$$

The magnetic field components can be found now from Eq. 9.21:

$$H_r(r,\theta) = -\frac{A\beta}{\omega\mu_0} J_1'(hr) \cos\theta;$$
$$H_\theta(r,\theta) = \frac{A\beta}{\omega\mu_0 r} J_1(hr) \sin\theta; \qquad (9.38)$$
$$H_z = -\frac{iAh^2}{\omega\mu_0} J_1(hr) \cos\theta.$$

The cut-off frequency for mode TE_{1n} is obtained by putting in Eq. 9.36 $\beta = 0$:

$$f_0 = \frac{c p_n}{2\pi b}; \qquad (9.39)$$

this gives $f_0 = 44$ GHz for a guide with 2 mm inner radius in TE_{11} mode. The cut-off frequency for the TM_{01} mode is 57 GHz, so that at 70 GHz both modes can propagate.

The phase velocity, guide wavelength and propagation impedance are for the TE_{11} mode:

$$v_p = \frac{\omega}{\beta},$$
$$\lambda_g = \frac{2\pi}{\beta} \text{ and} \qquad (9.40)$$
$$Z_g = \frac{253\,\Omega}{\sqrt{1 - \left(\frac{f_0}{f}\right)^2}}.$$

In the case of the TM_{01} mode, the Laplace equation can be again written in the form of Eq. 9.21, while noting that now the magnetic field has only the azimuthal component, with no radial nor axial component, and the electric field has no azimuthal component. This is simpler than in the above case of the TE_{11} mode.

Without going to the details, we just quote here the results for the TM_{01} mode:

$$\gamma = i\beta = i\sqrt{\omega^2 \mu_0 \varepsilon - \left(\frac{p_n}{b}\right)^2};$$

$$f_0 = \frac{cp_n}{2\pi b};$$

$$\beta = \sqrt{\left(\frac{\omega}{c}\right)^2 - \left(\frac{p_n}{b}\right)^2}; \qquad (9.41)$$

$$v_p = \frac{\omega}{\beta};$$

$$\lambda_g = \frac{2\pi}{\beta}.$$

The attenuation of the TM_{01} mode is approximately

$$\alpha = \frac{1}{b}\sqrt{\frac{\pi\varepsilon}{\sigma}} \sqrt{\frac{f}{1-\left(\frac{f_0}{f}\right)^2}}; \qquad (9.42)$$

the attenuation of the lowest mode TE_{11} has a similar magnitude and frequency dependence.

9.2.4 Waveguide Components

The microwave circuitry for a polarized target consists of

- the source and its control and measurement devices that can be conveniently mounted to an optical table in a control room;
- the oversized waveguide that brings the power to the target sitting in the beam area with controlled access;
- transition to a single mode guide and to a round guide, followed by a rotatable joint, the waveguide window and the round guide section and
- transition to a rectangular guide section and coupling to the target cavity.

The waveguides are connected together and to other components by standardized flanges that have precise alignment pins (dowel pins) to provide a good match of the rectangular openings of the guides. For V and E band guides the round flanges are most commonly designated UG-385 coming from the U.S. Military Standard. Other standardization organizations (IEC, EIA and RCSC) have different designations for essentially the same flange type, thus permitting their interconnection with minimal problems. Oversized X band guides use rectangular flanges with different type designations following the above standards.

The components include various types of passive and active devices; among the passive components are guide junctions, isolators, power absorbers, attenuators, phase shifters, tuners, couplers, shorts, antenna horns, switches, reflectometers and tapered transitions to other guide sizes and so on. Active devices include diode mounts, harmonic mixers, harmonic generators and amplifiers. The commercial suppliers provide test data sheets for their components that include insertion loss and often full scattering (S-) parameters measured in the specified frequency band.

If access to a fully equipped microwave lab with a vector network analyzers and spectrum analyzers is not conveniently available, the minimal instrumentation for the control of waveguide circuitry would include a search horn antenna and a tunable diode mount to search for possible leaks of the microwave power from the connections of the guides. Also, a reflectometer with a tuned diode mount and a calibrated attenuator allow the precise measurement of reflected power; this is useful for optimizing the round waveguide section and the coupling from the waveguide to the cavity inside the refrigerator. It is also desirable to control the reflections of the tapered transitions to and from the oversized guide sectors.

Transition from a single mode guide to a round guide should be equipped with a rotating flange on the round guide. This allows to twist the single mode guide so as to orient the electric field for optimum coupling to the target cavity.

9.2.5 Waveguide Windows

The waveguide must be equipped with at least one window that separates the RT section filled with air from the section traversing the isolating vacuum of the refrigerator. If the vacuum is sectorized so that the coldest part of the refrigerator has its independent vacuum, a second window must be installed in the guide. This is also required if the guide terminates in the part of the refrigerator that is exposed to helium.

The window is most easily installed in a round section of the guide. It is often made of FEP foil of 0.25 mm thickness, clamped between round guide flanges of the UG-385 size, shaped so that one side is equipped with a rounded knife, whose height is one half of the foil thickness. As such a flange may leak power out and also cause a reflection of power back towards the source, it is often equipped with a quarter-wavelength deep circular choke groove, which is about 0.4 mm wide and 1.2 mm deep. The short on the bottom of the groove ditch is thus transformed into a high impedance at its mouth. The groove should be at a distance of one quarter wavelength from the radius of the guide; thus the gap due to the FEP foil will transform the high impedance into a short circuit at the radius of the guide. Here we must note that the wavelength in this case refers to that in the FEP-filled gap, which is longer than the free-space wavelength.

It should be noted that the choke in the window flange is not needed if its design minimizes the power leakage, and if the reflected power does not cause problems for the microwave power source. Choked flanges are mainly used for high-power systems where pressurization of the guide is used for preventing arcing.

9.2.6 Target Cavity

Small target cavities can be fed by a conical antenna transition from the guide to the side or end of the cavity, as best fits for the layout of the refrigerator. For longer targets the coupling to the target cavity is usually made from the long side using a rectangular guide section.

When a single-mode guide is used, the coupling slots are machined to the short side of the guide. These transverse slots are 1.5 mm wide and inclined by about 15° from transverse orientation and spaced by one wavelength of the guide. If the coupling to a long cavity must be uniformly distributed, the coupling slots may be grouped, and the groups may be spaced so as to get progressively increased coupling along the guide. The slot width can be also used as an adjustable parameter.

If an oversized Ka band guide is used for the cavity feed, the coupling is made via a longitudinal slot machined to the center of the wide side of the guide. The slot can be tapered so as to obtain a better uniformity of the coupling along the length of the target; the tapered slot dimensions are best adjusted by measurement of the microwave field strength inside the cavity when it is loaded by a graphite absorber.

The target cavity can be best defined as a multimode resonant structure that is not tuned to a particular frequency or designed to favor a particular mode or field orientation. The walls are made of high-conductivity Cu sheet and leakage through cracks in the design is controlled by choke grooves and absorbers. The cavity can be cooled by the flow of helium so that its heat load due to microwaves is not absorbed by the target refrigeration cycle.

For a uniform irradiation of the target inside the untuned multimode cavity, the cavity should be as large as possible. However, as the cavity losses increase with its size, a compromise must be sought that also takes into account the requirement that heavy materials should be minimized in the target surroundings.

Microwave field strength inside the cavity can be monitored by a carbon composite resistance thermometer that acts as a very sensitive bolometer when the refrigerator is in normal operating mode. It serves for two useful purposes, with a fixed microwave frequency while scanning the magnetic field:

(1) Search of ESR absorption line edges, which appear as sudden reduction in the field strength when the magnetic field is scanned through the resonance condition.
(2) Observation of the cavity tune and mode switches when one scans the field outside the absorption line, where the dispersion part of the ESR line has a substantial frequency dependence.

The details of the ESR absorption line, however, cannot be resolved by such a bolometric detector, because the spin density is so high that the target appears as almost 'black' when its dimensions exceed the wavelength.

9.2.7 Frequency and Power Measurement

The microwave frequency can be measured by a cavity wavemeter and a diode detector. A calibrated micrometer dial of the wavemeter allows the direct reading of the frequency at the dip of the diode signal. As the microwave diodes have an extremely low capacitance, they are quickly destroyed by static charge if they are not connected to a readout device with a low impedance, such as a 10 µA moving coil meter.

The cavity wavemeter and diode can be replaced by a harmonic mixer that downconverts the microwave signal to the range of a direct frequency counter, which permits to relate the microwave frequency with that of a local oscillator phase locked to a crystal standard. Commercially available microwave counters determine which harmonic of the local oscillator has been selected for converting the frequency to the range of the counter and calculate directly the microwave frequency from the harmonic and downconverted signal frequencies.

Rectifying diode detectors mounted in a tunable cavity also give an idea of the microwave signal amplitude, if a calibration is available. A calibrated attenuator, based on Faraday rotation, permits to make accurate measurement of the relative power, based on the diode signal. The absolute power measurement, however, requires a calorimeter that is calibrated by electric power applied to the microwave absorber structure equipped with a heater and a thermometer.

The polarized target refrigerator can be also used as a calibrated calorimeter, if the cooling power has been accurately determined as a function of coolant temperature and its rate of circulation. This works well in evaporation refrigerators where the boiling pressure gives the temperature and cooling power. However, in a dilution refrigerator a resistance thermometer is the most practical thermometer, and as the microwave power heats it substantially, the power to be measured must be turned off and the record of the temperature reading has to be extrapolated back to the moment when the power was turned off. This extrapolated temperature then gives the heat load due to microwaves whose power was switched off just prior to the record [7].

As no powermeter can withstand the full output power of an EIO or a BWO tube, the power must be sampled by a directional coupler; this may be rated with a -20 dB or -40 dB coupling coefficient, depending on the power of the source and on the power handling capacity of the meter. Moreover, it is practical to place an adjustable attenuator between the coupler and the powermeter; this may be a blade attenuator or a calibrated Faraday rotation attenuator.

9.3 Iron Yoke Magnets with Warm Coils

First polarized targets used iron yoke magnets powered by water-cooled copper coils. They were operating in 1.8 T to 2.0 T field obtained between soft iron pole pieces, and the microwave frequency was around 70 GHz, because the target material was LMN single crystal with Nd^{3+} doping, oriented so that the g-factor had its maximum value.

The advent of high polarization in glassy alcohol and diol materials doped with free radicals or metallo-organic complexes led to the need of a higher magnetic field of 2.5 T, because the g-factors of these dopants are close to the free-electron value, with a fairly small anisotropy. It was easier and less costly to equip the magnets with cobalt-iron pole pieces rather than buying new microwave sources. Cobalt-iron alloys with about 50% cobalt have saturation magnetizations reaching 2.35 T, thus permitting to obtain 2.5 T in between pole pieces when the current coils add some tenths of tesla to the field strength. Such pole pieces need, however, to be annealed during 10 h at 800°C and then slowly cooled to reach maximum permeability. Figure 9.2 shows the permeability of annealed cobalt-iron alloy Vacoflux® 50.[1] This alloy is commonly used for 2.5 T magnets of polarized targets.

The soft iron yoke has usually C shape and a beam hole in the vertical return flux part, in order to maximize the angle of access to the target in the horizontal scattering plane. This allows to make transverse spin asymmetry measurements with vertical target spin orientations. Longitudinal spin orientation then necessitates frozen spin operation in the holding field created by a thin superconducting solenoid wound around the inner vacuum jacket covering the target. A typical C-yoke-polarized target magnet was pictured by Desportes in Ref. [8].

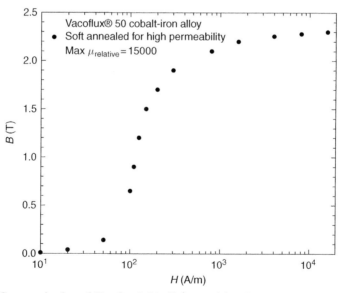

Figure 9.2 DC magnetization of Vacoflux® 50 alloy containing 49% Co, 49% Fe and 2% V. The material must be annealed for obtaining maximum relative permeability

---

[1] Vacuumschmelze GmbH.

A rule-of-thumb calculation of the current coils is based on the concept of 'reluctance' of the magnetic circuit, which has similarity with the resistance of a circuit with resistances in series. It is useful for the cases where the stray magnetic flux is small in comparison with the flux carried by the magnetic circuit, and in particular when the volume in the air gap(s) (or in media with low relative permeability) is small in comparison with the volume of the magnetic circuit with a high permeability.

9.3.1 Magnetic Susceptibility and Boundary Conditions

Let us assume that the magnetic parts are ideally permeable so that there is no permanent magnetic moment in the absence of true currents, and where the magnetic moment is produced only by the field produced by external true currents. The field equations are then [4]

$$\nabla \cdot \mathbf{B} = 0;$$
$$\oint \mathbf{H} \cdot d\mathbf{l} = J_{true}. \tag{9.43}$$

Assuming now that $\mathbf{M} = \chi_m \mathbf{H}$, we get

$$\mathbf{B} = \mu_0 (\chi_m + 1) \mathbf{H} = \mu_r \mu_0 \mathbf{H} = \mu \mathbf{H} \tag{9.44}$$

where μ_r is the relative permeability and μ is the absolute permeability. At a boundary between two linear media 1 and 2, denoting with $\mathbf{n}$ the unity vector normal to the interface, the relation between the normal components is

$$\mathbf{n} \cdot (\mathbf{B}_2 - \mathbf{B}_1) = \mathbf{n} \cdot (\mu_2 \mathbf{H}_2 - \mu_1 \mathbf{H}_1) = 0, \tag{9.45}$$

while for the tangential components

$$\mathbf{n} \times (\mathbf{H}_2 - \mathbf{H}_1) = \mathbf{n} \times (\nabla \cdot \phi_{m2} - \nabla \cdot \phi_{m1}) = \mathbf{n} \times \left(\frac{\mathbf{B}_2}{\mu_2} - \frac{\mathbf{B}_1}{\mu_1} \right) = \mathbf{K}, \tag{9.46}$$

where ϕ_m is the magnetic scalar potential and $\mathbf{K}$ is the true surface current between the two media, which is usually zero in magnetostatic problems. The normal component of $\mathbf{B}$ is thus continuous across the boundary, and the tangential component of $\mathbf{H}$ is continuous if there is no surface current. This also applies approximately in the saturated magnetic materials, if the saturation magnetization is not complete.

9.3.2 Magnetic Circuit and Reluctance

We can note that Eqs. 9.43 and 9.44 are mathematically identical with the equations of stationary flow of current in a continuous medium in the presence of nonconservative electromotive force, which leads to the concept of resistance of linear conductors in series [4], [9]. This analogy leads to the concept of the magnetic circuit, and the solution of linear magnetic media problems given by the expression for the magnetic flux Φ_m:

9.3 Iron Yoke Magnets with Warm Coils

$$\Phi_m = \int \mathbf{B} \cdot d\mathbf{S} = \frac{I}{\mathfrak{R}}, \qquad (9.47)$$

where

$$\mathfrak{R} = \sum_i \frac{l_i}{\mu_i S_i}$$

is the magnetic reluctance that has the units ampere/weber; here S_i is the cross-sectional area of the magnetic conductor, l_i is its length and μ_i is its magnetic permeability. Thus, the reluctances of components in series or in parallel are obtained analogously with electrical resistance from

$$\begin{aligned} \mathfrak{R} &= \sum_i \mathfrak{R}_i \quad \text{with } \mathfrak{R}_i \text{ in series, and} \\ \frac{1}{\mathfrak{R}} &= \sum_i \frac{1}{\mathfrak{R}_i} \quad \text{with } \mathfrak{R}_i \text{ in parallel.} \end{aligned} \qquad (9.48)$$

If the current conductor carrying I is a coil making N turns around the magnetic circuit's ring, the current must be understood as 'ampere turns', i.e. NI.

If a section of the magnet's ring is replaced by an air gap with shaped pole pieces, one obtains an electromagnet with reluctance [9]

$$\mathfrak{R} = \mathfrak{R}_{iron} + \mathfrak{R}_{air\,gap} = \mathfrak{R}_{iron} + \frac{l_{air}}{\mu_0 S'_{air}}, \qquad (9.49)$$

where l_{air} and S'_{air} are the length and effective cross-sectional area of the gap, respectively. The flux throughout the circuit is then obtained from Eq. 9.47:

$$\Phi_m = \frac{NI}{\mathfrak{R}} = \frac{NI}{\mathfrak{R}_{iron} + \dfrac{l_{air}}{\mu_0 S'_{air}}} \qquad (9.50)$$

and the magnetic flux density (= magnetic induction) in the air gap is

$$B_{air\,gap} = \frac{\Phi_m}{S'_{air}} = \frac{\mu_0 NI}{\mu_0 \mathfrak{R}_{iron} S'_{air} + l_{air}}. \qquad (9.51)$$

It should be noted that the magnetic permeability may be more than 10,000 at low magnetic field, while near saturation it can be two orders of magnitude lower. This happens in particular in the pole pieces shaped to concentrate the flux so as to reach 2.5 T flux density. In this case the reluctance of the pole pieces is not linearly related with the current of the coils, and an iterative process or simulation must be used to obtain the magnetic induction between the pole pieces.

In the case where the reluctance of the air gap dominates that of the magnetic circuit, Eq. 9.51 can be approximated by

$$B_{\text{air gap}} \cong \frac{\mu_0 NI}{l_{\text{air}}}, \tag{9.52}$$

which can be used to obtain a rough value of the magnetic induction in the gap at low current in the coils.

9.3.4 Magnetic Field Uniformity in the Air Gap

In Chapter 4 we have shown that the polarization can be increased if the magnetic field uniformity is made better. Improvement of DNP down to the range of 10^{-5} in field homogeneity can be seen when the dominant broadening mechanism is the anisotropy of the g-factor and the dynamic cooling of spin-spin interactions dominates other mechanisms of DNP.

The uniformity of 10^{-5} in an iron-core magnet can be reached by the shimming of the pole pieces. This is accomplished by attaching shaped foils of soft iron onto the pole pieces. The shape of the foils can be calculated by a numeric simulation of the magnetic circuit. The procedure is iterative: after measurement of the unshimmed field map in the air gap, the magnetic field is numerically simulated, and the simulation parameters are then adjusted so as to get a good fit with the measurement. The main adjustable parameters are those of the magnetic permeability curve and the effective density of the magnetic material in the pole pieces. The shape of the shims is then optimized by numeric simulation so that they give a desired improvement in the uniformity of the field within the planned volume of the target. They are fabricated and mounted for a new measurement of the field map. This iteration is then repeated until the desired field uniformity is reached.

The magnet modelling and numeric simulation are based on the potential field approach of Maxwell equations: in the case of static magnetic field, one determines the magnetic vector potential $\mathbf{A}$ from which the magnetic induction is readily obtained by $\mathbf{B} = \nabla \times \mathbf{A}$. In the general potential field approach Maxwell equations (see Appendix A.1.1) are

$$\nabla^2 \varphi + \frac{\partial}{\partial t}(\nabla \cdot \mathbf{A}) = -\frac{\rho}{\varepsilon_0},$$

$$\left(\nabla^2 \mathbf{A} - \frac{1}{c^2}\frac{\partial^2 \mathbf{A}}{\partial t^2}\right) - \nabla\left(\nabla \cdot \mathbf{A} + \frac{1}{c^2}\frac{\partial \varphi}{\partial t}\right) = -\mu_0 \mathbf{J}, \tag{9.53}$$

where φ is the scalar electric potential, ρ is the charge density and $\mathbf{J}$ is the current density. In the case of a static magnetic field, there is no explicit time dependence and the charge density is zero everywhere; moreover, for stationary currents we may put

$$\nabla \cdot \mathbf{A} = 0, \tag{9.54}$$

which involves no new physical assumptions [4]. With this simplifying assumption, the differential equation for $\mathbf{A}$ reduces to the vector form of Poisson's equation:

$$\nabla^2 \mathbf{A} = -\mu_0 \mathbf{J}, \tag{9.55}$$

which can be solved numerically by finite element (FE) computational methods that are commercially available. The commercial software can be used for the creation of a three-dimensional mesh in which Eq. 9.55 is solved for each node in the mesh, which can be of tetrahedral or of more complex shape. Simpler geometries can be solved in a two-dimensional mesh; these include PT magnets with circularly symmetric pole pieces. The non-linearity of the magnetic circuit materials is taken into account by the use of experimental data on the magnetic permeability as a function of magnetic field. The shimming in this case must be done at the field value chosen for the polarized target.

A large polarized target magnet, with cobalt-iron pole pieces of 25 cm pole tip diameter and 7.5 cm gap length, was successfully shimmed to reach the field uniformity of 10^{-5} at 2.5 T in the target volume of 12 cm length and 20 mm diameter [7]. The field modelling and simulation used the Poisson FE software of CERN program library [10]. Deuteron polarizations in excess of 40% were first time reached in a such a large target volume, which proved the usefulness of the improved field homogeneity. This target was then used in a series of polarized neutron charge exchange reactions using the polarized deuterated target and kaon and pion beams at CERN [11, 12].

9.4 Superconducting Magnets

Superconducting solenoids are ideally suitable for longitudinally polarized targets, but they can also be used for transverse polarization using a split coil design. The modelling and simulation is particularly easy if there are no magnetic materials in the proximity; otherwise it is necessary to use the FE software, especially if there are nearby magnetic shields or spectrometer magnets. The key issues arise in the winding of very regular coils that are needed for reaching the required field homogeneity:

- When using a round conductor, the wire valleys in the coil layers must be filled with epoxy before winding the next layer, so as to avoid winding irregularity at the crossing of the return layer wire with that of the underlying layer.
- With a rectangular conductor, the twist of the filaments ends up in a twist of the conductor after final drawing steps, unless bundles of filaments with opposite twist are used. If the twist remains under control, epoxy filling of the layers is not needed, and wet winding is alternative to impregnation after dry winding.
- Winding forces on the mandrel tend to bend and deform the solenoid into the shape of a 'banana'. A support mandrel is used to prevent this.
- Magnetic forces on the coils need to be supported by Al alloy tape or wire wound on top the coils.
- The coil assemblies must be supported inside the cryostat using cryogenic support rods that need heat sinking and low conductivity, while withstanding the applied magnetic and gravity forces.

The main configurations of axially symmetric superconducting coils were discussed already in 1971 by H. Desportes, with several examples of early magnets reaching over

9.4.1 Solenoid Field

To discuss field quality in detail, we must first estimate the magnetic field on the symmetry axis of an ideal thin solenoid shown in Figure. 9.3, with a winding radius a, total length l and total number of turns N. Using the notation of the figure, the field at point P is readily obtained by integration from the law of Ampère [9]:

$$H = \frac{NI}{2l}(\cos\beta_2 - \cos\beta_1). \tag{9.56}$$

At the center of a very long solenoid $\cos\beta_2 \cong 1$, $\cos\beta_1 \cong -1$ and the field is therefore close to

$$H \cong \frac{NI}{l} \tag{9.57}$$

and at the end it is one half of this:

$$H \cong \frac{NI}{2l}. \tag{9.58}$$

This suggests that the field at the end could be brought up to its value at the center by roughly doubling the number of turns towards the ends of the coil. This is done usually by adding compensating coils at the ends of the solenoid.

The field at the center of the uncompensated solenoid is approximately

$$H_0 \cong \frac{NI}{l}\left[1 - 2\left(\frac{a}{l}\right)^2\right], \tag{9.59}$$

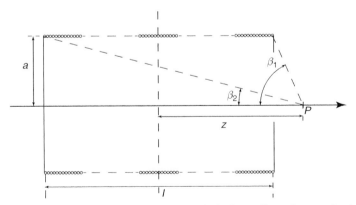

Figure 9.3 Model for a thin long solenoid of length l and winding radius a. See text for the evaluation of the field at point P on the axis

and if z is the axial distance from the center, the field on the axis drops as

$$H(z) \cong H_0 \left[1 - 2 \left(\frac{2a}{l} \right)^2 \left(\frac{z}{l} \right)^2 \right]. \tag{9.60}$$

Let us see numerically what this implies, by assuming a coil with $2a/l = 0.1$ and $z/l = 0.1$; these yield

$$H(z) \cong H_0 \left[1 - 2 \cdot 10^{-2} \left(\frac{z}{l} \right)^2 \right] = H_0 \left[1 - 2 \cdot 10^{-4} \right]. \tag{9.61}$$

Thus less than 20% of the length of this solenoid is usable for the DNP of a polarized target. Therefore, trimming coils are mounted along its whole length; these must have adjustable current so that winding irregularities can also be corrected.

9.4.2 Solenoid End Compensation

In a very long ideal solenoid the field in the central part is very homogeneous both axially and radially. In a real solenoid this is not the case, which is why we now evaluate the relationship between the radial and axial field gradients. This is most easily done by writing the Maxwell equation for the divergence of the magnetic induction explicitly in cylindrical coordinate system:

$$\nabla \cdot \mathbf{B} = 0 = \frac{1}{r} \frac{\partial (rB_r)}{\partial r} + \frac{1}{r} \frac{\partial B_\varphi}{\partial \varphi} + \frac{\partial B_z}{\partial z}. \tag{9.62}$$

Because the solenoid has axial symmetry, the azimuthal derivative is zero, and we find directly the relationship between the radial and axial derivatives

$$\frac{1}{r} \frac{\partial (rB_r)}{\partial r} = -\frac{\partial B_z}{\partial z}. \tag{9.63}$$

This is valid in the whole volume of the solenoid. The variation of the axial field is therefore directly related with that of the radial field: when the axial component drops, the radial one grows.

In order to keep the total length of the solenoid reasonable, end compensation coils are added to the main solenoid; these can be wound on top of the solenoid or at its ends. Additional trim coils, however, are required to get the field homogeneity to the level of 10^{-5}. These provide fine tuning of the end effects, and compensation for the inaccuracy of the main solenoid coil winding and that of the superconducting wire itself.

9.4.3 Winding Accuracy

In a well-compensated and trimmed solenoid, the residual field errors are mainly due to winding errors and inaccuracies. Let us now estimate the error due to the crossing of the wires in successive layers of a round wire, which happens when the surface of each layer

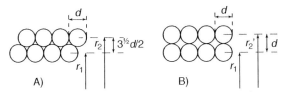

Figure 9.4 (A) The round wire of the second layer falls in the unfilled gaps of the first layer wire; (B) the round wire of the second layer crosses the wire of the first layer. This happens twice per turn on opposite sides of the coil

is not made flat by epoxy and/or by a sheet of rigid insulator. As shown in Figure 9.4, the second layer is lifted radially by

$$r_2 - r_1 = h = \frac{\sqrt{3}}{2} d \cong 0.866 \, d \tag{9.64}$$

when the wire of the next layer falls in between the turns underneath, but in the sector where the wires cross, the second layer is lifted by

$$r_2' - r_1 = d. \tag{9.65}$$

For one turn the coil error ('void' in the current density) can be estimated as from an opposite current loop with height

$$r_2' - r_2 = d\left(1 - \frac{\sqrt{3}}{2}\right) \cong 0.134 \, d, \tag{9.65'}$$

and length approximately $r/2$, carrying the same current as the coil wire. This gives rise to a fictitious dipole moment

$$\vec{\mu} = I\vec{A} = -\frac{Ird}{4}\left(1 - \frac{\sqrt{3}}{2}\right)\vec{z}_0, \tag{9.66}$$

which adds the dipole field component, from the upper limit of Eq. 2.4,

$$H_{dipole} = -2 \cdot \frac{\mu}{2\pi r^3} = -\frac{Id}{2r^2}\left(1 - \frac{\sqrt{3}}{2}\right) \tag{9.67}$$

on the axial field of the solenoid. In the above it is taken into account that there are two wire crossings per turn in a helically wound coil.

The wire crossings advance helically along the winding layer of the solenoid, with helically advancing field irregularity of the magnitude of Eq. 9.67, which can be compared with the field of Eq. 9.57:

$$\frac{H_{dipole}}{H_0} = -\frac{dl}{2r^2 N}\left(1 - \frac{\sqrt{3}}{2}\right) = -\frac{d^2}{2r^2}\left(1 - \frac{\sqrt{3}}{2}\right), \tag{9.67'}$$

where we have made use of the fact that $l/N = d$. As a numeric example, let us assume a wire diameter $d = 1.5$ mm and coil radius $r = 100$ mm, which yield for each turn

$$\frac{H_{dipole}}{H_0} = -1.5 \times 10^{-5}. \tag{9.67''}$$

As the dipole fields are inhomogeneous, their radial gradients add up to produce a helically advancing radial gradient that is multiplied by roughly $Nr/2l = r/2d = 67$. This may be partly compensated if the number of coil layers is large, but it turns out in practice that it is hard to wind coils without filling the gaps if the field homogeneity of 10^{-5} is required. The gap filling technique was used for winding the main solenoid of the EMC polarized target magnet [13, 14].

Alternatively the coil of a solenoid can be wound with a rectangular wire. This speeds up the fabrication because wet winding techniques can be used. Moreover, heat transport across the layers is improved, and there is more stabilizing copper in the coil. The solenoid of the Spin Muon Collaboration (SMC) polarized target was wound using such technique [15].

9.4.4 Control and Protection of Large Magnets

Large solenoids and dipole magnets are usually wound with a sizable wire or cable because the magnet is then easier to make accurate and its inductance is lower, which makes it easier to protect against quenches. Consequently, the coils must be supplied with a high current, and it becomes attractive to equip the magnet with a persistent mode switch and retractable current leads, which reduces the heat losses to the helium cryostat. This, in turn, then requires the addition of a sweep coil that permits the control of the field for the measurement of the baselines of the NMR Q-meters.

The use of a resistive shunt is possible for the measurement of current in small magnets, but is impractical and inaccurate when the current is several hundred amperes. The high current in a sizable magnet is best measured with a current transformer that adds no power dissipation in the main current circuit.

The quench protection of large magnets aims at dissipating the magnetic energy uniformly in the coil in the event of a transit of a part of the conductor from superconducting to normal state. For small magnets a part of the energy can also be dissipated in an external resistor, which is possible when the inductance and current are low. In large magnets this is impossible because of the high electric potential difference and electrical breakthrough between the coil layers. The coils of large magnets are therefore equipped with quench detection and quench heaters that warm up the entire coil system uniformly above the critical temperature.

9.4.5 Examples of Large Solenoids and Magnet Systems

In the following subsections we describe some of the leading developments in high-accuracy superconducting solenoids and magnet systems.

EMC Superconducting Solenoid Magnet

Rutherford Laboratory designed, built and tested the first major magnet of the large polarized target of the European Muon Collaboration at CERN. The target system is briefly described in Ref. [16]. The solenoid design and its test results were published in detail in Refs. [13, 14].

The polarized target of EMC was 1 m long, split into 40 cm long halves with a 20 cm gap that permitted unambiguous determination of the target half in which each event vertex was located. The main solenoid was 1.6 m long, with 12 trim coils mounted along its length; it was wound using a ⌀ 1.00 mm wire with Cu:SC ratio 2:1 of the 361 NbTi filaments of 30 μm diameter; the trim coils were wound of ⌀ 0.44 mm wire with 1:1 Cu:SC ratio and 1,000 filaments of 10 μm diameter. Both wires were insulated with 30 μm thick polyvinyl acetal layer. The trim coils had equal lengths and they were equally spaced over the main solenoid, with 1 mm gaps between the coil ends.

The main solenoid was wound on a Cu bore tube with 195 mm ID and 200 mm OD. During winding the bore tube was supported by a demountable mandrel of Al alloy. The trim coils were also wound on Cu bore tubes that were shrink fit onto the finished main solenoid.

Simulation using 3D GFUN FE program showed that wire cross-overs would cause a field error of 1.4×10^{-4}. Therefore a wet winding technique was adopted in which the valleys of each of the 12 layers of the main solenoid was filled by epoxy and then wiped flat before letting the epoxy harden. As the hardening involves shrinkage, the filling and wiping was repeated twice for each layer. The final trimmed field, measured by an NMR probe, was within ±8 parts in 10^5 in a volume of 1,000 mm length and 60 mm diameter.

The assembly of the main solenoid with the overlaid trim coils was finally wound over with an Al alloy wire, the differential dilatation of which provided support to the magnetic forces that tend to lift the coils off their bore tubes.

The Cu bore tubes were important parts of the quench protection system, because their induced current heating permitted fast quench propagation so as to distribute the resistive heating of the wire uniformly throughout the coils. The main solenoid current was 180 A for 2.5 T field, and its inductance was 8.63 H.

SMC Superconducting Solenoid and Dipole Magnet System

SMC has continued and perfected the polarized muon-proton and muon-deuteron experiments of EMC with a longer target and wider angle of acceptance; also, the beam was more intense after the redesign of the CERN M2 muon beam.

The magnet system of the SMC at CERN was described briefly in Ref. [17] and in detail in Ref. [15]. The target volume was 1,500 mm long and 50 mm in diameter; the main solenoid was 2,000 mm long, extended by 150 mm long compensation coils wound on the same supporting bore tube of Al alloy. The bore tube for the main and compensation coils was ⌀ 300 mm, and the outer dimension of the windings were ⌀ 326 mm and ⌀ 347 mm, respectively. The rectangular wire was approximately 1.5 mm in height and 2 mm in width. Wet winding technique was used for these coils.

Sixteen correction coils of 150 mm length were wound on top of the main solenoid, equally distributed along its length, and the whole assembly was wound with Al alloy tape for supporting the magnetic forces on the coils by differential dilatation.

An outer saddle coil dipole, with vertical field, was mounted on top of the coil assembly. This was wound, using Keystone Rutherford-type flat SC tape, on a support of $\varnothing$ 500 mm, and had a maximum field of 0.5 T, with 2% uniformity in the target volume. The coil was used for rotating the field of the main solenoid system so that zero crossing was avoided, and for operating the target in transverse spin mode. The field rotation permitted the reversal of the target spin orientation in about 10 minutes.

The nominal current of the main solenoid was 416 A at 2.5 T and inductance 5.6 H. The dipole current was 650 A for 0.5 T field and its inductance was 0.4 H.

The SMC magnet system was replaced by one with a larger bore and acceptance angle by the COMPASS collaboration [18]. Similar in principle with the SMC magnet system, it consists of a compensated a corrected solenoid with high winding density operating up to 2.5 T field, together with overlaid saddle-coil dipole reaching 0.63 T. The total length of the solenoid is just over 2 m and the homogeneous field length, in the target diameter of 3 cm, is about 1.4 m.

SLAC E143 Split-Coil Solenoid Magnet

The deep inelastic polarized electron-proton and electron-deuteron scattering experiments at SLAC used a longitudinal-field split-coil magnet system built into the vertical target refrigerator. The magnet, produced by Oxford Instruments, had a uniformity of 10^{-4} in a volume of 3 cm diameter and length [19, 20]. The coil split was 8 cm at the narrowest point with clear angles of acceptance of 34° horizontally and 50° vertically. The magnet was operated in persistent mode and its cross section is shown in Figure 8.7.

CEN-Saclay Magnet System for the Nucleon-Nucleon Experiments

A superconducting magnet system was designed for the Saturne II Nucleon-Nucleon scattering experiments with a polarized target and beam; this system consists of three coils and permits the orientation of the target spin in three orthogonal directions [21]. The orientability of the target spin makes it possible, in a single experiment, to perform complete sets of measurements allowing unique determination of the spin-dependent scattering amplitudes in two-body reactions of the nucleons, in this case polarized protons and neutrons.

The main coil with vertical homogeneous field of 2.5 T was lowered after DNP and freezing of the polarization, while one of the two larger holding coils was powered in order to hold the target polarization either vertically or in any direction of the horizontal scattering plane.

References

[1] L. R. Becerra, G. J. Gerfen, B. F. Bellew, et al., A spectrometer for dynamic nuclear polarization and electron paramagnetic resonance at high frequencies, *Journal of Magnetic Resonance, Series A* **117** (1995) 28–40.

[2] Y. Matsuki, H. Takahashi, K. Ueda, et al., Dynamic nuclear polarization experiments at 14.1 T for solid-state NMR, *Physical Chemistry Chemical Physics* **12** (2010) 5799–5803.

[3] M. Rosay, M. Blank, F. Engelke, Instrumentation for solid-state dynamic nuclear polarization with magic angle spinning NMR, *J. Magn. Res.* **264** (2016) 88–98.

[4] W. K. H. Panofsky, M. Phillips, *Classical Electricity and Magnetism*, Addison-Wesley, Reading, MA, 1962.

[5] J. M. Ziman, *Principles of the Theory of Solids*, Cambridge University Press, Cambridge, 1965.

[6] E. Jahnke, F. Emde, *Tables of Functions*, 4 ed., Dover Publications, Inc., New York, 1945.

[7] T. O. Niinikoski, Polarized targets at CERN, in: M.L. Marshak (ed.) *Int. Symp. on High Energy Physics with Polarized Beams and Targets*, American Institute of Physics, Argonne, 1976, pp. 458–484.

[8] H. Desportes, Superconducting magnets for polarized targets, in: G. Shapiro (Ed.) *2nd Int. Conf. on Polarized Targets*, LBL, University of California, Berkeley, 1971, pp. 57–68.

[9] H. E. Duckworth, *Electricity and Magnetism*, Holt, Rinehart and Winston, New York, 1961.

[10] C. Iselin, Solution of Poisson's or Laplace's Equation in Two-Dimensional Regions, CERN Program Library, CERN, Geneva, 1976.

[11] M. Fujisaki, M. Babou, J. Bystricky, et al., Polarization measurements in K^+n charge exchange at 6 and 12 GeV/c, in: G.H. Thomas (ed.) *High Energy Physics with Polarized Beams and Polarized Targets*, American Institute of Physics, Argonne, 1978, pp. 478–485.

[12] A. de Lesquen, L. van Rossum, M. Svec, et al., Measurement of the reaction $\pi^+ n\uparrow \to \pi^+ \pi^- p$ at 5.98 and 11.85 GeV/c using a transversely polarized deuteron target, *Phys. Rev. D* **32** (1984) 21.

[13] N. H. Cunliffe, P. Houzego, R. L. Roberts, A superconducting 2.6 T high accuracy magnet, *Sixth Int. Conference on Magnet Technology*, ALFA Bratislava, Bratislava, 1977, pp. 443–447.

[14] N. H. Cunliffe, R. L. Roberts, A superconducting uniform solenoidal field magnet for a large polarised target, *7th Int. Conf. on Cryogenic Engineering*, IPC Science and Technology Press, London, 1978, pp. 102.

[15] A. Dael, D. Cacaut, H. Desportes, et al., A Superconducting 2.5 T high accuracy solenoid and a large 0.5 T dipole magnet for the SMC target, *MT12*, Leningrad, URSS, 1991, pp. 560–563.

[16] S. C. Brown, G. R. Court, R. Gamet, et al., The EMC polarized target, in: W. Meyer (ed.) *Proc. 4th Int. Workshop on Polarized Target Materials and Techniques*, Physikalisches Institut, Universität Bonn, Bonn, 1984, pp. 102–111.

[17] Spin Muon Collaboration (SMC), D. Adams, B. Adeva, et al., The polarized double-cell target of the SMC, *Nucl. Instr. and Meth. in Phys. Res. A* **437** (1999) 23–67.

[18] E. R. Bielert, J. Bernhard, L. Deront, et al., Operational experience with the combined solenoid/dipole magnet system of the COMPASS experiment at CERN, *IEEE Transactions on Applied Superconductivity* **28** (2018) 4101805.

[19] D. G. Crabb, D. B. Day, The Virginia/Basel/SLAC polarized target: operation and performance during E143 experiment at SLAC, *Nucl. Instrum. and Meth. in Phys. Res. A* **356** (1995) 9–19.

[20] D. G. Crabb, D. B. Day, The Virginia/Basel/SLAC polarized target: operation and performance during experiment E143 at SLAC, in: H. Dutz, W. Meyer (eds.) *7th Int. Workshop on Polarized Target Materials and Techniques*, Elsevier, Amsterdam, 1994, pp. 11–19.

[21] R. Bernard, P. Chaumette, P. Chesny, et al., A frozen spin target with three orthogonal polarization directions, *Nucl. Instrum. and Methods A* **249** (1986) 176–184.

10
Other Methods of Nuclear Spin Polarization

Here we shall discuss other methods to produce high nuclear spin polarization, either in thermal equilibrium with the solid lattice or in dynamic equilibrium in a rotating frame. Optical pumping methods create a very high enhancement of the nuclear spin polarization based on spin-exchange collisions with atoms whose outer electron is polarized by circular polarized light. Some methods are also based on creating high non-equilibrium polarization that is then frozen in by increasing the spin-lattice relaxation time.

Chemical and biomedical research teams use the term 'hyperpolarization' to describe the general methods of spin polarization enhancement beyond thermal equilibrium; DNP methods belong clearly to these. Other methods include optical pumping and chemical polarization methods such as chemical-induced dynamic nuclear polarization (CIDNP) and parahydrogen-induced polarization (PHIP).

10.1 'Brute Force' Polarization

The equilibrium populations of spin states are given by the Boltzmann distribution in the absence of magnetic phase transitions, as was discussed in Section 1.2.1, based on simple and plausible arguments. When the spin-lattice relaxation is sufficiently fast in high magnetic field and low temperature, the thermal equilibrium polarization is reached in a reasonable time scale so that the target spin polarization, given by the Brillouin function of Eq. 1.63, can be used for experimentation. The spin-lattice relaxation was discussed in Section 2.3, and in better detail in Chapter 5 for nuclear spins; from these it was clear that nuclear spin relaxation due to paramagnetic impurities in dielectric materials is much too slow for getting thermal equilibrium in conditions favorable for high polarization.

The favorable conditions here mean that the Boltzmann factor of Eq. 1.63 satisfies roughly

$$x = \frac{\hbar \gamma B_0}{2 k_B T} \cong 24 \frac{f_{\mathrm{NMR}}/\mathrm{GHz}}{T/\mathrm{mK}} > 1. \qquad (10.1)$$

Assuming the gyromagnetic factor of free protons, for example, we find $x = 1.02$ in a magnetic field of 10 T at 10 mK temperature. Therefore, substantial spin polarizations are obtainable for a wide range of nuclei, and targets containing them can be used in experiments such as neutron transmission and scattering that deposit little heat in the target material.

The relaxation is fast enough in metallic materials and in quantum solids such as solid hydrogens HD and DT, and solid ^{3}He; these will be briefly discussed below. A review of the practical brute-force polarized target materials at TUNL[1] has been given by Seely et al. [1].

10.1.1 Metals

In many metals in a high magnetic field, the conduction electrons provide an efficient mechanism for nuclear spin-lattice relaxation. This obeys the Korringa law of Eq. 1.126, where κ is of order 1 sK for many common metals. Therefore, the relaxation time is in the range of 100 s at 10 mK temperature, which is quite manageable for a nuclear spin target.

Among the pure metals used in the TUNL brute-force polarized target are ^{27}Al, ^{93}Nb and ^{165}Ho. These and other relative abundant nuclei in metallic solids were compiled by Seely et al. in Ref. [1]. However, the Korringa constants are not known for many of these.

Metal hydrides such as TiH, TiH$_2$, ZrH$_2$, YH$_2$ and YH$_3$ may contain a large amount of hydrogen, with a hydrogen density comparable or larger (9.0×10^{22} free protons/cm^3 for TiH$_2$) than that of solid ammonia (8.2×10^{22} free protons/cm^3 for NH$_3$). Heeringa developed the fabrication method involving hydrogenation of Ti powder in a high-temperature oven to produce TiH$_2$ in the form of a sponge that was then compressed into a copper ring. This provides the cooling contact for the target [2]. Heeringa proved that the temperature measurement of such a target yielded a proton polarization value of 60% and confirmed the measurement by the known transmission difference of 1.2 MeV neutrons with spin parallel and antiparallel with that of the target [2]. The equilibrium polarization at 12 mK was reached with a time constant of a few days; this was determined to be due to the limited heat conductivity of the sponge material. The spin-lattice relaxation time at 3 mK had an upper limit of 5 min, thus showing Korringa-like behavior [2].

The metallic target temperature is most easily measured by ^{60}CoCo nuclear orientation thermometer that can work in the high field of the polarizing magnet but is more sensitive if located in a low-field region [1]. As it takes several hours of counting the gamma asymmetry of the nuclear orientation thermometer, TUNL have added a ^{3}He melting pressure thermometer to their target system.

Because the brute-force polarization method yields only positive spin temperatures and therefore only one target spin direction with respect to that of the high magnetic field, one has only two methods of polarization reversal:

- reversal of the magnetic field,
- adiabatic passage by RF field at low frequency.

The former must be done extremely slowly because of eddy current heating, and it is usable when only neutral particles are measured. The second is limited to studies of nuclear

---

[1] Triangle Universities Nuclear Laboratory, Durham, NC, USA.

magnetism, because the spin-lattice relaxation is fast in metallic materials, thus limiting severely the counting time with negative spin temperature. Consequently, the spin asymmetries must be obtained by comparing scattering from the target in polarized state with scattering from unpolarized (warm) target.

The rapid exchange of brute-force polarized TiH_2 target with a dummy Ti target was accomplished by a lifting mechanism that permits the target insert to be moved vertically inside the vertical dilution refrigerator [3]. The targets were cooled by a Cu rod with flexible thermal link to the mixing chamber of the dilution refrigerator.

10.1.2 HD

In H_2 the para ($J = 0$) spin state of the molecule has an energy substantially below that of the ortho ($J = 1$) state, as was discussed in Chapter 7, and therefore pure hydrogen cannot be polarized in its ground state. The HD molecules are not limited by this because the H and D nuclei are not identical, and their polarizations are independent as there is no efficient cross-relaxation mechanism. However, if the solid HD is contaminated by some H_2 in ortho state, this provides a mechanism for spin-lattice relaxation of the proton spins even close to 1 K temperature, as was demonstrated by Hardy and Gaines in 1966 [4]. Consequently, Honig [5] proposed to use the ortho–para conversion of H_2 as a switch between the polarization and frozen spin modes of an HD target.

It turned out quickly that in the frozen spin state the proton and deuteron spins of HD were quite efficiently decoupled from each other, and that the D spins took very much longer to polarize by the brute force method. The only method to polarize the D spins then remained RF mixing with the proton spins, which worked well [6].

As the thermal conductivity of solid HD is quite limited at millikelvin temperatures, the heat from spin relaxation (and from the possible beam) must be carried away by metallic wires embedded in the solid target. The background material thus added can be minimized by making the wires out of high-purity Al [6].

A polarized solid HD target was recently used successfully in a polarized photon beam to study inclusive π production [7].

10.1.3 Solid ^3He

^3He has the peculiar behavior of exhibiting a minimum in the melting curve at 315.24 mK that prevents compression into the solid phase by external pressure. In the liquid phase, the Fermi liquid properties limit the polarization to rather low values, but the solid stays paramagnetic in a high field, permitting brute force polarization of the nuclear spins.

The use of solid ^3He as a polarized target was already discussed by J. Wheatley in 1971 [8]; in the early phase it was proposed that the technique of Pomeranchuk cooling be used for cooling the ^3He in a high field. A more practical technique was developed for the TUNL brute-force polarized target facility whereby the compressed liquid is cooled at a density of 0.125 g/cm² that corresponds to a pressure higher than the minimum of the melting curve, and at this pressure and density the liquid solidifies upon cooling into the paramagnetic phase in a high magnetic field [1]. The exchange interaction in the quantum solid is rather strong, and its effect is that the Curie law must be corrected to read

$$P(^3\text{He}) = \tanh \frac{\mu B + \Theta P(^3\text{He}) + KP^3(^3\text{He})}{kT}, \qquad (10.2)$$

where the spin exchange constants are $\Theta/k = -1.18\,\text{mK}$ and $K/k = -1.962\,\text{mK}$. We recall that the nuclear magnetic moment of ^{3}He is negative and therefore the spin polarization vector is antiparallel with the magnetic field when the spin temperature is positive. Thus, the effect of the two spin exchange terms is to reduce the polarization below that given by the Curie law.

10.2 Spin Refrigerator

Spin refrigerator technique is based on a paramagnetic crystal, with an anisotropic g-factor, rotated in a steady magnetic field at a low temperature. The case best studied is yttrium ethyl sulphate (YES), $Y(C_2H_5SO_4)_3 \cdot 9H_2O$, in which ≤ 0.04 at.% of the non-magnetic Y^{3+} ions are substituted by ytterbium Yb^{3+} ions that has in its lowest Kramers doublet a very anisotropic g-factor [9]

$$g(\theta) = \left(g_\parallel^2 \cos^2\theta + g_\perp^2 \sin^2\theta\right)^{\frac{1}{2}}, \qquad (10.3)$$

where θ is the angle between the applied field and the c-axis of the crystal:

$$g_\parallel = 3.33$$
$$g_\perp = 0.00302.$$

The relaxation rate of the ytterbium ion varies approximately as $\sin^2\theta \cos^2\theta$ [9]. If the crystal is rotated in a steady field at a rate greater than the relaxation rate, the Yb^{3+} ion reaches an equilibrium polarization that corresponds roughly to $g(45°)$, or

$$P(\text{Yb}) = \tanh\left(\frac{2.4\mu_B B}{kT}\right). \qquad (10.4)$$

At $\theta = 90°$, the doublet splitting approaches that of the proton, which makes the proton spin system cool by cross-relaxation towards that of the Yb^{3+} ions that were strongly cooled due to the reduction of the g-factor. As the angle progresses beyond 90°, the proton relaxation rate drops rapidly, and the protons become isolated from the paramagnetic ion system.

We note that the rotation at speeds greater than the relaxation rate keeps at times the spin systems adiabatic, while at 90° there can be thermal mixing between the spin systems. Close to 45° the rapid spin-lattice relaxation of the paramagnetic ions repolarizes them. The spin refrigerator can therefore 'pump' magnetic energy from the protons to the lattice.

Simplistically one might say that the spin refrigeration works best with oriented single crystals, but it turns out that high proton polarizations can also be reached in powder samples. This is mainly because the g-factor is nearly zero in the whole plane perpendicular to the c-axis and this can be reached during the rotation whatever is the orientation of the crystallite with respect to the rotation axis, unless the c-axis is parallel to the axis of rotation. The first confirmation was when proton polarization of approximately 65% was achieved with the University of Massachusetts spin refrigerator in a non-uniform 1.07 T magnetic field at 1.25 K temperature. The free protons in YES were polarized in a sample rotated at speeds between 100/s and 200/s [10]. The target was successfully used in an experiment at BNL in secondary kaon and pion beams [11].

Other uses of the spin refrigerator are in spin filtering of neutron beams. The method has the great simplicity that only moderate magnetic field is required, and the field can be very inhomogeneous. The advantage of using polarized protons in such filters is that it is the only method to filter truly white neutron beams up to keV neutron energies.

10.3 Chemical Spin Polarization

10.3.1 Atomic Hydrogen

Chemical reactions are sensitive to the spin states of the atoms in the reacting molecules. Atomic hydrogen is an extreme case of this, and this exotic gas has been proposed to be used in polarized targets and beam sources [12]. The atoms populate the four spin states labelled a, b, c and d from the lowest to the highest magnetic energy, as shown by Figure 10.1 schematically. In the figure the magnetic energy is scaled by the zero-field hyperfine splitting of hydrogen

$$E_{HFS}(B=0) = a_H \hbar^2,$$

and the magnetic field is scaled by

$$B_{HFS} = \frac{a_H \hbar^2}{g_e \mu_B} = 50.7 \text{ mT}.$$

Hydrogen atoms on superfluid helium have an anomalously low surface adsorption energy of $\varepsilon_a / k = 0.9$ K on ^{4}He and $\varepsilon_a / k = 0.3$ K on ^{3}He-^{4}He mixture [13, 14]. This and the electron spin alignment by magnetic selection reduce the surface recombination reaction speed and allow the gas phase density n_v to be increased by magnetic compression. The surface density

$$n_s = \lambda n_v \exp(\varepsilon_a / kT);$$

$$\lambda = \left(\frac{2\pi\hbar^2}{mkT}\right)^{\frac{1}{2}}, \tag{10.5}$$

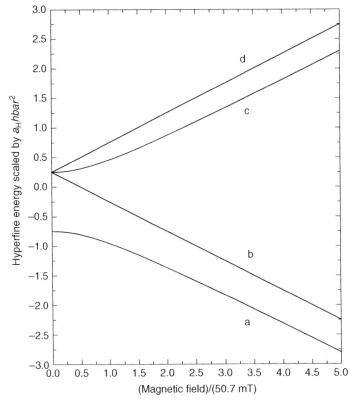

Figure 10.1 Magnetic energy (scaled by the zero-field hyperfine splitting) of the hyperfine states a, b, c and d of atomic hydrogen, as a function of the static magnetic field scaled by the hyperfine splitting divided by the magnetic moment of the free electron

where λ is the thermal de Broglie wavelength, increases rapidly when kT goes below the adsorption energy and approaches then the saturation limit

$$n_s^{sat} = \frac{\varepsilon_a}{2V_{int}^s}, \qquad (10.6)$$

where V_{int}^s is the interaction potential of the 2-d gas of atomic hydrogen H⇓ with electron spins aligned. In practice it is difficult to reach this density, because recombination heats up the superfluid film on the surface.

In a high trap field $B > 8\,\mathrm{T}$, the thermal leak time constant is

$$\tau = \frac{4c_{He}c_M V_{eff}}{K\,\bar{v}A}, \qquad (10.7)$$

where V_{eff} is the effective volume, c_{He} and c_M are the compression ratios due to the helium vapor pumping effect and magnetic attraction, K is the Clausing flow conductance factor and A the filling tube cross section. This time constant becomes substantially longer ($>10^3$ s) than the decay time constant due to the surface recombination on ^{4}He. The addition of ^{3}He on the film surface increases the relaxation time $a \leftrightarrow b$ to about one hour, which leads to a large enhancement of the pure b-state because the mixed a-state has a higher recombination speed. This leads to gas densities of the order 10^{17} cm^{-3} of the atomic gas, with nearly complete electron spin polarization and high proton spin polarization[15]. The proton spin polarization, however, behaves in a non-linear manner when exiting the transverse spin by a resonant transverse magnetic field that gives rise to spin waves due to the oscillatory behavior in the spin diffusion. This invalidates the linear response hypothesis discussed in Section 2.2.2 that is the basis of the linear relation of Eqs. 2.64, 2.65 and 2.81; consequently the integral of the NMR absorption signal cannot be assumed to have a linear relationship with the nuclear spin polarization. The spin waves also influence strongly the NMR lineshape [15] and complicates further the NMR measurement of proton spin polarization.

10.3.2 Chemical-Induced Dynamic Nuclear Polarization (CIDNP)

The dynamics of the spin states in stable atomic hydrogen is an example of chemical spin polarization or, more generally, CIDNP that relies on the ability of nuclear spin interactions to alter the recombination probability in chemical reactions that proceed through radical pairs [16]. In photochemical systems, detected as enhanced signals in the NMR spectra of the reaction products, CIDNP has been used to characterize transient free radicals and their reaction mechanisms. In certain cases, CIDNP also offers the possibility of large improvements in NMR sensitivity. The principal application of this photo-CIDNP technique, as devised by Kaptein [17] in 1978, has been in the studies of proteins and their chemical reactions.

10.3.3 Parahydrogen-Induced Polarization (PHIP)

PHIP is a technique used in magnetic resonance imaging. The technique relies on the incorporation of hyperpolarized H$_2$ into molecules, usually by hydrogenation [18]. While in principle hyperpolarized proton sites can be used directly for imaging applications, the spin-lattice relaxation time of hyperpolarized protons is frequently too short for biomedical use (unless long-lived spin states are created), and polarization from nascent parahydrogen protons must be transferred intramolecularly to longer-lived ^{13}C or ^{15}N spins using spin-spin couplings [19].

The CIDNP and PHIP methods are demonstrably successful, but they are often system specific.

10.4 Overhauser Effect
10.4.1 Metals and liquids

Historically this is the first DNP effect, proposed in 1953 by Albert W. Overhauser [20] and experimentally verified by Carver and Slichter in the same year [21]. The method was already described in Section 4.7 [22] and here we shall just give a brief overview.

In non-magnetic metals the nuclear spins couple with the conduction electrons that have a fast spin-lattice relaxation time of order $\geq 10^{-10}$ s. The conduction electrons exchange the kinetic energy with the lattice phonons even faster, with τ of the order 10^{-13} s. The nuclear spin-lattice relaxation obeys the Korringa 1/T-law, Eq. 1.126, at and below helium temperatures, and the nuclear Larmor frequency is shifted upwards (mostly) by an almost constant amount; this is called Knight shift and was discussed in Section 2.1.5.

The conduction electrons obey Fermi statistics, which means that at low temperatures most of the electrons have their spins paired; only close to the Fermi surface E_F there are electrons thermally excited above the surface, with a corresponding hole below the surface. It is these electrons and holes that exhibit paramagnetism and that also couple with the nuclear spins. The equilibrium Fermi distribution functions in a magnetic field are, for the two orientations of the electron spin $S_z = \pm \frac{1}{2}$,

$$f_\pm = \left[1 + \exp \frac{E \pm \frac{1}{2}\hbar\omega_e - E_F^\pm}{kT}\right]^{-1}, \quad (10.8)$$

where E is the kinetic energy of the electron; in equilibrium these distributions yield the temperature-independent equilibrium electron polarization

$$S_0 = \frac{-\hbar\omega_e \rho(E_F)}{2N}. \quad (10.9)$$

Here $\rho(E_F)$ is the density of states on the Fermi surface and N is the number density of conduction electrons.

Electron spin resonance saturation tends to make disappear the electron polarization that corresponds to $f_+ = f_-$ and which can be quantified by the saturation parameter s:

$$E_F^+ - E_F^- = \hbar\omega_e \frac{S_0 - \langle S_z \rangle}{S_0} = s\hbar\omega_e. \quad (10.10)$$

The introduction of nuclear spins requires now to write the rate equations for their spin-up and spin-down populations n_+ and n_- (assuming nuclei with spin ½); this is beyond our present scope and can be found in Ref. [23], p. 367. These yield at the end the Overhauser effect

$$\frac{n_+}{n_-} = \exp\left[\frac{\hbar(s\omega_e - \omega_n)}{kT}\right], \quad (10.11)$$

where $0 < s < 1$ describes the saturation. If the nuclear gyromagnetic factor is positive, the nuclear polarization is positive, and has the opposite sign if γ_n is negative; unlike DNP in solids with off-resonance saturation, only one sign of polarization is possible.

In practice the Overhauser effect in metals is limited by the skin effect that attenuates the microwave field inside the metallic grains. Nuclear polarization enhancements of several hundreds have been observed with nanometric grains.

The Overhauser effect also works in liquids where paramagnetic centers can be mobile, such as Na dissolved in ammonia [22] and liquids with protons and other nuclei in diamagnetic molecules and protons within free radicals [24].

10.4.2 Nuclear Overhauser Effect

The nuclear Overhauser effect (NOE) is the transfer of nuclear spin polarization from one spin bath to another spin bath via cross-relaxation. In nuclear magnetic resonance spectroscopy, NOE can be used to help resolve the structures of organic compounds.

In the NOE technique, the magnetization on one of the spins is reversed by applying a selective pulse sequence; the sequence may select a particular chemical site of a complex organic molecule. At short times then, the resulting magnetization of another nearby spin evolves in a way described by the differential equations of Solomon [25], with a significant enhancement of the other spin in the spectrum.

The technique is sensitive to the distance between the spins and offers a powerful method for the study of complex biochemical molecules. Electron spins are not involved in the NOE studies.

10.5 DNP Using Nuclear Orientation Via Electron Spin Locking (NOVEL)

In some photochemical reactions yielding triplet-state paramagnetic centers the DNP mechanisms can be shown to work. A clean case is solid naphthalene doped with pentacene, where photo-excited triplet states of pentacene and the protons of naphthalene were polarized using pulse sequences called nuclear orientation via electron spin locking (NOVEL) [26]. The effect was first observed in an Si single crystal doped with boron acceptors; the paramagnetic centers are the holes bound to these acceptors [27].

This NOVEL technique developed by the team of Wenckebach uses a microwave pulse sequence adapted from that of Hartmann and Hahn, which applies to the transfer of polarization between two nuclear spin species and to the detection of the resonance of very rare nuclear spin species [28]. The NOVEL sequence consists of a 90° microwave pulse that orients the electron spins (polarized in the beginning along the steady field in z-direction) along the y-axis in the rotating frame; at the end of this pulse the microwave field is then phase shifted by 90° and has the effect of orienting the effective rotating field exactly parallel to the rotating polarization vector along the y-axis. This concerns only one spin packet of the inhomogeneously broadened line. By choosing the amplitude of the phase-shifted rotating effective field, one can make the nuclear spin resonance frequency in the rotating frame coincide with that of the electrons, which results in a rapid transfer of the electron spin polarization to that of the nuclei. We note that this involves a rather high microwave field strength and therefore a high power dissipation.

For a complete mathematical description, with comprehensive illustrations, we refer to the recent book of Wenckebach [29], pp. 255–259.

10.6 Gas Targets

Gaseous polarized targets are much less sensitive to radiation damage than solid targets, which is why they can be used as internal targets in an accelerator or in a storage ring. Thanks to new optical pumping methods, their density has also increased so much that they can be used in extracted primary beams. We shall briefly describe here the magnetic spin selection, optical pumping and cryogenic stable atomic hydrogen techniques.

10.6.1 Spin Selection in an Atomic Beam

Historically this is the first method to obtain spin-polarized neutral hydrogen. A beam of neutral ground-state hydrogen atoms is formed at the orifice of an accommodator after dissociation by an RF discharge. The dissociator works at a pressure of several millibar and its nozzle may be cooled to 80 K [30]. The flow of H_2 gas into the dissociator is up to 5 mbar×l/s [31]. The atoms effusing from the exit orifice of an accommodator undergo an accelerating expansion into vacuum and cool down until reaching the radius at which the density is so low that the atoms undergo few collisions; this is called the 'freezing radius' because beyond this point the velocity distribution remains constant. The typical Mach numbers reached in this expansion are 2 to 5 and the mean velocity is 1000 m/s to 1500 m/s, if the accommodator cools the atoms to about 35 K [32, 33]; [34]. Accommodator temperatures below 30 K result in the loss of density in the atomic beam and in an unstable operation; the flow in the accommodator with 3 mm diameter and 2 cm length is laminar and the heat transfer efficiency is better than 99.9% [34]. The most likely cause of poor accommodator performance below 35 K is the recombination on its surface that becomes covered with molecular hydrogen; the addition of a few percent of N_2 and/or H_2O in the hydrogen input flow helps getting a stable adsorbed surface layer that yields a high and stable atomic beam density at 35 K accommodator temperature [35].

Subsequently a narrow beam is formed by a skimmer with a possible collimator and with differential pumping, followed by the selection of the higher two hyperfine states (c and d) of the atoms by focusing them in a hexapole magnet using the Stern-Gerlach effect. The skimmer position and the pumping speed are critical parameters because the supersonic flow may result in a shock wave starting at the input orifice of the skimmer [36]. Obviously, the shock wave gives a higher density inside the skimmer and the loss of the molecular flow regime locally.

The flow of polarized atoms from the source is approaching 10^{17} s^{-1} and the atomic thickness of a pointlike jet target is in the range of 10^{12} cm^{-2} [37]. This, however, does not compare favorably with experiments with extracted polarized beams and solid polarized targets. Therefore, the density of the atomic beam target must be increased by storing the atomic gas in a storage cell that works with the principle of a compression tube while also extending the target length along the beam axis. By cooling the storage cell to 100 K the thickness of proton and deuteron spin polarized atomic hydrogen has been increased to 7×10^{13} cm^{-2} [38] while maintaining a high polarization above 0.8 for both H and D nuclei [39]. To reach this, a substantial R&D program targeted the atomic beam and spin selection stages, and the density and nuclear spin relaxation in the storage cell [40]. The polarized atoms make several hundred collisions with the cell walls (pure Al coated by Dryfilm) before escaping; they lose a minor amount of polarization in this process and thermalize to the cell wall temperature in a few collisions. During their sojourn in the cell, they undergo from 1.5 to 15 spin exchange collisions with other atoms in the gas phase and lose no more than 1.5% of their polarization.

The nuclear spin polarization of the HERMES target is measured by a Breit-Rabi polarimeter that uses a small amount of the stored atoms escaping through another pipe attached to the cell. These escaping atoms pass through another series of hexapole focusing magnets and adiabatic RF transitions that filter the spin states of the beam before measuring their flux by a sensitive quadrupole mass spectrometer mounted behind a chopper [40].

The depolarization due to the electromagnetic fields generated by the intense bunches of electrons and positrons of the HERA storage ring was studied by the HERMES group. Only during the injection, with collimators open, a measurable effect on the target polarization was observed, but during data taking with collimators closed the target polarization remained stable at 92% while the beam was decaying exponentially from 40 mA to 13 mA for 12 h [40].

The nuclear spin polarization of the HERA targets can be reversed rapidly by selecting the right transition in the atomic beam adiabatic transitions, and the polarization can be measured by the Breit-Rabi polarimeter in some 10 s [40].

10.6.2 Optical Pumping in Gas Phase

Optical pumping can be applied to ground state and metastable atoms and ions. When a transition of an atom is saturated by circular polarized light, the angular momentum of the polarized photons is transferred to the excited electron that gets polarized in the process. The electron polarization is then transferred to noble-gas nuclei by a weak hyperfine interaction in the rubidium spin-exchange optical pumping (Rb-SEOP) method, or by a metastability-exchange (ME) collision in the ME-optical pumping (MEOP) method. The latter requires the creation of long-lived metastable atoms by an electric discharge; these have a much larger spin-exchange collision cross section and yield a higher pumping speed but apply only to ^{3}He.

The first ^{3}He targets polarized by optical pumping were based on the spin transfer by ME between the optically pumped 2^3S_1 polarized ^{3}He atoms and the ground-state 1^1S_0 atoms.

The optical pumping in these early studies was made by circularly polarized resonance radiation from a helium discharge lamp directed along the magnetic field, absorbed by the ^{3}He atoms that were optically excited from the metastable 2^3S_1 state to 2^3P states, followed by the spontaneous decay back to one of the 2^3S_1 Zeeman sublevels [41]. A weak discharge maintained a suitable concentration of the metastable atoms and the optimum pressure was around a few millibar, which yields low target thickness. It was known in 1966 that also the vapor of alkali metals such as rubidium can be used for the optical pumping of dense ^{3}He, but there was no suitable source of circular polarized light at that time [42].

Denser gaseous ^{3}He-polarized targets have been developed using the technique of collisional SEOP. This is based on the laser excitation of alkali-metal vapor, typically rubidium, via the strong resonance line connecting the Rb $5s^2S_{1/2}$ ground state to the $5p^2P_{1/2}$ excited states; such target can operate at several bar pressure at room temperature, and in 1987 Chupp et al. showed that the nuclear spin polarization is not sensitive to 360 nA beam of 18 MeV α-particles [43]. A review of the successful operation of such a target in SLAC experiment E142 is given in Ref. [44]. This development relied on the application of powerful Ti:sapphire lasers at 795 nm and permitted to operate a two-chamber target cell at several bar pressure with 35% polarization of the ^{3}He nuclei.

Further improvements were made by introducing powerful spectrally narrowed laser diode arrays whose emission linewidth of 2 nm matches better the pressure-broadened absorption linewidth of 0.3 nm in Rb vapor [45]. This and the use of a mix of K and Rb vapor have led to over 70% polarization of the ^{3}He nuclear spins [45]. The polarization of the ^{3}He nuclei at the 30 mT field was measured by NMR calibrated with water filling the target cell.

The high-power laser diodes at 1,083 nm wavelength also improved the performance of the MEOP of ^{3}He. These and the application of higher magnetic field (up to 120 mT instead of a few mT) permitted to increase the cell pressure to 32 mbar while obtaining 70% nuclear spin polarization [46]. The polarized gas has a long relaxation time in contact with walls that are suitably coated, so that in can be compressed to pressures above atmospheric and then transferred to an experiment in a particle beam or to a patient in an MRI set-up [47]. Transferred targets were successfully used in a tagged photon beam in the Mainz Microtron (MAMI); the target polarization was continuously monitored by pulsed NMR [48]. This was accurately calibrated by a flux gate magnetometer that measured the nuclear spin magnetization change after reversing the polarization by adiabatic fast passage [48].

Optical pumping does not enable to produce opposite sign of polarization, but because the guide field has the strength of only a few mT, the field orientation can be reversed in most experiments, and also transverse, longitudinal and any intermediate target spin orientation can be easily obtained.

Alkali-SEOP techniques can also be used for polarizing the nuclei of atomic hydrogen and deuterium at a low density. This technique can be used in polarized proton and deuteron ion sources, which are beyond our present scope.

10.6.3 Stable Atomic Hydrogen as Polarized Electron Target

This was already discussed in Section 10.1.3, but we wish here to point out that the stable atomic hydrogen can also be used as a polarized electron target; such targets traditionally use ferromagnetic foils magnetized to saturation. These, however, have more unpolarized than polarized electrons, and their contribution and that of the orbital moments depends on the applied external field. Therefore, the measurement of the effective electron polarization of a magnetized foil requires probing it by particle scattering.

In stable atomic hydrogen all electrons are polarized to very near 100%, if the stabilization cell temperature is such that the vapor pressure of the helium coating is sufficiently low. Without ^{3}He the nuclear spin polarization remains low, and in most applications requiring polarized electrons the scattering off the electrons can be kinematically separated from that off the protons.

Stable atomic hydrogen-polarized electron target was proposed for the measurement of electron beam polarization by Møller scattering in intense high-energy electron beams with a theoretical accuracy of about 0.5% [49, 50]. The team of the P2 experiment at Mainz MAMI plans to apply this technique for their beam polarization measurement [51].

References

[1] M. L. Seely, C. R. Gould, D. G. Haase, et al., Polarized targets at triangle universities nuclear laboratory, *Nuclear Instruments and Methods in Physics Research* **A356** (1995) 142–147.

[2] W. Heeringa, A brute-force polarized proton target, in: W. Meyer (ed.) *Proc. 4th Int. Workshop on Polarized Target Materials and Techniques*, Physikalisches Institut, Universität Bonn, Bonn, 1984, pp. 129–135.

[3] W. Heeringa, C. Maier, H. Skacel, Exchange of polarized targets in a high magnetic field and at mK temperatures, in: S. Jaccard, S. Mango (eds.) *Proc. Int. Workshop on Polarized Sources and Targets*, 1986, pp. 795–798.

[4] W. N. Hardy, J. R. Gaines, Nuclear spin relaxation in solid HD with H_2 impurity, *Phys. Rev. Lett.* **17** (1966) 1278–1281.

[5] A. Honig, Highly spin-polarized proton samples-large, accessible, and simply produced, *Phys. Rev. Lett.* **19** (1967) 1009–1010.

[6] A. Honig, Q. Fan, X. Wei, A. M. Sandorfi, C. S. Whisnant, New investigations of polarized solid HD targets, in: H. Dutz, W. Meyer (eds.) *Proc. 7th Int. Workshop on Polarized Target Materials and Techniques*, Elsevier, Amsterdam, 1995, pp. 39–46.

[7] LEGS-Spin Collaboration, S. Hoblit, A. M. Sandorfi, et al., Measurements of $\vec{\text{HD}}(\vec{\gamma}\pi)$ and implications for the convergence of the Gerasimov-Drell-Hearn integral, *Phys. Rev. Lett.* **102** (2009) 172002.

[8] J. C. Wheatley, Possible polarized He^3 targets using the adiabatic compression method, in: G. Shapiro (ed.) *Proc. 2nd Int. Conf. on Polarized Targets*, LBL, University of California, Berkeley, Berkeley, 1971, pp. 73–76.

[9] K. H. Langley, C. D. Jeffries, Theory and operation of a proton-spin refrigerator, *Phys. Rev.* **159** (1966) 358–376.

[10] J. Button-Shafer, R. L. Lichti, W. H. Potter, High proton polarization achieved with a $(Yb, Y)(C_2H_5SO_4)_3 \cdot 9H_2O$ spin refrigerator in a nonuniform magnetic field, *Phys. Rev. Lett.* **39** (1977) 677–680.

[11] J. Button-Shafer, The University of Massachusetts spin refrigerator and strange particle physics with the Brookhaven multiparticle spectrometer, in: G.H. Thomas (ed.) *High Energy Physics with Polarized Beams and Polarized Targets*, American Institute of Physics, New York, 1979, pp. 41–47.

[12] T. O. Niinikoski, S. Penttila, J.-M. Rieubland, Stable atomic hydrogen: possible applications in intense polarized sources, in: G. M. Bunce (ed.) *Proc. 5th Int. Symp. on High-Energy Spin Physics*, American Institute of Physics, New York, 1983, pp. 597–600.

[13] A. T. M. Matthey, J. T. M. Walraven, I. F. Silvera, Measurement of pressure of gaseous H↓: adsorption energies and surface recombination rates on helium, *Phys. Rev. Lett.* **46** (1981) 668.

[14] G. van Yperen, A. T. M. Matthey, J. T. M. Walraven, I. F. Silvera, Adsorption energy and nuclear relaxation of H↓ on ^{3}He-^{4}HeMixtures, *Phys. Rev. Lett.* **47** (1981) 800–803.

[15] S. Penttilä, T. O. Niinikoski, J.-M. Rieubland, A. Rijllart, Continuous-wave NMR in spin-polarized atomic hydrogen, *Phys. Rev. B* **36** (1987) 3577–3582.

[16] R. G. Griffin, T. F. Prisner, High field dynamic nuclear polarization—the renaissance, *Physical Chemistry Chemical Physics* **12** (2010) 5737–5740.

[17] R. Kaptein, Photo-CIDNP studies of proteins, *Biol. Magn. Res.* **4** (1982) 145–191.

[18] T. C. Eisenschmid, R. U. Kirss, P. P. Deutsch, et al., Para hydrogen induced polarization in hydrogenation reactions, *J. Am. Chem. Soc.* **109** (1987) 8089–8091.

[19] K. Golman, O. Axelsson, H. Johannesson, et al., Parahydrogen-induced polarization in imaging: subsecond C-13 angiography, *Magn. Reson. Med.* **46** (2001) 1–5.

[20] A. W. Overhauser, Polarization of nuclei in metals, *Phys. Rev.* **92** (1953) 411–415.

[21] T. R. Carver, C. P. Slichter, Polarization of nuclear spins in metals, *Phys. Rev.* **92** (1953) 212–213.

[22] T. R. Carver, C. P. Slichter, Experimental verification of the Overhauser nuclear polarization effect, *Phys. Rev.* **102** (1956) 975–980.

[23] A. Abragam, *The Principles of Nuclear Magnetism*, Clarendon Press, Oxford, 1961.

[24] K. H. Hausser, D. Stehlik, Dynamic nuclear polarization in liquids, *Advances in Magnetic and Optical Resonance* **3** (1968) 79.

[25] I. Solomon, Relaxation processes in a system of two spins, *Phys. Rev.* **99** (1955) 559–565.

[26] A. Henstra, W. T. Wenckebach, The theory of nuclear orientation via electron spin locking (NOVEL), *Molecular Physics* **106** (2008) 859–871.

[27] A. Henstra, P. Dirksen, J. Schmidt, W. T. Wenckebach, Nuclear spin orientation via electron spin locking (NOVEL), *J. Magn. Reson.* **77** (1988) 389–393.

[28] S. R. Hartmann, E. L. Hahn, Nuclear double resonance in the rotating frame, *Phys. Rev.* **128** (1962) 2042–2053.

[29] W. T. Wenckebach, *Essentials of Dynamic Nuclear Polarization*, Spindrift Publications, The Netherlands, 2016.

[30] W. Haeberli, Review of operating atomic beam sources, in: S. Jaccard, S. Mango (eds.) *Proc. Int. Workshop on Polarized Sources and Targets*, Birkhäuser, Montana, Switzerland, 1986, pp. 513–525.

[31] W. Korsch, Intensity measurements on the FILTEX atomic beam source, in: W. Meyer, et al. (eds.) *High Energy Spin Physics*, Springer Verlag, Heidelberg, Bonn, 1990, pp. 168–172.

[32] W. Grüebler, P. A. Schmelzbach, D. Singy, W. Z. Zhang, Polarized atomic beams for targets, in: W. Meyer (ed.) *Proc. 4th Int. Workshop on Polarized Target Materials and Techniques*, Physikalisches Institut, Universität Bonn, Bonn, 1984, pp. 193–201.

[33] W. Grüebler, P. A. Schmelzbach, D. Singy, W. Z. Zhang, Progress report on the cooled ETH polarized ion source, in: S. Jaccard, S. Mango (eds.) *Proc. Int. Workshop on Polarized Sources and Targets*, Montana, Switzerland, 1986, pp. 568–572.

[34] A. Herschcovitch, A. Kponou, T. O. Niinikoski, Cooling of high-intensity atomic beams to liquid helium temperatures, in: S. Jaccard, S. Mango (eds.) *Proc. Int. Workshop on Polarized Sources and Targets*, Montana, Switzerland, 1986, pp. 526–538.

[35] P. A. Schmelzbach, Recombination problems between 4 and 100 K, in: S. Jaccard, S. Mango (eds.) *Proc. Int. Workshop on Polarized Sources and Targets*, Montana, Switzerland, 1986, pp. 539–546.

[36] A. V. Sukhanov, D. K. Torpakov, The pumping speed limitations of the atomic beam intensity, in: W. Meyer, et al. (eds.) *High Energy Spin Physics*, Springer Verlag, Heidelberg, Bonn, 1990, pp. 173–177.

[37] E. Steffens, Workshop report: polarized gas targets for storage rings, Heidelberg 1991, in T. Hasegawa et al. (Eds.) *Proc. 10th Int Symp. On High-Energy Spin Physics*, Universal Academy Press, Inc., Tokyo, 1992, pp. 259–268.

[38] A. Golendoukhin, for the HERMES Collaboration, The HERMES polarized proton target at HERA, in: C. W. de Jager, et al. (eds.) *12th Int. Symp. on High-Energy Spin Physics*, World Scientific, Singapore, 1996, pp. 331–333.

[39] A. Airapetian, N. Akopov, Z. Akopov, et al., The HERMES polarized hydrogen and deuterium gas target in the HERA electron storage ring, *Nuclear instruments and methods in physics research A* **540** (2005) 68–101.

[40] B. Braun, Polarization of the HERMES hydrogen target, in: C. W. de Jager, et al. (eds.) *12th Int. Symp. on High-Energy Spin Physics*, World Scientific, Singapore, 1996, pp. 241–243.

[41] G. K. Walters, Polarized ^{3}He targets and ion sources by optical pumping, *Polarized Targets and Ion Sources*, La Direction de la Physique, CEN Saclay, Saclay, France, 1966, pp. 201–214.

[42] T. R. Carver, Some general problems in producing dense polarized ^{3}He targets, *Polarized Targets and Ion Sources*, La Direction de la Physique, CEN Saclay, Saclay, France, 1966, pp. 191–199.

[43] T. E. Chupp, M. E. Wagshul, K. P. Coulter, A. B. McDonald, W. Happer, Polarized, high-density, gaseous ^{3}He targets, *Physical Review C* **36** (1987) 2244–2251.

[44] J. R. Johnson, A. K. Thompson, T. E. Chupp, et al., The SLAC high-density gaseous polarized ^{3}He target, in: H. Dutz, W. Meyer (eds.) *7th Int. Workshop on Polarized Target Materials and Techniques*, Elsevier, Amsterdam, 1994, pp. 148–152.

[45] J. Singh, P. Dolph, K. Mooney, et al., Recent advances in polarized He-3 targets, in: D. G. Crabb, et al. (eds.) *18th International Spin Physics Symposium*, American Institute of Physics, New York, 2008, pp. 823–832.

[46] P.-J. Nacher, E. Courtade, M. Abboud, et al., Optical pumping of helium-3 at high pressure and magnetic field, https://hal.archives-ouvertes.fr/hal-00002223/ 2002, pp. 2225–2236.

[47] E. W. Otten, Take a breath of polarized noble gas, *Europhysics News* **35** (2004) 16–20.

[48] J. Krimmer, W. Heil, S. Karpuk, Z. Sahli, Polarized ^{3}He targets at MAMI, in: D.G. Crabb, et al. (eds.) *18th International Spin Physics Symposium*, American Institute of Physics, New York, 2008, pp. 829–832.

[49] E. Chudakov, V. Luppov, Møller polarimetry with atomic hydrogen targets, *IEEE Trans. Nucl. Sci.* **51** (2004) 1533–1540.

[50] E. Chudakov, V. Luppov, Møller polarimetry with atomic hydrogen targets, *The European Physical Journal A* **24** (2005) 123–126.

[51] D. Becker, R. Bucoveanu, C. Grzesik, et al., The P2 experiment – a future high-precision measurement of the electroweak mixing angle at low momentum transfer, *arXiv*, 2018, pp. 1–64.

11

Design and Optimization of Polarized Target Experiments

The figure of merit is defined for some scattering applications; this figure permits the objective comparison of various target types and polarization methods. The optimization of the polarized target operation in particle physics experiments is briefly discussed before treating the sources of possible false asymmetries due to the target.

Finally a series of uses of polarized target techniques beyond particle and nuclear physics experiments is presented. These include notably the coherent small-angle neutron scattering (SANS) used in the studies of biological macromolecules, time-resolved SANS, pseudomagnetism, nuclear magnetic ordering, dynamic nuclear polarization (DNP) enhancement of high-resolution NMR spectroscopy, particularly in solid state using the magic angle spinning (MAS) techniques. The sensitivity and contrast enhancement are briefly discussed for magnetic resonance imaging (MRI) techniques. These use various DNP techniques and radical-free injectable polarized fluid methods, as well as the dissolution DNP (dDNP) techniques.

11.1 Particle and Nuclear Physics Experiments

11.1.1 Figure of Merit for an Experiment with a Single Target

As described in Chapter 7, the optimization of the particle-scattering experiment consists of optimizing the statistical accuracy to which the desired polarization asymmetry can be determined during the allocated beam time. The target asymmetry

$$A = \frac{\Delta\sigma}{\sigma_0} \tag{11.1}$$

is determined, in each kinematical bin, from the number of counts $N_\pm$ with target polarization along or opposite to the magnetic field:

$$N_+ = \Phi_+ a n_t (\sigma_0 + f P_+ P_b \Delta\sigma),$$
$$N_- = \Phi_- a n_t (\sigma_0 + f P_- P_b \Delta\sigma). \tag{11.2}$$

Here $\Phi_\pm$ are the integrated beam fluxes through the target with average polarizations, $P\pm$ and $N_\pm$ are the corresponding number of counts in the bin, a is the acceptance of the detector in the kinematical bin, σ_0 is the unpolarized cross section, $\Delta\sigma$ is the cross-section difference with opposite orientations of the target nuclear spins, n_t is the target thickness (number of nucleons or nuclei per cm^2), P_b is the beam polarization and f is the target dilution factor:

$$f = \frac{n_p \sigma_p}{n_t \sigma_0} = \frac{n_p \sigma_p}{n_p \sigma_p + \sum_i n_i \sigma_i}, \qquad (11.3)$$

where the indexes p refer to the polarizable nucleons and i to unpolarizable background nucleons. Clearly f is different for each kinematic bin as the cross sections depend on the kinematics.

By scaling the numbers of counts by the integrated beam fluxes, we get an equation from which the asymmetry A can be solved by iteration:

$$\frac{N_+/\Phi_+ - N_-/\Phi_-}{N_+/\Phi_+ + N_-/\Phi_-} = f P_b A \frac{P_+ - P_-}{2 + f P_b A(P_+ + P_-)}. \qquad (11.4)$$

The solution converges fast because the mean target polarizations are nearly equal but opposite so that $P_+ + P_- \ll 1$ and often A is also small. Based on this equation and on only the counting statistics, an estimate for the statistical error of the asymmetry was obtained in Chapter 7:

$$\delta A \cong \frac{1}{\mathcal{M}_t} \frac{1}{P_b \sqrt{2\Phi a \sigma_0}}. \qquad (11.5)$$

Here we have defined the figure of merit of the target

$$\mathcal{M}_t = f \bar{P} \sqrt{n_t} \;\; ; \;\; \bar{P} = \frac{P^+ - P^-}{2}, \qquad (11.6)$$

which can be maximized so as to guide in the choice of the target material and thickness. The target dilution factor f is to be understood here as a mean value of Eq. 11.3 that covers the kinematic range of the experiment.

If the target length is determined by the space available or by the detector requirements, rather than by beam attenuation or multiple scattering of the beam or the secondary particles, the choice of the material can be further simplified by writing the target nucleon thickness in the terms of its average density, length and nucleon mass:

$$n_t = \frac{\rho_t V_t}{m_n} \frac{1}{A_t} = \rho_t \frac{L_t}{m_n}, \qquad (11.7)$$

where the liquid helium coolant filling the voids between the target beads must also be taken into account. The figure of merit now reads

$$\mathcal{M}_t = f \bar{P} \sqrt{\rho_t} \sqrt{\frac{L_t}{m_n}} \qquad (11.8)$$

and the material-dependent part can be determined for a substance once its filling factor and average polarization are known.

The polarization may evolve during the data taking, because its frequent reversal is often required for reducing systematic errors due to the slow drift of the beam or the detector acceptance, and possibly because of the radiation damage of the target material. The average polarization in Eq. 11.6 is then obtained from the square roots of the time averages of the squared polarizations, which can be determined when the time evolution of the polarization during DNP, and the dose dependence of the reduction of the polarization are known.

In a high-intensity beam, the polarization of the target may be reduced by the direct heating of the material by the beam and by the radiation damage, which gradually accumulates during the experiment. In this case the figure of merit of the experiment also follows from the minimization of the statistical uncertainty of the target asymmetry of Eq. 11.5, which requires the maximization of

$$\sqrt{\Phi}\mathcal{M}_t = t_{exp}\sqrt{I}\sqrt{f^2\langle P^2\rangle}. \tag{11.9}$$

Here t_{exp} is the effective duration of the data taking excluding time needed for target annealing or change, I is the beam intensity and the time average of the polarization needs the knowledge of the polarization build-up during reversal and the reduction of polarization due to the accumulated dose and due to the material heating that depends on the intensity I. It is clear that these parameters can only be obtained by direct measurement, and that also the cooling system will strongly influence the maximization of expression 11.9. These factors were discussed in Section 7.5.

If multiple scattering limits the length of the target, the best material is one that has the highest material-dependent figure of merit and has a low relative number of heavier nuclei so that the length can be increased. The criteria related with multiple scattering unfortunately cannot be written in simple analytic form and the judgement between materials of roughly equal and high figure of merit must be based on their relative heavy-element contents. The parameter relevant for multiple scattering is the radiation length X_0, which was discussed in Section 7.1.1.

If some of the heavy nuclei also become polarized, their contribution to the scattering asymmetry must be estimated. This requires the estimation or measurement of their polarization. The errors related with these procedures are usually taken into account in the systematic error analysis, because they are usually dominated by the incomplete knowledge of the nuclear structure of the heavy nuclei.

The statistical accuracy can be improved by taking more data to the point that other errors begin to dominate. Among these are the accuracy of the beam normalization, the drift in the acceptance a and variations in the effective target thickness n_t due to the microwave power for DNP. These can give rise to false asymmetries that can be evaluated and optimized based on Eq. 11.2.

11.1.2 False Asymmetry Estimation and Mitigation

Beam Normalization

We focus here on spin asymmetry measurements with extracted beams and solid polarized targets.

Accurate spin asymmetry measurements require high statistics and therefore long experimental runs. The stability of the beam source and of the beam monitoring system therefore becomes a major concern.

Beam normalization is often the leading source of false asymmetry, because the beam control and counting cannot be done to much better than 5% over several month periods of time. The leading causes for this are the limitations in the stability of the accelerator and of the beamline components, in addition to the drifts in the stability of the beam detector and monitor systems.

Given the uncertainty of the beam normalization over extended run periods, the main method of mitigation is the frequent reversal of the target polarization. If the beam stability and its main causes of instability are monitored, we can use the spectrum of its time variations to optimize the frequency of the polarization reversal. This will be discussed in Section 11.1.2.

Another way to eliminate the beam normalization errors is to make the experiment with opposite polarizations with two oppositely polarized targets placed in the same beam. This will be discussed in Section 11.1.3.

In elastic scattering experiments, the kinematics of the reaction can be fully determined so that the elastic peak due to scattering off protons is well resolved from the quasi-elastic background due to scattering off nucleons in complex nuclei, because the elastic peak in the latter case is much broadened due to the Fermi motion inside the nuclear potential well. This background is unpolarized and can therefore be used as a method of beam normalization. Obviously, the complex nuclei should be consist of as few as possible other nuclei than spin 0 species such as ^{12}C and ^{16}O.

Acceptance Drift

The parameter 'acceptance' = a in Eq. 11.2 consists of the geometric acceptance of the detector in each kinematic bin and of the efficiency and dead time parameters of the detector. While the geometric acceptance stays constant, the detector threshold and efficiency depend on many poorly controlled parameters. Among these are notably the efficiency drift of gas-filled tracker detectors due to the variation of the ambient temperature and humidity, and the dependence of the dead time on the beam intensity.

Frequent polarization reversal is again an effective countermeasure to mitigate the effect of the slow drift of the acceptance, because acceptance has, in principle, no correlation with the sign of the target polarization.

Target Thickness Variation

The heat load distribution and its variation with the microwave power and beam current influence the coolant density in the target volume. This can give rise to a false asymmetry if the applied heat load is systematically different for the positive and negative polarization states of the target. The thickness variation depends on the cooling method:

- In evaporative cooling methods the higher heat load entails higher quality factor (vapor fraction) of the coolant fluid, both spatially and in time.
- In dilution refrigerator the higher heat load leads to spatially slightly higher coolant density owing to osmotic pressure, and higher overall heat load leads to slightly lower coolant fluid density because of higher solubility of ^{3}He in the dilute phase at the higher temperature.

In the case of cooling by dilution refrigerator, the target is always submerged in the dilute phase with ^{3}He concentration close to its solubility limit. The temperature dependence of the phase diagram is discussed in Appendix A5.1. The density of the coolant fluid can be monitored by recording the coolant temperature, in principle. In practice, however, this is not accurate during DNP because the thermometer sensors are heated up by the microwave power losses in them, thus monitoring bolometrically the microwave field rather than the fluid temperature. The simplest way to measure the coolant temperature consists of turning off the microwaves and recording the transient in the thermometer reading, which can then be extrapolated back to the moment of power turnoff. The ^{3}He vapor pressure thermometer bulb (used for TE calibration temperature measurement) is less sensitive to microwave power absorption than resistance thermometers, and it may serve as a monitor of the coolant temperature once the heating effect is calibrated.

During frozen spin operation, the coolant temperature can be accurately measured by resistance thermometers and its effect on the coolant density is also smaller because the solubility flattens below 50 mK temperature.

Because the vapor density is much lower than the liquid phase density in evaporative cooling both by ^{4}He and by ^{3}He, the likelihood of false asymmetry from the coolant of the target is higher in the evaporative cooling methods. Therefore, it is recommendable to monitor the amount of coolant in the target in order to be able to correlate and correct any possible false asymmetry due to the effective density of the coolant.

In ^{3}He refrigerators the coolant density in the target volume can be monitored by recording the condensation pressure while keeping the amount of gas in the closed-loop system constant. This obviously requires that the condenser operates in steady conditions and has a small volume compared with that of the target cavity.

In continuous-flow ^{4}He refrigerators, the monitoring of the amount of coolant is not possible in this way because the cooling circuit is open. Therefore it is important to keep the microwave power absorption in the target equal in both polarization states of the target, and to monitor all parameters of the refrigeration system so as to keep the as constant as possible while reversing the polarization by DNP.

The target thickness can, in principle, be also monitored by using a scattering reaction with high statistics and no spin effects, in parallel with the polarization asymmetry measurement.

Dummy Target Tests

The sources of false asymmetries are often studied by replacing the polarized target with an unpolarized dummy containing polyethylene or graphite beads. The data collected with such a target should therefore yield zero asymmetry, and in the data with the graphite target the elastic peak should be absent.

11.1.3 Methods and Frequency of Polarization Reversal

Reversal by DNP

It is often stated that DNP has the benefit of enabling the reversal of the target spin polarization with nothing but a small change in the saturating microwave frequency. This is true despite of the fact that the microwave power may be systematically slightly different for the two polarization states, thus producing a small false asymmetry effect on the background due to scattering off the coolant nuclei.

The speed of polarization depends on the spin-lattice relaxation time of the electron spins and on the power and frequency of the applied microwave irradiation, as was discussed in Chapter 4. Zero-crossing takes place typically in a few minutes, while 70% of the final polarization is reached in 30 minutes at 0.5 K and 10 minutes when using ^{4}He as a coolant at 1 K.

The polarization speed depends on the density of the paramagnetic electrons in the target. In heavily irradiated targets, the polarizing time constants also depends on the radiation damage accumulated during the experiment.

Reversal by Spin Rotation

In some experiments the magnetic field can be reversed without noticeable effects on the kinematics, or with effects that are well controlled and compensated. This is often the case with neutral beam particles such as neutrons and gammas, in particular, when the field is axial with the beam and when also the scattered particles are neutral. Then the target spin can be reversed as rapidly as the field can be rotated.

The field can be most easily rotated by powering a magnet that produces a field orthogonal with the main field. If the target is operating in the frozen spin mode, the transverse field can have a value of about 0.5 T, where the spin-lattice relaxation time at 50 mK temperature is typically around 200 h (see Eq. 5.97 and Figure 5.8). Eddy current heating limits the rotation speed so that it is usually accomplished in about 10 minutes.

The field rotation can be done with either positive or negative polarization of the target. Often both polarities are used in order to compensate systematic effects due to the change of the field direction. In the case of negative polarization of protons, care must be taken to avoid superradiance of the NMR hybrid resonant circuits, as was discussed in Section 6.3.6. If superradiance is suspected, it can be avoided by making the magnetic field sufficiently inhomogeneous during its ramp down and up.

With target in the frozen spin operation mode, the magnetic field can be rotated in any orientation that the experiment may require, as long as the field value is about 0.3 T or higher. The superconducting magnet system of Saturne II Nucleon-Nucleon scattering experiments was briefly described in Chapter 9; this system consists of three coils and permits the orientation of the target spin in three orthogonal directions [1]. The orientability of the target spin makes it possible, in a single experiment, to perform complete sets of measurements allowing unique determination of the spin-dependent scattering amplitudes in two-body reactions of the nucleons, in this case polarized protons and neutrons [2]. The polarization was also reversed by field rotation in the cases where the experiment permitted it.

CERN PS199 experiment using antiproton beams from the LEAR ring was designed to measure the spin-dependent scattering amplitudes in the $\bar{p}p$ system [3, 4]. The 12 cm long target was polarized in the 2.5 T homogeneous vertical field of an iron-yoke magnet. After reaching 90% polarization, this main field was lowered to rotate the field along the axis of the target, obtained by powering a thin holding coil wound around the thin-walled vacuum inner chamber of the target refrigerator [5]. The holding coil could generate a 0.3785 T field, which was needed for transporting the low-energy antiprotons to the target along the axis of the refrigerator, but could also be used for polarization rotation.

Polarization Reversal by Adiabatic Passage in Rotating Frame (APRF)

The adiabatic passage in the rotating frame can be done as fast as in a few seconds in a small target. This is most easily accomplished by a transverse field coil driven by a constant frequency RF source, while sweeping the main field through the resonance condition with the Larmor frequency. The transverse RF field does not need to be very homogeneous, but its field needs to cover well enough the target volume. In the case of polarized protons, this limits the length of the target to about 5 cm because otherwise the coil itself has resonant frequencies that are too close to the Larmor frequency.

The efficiency of reversal by APRF is limited by the speed at which thermal mixing between the Zeeman and dipolar reservoirs can be obtained, and by the fact that there are electron spins and possible other nuclear spin species in the target material.

Using classical equations of motion for the magnetization due to free spins, Abragam derives the adiabatic condition [6]

$$A = \frac{\gamma B_{eff}^2}{\dot{B}_0 \sin\theta} \gg 1 \qquad (11.10)$$

which is strongest at resonance where the effective field is perpendicular to the steady field so that $\sin\theta = 1$ and $B_{eff} = B_1$. This ignores any relaxation phenomena and complications arising from the contact with the paramagnetic electron spin system.

The internal thermal equilibrium of the Zeeman and dipolar reservoirs is established in a time scale of T_2, the decay time of the free-precession NMR signal, whereas the Zeeman and dipolar reservoirs relax towards the lattice temperature at very much slower rates T_{1Z}

and $T_{1D} \approx T_{1z}/3$, respectively. When the magnetic field is changed, the Zeeman reservoir is cooled or heated if the field change happens at a constant entropy, whereas the dipolar reservoir is not influenced by the field change. Similarly, if an adiabatic demagnetization or magnetization is made in the rotating frame at high enough effective field, the dipolar reservoir is unaffected whereas the Zeeman reservoir is cooled or heated. The two reservoirs reach a common temperature when the effective field has a low enough value, called mixing field.

Goldman [7] and Abragam and Goldman [8] discuss the quasi-equilibrium of the spin systems at high fields and make the reasonable assumption, well verified by several types of experiments, that the systems can be described by the density matrix with two temperatures:

$$\rho = \frac{\exp(-\beta_z \omega_0 I_z - \beta \mathcal{H}'_D)}{\text{Tr}\{\exp(-\beta_z \omega_0 I_z - \beta \mathcal{H}'_D)\}}. \tag{11.11}$$

In the high field $\mathcal{H}_D$ is replaced by the so-called secular part of the dipolar Hamiltonian $\mathcal{H}'_D$, which has the important property that it commutes with the Zeeman Hamiltonian. The reasons for this were discussed in Chapter 2. In the presence of a transverse RF field, the Hamiltonian in the rotating frame becomes

$$\mathcal{H}_{\text{eff}} = -\hbar(\omega_0 - \omega)I_z + \mathcal{H}'_D + \hbar\omega_1 I_x \tag{11.12}$$

where $\omega_1 = B_{1y}$ is the amplitude of the transverse rotating field. When all of the three terms of Eq. 11.12 are of the same order of magnitude, the relaxation towards a common spin temperature happens fast, with a time constant of the order of T_2, which is roughly the inverse of the dipolar linewidth and is of the order of 1 ms. This relaxation also depends on the magnitude of the transverse rotating field in a complicated way.

Let us now perform adiabatic demagnetization in rotating frame (ADRF) from such a large effective field that no thermal mixing occurs, until arriving to the mixing field

$$b_m = B_0 + \frac{\omega_m}{\gamma} = \frac{\omega_m - \omega_0}{\gamma}, \tag{11.13}$$

where the relaxation between the Zeeman and dipolar interaction energy reservoirs is reasonably fast in comparison with the spin-lattice relaxation. The relaxation takes place at constant total energy, and the final inverse temperature, in the high-temperature approximation, is analogous with that of Eq. 1.121:

$$\beta_f = \beta_i \frac{B_i}{\sqrt{b_m^2 + B_1^2 + B'^2_{loc}}} \times \frac{\sqrt{b_m^2 + B_1^2}}{\sqrt{b_m^2 + B_1^2 + B'^2_{loc}}}, \tag{11.14}$$

where the second term on the right describes again the non-adiabaticity due to the mixing. The losses in entropy can be of the order of 1% in favorable cases. If the effective field is further reduced so slowly that the dipolar and Zeeman temperatures are always in good equilibrium with each other, the common inverse temperature undergoes reversible changes with the effective field:

$$\beta_f = \beta_i \frac{B_i}{\sqrt{b^2 + B_1^2 + B'^2_{loc}}} \times \frac{b_m}{\sqrt{b_m^2 + B'^2_{loc}}}, \qquad (11.15)$$

where $b = (\omega - \omega_0)/\gamma$ and where it was taken into account that the rotating field B_1 must be small in comparison with both the local field and the mixing field.

The APRF can be stopped at any effective field and the transverse field can be adiabatically reduced to zero. If this is done at the frequency ω_0, all available Zeeman order is transformed to dipolar order, and if the initial polarization is sufficiently high, magnetic ordering may take place in the nuclear spin system. This is very interesting in its own right, but polarized target materials are not optimized for such because usually the electron spin density is too high and the shape of the target is not an ellipsoid. These would cause the smearing of the transition and make it difficult to interpret the domain structure of the magnetic phase. However, such a transition is likely to cause loss of reversibility in the APRF.

If the frequency or field sweep is continued through zero effective longitudinal field b, the Zeeman temperature will be adiabatically reversed. In this process there is no adiabatic loss equivalent to that of the field flip given by Eq. 1.134, because in the rotating frame one can proceed adiabatically through zero effective longitudinal field without losses due to relaxation, provided that the spin-lattice relaxation times are much longer than the time spent in the passage.

The above formulas are derived using the high-temperature approximation because there is no equivalent and simple model for the APRF covering the case of low temperatures. Thus these theoretical results are at best qualitatively correct, with no clear way of optimization for the strength of the rotating field and the field sweep rate.

In practice the strength of the transverse field and the sweep rate of the steady field are experimentally optimized so that the losses due to relaxation are minimized. The polarization loss in the reversal by APRF is then reduced to that due to the loss of entropy when performing the thermal mixing, and to the loss due to other spins whose temperature remains untouched during the passage. At very high polarization the above results based on the high-temperature approximation are qualitatively valid, but nuclear magnetic phase transition phenomena may reduce the efficiency of the polarization reversal.

The reversal can be performed starting from positive or negative polarization and spin temperature, and the sweep of the frequency or field can be started from above or below, with approximatively same results. To reduce losses due to thermal mixing, however, it is best to perform thermal mixing at positive Zeeman temperature when the initial polarization is positive, and at negative Zeeman temperature when the initial polarization is negative.

To our knowledge the APRF reversal has never been applied in polarized proton targets during operation in a particle physics experiment. Among the rare available reports, those of Patrick Hautle are probably the most systematic. He describes the problems discussed above and finds the experimental reversal efficiency $P_{final}/P_{initial} = -0.36$ for protons in

1-butanol doped with 4×10^{19} spins/cm^3 of porphyrexide [9, 10]. This best efficiency was reached at $A = 20$. The experimental results of Hautle [11] confirm earlier work of Parfenov and Prudkoglyad [12].

The foregoing discussion applies to spin ½ nuclei. The tests were performed with a frequency sweep rather than field scan. For deuterons with a quadrupole broadened NMR line, an additional complication arises because in a single APRF through the resonance line a part of the spins are reversed twice. Hautle remedied this by a second passage that was stopped in the middle of the deuteron resonance line [9, 10]. This technique permitted to reach reversal efficiencies as high as $P_{final}/P_{initial} = -0.92$. The optimization of the frequency sweep rate and of the transverse field strength yielded best reversal efficiencies in the range 0.36 T/s $\leq dB/dt \leq 1.28$ T/s and RF field amplitudes around $2B_1 = 0.1$ mT.

The reports on the APRF reversal of proton polarization fail to take into account the internal field due to the magnetization of the highly polarized proton spins. This is of the same magnitude as the dipolar field and is higher than the rotating field due to the RF drive coil. When the magnetization is negative, it induces a large current in the coil and may lead to superradiance unless the resonance of the tuned circuit is highly damped. The superradiance will happen once in each reversal, irrespective of the direction of the sweep or initial sense of polarization. This will limit the final polarization to the value at which the superradiance conditions are met, as was described in Chapter 6. It remains to be experimentally verified if superradiance can explain the poor results of APRF in polarized proton targets.

While for protons the classical adiabatic condition $A = 20$ for best efficiency seems to verify Eq. 11.10, the loss of polarization is higher than predicted and remains unexplained. On the other hand, the reversal efficiency for deuterons is more in conformity with the prediction, while it is achieved with $A < 1$ in contradiction with the classical picture of APRF. These facts remain to be clarified by further experimental and theoretical work.

Optimization of the Reversal Frequency

The accuracy of the beam normalization, the drift in the acceptance a and variations in the effective target thickness n_t were discussed above and it was stated that frequent polarization reversal is the best mitigation strategy. However, if the period of operation with a given polarity is pushed too short, the mean polarization of Eq. 11.6 remains low, which limits the accuracy of asymmetry measurement. Thus there must be an optimum reversal frequency.

In polarization asymmetry measurements of inelastic and inclusive reactions, there is no elastic peak that can be normalized with the unpolarized background scattering events off complex nuclei. The experimental asymmetry is therefore limited by the dilution factor of Eq. 11.3 and it may remain buried in the 'noise' due to the drift of the above parameters.

Chabaud and Kuroda developed a time-dependent analysis in which the slow drift was modelled by a polynomial modulated by the target polarization via the scattering asymmetry in each kinematic bin [13]. The polynomial was chosen so that it does not fit away the polarization asymmetry but permits to detect the rather rapid change of the polarization sign at each reversal, even when buried in the statistical noise.

The above method is based on the assumptions that the drift is multiplicative and continuous in time. Unfortunately it fails to yield a reliable estimate of the residual error in the resulting asymmetry. This is mainly due to the fact that the fitting process is equivalent to a non-linear filtering operation. Moreover, the polynomial is likely to contain Fourier components at the frequency of polarization reversal; these are indistinguishable from the 'signal' due to the polarization reversals.

In order to optimize the polarization reversal frequency and to obtain a well-understood estimate for the residual error in asymmetry, it was proposed to use the optimum linear filter theory [14], which is well known from communication theory [15]. The optimum linear filter has a frequency response function shaped like the complex conjugate of the Fourier transform of the signal, weighted with the reciprocal of the noise power spectrum. Thus, as the signal is the replica of the time evolution of the target polarization, it is just required to experimentally find the noise power spectrum in order to compute the filter. The remaining optimization consists of placing the polarization signal optimally in the frequency domain, so as to minimize the resulting error in the estimate of the polarization asymmetry.

The noises due to the drifts of the counting rate signal are modelled by multiplicative and additive components and by statistical fluctuations. The time sequence of counts in each kinematic bin is approximated by a continuous signal

$$\dot{N}(t) = \frac{N_r}{t_r}[1+n_1(t)] \cdot [1+n_2(t)+\varepsilon P_t(t)], \qquad (11.16)$$

where N_r is the total number of counts in an experimental run lasting during the time t_r of the run, $n_1(t)$ is the multiplicative part of the noise (mainly the drift of the counting efficiency and of the acceptance) and $n_2(t)$ is the additive part of the noise coming from statistical fluctuation and fluctuation of the target density. The experimental asymmetry ε is modulated by the target polarization $P(t)$, which is assumed to evolve exponentially during each reversal period:

$$P_t(t) = P_\infty(1-e^{-t/\tau}) \qquad (11.17)$$

where τ is typically 5 min to 10 min in materials such as propanediol with PD-Cr(V) complexes. It turns out that if the asymmetry signal is small, all noise sources can be considered to be additive.

The optimum filter operation consists of pre-filtering the raw data with the reciprocal of the noise power spectrum, followed by the calculation of the cross correlator of the counting asymmetry and polarization signals:

$$\frac{1}{t_r}\int_0^{t_r} \dot{N}(t)P_t(t)dt \cong \frac{N_r}{t_r^2}\int_0^{t_r} dt P_t(t)\left[n(t)+\varepsilon P_t(t)\right]. \qquad (11.18)$$

Here second-order terms are assumed to be small and the noise signal is

$$n(t) = n_1(t)+n_2(t)+n_1(t)n_2(t) \qquad (11.19)$$

11.1 Particle and Nuclear Physics Experiments

which is Gaussian if both $n_1(t)$ and $n_2(t)$ are Gaussian. With sub-runs of length Δt and number of counts ΔN, the statistical counting noise is white with amplitude and bandwidth

$$G_1 = \frac{\sigma_n^2}{B_n} = \frac{\Delta t}{\Delta N} = \dot{N}^{-1};$$

$$B_n = \frac{1}{\Delta t}.$$
(11.20)

Furthermore, if the drift of the counting efficiency can be represented by a noise with $1/f$ spectral density, the total noise spectrum is

$$G_n(\omega) = \frac{G_0}{1+\omega^2 T_c^2} + G_1.$$
(11.21)

The RMS variation of the estimate for the polarization asymmetry is then [14]

$$(\delta\varepsilon)_{rms} \cong \frac{1+(T/T_0)}{\sqrt{N_r \langle P_t^2 \rangle}},$$
(11.22)

where T is the length of time between polarization reversals and

$$T_0^{-1} = (2\pi)^{-1} \left[\dot{N} \frac{G_0}{T_c^2} \right]^{\frac{1}{2}}$$
(11.23)

is the frequency at which the $1/f$-noise is equal to the white statistical noise.

The minimization of Eq. 11.22 with respect to the reversal period T, while keeping the total number of counts N_r constant, can now be obtained by requiring that its partial derivative with respect to T is zero. To do this we find first write

$$\langle P_t^2 \rangle = P_\infty^2 \left(1 - \frac{2}{x} \tanh \frac{x}{2} \right); \quad x = \frac{T}{2\tau}$$
(11.24)

and write the nominator of Eq. 11.22 as

$$1 + \left(\frac{T}{T_0} \right)^2 = 1 + \left(\frac{x}{x_0} \right)^2; \quad x_0 = \frac{T_0}{2\tau}.$$
(11.25)

These yield

$$\frac{\partial}{\partial x}(\delta\varepsilon)_{rms} = \frac{\partial}{\partial x} \frac{1+\left(\frac{x}{x_0}\right)^2}{P_\infty \left[1 - \frac{2}{x} \tanh \frac{x}{2} \right]^{\frac{1}{2}}} = 0,$$
(11.26)

which can be solved numerically for x. This was done in Ref. [14] using the experimental conditions of Ref. [16] in intense secondary beams of $\pi^\pm$ and p at 8 GeV/c at CERN PS, with $T_0 \approx 100$ minutes. The solution shown in Figure 11.1 suggests that the optimum reversal period should be about $T = 45$ minutes, while the experiment was run with reversals each 5.9 h on the average. It is clear that this was sub-optimal but, on the other hand, the target operation was manual rather than automated as required when reversing the polarization each 45 minutes.

Many inclusive spin asymmetry measurements have taken advantage of the shortened polarization reversal periods. There is, however, the technique of using two oppositely polarized targets placed in the same beam. In this case the slow drift of the beam and acceptance can be eliminated to a large extent, which will be described below.

11.1.4 Multiple Targets in the Same Beam

If the scattering events are reconstructed so that the event vertex can be located in the cell of the polarized target, two target cells with opposite polarizations can be placed in the same beam so as to eliminate to a large extent the effects of the drift in the beam and in the acceptance. This technique was used for the first time by the European Muon Collaboration (EMC) in their investigation of deep inelastic scattering of polarized muons on polarized protons in 1984 and 1985 [17],[18]. The M2 muon beam of the CERN SPS accelerator was

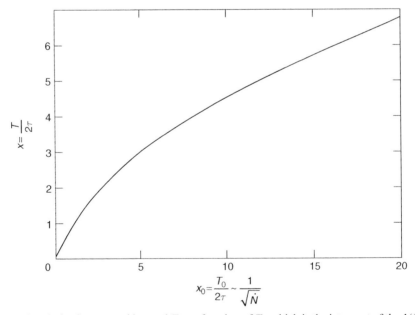

Figure 11.1 Polarization reversal interval T as a function of T_0, which is the intercept of the $1/f$ noise with the statistical noise (see the text for the precise definition). Reprinted from Ref. [14] with minor improvements, with permission from Elsevier

instrumented with a fast beam tracker that allowed the beam phase space cuts to be made so that the same beam traversed the two target cells. The cells were of 360 mm long and of 1 L volume; they were separated by a 220 mm gap. The vertex distribution in the two cells along the beam direction is shown in Figure 11.2.

The spin-dependent differential cross sections give rise to the free proton asymmetry

$$A = \frac{d\sigma^{\uparrow\downarrow} - d\sigma^{\uparrow\uparrow}}{d\sigma^{\uparrow\downarrow} + d\sigma^{\uparrow\uparrow}} \qquad (11.27)$$

and the measured event yields from the two target cells are related with this by

$$N_u = \Phi n_u a_u \sigma_0 (1 - f P_b P_u A), \quad N_d = \Phi n_d a_d \sigma_0 (1 - f P_b P_d A), \qquad (11.28)$$

where the subscripts u and d refer to the upstream and downstream target halves, n is the number density of the target nucleons, Φ is the beam flux, a is the acceptance of the detector system, σ_0 is the unpolarized cross section, f is the fraction of the event yield from the polarized protons of the target, P_b is the beam polarization and $P_{u(d)}$ are the polarizations of the upstream (downstream) target halves. As the polarization of the positive muon beam was fixed and opposite to the beam direction, it amounted between –0.77 and –0.82 depending on the energy that was varied from 100 to 200 GeV.

For an experimental run with P_u initially in the direction of the beam and P_d opposite the measured asymmetries A_m are

$$A_m = \frac{N_u - N_d}{N_u + N_d}; \quad A'_m = \frac{N'_d - N'_u}{N'_d + N'_u}, \qquad (11.29)$$

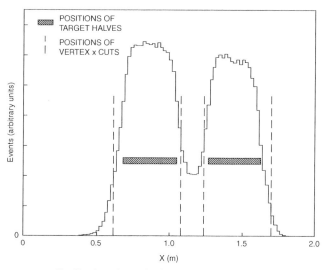

Figure 11.2 Event vertex distribution along the beam direction in the double-cell polarized target of the EMC [18]. The positions of the target edges and the applied cuts are shown. The slope of the number of counts in the two targets is due to the their different acceptances seen by the scattered muon spectrometer

where the unprimed (primed) quantities refer to the event yields before (after) the reversal of the polarizations of the target halves. The free proton asymmetry A is related to the measured asymmetries by

$$\bar{A}_m = \frac{1}{2}(A_m + A'_m) = -AfP_b\left(\left|P_u\right| + \left|P_d\right| + \left|P'_u\right| + \left|P'_d\right|\right)/4. \qquad (11.30)$$

The measured asymmetries were always less than 0.02 so that the experiment requires high statistics and stable equipment, even if the false asymmetries exactly cancel from Eq. 11.30. However, if the acceptance drifts slightly, a false asymmetry may show up if the ratio of the downstream and upstream acceptance ratios before and after reversal

$$K = \frac{a_u/a_d}{a'_u/a'_d} \qquad (11.31)$$

deviates from unity. This turned out to be the leading systematic source of error in the EMC result on the Ellis-Jaffe sum rule [19], and requires the precise monitoring of the acceptances in the improved experiments in order to be able control and correct them. Such improvements were implemented by the Spin Muon Collaboration along with their new 50% longer polarized target [20]: the contribution of the acceptance drift to the errors of the integrals of the spin structure functions was reduced by a factor of almost 4 by more frequent polarization reversals and by introducing the field rotation as an additional method of reversal. The new magnet system permitting the field rotation included dipole coils with 0.5 T transverse field, which also allowed target operation with transverse spin orientations [21].

The technique of placing several different targets in the same muon beam was also used for the measurement of precise deep-inelastic cross-section ratios on nucleons in nuclei of different sizes, in order to verify the differences in the nucleon structure functions between the quasi-free nucleons of deuteron and those bound inside heavier nuclei [22].

The two-cell target method was developed further by the CERN COMPASS collaboration by splitting the SMC target into three sections, with a central section twice longer than the end sections [23]. The central section was polarized in direction opposite to that of the end sections. In this way the mean acceptances of the oppositely polarized sections were more equal, and the adverse effects of the slow drift were reduced beyond what was possible with the two-cell targets.

11.2 Spin Filtering of Neutron Beams

Spin filtering of a neutron beam by polarized protons was one of the early applications of the dynamically polarized LMN targets [24]. In the neutron beam energy interval 10 eV to 100 keV, this is the most efficient technique, because the neutron-nucleus cross sections are large and they have a large and calculable spin-dependent part determined by the s-wave scattering. In s-wave neutron-nucleus scattering, the total cross section can be written in the general form

11.2 Spin Filtering of Neutron Beams

$$\sigma = \sigma_0 - P_n P_N \sigma_P \tag{11.32}$$

where σ_0 is the total cross section with unpolarized particles, $P_{n(N)}$ are the polarizations of the neutron (nucleus) and σ_p is the so-called polarization cross section:

$$\sigma_p = \frac{I}{2I+1} 4\pi(a_-^2 + a_+^2) + \frac{I}{2I+1}(\sigma_{c+} - \sigma_{c-}). \tag{11.33}$$

Here I is the spin of the target nucleus and $a_\pm$, $\sigma_{c\pm}$ are the scattering lengths and capture cross sections for neutron-nucleus collisions with total spin $I + \frac{1}{2}$ and $I - \frac{1}{2}$, respectively.

The transmission of the target for an unpolarized incident neutron beam is

$$T = \frac{T_+ + T_-}{2} = e^{-n\sigma_0 t} \cosh(P_N n \sigma_p t) \tag{11.34}$$

and the transmitted beam polarization is

$$P_n = \frac{T_+ - T_-}{T_+ + T_-} = \tanh(P_N n \sigma_p t) \tag{11.35}$$

For 1 eV to 10 keV neutrons scattering on polarized protons has $\sigma_0 = 20.4$ b and $\sigma_p = 16.7$ b.[1] Targets of a few cm thickness can thus polarize transmitted neutrons almost completely by spin filtering without losing more than 90% of the beam intensity [24].

At very low neutron energy the cross sections entering in Eqs. 11.34 and 11.35 are

$$\sigma_0' = \sigma_{inc}\left(1 - \frac{I}{I+1}P_N^2\right) + \sigma_{c0};$$

$$\sigma_p' = \frac{I}{I+1}\sigma_{inc} + \frac{1}{2I+1}(\sigma_{c-} - \sigma_{c+}) \tag{11.36}$$

Here σ_{inc} is the incoherent scattering cross section and σ_{c0} is the unpolarized capture cross section. For slow neutrons with energy $\ll 1$ eV on polarized protons, we have

$$\sigma_0' = 80\left(1 - \frac{1}{3}P_N^2\right) \text{ b};$$

$$\sigma_p' = 53 \text{ b}. \tag{11.36'}$$

The main advantages of polarizing filters are [25]:
- They can accept wide beam divergence.
- They can be designed to have high transmittances and polarizing efficiencies.
- The thickness of the polarizing filter controls the above parameters via Eqs. 11.34 and 11.35.
- They operate efficiently over a broad neutron energy range.

---

[1] The unit barn (b) is used for the cross sections in nuclear physics; 1 b = 10^{-28} m².

For these reasons the polarized proton filters are the only means of polarizing truly white neutron beams such as those in the spallation neutron sources. These filters are usually cooled by evaporation of ^{4}He because ^{3}He has a large absorption cross section for slow neutrons.

Other materials for neutron spin filtering include samarium, ^{3}He and TiH_2. The samarium filter [25] is based on selective neutron absorption by brute force polarized samarium nuclei in a saturated ferromagnet cooled to LHe temperatures. The brute force polarized ^{3}He and TiH_2 were briefly discussed in Chapter 10. More recently optically pumped ^{3}He has been proposed for neutron spin filtering [26].

Fast polarized neutrons are used for the studies of nuclear structure and resonances. DNP may be used for polarizing the spins of the target nuclei used in elastic and inelastic scattering experiments. In these studies often very thin targets are required, whose preparation was briefly discussed in Chapter 7.

For ultracold and cold neutrons in the wavelength range 4–20 Å multilayer magnetic supermirrors are the most efficient way of getting the neutron beam highly polarized [27–29]. Such slow polarized neutrons are useful for solid-state research, in particular, for the material studies using coherent scattering, which is the topic of next section.

11.3 Coherent Slow Neutron Scattering

11.3.1 Born Approximation and Fermi Pseudopotential

Neutrons thermalized at 293 K have the de Broglie wavelength

$$\lambda = \frac{h}{m\sqrt{\langle v^2 \rangle}} = \frac{h}{3kT} = 0.147 \, \text{nm}, \tag{11.37}$$

which is in the range of atomic distances in solids and in complex molecules. Thermal and cold neutrons are therefore suitable for the structural studies of large molecules such as proteins, carbohydrates, lipids, enzymes and biopolymers such as RNA and DNA. In comparing with X-rays and electrons, the fact that all biomolecules contain hydrogen makes the neutrons particularly suitable for these studies due to the fact that protons scatter neutrons very efficiently, while neutrons are almost insensitive to the atomic electrons and feel much less the heavier nuclei.

The Born approximation describes well the slow neutron scattering by a nuclear potential, by considering the incoming neutron as a plane wave and by taking into account that the interaction potential well is shallow. In this case the scattering amplitude is [8]

$$f(\theta) = -\frac{M}{2\pi\hbar^2} \int \exp(i(\mathbf{k} - \mathbf{k}') \cdot \mathbf{r}') V(\mathbf{r}') d^3 r', \tag{11.38}$$

where M is the neutron mass, $\mathbf{k}$ is its incoming wave vector and $\mathbf{k}'$ is the outgoing wave vector. Moreover, when the neutron wavelength of Eq. 11.37 is very long in comparison with the size of the scattering potential well, the scattering becomes isotropic and the scattering amplitude of Eq. 11.38 has negligible imaginary part for nuclei below mass 17, with the exception of ^{3}He and ^{6}Li.

If in addition the interaction potential is approximated by the Fermi pseudopotential

$$V_F(\mathbf{r}) = \frac{2\pi\hbar^2}{M} a\delta(\mathbf{r}), \qquad (11.39)$$

the scattering amplitude becomes

$$f(\theta) = -\frac{M}{2\pi\hbar^2} \int \exp\left(i(\mathbf{k}-\mathbf{k}')\cdot\mathbf{r}'\right) V_F(\mathbf{r}') d^3r' = -a. \qquad (11.40)$$

If furthermore the nuclear spin is considered, the scattering amplitude can be expressed in operator form as

$$a = b + 2B\mathbf{I}\cdot\mathbf{s}, \qquad (11.41)$$

where $\mathbf{I}$ is the nuclear spin operator and $\mathbf{s}$ is the neutron spin operator; b, B are constants determined by the two eigenvalues b_+ and b_- of the operator $\mathbf{I}\cdot\mathbf{s}$ in the states $J_\pm = I \pm \frac{1}{2}$ [8]:

$$b = \frac{(I+1)b_+ + Ib_-}{2I+1} \;;\; B = \frac{b_+ - b_-}{2I+1}. \qquad (11.42)$$

If the neutron can be absorbed by the nucleus, the scattering amplitude can also have an imaginary part, but in the case of cold neutrons the absorption cross section is small for all light nuclei with the exception of ^{3}He and ^{6}Li.

There is no theoretical method to determine these amplitudes accurately from nuclear models, but they have been experimentally measured for most nuclei. Table 11.1 lists the parameters b and B for nuclei of interest in biochemical molecules. More recent tables of neutron scattering lengths for a wide range of nuclei are available in Ref. [30].

In targets with many dense clusters of nuclei scattering in the forward direction has an enhancement due to the coherence of the waves scattered from nearby nuclei. The scattering amplitudes of the clusters and solvent are obtained by adding the spherically symmetric

Table 11.1 Parameters b and B that determine the spin-dependent scattering amplitudes of slow neutrons on various nuclei common in biological macromolecules, from Ref. [31].

Nucleus	I	b (10^{-15} m)	B (10^{-15} m)
^{1}H	½	3.74	29.12
^{2}H	1	6.67	2.85
^{12}C	0	6.65	0
^{13}C	½	6.29	−0.6
^{14}N	1	9.37	1.4
^{16}O	0	5.80	0

elastically scattered waves. In such forward elastic scattering, the recoil mass is that of the whole crystal or cluster in the case of cold neutrons. For faster neutrons, phonons may be created but these are not of interest in the study of macromolecular structures.

11.3.2 Spin Contrast Variation in Small-Angle Neutron Scattering

Polarization of Nuclei in Large Biomolecules

It is not self-evident that proton or deuteron spin polarization can diffuse rapidly from nuclei in the solvent matrix into the large biomolecules, even if the liquid solution preserves its random distribution in the sample vitrification process.

Stuhrmann and coworkers polarized dynamically various samples containing an enzyme (hen egg white lysozyme) dissolved in a mixture of heavy water and deuterated propanediol doped with deuterated HMPA-Cr(V), containing about 7% of unsubstituted protons [32]. The conclusions were based on the analysis of the proton NMR signal shape and time evolution, and on the tests of the spin temperature equilibrium between the deuteron and proton spin systems. It turns out that the dense clusters of the 700 protons in the lysozyme structures are polarized at almost the same speed as the unsubstituted hydrogens of the solvent, and that the protons reached the same spin temperature as the solvent deuterons, within the experimental accuracy.

Nuclear Spin-Dependent Amplitudes

Using the scattering length operators of Eq. 11.41, we may define two scattering amplitudes for a macromolecule [33, 34]

$$U(\mathbf{Q}) = \sum_{j=1}^{M} b_j \exp\left[i(\mathbf{Q} \cdot \mathbf{r}_j)\right],$$
$$V(\mathbf{Q}) = \sum_{j=1}^{N} P_j I_j B_j \exp\left[i(\mathbf{Q} \cdot \mathbf{r}_j)\right],$$
(11.43)

where U is the invariant and V is the polarization-dependent amplitude, $\mathbf{Q} = \mathbf{k} - \mathbf{k}'$ is the momentum transfer vector and $\mathbf{r}_j$ is the position of the nucleus j. The macromolecule has M nuclei of which N have a non-zero spin. The corresponding parts of the macromolecular structure $\rho(\mathbf{r})$ are

$$\rho_U(\mathbf{r}) = \sum_{j=1}^{M} b_j \delta(\mathbf{r} - \mathbf{r}_j),$$
$$\rho_V(\mathbf{r}) = \sum_{j=1}^{N} b_j \delta(\mathbf{r} - \mathbf{r}_j).$$
(11.44)

The elastic coherent scattering cross section is given by

11.3 Coherent Slow Neutron Scattering

$$S = \left(\frac{d\sigma}{d\Omega}\right)_{coh} = UU^* + P_n(UV^* + VU^*) + VV^*, \quad (11.45)$$

where the polarization P_n of the incident neutron is along the same common axis as that of the target nuclei I_p, i.e. along the static magnetic field.

While the s-wave scattering of the neutron on a single nucleus is isotropic, the elastic coherent scattering is strongly peaked in the forward direction, because the sums of Eqs. 11.43 get large under the condition that **Q** be small. It is these forward peaks in the polarized scattering functions that are of interest in the study of macromolecules.

Assuming that only one nuclear species with polarization P is contributing to the amplitude $V(\mathbf{Q})$ of Eq. 11.43, the coherent scattering can be expressed as

$$S(\mathbf{Q}) = S_U(\mathbf{Q}) + P_n P S_{UV}(\mathbf{Q}) + P^2 S_V(\mathbf{Q}), \quad (11.46)$$

where S_U, S_{UV} and S_V are the basic scattering functions of spin-contrast variation; $S_U(\mathbf{Q})$ is the scattering function of the unpolarized target with structure $\rho_U(\mathbf{r})$. The polarized nuclei with structure $\rho_V(\mathbf{r})$ give rise to $S_V(\mathbf{Q})$, and the cross-term $S_{UV}(\mathbf{Q})$ is due to the convolution of $\rho_U(\mathbf{r})$ with $\rho_V(\mathbf{r})$.

The experimental determination of the basic scattering functions involves a series of measurements of $S(\mathbf{Q})$ with different P and P_n; $S_U(\mathbf{Q})$ is first measured with the zero polarization of the target $P = 0$. As the spin of the neutron beam can be quickly inverted by a flip coil, the functions with opposite neutron polarizations P_n and $-P_n$ are then measured:

$$\begin{aligned} S_{\uparrow\uparrow} &= S_U + P_n P S_{UV} + P^2 S_V, \\ S_{\uparrow\downarrow} &= S_U + P_n P S_{UV} + P^2 S_V. \end{aligned} \quad (11.47)$$

From the difference of these we obtain

$$S_{UV} = \frac{S_{\uparrow\uparrow} - S_{\uparrow\downarrow}}{2 P_n P}. \quad (11.48)$$

The sum of these yields

$$S_{\uparrow\uparrow} + S_{\uparrow\downarrow} = 2 S_U + 2 P^2 S_V, \quad (11.49)$$

from which S_V can be extracted because S_U was already determined from the measurement of the scattering function with unpolarized target.

The experimental scattering functions were determined for several large biomolecules, from which the proton spin densities and molecular sizes were directly determined. The scattering functions determined for the biomolecules using different deuterated labels yielded important information for the relative positions of the labels and therefore the shapes of the molecules.

Dynamically polarized protons of a crystal of lanthanum magnesium nitrate (LMN) ^{142}Nd^{3+} were first time studied by polarized slow neutron diffraction by Hayter, Jenkin and White in 1974 [35]. The positions of the H atoms were obtained from a limited set of reflections. The non-uniform proton polarization was one of the reasons for the unexpectedly poor quality of the polarized proton-density map [36]. The work on LMN crystals was not pursued at that time.

In the mid-80s, several groups embarked on experiments of polarized SANS from dynamically polarized proton spins. Demonstration-quality results were obtained in crown ethers by Kohgi et al. [37], polymers by Glättli et al. [38] and biological macromolecules by Knop et al. [33, 39]. All of them took advantage of the new type of glassy polarized target materials using stable Cr(V), which had been developed for high-energy physics experiments. Its preparation for the purpose of neutron scattering from hydrophilic macromolecules is simple. A small amount of EHBA-CrV together with the biological macromolecule of interest is added to a glycerol/water mixture and rapidly frozen to a glassy platelet in a liquid-nitrogen-cooled copper mould.

Most of these tests were made using ^{4}He evaporation refrigerators, which eliminates the neutron absorption by ^{3}He. Knop and coworkers used first a fast-loading dilution refrigerator with quartz inserts to reduce the amount of ^{3}He in the beam path through the target, and then developed a dilution refrigerator with a ^{4}He-filled sample cell [40]. The advantage of a dilution refrigerator is that it enables spin freezing and selective depolarization of protons or deuterons.

The recent progress in the applications of SANS using polarized targets is reviewed by Stuhrmann [41, 42].

Time-Resolved Small-Angle Neutron Scattering (SANS)

The paramagnetic molecules used for DNP may be rather large and it is often asked how the nuclear spins in these molecules behave during DNP and when the polarization is frozen. This was discussed in the terms of the diffusion barrier in Chapter 5. Van den Brandt and coworkers addressed the issue in the case of EHBA-Cr(V) by developing a technique to follow the time evolution of the small-angle scattering during the reversal of polarization in glassy glycerol-water samples of different degrees of deuteration [43]. The time constants that describe the build-up of polarization around the paramagnetic center and the subsequent diffusion of polarization in the solvent were determined by analysing the temporal evolution of the nuclear polarization, which in turn was obtained by fitting a core-shell model to the time-dependent SANS curves. The results on the spin dynamics obtained using the scattering function of a core-shell could be independently confirmed by evaluating the integrated SANS intensity. A thermodynamic one-center model was presented, which is able to reproduce the observed dependence of the proton polarization times on the proton concentration of the solvent.

The simultaneous global fitting of typically 200 time-resolved differential cross-section data sets provided the time dependence $P(t)$ of the close proton polarization. Two time constants could be identified in the dynamics of the polarization process. The short time constant τ_1, typically of the order of 1 s, can be related to the relatively fast polarization build-up of the protons close to the paramagnetic center, whereas the long-time constant τ_2 ~10 s was interpreted as describing the slower build-up of the polarization of the distant bulk protons of the solvent [43]. A thermodynamic model expressing the flow of polarization between three reservoirs coupled in series is able to reproduce the concentration dependence of the experimental time constants.

Paramagnetic electron spin centers may also occur naturally in biomolecules, and these may be used for DNP in order to complement EPR results with time-resolved SANS. A differentiation between dynamic polarized protons close to tyrosyl radical sites in a large biomolecule bovine liver catalase and those of the bulk was achieved by time-resolved polarized neutron scattering [44]. This was possible despite of the polarization of 3% reached with the low concentration of the tyrosyl radical. Three radical sites, all of them being close to the molecular center and the haem,[2] appear to be equally possible. Among these is tyr-369, the radial site of which had previously been proven by EPR [44].

Future Neutron Beams for Molecular Studies

The spallation neutron sources produce intense pulses of neutrons from high-energy proton bunches hitting a liquid mercury jet target. The produced neutrons can be slowed down in moderators operating at different temperatures, before collimating them into neutron guides. The velocity of each neutron can be obtained from the time of flight between the pulse and the detector. Such pulsed neutron beams have the advantage that the velocity of the neutrons in the beam can be recorded, event by event. Thus the intensity is not reduced by a velocity selector. Moreover, as the beam velocity range can be rather wide, the range of momentum transfer **Q** can also be wide. This gives an extra parameter that may be used in the studies of molecular structures.

These pulsed neutron facilities are predicted to gain considerable importance in time-resolved SANS in the future [45]. An example is the iMATERIA spectrometer at the J-PARC spallation neutron source, which features a polarized target with a 7 T magnet and 1.2 K ^{4}He refrigerator [46]. The facility is planned for the studies of microstructures of polymers and elastomers.

11.3.3 Pseudomagnetism

When polarized particles (neutrons, electrons or photons, for example) pass through a polarized target, the interaction of the particle with the matter can also be described as waves that become superposed and therefore interfere. This was predicted for neutrons by Barychevskii and Podgoretskii [47] who found that slow polarized neutrons will precess substantially differently about the nuclear polarization vector when the polarization is non-zero.

---

[2] Haem are recognized as the components of haemoglobin but are also found in a number of other biologically important haemoproteins such as catalases and myoglobin. Haem is a coordination complex consisting of an iron ion coordinated to a porphyrin and one or two axial ligands.

As was described in Section 11.3.1, only isotropic s-wave scattering is substantial for thermal neutrons, whose wavelength is several orders of magnitude longer that the range of the nuclear force. Following Tsulaia [48], the propagation of slow neutrons through matter can be described by an index of refraction n

$$n^2 = 1 + \frac{4\pi}{k^2} \sum_i N_i A_i(0), \tag{11.50}$$

where $k = 2\pi/\lambda$ is the wave vector, $A_i(0)$ is the amplitude of the non-spin-flip forward elastic scattering of nuclear species i and N_i is the number density of such nuclei.

The Fermi pseudopotential of Eq. 11.39 can be written, using the amplitudes of Eq. 11.42

$$V_F(\mathbf{r}) = \frac{4\pi\hbar^2}{M} \left[\frac{I+1+\mathbf{I}\cdot\mathbf{s}}{2I+1} f^+ + \frac{I-2\mathbf{I}\cdot\mathbf{s}}{2I+1} f^- \right] \delta(\mathbf{r}), \tag{11.51}$$

where M is the reduced mass of the neutron-nucleus system, f^+ and f^- are the amplitudes for the cases of total angular momentum $J = I \pm \frac{1}{2}$, respectively.

The angular brackets of Eq. 11.51 express the total scattering amplitude

$$A(0) = \left[\frac{I+1+\mathbf{I}\cdot\mathbf{s}}{2I+1} f^+ + \frac{I-2\mathbf{I}\cdot\mathbf{s}}{2I+1} f^- \right] \tag{11.52}$$

from where one can immediately see that when the neutron and target spins are parallel, we have the forward scattering amplitude

$$A_1(0) = f^+, \tag{11.53}$$

while in the non-flip forward scattering of a wave whose polarization is antiparallel with the polarization of the target nuclei, the scattering amplitude is

$$A_2(0) = \frac{1}{2I+1} f^+ + \frac{2I}{2I+1} f^-. \tag{11.54}$$

Therefore the index of refraction of Eq. 11.50 has different values for the two spin states of the neutron-nucleus system:

$$n_1^2 = 1 + \frac{4\pi N}{k^2} f^+. \tag{11.55}$$

$$n_2^2 = 1 + \frac{4\pi N}{k^2} \left(\frac{1}{2I+1} f^+ + \frac{2I}{2I+1} f^- \right). \tag{11.55a}$$

Given the fact that both of these deviate little from 1, we may linearize and write them as

$$n_1 = 1 + \frac{2\pi N}{k^2} f^+, \tag{11.56}$$

$$n_2 = 1 + \frac{2\pi N}{k^2}\left(\frac{1}{2I+1}f^+ + \frac{2I}{2I+1}f^-\right), \quad (11.56b)$$

with the difference

$$\Delta n = n_1 - n_2 = \frac{4\pi N}{k^2}\frac{I}{2I+1}\left(f^+ - f^-\right). \quad (11.57)$$

It follows from the above that a neutron propagating at velocity v through a nuclear polarized medium will see its spin component perpendicular to the nuclear polarization precess at angular frequency that differs from the precession due to the magnetic induction by

$$\Delta\omega_n = k\Delta nv = kv\frac{4\pi N}{k^2}\frac{I}{2I+1}\left(f^+ - f^-\right). \quad (11.58)$$

The change in the state of polarization of the particles, passing through the target is described phenomenologically in terms of their precession in an additional pseudomagnetic field[3] B^*:

$$\Delta\omega_n = \gamma_n B^*. \quad (11.59)$$

Such neutron spin rotation, while moving in matter with polarized nuclei, is caused by strong interactions, and can be substantially faster than that caused by the sum of the external and internal magnetic fields.

The outline presented above for the example of a beam of slow monoenergetic neutrons passing through a polarized nuclear target can be generalized to beams of other particles. Results of experiments with neutrons, electrons and γ-ray quanta have been presented by Pokazan'ev [49].

The concept of a nuclear pseudomagnetic field and the phenomenon of neutron spin precession in matter with polarized nuclei were experimentally verified first time by Abragam's group in France in 1972 [50]. The accumulated results were reviewed by Glättli et al. in 1979 [51]. Abragam and Goldman have reviewed the method and the pseudomagnetic moments of more than 30 nuclei in Chapter 7 of Ref. [8].

The neutron precession in a nuclear spin polarized sample has evolved into a convenient tool for measuring spin-dependent scattering lengths of slow neutrons on nuclei. The method is, according to Glättli et al. [51], straightforward when

- the sample has only one nuclear species with non-zero spin;
- the number density of the nuclei with spin is at least 10^{22} cm^{-3};
- the sample does not absorb thermal neutrons too strongly;
- nuclear polarization > 1% can be obtained and measured.

---

[3] We use the SI units; the original papers by Abragam and coworkers are written with CGS units, where the magnetic field H is used instead of the magnetic induction B (see Appendix A.1.2.).

11.4 Solid-State Physics, Chemistry and Biomedical Applications

DNP was developed originally for the needs of particle and nuclear physics, but it was soon realized that it enabled new research and expanded the existing methods in solid-state physics, chemistry and biomedical chemistry. Among these are nuclear magnetic transitions in solids, high-resolution NMR spectroscopy in solids and sensitivity enhancement in NMR and MRI using rare nuclei.

The overview below is brief and does not attempt to give exhaustive coverage because the topics are currently rapidly expanding. Moreover, these topics are not in the main focus of this book.

Chemists use the generic term hyperpolarization to describe all methods to enhance spin polarization beyond its thermal equilibrium value.

11.3.4 Nuclear Dipolar Magnetic Ordering

We briefly include this topic for two reasons:

(1) The polarization reversal by adiabatic passage in rotating frame (APRF) meets with the conditions under which nuclear dipolar ordering has been observed in several single crystal materials.
(2) The poor success of APRF in polarized proton targets remains to be explained and, if possible, cured.

Magnetic ordering of nuclear dipole moments has been predicted to happen at a critical temperature T_c when the thermal energy per spin is of the same order as the dipolar energy:

$$k_B T_c \cong \hbar \gamma B_{loc}. \tag{11.60}$$

The transition temperature is therefore in or near the range of 10^{-7} K to 10^{-6} K for most solids with a high density of nuclear spins. As DNP in a 2.5 T field can yield spin temperatures of a few mK (see Chapter 4), adiabatic demagnetization to 0.25 mT field will then yield a spin temperature of a few 10^{-7} K, comfortably in the required range of the dipolar fields in solids.

In Section 11.1.3 we discussed briefly the loss of polarization of protons upon reversal by adiabatic passage in the rotating frame. Magnetic ordering during the passage can cause loss of entropy and therefore loss of polarization that has been observed in many experiments. The conditions under which this can happen in a polarized target are, however, so poorly defined that the possible magnetic phase transition is likely to be too diffuse to detect in anything else than in the loss of entropy that occurs due to the relaxation of the dipolar energy when the effective field in the rotating frame is perpendicular to the main steady field. The relation between entropy and polarization was discussed in Sections 1.3.2 and 1.3.3.

Abragam and Goldman [8] discuss the conditions required for the unambiguous observation of the magnetic phases in the rotating frame, listing the problems that needed to be addressed beyond the straightforward polarized target techniques:

- The samples need to be oriented single crystals.
- The sample must be spherical or ellipsoidal so that the demagnetizing field is homogeneous.
- The concentration of the paramagnetic dopant must be low because ordering may not be observable within the dipolar field of the unpaired electron.
- The external steady field must be much more homogeneous than 10^{-4} so that the demagnetization to zero effective field is met in the whole sample.
- The DNP must yield a high and homogeneous nuclear spin polarization.

Once the sample material and the suitable paramagnetic center and its concentration have been optimized, the nuclear dipolar ordering consists of obtaining a high DNP in the sample, followed by ADRF to zero effective magnetic field. The rotating field can then be reduced to zero and one may begin magnetic measurements or neutron diffraction experiments (in the case of ordering in the ^{1}H spins) in view of studying the magnetic phase diagram.

Antiferromagnetic and ferromagnetic phases in the rotating frame have been observed in spin systems of ^{1}H, ^{6}Li, ^{7}Li and ^{19}F nuclear spins. These were chemically bound in single crystals of LiH, LiF, CaF_2 and $Ca(OH)_2$. The two last ones also contained ^{43}Ca spins. The paramagnetic F-centers, created by various radiation sources, were used in LiH and LiF. CaF_2 was doped with U^{3+} or Tm^{2+} to get DNP by solid effect [8]. $Ca(OH)_2$ was doped with paramagnetic OH^- centers created by electron irradiation [52].

The nuclear magnetic ordering in the rotating frame was first observed in CaF_2 by the group of Abragam in Saclay in 1969 [53]; [54]. A transition to ordered state was observed in the NMR dispersion signal when ADRF was performed to zero effective field, with initial polarizations $-P \gg 0.3$. It was not possible to determine the transition temperature in these early experiments, but it was found that when the steady field was along the |100| crystal axis, antiferromagnetic order resulted at $T < 0$, whereas a ferromagnetic order with sandwich domain structure was found for |111| direction. For positive temperatures a transition to helical order was found with field along the |111| axis. Neutron scattering experiments at Saclay with LiH similarly revealed an antiferromagnetic transition, with $T_N = -1.1$ µK [55]; [56]; [57]. The Leiden group [52] studied $Ca(OH)_2$ and found a transition to ferromagnetic state with $T_c = -0.9 \pm 0.2$ µK [58]. This was also obtained at initial polarizations $-P > 0.3$ before ADRF.

In glassy polarized target materials, the observation of a possible magnetic phase transition is obscured by the high electron spin concentration and by the shape of the sample. The latter, however, can be cured by preparing a spherical sample 3–5 mm in diameter using the mould techniques developed by Stuhrmann and coworkers (see Chapter 7). In analogy with the spin glass transition in dilute electron spin systems, one could possibly observe a transition in the nuclear spin glass state in a polarized proton target material, by ADRF to zero effective field.

11.3.5 DNP MAS NMR Spectroscopy

MAS is a powerful technique that enables narrow NMR signals to be measured with high-resolution in solid samples, where the motional narrowing is absent. The high-resolution NMR technique relies on pulsed NMR, the details of which are beyond the scope of this book.

The dipolar broadening is due to the terms A and B of Eqs. 2.11a and 2.11b, which need to be averaged over the internuclear polar angles θ_{jk} to get the dipolar lineshape in a solid powder sample. In the case of rapid molecular rotation in a solid, Gutowsky and Pake have shown that the average of the angular factor $1-3\cos^2\theta_{jk}$ should be replaced by [59]

$$\left\langle 1-3\cos^2\theta_{jk}\right\rangle_{av} = \left(1-3\cos^2\theta'\right)\left(\frac{3\cos^2\eta_{jk}-1}{2}\right), \tag{11.61}$$

where θ' is the angle the molecular rotation makes with respect to the static field and η_{jk} is the angle between the rotation axis and the internuclear vector r_{jk}. Such a linewidth transition occurs in solid ammonia around 40 K, for example. This led Lowe [60] and Andrew [61] to propose the rotation of the entire sample about an axis that makes the angle θ' with respect to the magnetic field so that

$$1-3\cos^2\theta' = 0, \tag{11.62}$$

which would make the dipolar broadening vanish. The angle arc cos $\sqrt{1/3} = 54.7°$ became soon known as the magic angle, and it was quickly understood that it could eliminate broadening due to chemical shift anisotropy and first-order quadrupole splitting as well. A thorough introduction to MAS and also to pulse sequences for spin-flip narrowing is presented by Slichter in Chapter 8 of Ref. [62].

Beyond this, the second-order quadrupole splittings and other second-order broadenings can be eliminated by rotating the sample simultaneously about two axes inclined at the zeros (= magic angles) of the $l = 2$ and $l = 4$ Legendre polynomials [63].

In pulsed NMR the free-induction decay signal is Fourier transformed to get the spectrum of the Larmor frequencies of the chemically shifted absorption signals. The rotation frequency of the sample is seen as sidebands to each of the narrow lines of this spectrum. Therefore, a high spinning speed is desirable to avoid overlap of the chemical shift spectra.

The MAS NMR is particularly applicable to large complex chemical entities of structural biology, which cannot exhibit motional narrowing and which also escape X-ray studies. Also a wide range of topics in the materials science, in particular soft materials, have profited of the new method. The first pioneering work was done on ^{13}C spectroscopy of coal in Delft in 1985 [64].

MAS uses miniature gas turbines to rotate the sample at speed up to 100 kHz or even higher. The gas also cools the sample and acts as a gas-dynamic bearing for the rotor, similar to turbine expanders of liquefiers. The gas may be N_2 cooled by an LN_2 bath or a cryocooler, or He cooled by a cryocooler or by exchanger using evaporation from an LHe bath. In the former case temperatures down to 100 K are easily controlled, limited by the increasing viscosity of N_2 and by irregular operation when the gas begins to liquefy, while with He temperatures down to 30 K or even 10 K are available. In the latter case the electron spins are almost completely polarized similar to polarized targets, in particular when 14 T field is applied.

The field sweep of the high-field magnets is produced with a separate sweep coil system. The sweep must be done very slowly because the metallic mandrels and other support structures, in addition to the conductor itself, have eddy currents that decay slowly. These distort the field and destroy its homogeneity until the currents have decayed.

The cooling of the rotating sample is mainly by conduction through the probe cell walls and by forced convection of the turbine gas. As rather large temperature differences can be used, the microwave power absorption can be several W/cm^3, which is required for the solid effect or differential solid effect transitions to be well saturated. In the case of the use of cross-effect or Overhauser effect, less power is required because the first-order allowed transition is more easily saturated. Although the frequency modulation effects are not well understood, experimental findings suggest that the above DNP mechanisms all profit from the frequency modulation. On the other hand, any anisotropy in the g-factor or hyperfine splitting also turns into an effective modulation of the transitions, because the axis of rotation is not parallel to the steady field. But the rotation alone cannot produce nuclear polarization, because the anisotropy is small and the magnetic field is very high, in contrast with the spin refrigerator that was briefly described in Section 10.2. On the other hand, in the theoretical estimates of microwave-induced DNP, the sample rotation certainly plays a significant role and cannot be ignored. Therefore the experimental effect of the sample rotation speed Ω is an interesting item to be studied and understood.

The microwave source can be gyrotron (for highest power), EIO or BWO; these were described briefly in Chapter 9. In the case of a gyrotron and EIO, the sources are tunable electrically only over a small range, which requires that the magnet system must include a sweep coil to be able to match the field to the resonance conditions required by the particular free radical used in the experiment. However, an electrical control of the frequency is required for the modulation of the microwave frequency, in order to make use of the beneficial effects of the modulation. This was discussed in Chapters 3 and 4 from the point of view of polarized target applications.

In the present DNP MAS systems, the microwaves are launched to the probe directly from the waveguide, with no reflecting structures. As a tunable resonating cavity is impractical at millimeter wavelength, similar to the polarized targets, the best solution is to use a multimode cavity with a suitable coupling structure such as a horn antenna. The present systems launch the microwaves directly from a corrugated oversized waveguide.

As the MAS DNP applications have expanded rapidly, equipment have become commercially available from several sources.[4] A recent review gives a good overview of the subject [65].

---

[4] Bruker Inc. delivers complete systems for DNP MAS NMR up to 593 GHz (21 T). Gyrotrons and other millimeter wave sources are also available from CPI International Inc. (ancient Varian) and Thales Group (ancient Thomson CSF).

A major development has taken place in the free radicals that are usable in the conditions of high temperature (10 K to 100 K) and high field (5 T to 21 T) of the MAS DNP. These are also likely to profit future polarized target materials. The concentrations are in the range of 10 mM to 40 mM,[5] which corresponds to the number density range 6×10^{17} cm^{-3} to 2.8×10^{18} cm^{-3}. In the lower limit the cross-effect, solid effect or differential solid effect is likely to dominate in DNP, whereas in the higher limit the cross-effect or solid effect should be effective, depending on the linewidth of the radical in the solid. Dynamic cooling of the spin-spin interactions would require about 10 times higher number density of the free radicals, as was discussed in Chapter 7. Such high electron spin density may not be desirable in the MAS NMR samples.

The free radicals are called by the generic name 'polarizing agent' (PA) by the chemists. Among these are notably BDPA- and trityl-based radicals, including notably sulphonated BDPA (SA-BDPA) and sulphonamide BDPA (SN-BDPA), and trityl derivatives OX063 and CT-03. The latter are water soluble and relatively stable in air, which greatly facilitates their use in biochemical samples.

Among the nitroxide radicals, TEMPO was the first one to demonstrate the enhancement of ^{1}H and ^{13}C by DNP under NMR conditions of 5 T field and at a temperature below 100 K by the group of Griffin at MIT magnet laboratory [66, 67]. The group went on to introduce biradicals that combine two interconnected TEMPO molecules by adjustable-length glycol chains, a whole series of new radical pairs was developed to match the requirements of cross-effect DNP that works with low enough electron spin concentration. The most successful of these is called TOTAPOL [68].

Beyond these, a wide range of biradicals, triradicals and more complex mixtures have been developed. These are reviewed by the group of Corzilius in Ref. [65].

11.3.6 MRI Sensitivity and Contrast Enhancement

Overhauser effect can be, in principle, used for nuclear spin polarization enhancement in liquid state in situ inside an imaging magnet. This was demonstrated inside the bore of a 1.5 T MRI magnet where proton polarization in a water sample was enhanced by a factor up to 98 before letting it flow into a chamber where water was in thermal equilibrium at RT. The polarization was preserved and imaging could be demonstrated with high resolution of the contrast in the polarization [69].

Pure water with polarized protons has been proposed already earlier as a contrast-agent-free contrast medium to visualize its macroscopic evolution in aqueous media by MRI by McCarney et al. [70]. The method is called remotely enhanced liquids for image contrast and it utilizes the proton signal of water that is enhanced outside the sample in continuous-flow mode and immediately delivered to the sample to obtain maximum contrast between entering and bulk fluids. The approach features enhancement of the proton MRI signal by up to two orders of magnitude through the Overhauser effect under ambient

---

[5] 1 M = 1 mol/L is the unit used by chemists to indicate the number density of the solute molecules in the solution.

conditions at 0.35 T by using spin-polarized electrons of TEMPO molecules that are covalently immobilized onto a porous, water-saturated gel matrix. The continuous polarization of radical-free flowing water allowed to distinctively visualize vortices in model reactors and dispersion patterns through porous media. A proton signal enhancement of water by a factor of −10 and −100 provided for an observation time of 4 s and 7 s, respectively, upon its injection into fluids with a relaxation time $T_1 = 1.5$ s [70]. The team used an agarose-based porous media that was covalently spin labelled with the stable nitroxyl radicals TEMPO. The loading of solvent-accessible radical is sufficiently high and their mobility approximates that in solution, which ensures high efficiency for Overhauser mechanism induced DNP without physically releasing any measurable radical into the solution. Under ambient conditions at 0.35 T magnetic field, the DNP enhancement efficiency of proton signal of stagnant and continuously flowing water compares favorably with the performance of freely dissolved radicals [71].

Ebert et al. [72] have implemented a similar flow system into a mobile DNP polarizer, where radicals are immobilized in a gel matrix and the hyperpolarized radical-free fluid is subsequently directly pumped into the MRI scanner. They showed that even at flow conditions, the NMR signal is enhanced due to Overhauser DNP in the 0.35 T magnet as well as in the MRI scanner (4.7 T) at a distance of 1.4 m. Acquired images demonstrate the use of enhanced and the inverted NMR signals, which provide an excellent MRI contrast even for small enhancements.

Such enhanced signals may also prove useful when using ^{13}C, ^{15}N or ^{31}P nuclei for MRI; clearly this requires the gel technique to be adapted to the fluid in question.

11.3.7 Dissolution DNP

dDNP is a method to create solutions of molecules with nuclear spin polarization close to unity. The many orders of magnitude signal enhancement have enabled new applications, particularly in in vivo MR metabolic imaging. The method relies on solid-state DNP at low temperature (1 K, 5 T), followed by rapid dissolution to produce the room temperature solution of highly polarized spins. The technique was recently reviewed by Ardenkjaer-Larsen [73] who has originally developed the method to get the sample from 1 K to the liquid state in a matter of seconds.

dDNP is mainly used for low abundance NMR-active nuclei such as ^{13}C, ^{15}N in molecules of interest, thereby enhancing the signal-to-noise ratio by up to three orders of magnitude. The signal enhancement makes it possible to study biochemical phenomena and diseases in real time in vivo by ^{13}C NMR/MRI.

Typically, a sample containing the metabolite of interest (e.g. pyruvate, lactate) and stable radical is frozen, and then irradiated by microwave radiation in the DNP apparatus at ≤1.2 K for approximately 1 h. The sample is then rapidly melted and transferred to an automatic injection device using a heated buffer solution (e.g. water), where it can be introduced into the tissue, organ or cell culture to be studied. The whole process from the end of DNP until sample melting and dissolution takes about 10 s. The hyperpolarized signal can then be observed for up to 5 minutes after the dissolution process has been initiated.

In ETH Zurich dDNP has been developed for operation at 7 T static magnetic field and a temperature of 1.4 K [74]. The equipment was optimized for trityl-based samples. In [1-^{13}C]-pyruvic acid polarization levels of about 56% of ^{13}C are achieved compared to typical polarization levels of about 35%–45% at a standard field of 3.4 T. At the same time, the polarization build-up time increased significantly from about 670 s at 3.4 T to around 1300 s to 1900 s at 7 T, depending on the trityl derivate used. Adding trace amounts of Gd^{3+} to two samples was studied, yielding in one trityl compound no benefit, while the other profits significantly, boosting achievable polarization by 6%.

In the AMRIS Facility of National High Magnetic Field Laboratory in Gainesville, FL, a dDNP polarizer operating at 5 T/140 GHz has been built for in vivo, ex vivo and in vitro MRI and spectroscopy (MRI/S) studies. The custom-built microwave source features a tunable diode microwave source, which also allows for solid-state studies of the polarization dynamics. It has been designed as a platform for both methodology development and in vivo metabolic studies via MRI/S at both 4.7 T and 11 T, and ex vivo/in vitro studies at 14 T.

As most free radicals are likely to interact unfavorably with biochemical substances, it is highly desirable to remove them in the process of dissolution, in which case also concentrations in the range of polarized targets can be utilized. This can be realized by the use of radiolytic free radicals that react and anneal away quickly upon warmup of the sample.

To this end, Capozzi and coworkers used free radicals generated by UV-light irradiation of a frozen solution containing a fraction of pyruvic acid. They have demonstrated their dDNP potential, by providing up to 30 % [1-^{13}C]pyruvic acid liquid-state polarization [75]. Moreover, their labile nature has proven to pave a way to nuclear polarization storage and transport. Herein, different from the case of pyruvic acid, the issue of providing dDNP UV-radical precursors (trimethylpyruvic acid and its methyl-deuterated form) not involved in any metabolic pathway was investigated. The ^{13}C dDNP performance was evaluated for hyperpolarization of ^{13}C6 labelled deuterated glucose. The generated UV radicals proved to be versatile and highly efficient polarizing agents, providing, after dissolution and transfer (10 s), a ^{13}C liquid-state polarization of up to 32 %.

Complete equipment for dDNP have become recently available under the trade name Spinlab.[6] In-house developed dDNP equipment exist in numerous laboratories:

- Technical University of Denmark, Center for Magnetic Resonance led by Prof. J.H. Ardenkjaer-Larsen (dDNP);
- ETH Zurich, group of Prof. B. Meier (solid-state NMR, dDNP);
- EPF Lausanne, group of Prof. L. Emsley, Laboratory of Magnetic Resonance (solid-state NMR, dDNP);
- PSI Villigen, LMD Department, group of B. van den Brandt (dDNP, polarized targets);
- National High Magnetic Field Laboratory, Gainesville (FL) (dDNP, MRI imaging).

---

[6] Supplier: GE Healthcare.

It is impressive that the number of research teams using DNP for the purposes of biochemistry and biomedical applications now compares with the number of teams active in polarized targets for particle and nuclear physics, and that complete equipment for MAS DNP, DNP-MRI and dDNP have become commercially available.

References

[1] R. Bernard, P. Chaumette, P. Chesny, et al., A frozen spin target with three orthogonal polarization directions, *Nucl. Instrum. and Methods A* **249** (1986) 176–184.
[2] J. Bystricky, P. Chaumette, J. Deregel, et al., Measurement of the spin correlation parameter A_{oonn} and of the analyzing power for pp elastic scattering in the energy range from 0.5 to 0.8 GeV, *Nucl. Phys. B* **262** (1985) 727–743.
[3] R. A. Kunne, C. I. Beard, R. Birsa, et al., First measurement of the D_{onon} in p(bar)p elastic scattering, *Phys. Lett. B* 261 (1991) 191–196.
[4] R. Birsa, F. Bradamante, A. Bressan, et al., Measurement of the analysing power of the charge-exchange p(bar)p → n(bar)n reaction in the momentum range 546–875 MeV/c at LEAR, *Phys. Lett. B* **273** (1991) 533–539.
[5] T. O. Niinikoski, F. Udo, "Frozen spin" polarized target, *Nucl. Instr. and Meth.* **134** (1976) 219–233.
[6] A. Abragam, *The Principles of Nuclear Magnetism*, Clarendon Press, Oxford, 1961.
[7] M. Goldman, *Spin Temperature and Nuclear Magnetic Resonance in Solids*, Clarendon Press, Oxford, 1970.
[8] A. Abragam, M. Goldman, *Nuclear Magnetism: Order and Disorder*, Clarendon Press, Oxford, 1982.
[9] P. Hautle, W. Grüebler, B. van den Brandt, et al., Polarization reversal by adiabatic fast passage in a deuterated alcohol, *Phys. Rev. B* **46** (1992) 6596–6599.
[10] P. Hautle, W. Grüebler, B. van den Brant, et al., Fast polarization reversal in deuterated alcohol targets by the use of the adiabatic fast passage mechanism, in: T. Hasegawa, et al. (eds.) *Proc. 10th Int. Symp. on High-Energy Spin Physics*, Universal Academy Press, Inc., Tokyo, 1993, pp. 371–474.
[11] P. Hautle, W. Grüebler, J. A. Konter, et al., Polarization reversal by adiabatic fast passage in deuterated alcohols, in: E. Steffens, et al. (eds.) *Proc. 6th Workshop on Polarized Solid Targets*, Springer Verlag, Berlin, Bonn, 1990, pp. 364–368.
[12] L. B. Parfenov, A. F. Prudkoglyad, Rapid reversal of sign of dynamically enhanced polarization in polarized targets, *Sov. Phys. JETP* **60** (1984) 123–127.
[13] V. Chabaud, K. Kuroda, A time-dependent analysis of asymmetry measurements in polarized-target experiments, *Nucl. Instrum. Methods* **125** (1975) 119–124.
[14] T. O. Niinikoski, Optimum measurement and analysis of small polarization asymmetry in high-energy inelastic scattering using a polarized target, *Nucl. Instrum. and Meth.* **134** (1976) 235–241.
[15] A. B. Williams, F .J. Taylor, *Electronic Filter Design Handbook*, McGraw-Hill, New York, 1995.
[16] L. Dick, A. Gonidec, A. Gsponer, et al., Spin effects in the inclusive reactions π± + p(↑) → π± + anything at 8 GeV/c, *Physics Letters B* **57** (1975) 93–96.
[17] European Muon Collaboration (EMC), J. Ashman, B. Badelek, et al., A measurement of the spin asymmetry and determination of the structure function g_1 in deep inelastic muon proton scattering, *Phys. Lett. B* **206** (1988) 364–370.

[18] European Muon Collaboration (EMC), J. Ashman, B. Badelek, et al., An investigation of the spin structure of the proton in deep inelastic scattering of polarized muons on polarized protons, *Nucl. Phys. B* **328** (1989) 1–35.

[19] J. Ashman, B. Badelek, G. Baum, et al., An investigation of the spin structure of the proton in deep inelastic scattering of polarized muons on polarized protons, *Nucl. Phys. B* **328** (1989) 1–35.

[20] Spin Muon Collaboration (SMC), D. Adams, B. Adeva, et al., The polarized double-cell target of the SMC, *Nucl. Instr. and Meth. A* **437** (1999) 23–67.

[21] Spin Muon Collaboration (SMC), B. Adeva, S. Ahmad, et al., Measurement of the spin-dependent structure function $g_1(x)$ of the proton, *Phys. Lett. B* **329** (1994) 399–406.

[22] European Muon Collaboration (EMC), J. Ashman, B. Badelek, et al., A measurement of the ratio of the nucleon structure function in copper and deuterium, *Z. Physik C* **57** (1993) 211–218.

[23] P. Abbon, E. Albrecht, V. Y. Alexakhin, et al., The COMPASS experiment at CERN, *Nuclear Instruments and Methods in Physics Research Section A: Accelerators, Spectrometers, Detectors and Associated Equipment* **577** (2007) 455–518.

[24] F. L. Shapiro, Polarized nuclei and neutrons, *Polarized Targets and Ion Sources*, La Direction de la Physique, CEN Saclay, Saclay, France, 1966, pp. 339–356.

[25] W. G. Williams, Polarizing filters, in: G. R. Court, et al. (eds.) *Second Workshop on Polarized Target Materials*, SRC, Rutherford Laboratory, Cosener's House, Abingdon, Chilton, Didcot, UK, 1979, pp. 102–105.

[26] D. R. Rich, T. R. Gentile, T. B. Smith, A. K. Thompson, G. L. Jones, Spin exchange optical pumping at pressures near 1 bar for neutron spin filters, *Appl. Phys. Lett.* **80** (2002) 2210–2212.

[27] O. Schaerpf, Comparison of theoretical and experimental behaviour of supermirrors and discussion of limitations, *Physica B: Condensed Matter* **156–157** (1989) 631–638.

[28] S. Masalovich, Analysis and design of multilayer structures for neutron monochromators and supermirrors, *Nuclear Instruments and Methods in Physics Research Section A: Accelerators, Spectrometers, Detectors and Associated Equipment* **722** (2013) 71–81.

[29] V. G. Syromyatnikov, V. M. Pusenkov, New compact neutron supermirror transmission polarizer, *Journal of Physics: Conference Series* **862** (2017) 012028.

[30] J. Dawidowski, J. R. Granada, J. R. Santisteban, F. Cantargi, L. A. R. Palomino, Appendix – neutron scattering lengths and cross sections, in: F. Fernandez-Alonso, D. L. Price (eds.) *Experimental Methods in the Physical Sciences*, Academic Press, 2013, pp. 471–528.

[31] H. Glättli, M. Goldman, Neutron scattering, in: K. Sköld, D. L. Price (eds.) *Methods of Experimental Physics Vol. 23 C*, Academic Press, New York, 1987.

[32] H. B. Stuhrmann, O. Scharpf, M. Krumpolc, et al., Dynamic nuclear polarization of biological matter, *Eur. Biophys. J.* **14** (1986) 1–6.

[33] W. Knop, K. H. Nierhaus, V. Novotny, et al., Polarised neutron scattering from dynamic polarised targets of biological origin, in: S. Jaccard, S. Mango (eds.) *Proc. Int. Workshop on Polarized Sources and Targets*, 1986, pp. 741–746.

[34] W. Knop, H.-J. Schink, H. B. Stuhrmann, et al., Polarized neutron scattering by polarized protons of bovine serum albumin in deuterated solvent, *J. Appl. Crystallography* **22** (1989) 352–362.

[35] J. B. Hayter, G. T. Jenkin, J. W. White, Polarized-neutron diffraction from spin-polarized protons: a tool in structure determination?, *Phys. Rev. Lett.* **33** (1974) 696–699.

[36] M. Leslie, Crystallographic studies of polarized lanthanum magnesium nitrate (LMN) using polarized neutrons, in: G. R. Court, et al. (eds.) *Second Workshop on Polarized Target Materials*, SRC, Rutherford Laboratory, Cosener's House, Abingdon, Chilton, Didcot, UK, 1979, pp. 106–111.
[37] M. Kohgi, M. Ishida, Y. Ishikawa, et al., Small-angle neutron scattering from dynamically polarized hydrogenous materials, *Journal of the Physical Society of Japan* **56** (1987) 2681–2688.
[38] H. Glättli, C. Fermon, M. Eisenkremer, M. Pinot, Small angle neutron scattering with nuclear polarization on polymers, *J. Phys. France* **50** (1989) 2375–2387.
[39] W. Knop, K. H. Nierhaus, V. Novotny, et al., *Spin Contrast Variation – a New Tool in Macromolecular Structure Research*, GKSS Geesthacht Report 1988.
[40] W. Knop, M. Hirai, H.-J. Schink, et al., A new polarized target for neutron scattering studies on biomolecules: first results on apoferritin and the deuterated 50S subunit of ribosomes, *J. Appl. Crystallography* **25** (1992) 155–165.
[41] H. Stuhrmann, Contrast variation in X-ray and neutron scattering, *Journal of Applied Crystallography* **40** (2007) 23–27.
[42] H. B. Stuhrmann, Contrast variation application in small-angle neutron scattering experiments, *Journal of Physics: Conference Series* **351** (2012) 012002.
[43] B. van den Brandt, H. Glättli, I. Grillo, et al., Time-resolved nuclear spin-dependent small-angle neutron scattering from polarised proton domains in deuterated solutions, *The European Physical Journal B – Condensed Matter and Complex Systems* **49** (2006) 157–165.
[44] Z. Oliver, M. J. Hélène, B. S. Heinrich, Time-resolved proton polarisation (TPP) images tyrosyl radical sites in bovine liver catalase, *Journal of Physics: Conference Series* **848** (2017) 012002.
[45] H. B. Stuhrmann, Time-resolved polarized neutron scattering from dynamic polarized nuclear spin targets, *Journal of Physics: Conference Series* **351** (2012) 012003.
[46] Y. Noda, S. Koizumi, Dynamic nuclear polarization apparatus for contrast variation neutron scattering experiments on iMATERIA spectrometer at J-PARC, *Nuclear Instruments and Methods in Physics Research Section A: Accelerators, Spectrometers, Detectors and Associated Equipment* **923** (2019) 127–133.
[47] V. G. Baryshevskii, M. I. Podgoretskii, Nuclear precession of neutrons, *Zh. Eksp.Teor. Fiz. (Sov. Phys. JETP 20 (1965) 704)* **47** (1964) 1050–1054.
[48] M. I. Tsulaia, Neutron nuclear precession—nuclear pseudomagnetism, *Phys. Atom. Nuclei* **77** (2014) 1321–1333.
[49] V. G. Pokazan'ev, G. V. Skrotskiĭ, Pseudomagnetism, *Soviet Physics Uspekhi* **22** (1979) 943–959.
[50] A. Abragam, G. L. Bacchella, H. Glättli, et al., Résonance nucléaire "pseudo-magnétique" du neutron induite par un champ nucléaire de radiofréquence, *C.R. Acad. Sci. B* **274** (1972) 423–433.
[51] H. Glättli, G. L. Bacchella, M. Fourmond, et al., Experimental values of spin dependent nuclear scattering lengths of slow neutrons, *Journal de Physique* **40** (1979) 629–634.
[52] J. Marks, W. T. Wenckebach, N. J. Poulis, Magnetic ordering of proton spins in $Ca(OH)_2$, *Physica B* **96** (1979) 337–340.
[53] M. Chapellier, M. Goldman, V. H. Chau, A. Abragam, Production et observation d'un état antiferromagnétique nucléaire, *C.R. Acad. Sci.* **268** (1969) 1530–1533.
[54] M. Chapellier, M. Goldman, V. H. Chau, A. Abragam, Production and observation of a nuclear antiferromagnetic state, *J. Appl. Phys.* **41** (1970) 849–853.
[55] Y. Roinel, V. Bouffard, C. Fermon, et al., Phase diagrams in ordered nuclear spins in LiH: a new phase at positive temperature?, *J. Physique* **48** (1987) 837–845.

[56] Y. Roinel, G. L. Bachella, O. Avenel, et al., Neutron diffraction study of nuclear magnetic ordered phases and domains in lithium hydride, *J. Physique Lett.* **41** (1980) 123–125.

[57] Y. Roinel, V. Bouffard, G. L. Bachella, et al., First study of nuclear antiferromagnetism by neutron diffraction, *Phys. Rev. Lett.* **41** (1978) 1572–1574.

[58] C. M. Van der Zon, G. D. Van Velzen, W. T. Wenckebach, Nuclear magnetic ordering in Ca(OH)$_2$. III. Experimental determination of the critical temperature, *J. Physique* **51** (1990) 1479–1488.

[59] H. S. Gutowsky, G. E. Pake, Structural investigations by means of nuclear magnetism. II. hindered rotation in solids, *J. Chem. Phys.* **18** (1950) 162–170.

[60] I. J. Lowe, Free induction decays of rotating solids, *Phys. Rev. Lett.* **285** (1959) 285–287.

[61] E. R. Andrew, A. Bradbury, R. G. Eades, Nuclear magnetic resonance spectra from a crystal rotated at high speed, *Nature* **182** (1958) 1659–1659.

[62] C. P. Slichter, *Principles of Magnetic Resonance*, 3rd ed., Springer-Verlag, Berlin, 1990.

[63] A. Samoson, E. Lippmaa, A. Pines, High resolution solid-state N.M.R., *Molecular Physics* **65** (1988) 1013–1018.

[64] R. A. Wind, M. J. Duijvestijn, C. van der Lugt, A. Manenschijn, J. Vriend, Applications of dynamic nuclear polarization in 13 C NMR in solids, *Progress in Nuclear Magnetic Resonance Spectroscopy* **17** (1985) 33–67.

[65] A. S. Lilly Thankamony, J. J. Wittmann, M. Kaushik, B. Corzilius, Dynamic nuclear polarization for sensitivity enhancement in modern solid-state NMR, *Progress in Nuclear Magnetic Resonance Spectroscopy* **102–103** (2017) 120–195.

[66] L. R. Becerra, G. J. Gerfen, B. F. Bellew, et al., A spectrometer for dynamic nuclear polarization and electron paramagnetic resonance at high frequencies, *Journal of Magnetic Resonance, Series A* **117** (1995) 28–40.

[67] G. J. Gerfen, L. R. Becerra, D. A. Hall, et al., High frequency (140 GHz) dynamic nuclear polarization: polarization transfer to a solute in frozen aqueous solution, *The Journal of Chemical Physics* **102** (1995) 9494–9497.

[68] C. Song, K.-N. Hu, C.-G. Joo, T. M. Swager, R. G. Griffin, TOTAPOL: a biradical polarizing agent for dynamic nuclear polarization experiments in aqueous media, *Journal of the American Chemical Society* **128** (2006) 11385–11390.

[69] J. G. Krummenacker, V. P. Denysenkov, M. Terekhov, L. M. Schreiber, T. F. Prisner, DNP in MRI: an in-bore approach at 1.5 T, *J. Magn. Res.* **215** (2012) 94–99.

[70] E. R. McCarney, B. D. Armstrong, M. D. Lingwood, S. Han, Hyperpolarized water as an authentic magnetic resonance imaging contrast agent, *Proceedings of the National Academy of Sciences* **104** (2007) 1754–1759.

[71] E. R. McCarney, S. Han, Spin-labeled gel for the production of radical-free dynamic nuclear polarization enhanced molecules for NMR spectroscopy and imaging, *J. Magn. Res.* **190** (2008) 307–315.

[72] S. Ebert, A. Amar, C. Bauer, et al., A mobile DNP polarizer for continuous flow applications, *Applied Magnetic Resonance* **43** (2012) 195–206.

[73] J. H. Ardenkjaer-Larsen, On the present and future of dissolution-DNP, *J. Magn. Res.* **264** (2016) 3–12.

[74] F. Jähnig, G. Kwiatkowski, A. Däpp, et al., Dissolution DNP using trityl radicals at 7 T field, *Physical Chemistry Chemical Physics* **19** (2017) 19196–19204.

[75] A. Capozzi, S. Patel, C. P. Gunnarsson, et al., Efficient hyperpolarization of U-13 C-glucose using narrow-line UV-generated labile free radicals, *Angewandte Chemie International Edition* **58** 1334–1339.

Appendices

A.1 Units, Variables, Symbols and Constants

A.1.1 International System (SI) of Units Compared with CGS Gaussian Units

The International System [1, 2] defines units for length (m), mass (kg), time (s), current (A), temperature (K), amount of substance (mole = mol) and luminous intensity (candela = cd). All other units are then expressed using these basic definitions and rules derived from constitutive relations. For mechanical and electrodynamic relations alone, four basic units (m, kg, s, A) are defined, and these constitute a self-consistent base of units for the variables needed in the description of the laws of nature.

In the CGS Gaussian system (also called CGS Electromagnetic system), the units of length (cm), mass (g) and time (s) are defined; the charge unit (esu) is linked to these based on the Coulomb force between two charges, and on the definition that the dielectric constant (permittivity) of free space is chosen to be unity. Other units are then derived from this set, in a way similar to the SI units. The charge unit in SI system, 1 coulomb = 1 A·s, is converted to CGS Gaussian unit by a multiplying factor, which amounts to 10 times the velocity of light in vacuum expressed in m/s, i.e. $10\,c/(m/s)$.

There is the fundamental difference that the CGS Gaussian system is three dimensional [1, 2], and therefore cannot be compared with the SI units, which has four dimensions when mechanical and electrical quantities are considered.

Table A1.1 lists the conversion factors for the main units in the SI and CGS Gaussian systems. It should be noted that the magnetic flux density and magnetic field are in CGS Gaussian units, which cannot be compared and converted directly to the corresponding SI units. This follows from the fact that the constitutive relations of electric polarization and magnetic induction are different in the two systems. Therefore, the relations linking mechanical forces and electromagnetic variables and other basic electromagnetic and electrodynamic relations are also different in the two systems.

We notice that the velocity of light enters in the conversion factors between some of the electromagnetic units in the two systems. This follows from the definition that the permeability of free space is also defined as unity in the CGS system.

The electromagnetic equations in vacuum, derived in the SI units, can sometimes be transformed to the CGS Gaussian system by substitutions, and vice versa, but these depend on the way how the equations are written. It is therefore best that the equations are re-derived from the basic constitutive and other basic electromagnetic relations in the other systems of units. The main relations are listed below:

Table A1.1 Comparison of units in the International System and CGS Gaussian system.

Quantity	SI unit	In CGS Gaussian units
Energy	1 joule (J)	$= 10^7$ erg (erg = g·cm^2·s^{-2})
Force	1 newton (N)	$= 10^5$ dyne (dyn = abampere^2 = g·cm·s^{-2})
Charge	1 coulomb (C)	$= 2.99792458 \cdot 10^9$ esu
Current	1 ampere (A)	$= 0.1$ abampere
Electric potential	1 volt (V = J/C)	$= 1/299.792458$ statvolt (statvolt = erg/esu)
Electrical resistance	1 ohm (Ω)	$= 1/29.9792458$ statvolt/abampere
Magnetic flux density	1 tesla (T)	$= 10^4$ gauss (= dyne/esu)
Magnetic field	1 A/m	$= 10^{-3} \cdot 4\pi$ oersted

Table A1.2 Electromagnetic equations in SI and CGS Gaussian systems.

	SI system	CGS Gaussian system
Permeability of free space:	$\mu_0 = 4\pi \cdot 10^{-7}$ Vs/Am	$\mu_0 = 1$
Permittivity of free space:	$\varepsilon_0 = 8.854187 \ldots \cdot 10^{-12}$ F/m	$\varepsilon_0 = 1$
Electric polarization:	$\mathbf{D} = \varepsilon_0 \mathbf{E} + \mathbf{P}$	$\mathbf{D} = \mathbf{E} + 4\pi \mathbf{P}$
Magnetic induction:	$\mathbf{H} = \dfrac{\mathbf{B}}{\mu_0} - \mathbf{M}$	$\mathbf{H} = \mathbf{B} - 4\pi \mathbf{M}$
Coulomb force:	$\mathbf{F} = \dfrac{q_1 q_2}{4\pi\varepsilon_0} \dfrac{\mathbf{r}_{12}}{r_{12}^3}$	$\mathbf{F} = q_1 q_2 \dfrac{\mathbf{r}_{12}}{r_{12}^3}$
Lorentz force:	$\mathbf{F} = q(\mathbf{E} + \mathbf{v} \times \mathbf{B})$	$\mathbf{F} = q\left(\mathbf{E} + \dfrac{\mathbf{v}}{c} \times \mathbf{B}\right)$
Maxwell equations:	$\nabla \cdot \mathbf{D} = \rho$	$\nabla \cdot \mathbf{D} = 4\pi\rho$
	$\nabla \cdot \mathbf{B} = 0$	$\nabla \cdot \mathbf{B} = 0$
	$\nabla \times \mathbf{H} - \dfrac{\partial \mathbf{D}}{\partial t} = \mathbf{J}$	$\nabla \times \mathbf{H} - \dfrac{1}{c}\dfrac{\partial \mathbf{D}}{\partial t} = 4\pi \mathbf{J}$
	$\nabla \times \mathbf{E} - \dfrac{\partial \mathbf{B}}{\partial t} = 0$	$\nabla \times \mathbf{E} - \dfrac{1}{c}\dfrac{\partial \mathbf{B}}{\partial t} = 0$
Static potentials:	$V(\mathbf{r}) = \dfrac{1}{4\pi\varepsilon_0} \int \dfrac{\rho(\mathbf{r}')}{|\mathbf{r}-\mathbf{r}'|} d^3x'$	$V(\mathbf{r}) = \int \dfrac{\rho(\mathbf{r}')}{|\mathbf{r}-\mathbf{r}'|} d^3x'$
	$\mathbf{A}(\mathbf{r}) = \dfrac{\mu_0}{4\pi} \int \dfrac{\mathbf{J}(\mathbf{r}')}{|\mathbf{r}-\mathbf{r}'|} d^3x'$	$\mathbf{A}(\mathbf{r}) = \dfrac{1}{c} \int \dfrac{\mathbf{J}(\mathbf{r}')}{|\mathbf{r}-\mathbf{r}'|} d^3x'$

Table A1.2 (cont.)

	SI system	CGS Gaussian system
Fields from potentials:	$\mathbf{E} = -\nabla \bullet V - \dfrac{\partial \mathbf{A}}{\partial t}$	$\mathbf{E} = -\nabla \bullet V - \dfrac{1}{c}\dfrac{\partial \mathbf{A}}{\partial t}$
	$\mathbf{B} = \nabla \times \mathbf{A}$	$\mathbf{B} = \nabla \times \mathbf{A}$

Some older sources use the Heaviside-Lorentz system of units, which is a CGS system rationalized in the same way as the MKSA system, such that the factor 4π appears in the Coulomb and Ampère laws explicitly. The electromagnetic relations in this system are compared with the MKSA (same as SI) and CGS Gaussian systems in Ref. [3].

Because of the inconvenient and somewhat unnatural CGS units for charge, current, potentials and fields, several branches of physics, notably in solid state physics, have moved to a hybrid CGS system, where the units of coulomb and ampere were used for replacing the units statvolt and abampere. Sometimes the system of units is not clearly stated in publications, which may result in a considerable confusion. In these cases we have re-derived the equations using the basic SI relations listed above.

A.1.2 Definition of Fundamental Physical Quantities and Variables

In the following table we shall give the definition and values of important fundamental quantities and parameters, relevant in magnetic resonance and other fields of the physics of polarized targets, in the SI system of units:

Variable	Symbol	Definition	SI unit
Magnetic field	$\mathbf{H}$	$\oint_a \mathbf{H} \cdot d\mathbf{l} = I_a$	A/m
Magnetization	$\mathbf{M}$	$= n\boldsymbol{\mu}$	A/m
Magnetic induction = magnetic flux density	$\mathbf{B}$	$= \mu_0(\mathbf{H} + \mathbf{M}) = \nabla \times \mathbf{A}$ $= -\nabla V_m$	Vs/m^2 = T (tesla)
Magnetic flux	Φ_m		Vs = Wb (weber)
Vector potential	$\mathbf{A}$	$= (\mu_0 I / 4\pi)\oint_a d\mathbf{l}/r$ $= (\mu_0 I/4\pi)\,\boldsymbol{\mu}\times(\nabla(1/r))$	Vs/m
Magnetic scalar potential	V_m	$= (\mu_0/4\pi r^3)\,\boldsymbol{\mu}\cdot\mathbf{r}$	Vs/m
Magnetic moment	$\boldsymbol{\mu}$		A m^2 = J/T
Angular momentum	$\mathbf{J}$		kg m^2/s
Gyromagnetic ratio	Γ	$= |\boldsymbol{\mu}|/|\mathbf{J}|$	A s/kg = m^2/(V s^2)
Number density	N		m^{-3}
Larmor frequency	ν_L	$= -\gamma B/2\pi$	s^{-1} = Hz (hertz)
(in angular units)	ω_L	$= -\gamma B$	radians/s

In this book, for simplicity, we use the term 'magnetic field' for both **H** and **B** if the meaning appears clear from the context. This seems to be a common practice in modern literature, although it can be misleading in some cases.

A.1.3 Physical Constants and SI Base Units Redefined in 2019

The table below gives the fundamental constants that are relevant in the physics of polarized targets. While the numeric values of Particle Data Group of Ref. [4] are accurate for most purposes, the table below reflects also the changes of the new SI definitions introduced in 2019 and described by Ref. [5].

Quantity	Symbol, Equation	Value in SI Units		SI Unit	Uncertainty (ppb)
Speed of light	c	299 792 458		m/s	exact
Permeability of free space	μ_0	4π	$\cdot 10^{-7}$	Vs/Am	exact
Permittivity of free space	$\varepsilon_0 = (c^2 \mu_0)^{-1}$	8.854 187 817	$\cdot 10^{-12}$	F/m	exact
Impedance of free space	$Z_{free} = \mu_0 c$	119.916 983 2 π		Ω	exact
Planck constant	h	6.626 070 15	$\cdot 10^{-34}$	J·s	exact
Planck constant/2π	$h/2\pi$	1.054 571 800	$\cdot 10^{-34}$	J·s	exact
		= 6.582 119 514	$\cdot 10^{-22}$	MeV s	6.1
Unit charge	e	1.602 176 634	$\cdot 10^{-19}$	C = A·s	exact
Electron mass	m_e	0.510 998 9461		MeV/c²	6.2
		= 9.109 383 56	$\cdot 10^{-31}$	kg	12
Bohr magneton	$\mu_B = \beta$	5.788 381 8012	$\cdot 10^{-11}$	MeV/T	0.45
Proton mass	m_p	938.272 0813		MeV/c²	6.2
Nuclear magneton	$\mu_N = he/4\pi m_p$	3.152 451 2550	$\cdot 10^{-14}$	MeV/T	0.46
Deuteron mass	m_d	1875.612 928		MeV/c²	6.2
Avogadro number	N_A	6.022 140 76	$\cdot 10^{23}$	mol⁻¹	exact
Atomic mass unit	$m(^{12}C)/12$	931.494 0954		MeV/c²	6.2
	$= (10^{-3}\,\text{kg})/N_A$	= 1.660 539 040	$\cdot 10^{-27}$	kg	12
Boltzmann constant	k	1.380 649	$\cdot 10^{-23}$	J/K	exact
		= 8.617 3330	$\cdot 10^{-5}$	eV/K	570
Gas constant	$R = kN_A$	8.314 459 86		J/(mol·K)	exact
Standard temperature	T_{STP}	273.15		K	definition
Standard pressure	p_{STP}	1.013 25	$\cdot 10^5$	N/m²	definition
Mol. volume STP	$N_A k T_{STP}/p_{STP}$	22.413 962		l/mol	definition
Spin temp. constant	h/k	47.992 430 734		mK/GHz	exact

A.2 Paramagnetic Compounds, Ions and Free Radicals

Table A2.1 is reproduced from M. Borghini's report listing the spectral widths of the hyperfine lines measured in diluted liquid samples at RT [6], obtained mainly from Ref. [7].

Table A2.1 Extreme hyperfine structure (HFS) of free radicals listed by Borghini in 1966. The data is mainly from Ref. [7]. The hyperfine widths were measured mainly in diluted liquid samples at RT, and they may miss weak HF lines due to ^{13}C and other rare spins. The g-factors were not accurately measured and anisotropies were not measured in glassy solid state. No solvent effects were recorded.

Free radical	Extreme HFS (10^{-4} T)	Abbreviation
Bianthrone	10	
1,2 Bis-diphenylene-1-phenylallyl	7.1	BDPA
Dianisyl nitric oxide	30	
1,1-Diphenyl-2-picryl hydrazyl	58	DPPH
Diphenylene-triphenyl-ethyl	5	
Galvinoxyl		
Kenyon-Banfield radical	8.9	
Pentaphenylcyclopentadienyl		
Picryl-aminocarbazyl	60	
Porphyrexide	17	PX
Porphyrindene	10.7	
Tetramethyl benzidine formate	3.4	
Triacetonamyl		
Tri-p-anisylaminium perchlorate	0.68	
Tri-p-aminophenylaminium perchlorate	0.33	
Tri-t-butyl phenoxyl	7.7	
Tri-p-nitrophenylmethyl	0.7	
Tri-phenylamine perchlorate	2	
Tri-p-xenylmethyl	5.7	
Violanthrene	26	
Violanthrone	30	
Würster's blue perchlorate	2.7	
P-benzosemiquinone	9.5	
Mono-methylquinone	14	
Tetra-methylquinone	23	
Mono-chloroquinone	6.0	
Tri-chloroquinone	2.1	
Tetra-chloroquinone	0.4	
2.5 Di-t-butylquinone	4.3	
1.4-Naphtosemiquinone	8	
2.3-Dimethylquinone	12	

Table A2.1 (cont.)

	Extreme HFS
0-Benzosemiquinone	10
4-Tert-butylquinone	6
3-Phenylquinone	8

Table A2.2 Values of g-tensor for some organic free radicals used for DNP. Most of these radicals are commercially available from Sigma Aldrich (now Merck KGaA).

Organic free radicals	g_1	g_2	g_3		Note
1,1-Diphenyl-2-picryl hydrazyl	2.005	2.010	2.010	DPPH	
Porphyrexide	1.9997	2.0047	2.0097	PX	
Li-7,7,8,8-tetracyanoquinodimethane	2.00252	$\Delta g <$	0.0018	TCNQ-Li	
BDPA-derived radicals					
1,2 bis-diphenylene-1-phenyl allyl	2.0035	$\Delta g <$	0.00015	BDPA	1
Sulphonated BDPA				SA-BDPA	2
Sulphonamide BDPA				SN-BDPA	2
Nitroxide radicals and biradicals					
2,2,6,6-Tetramethylpiperidin-1-oxyl	2.002	2.005	2.009	TEMPO	3
4-Hydroxy-2,2,6,6-tetramethylpiperidin-1-oxyl				TEMPOL	4
4-Oxo-2,2,6,6-tetramethyl-1-piperidinyloxy				4-oxo-TEMPO	4
4-Amino-2,2,6,6-tetramethylpiperidine-1-oxyl				4-amino-TEMPO	4
Biradicals based on TEMPO and others					5
Trityl radicals					
Triphenylmethyl (triarylmethyl)				TAM	
Trityl OX063	2.005	$\Delta g <$	0.0007	OX063	
Finland trityl CT-03				CT-03	

Notes: (1) BDPA is commonly used as EPR marker for magnetic field strength; soluble in organic solvents such as styrene. (2) These are water soluble versions of BDPA. (3) TEMPO is water soluble and sublimable, with melting point close to 37 °C. (4) These versions of TEMPO enter into the formulation of biradicals. (5) Biradicals are two radicals linked together chemically; for a review, see Ref. [8].

A.3 Nuclear Moments and NMR Frequencies

Table A3.1 gives the spins, magnetic moments and quadrupole moments of the nuclei of stable isotopes of some elements in their ground state. Included are light elements up to $Z = 30$ (^{67}Zn) and some heavier ones, which are met in polarized targets. In the table we have

A.3 Nuclear Moments and NMR Frequencies

also included the neutron and tritium, as these are important for the tests of the theoretical models of the nuclear moments.

The light elements up to $Z = 21$ are almost all dynamically polarized in target materials, with the exception of Be, Ne and Sc. The transition metals of the iron group Ti-Zn form paramagnetic metallo-organic complexes, and are also paramagnetic in many of their ionic salts and minerals; the most common of these is Cr. Their nuclear moments influence the EPR line via hyperfine interactions, and their knowledge is therefore vital. From the rare earth group only Nd and Tm are represented; they are well-known dopants of the LMN and CaF_2, for instance. More details and examples on the paramagnetic atoms and ions are given in Chapter 3.

Included in Table A3.1 are also the moments of the two isotopes of Ag, which were used in the work on the nuclear magnetism in this metallic element, described in Chapter 1. ^{3}He and Xe have been optically pumped to a high polarization, as described in Chapter 10.

The magnetic dipole moments in the table are scaled by the nuclear magneton $\mu_N = 3.1524512550 \times 10^{-14}$ MeV/T from table of Section A1.3. This parameter is known to 0.46 ppb from measurements on protons in a Penning trap. The magnetic moments of the proton and deuteron are obtained from similar measurements to a high accuracy. Neutron beam measurements yield the most precise value for its magnetic dipole moment. Most other magnetic moments are determined by NMR or other spectroscopic methods in condensed matter in gas or liquid phase. Such measurements are subject to shifts due to interactions with the electronic system, in particular the chemical shift that increases with the atomic number. The relative accuracy therefore has a tendency to decrease with increasing Z.

The table lists the NMR frequencies of the nuclei at 2.5 T magnetic field, calculated directly from the tabulated nuclear moments using the Larmor precession frequency defined by Eq. 1.39:

$$f_{NMR} = \frac{|\omega_0|}{2\pi} = \frac{|\gamma B_0|}{2\pi}. \tag{A3.1}$$

The experimental frequency in solid materials may be somewhat different, due to the chemical shift in dielectric materials, and due to the Knight shift in metals. The gyromagnetic ratios are calculated from Eqs. (1.5) and (5.1):

$$\gamma = \frac{\mu}{J} = \frac{\mu}{\mu_N} \frac{\mu_N}{\hbar I}, \tag{A3.2}$$

and the dipole moment μ is scaled by dividing with μ_N of Eq. 5.2, giving $\mu/\mu_N = gI$. The natural abundances vary depending on the origin of the sample. A broader list including unstable and heavier isotopes can be found in *CRC Handbook of Chemistry and Physics*.

Table A3.1 Nuclear moments of light stable isotopes; see text for explanations.

Z	Elem.	A	I	f_{NMR} at 2.5 T (MHz)	Gyromagnetic ratio γ (s^{-1}/T)	Dipole moment $\mu/\mu_N = gI$	Abundance (%)	Quadrupole moment (barn)
0	n	1	½	72.915804	−1.832574E+08	−1.913042	0.000	
1	^{1}H	1	½	106.443672	2.675221E+08	2.792847	99.9885	
1	^{2}H	2	1	16.338777	4.106383E+07	0.857387	0.0115	2.73E−03
1	^{3}H	3	½	113.529733	2.853313E+08	2.978770	0.000	
2	^{3}He	3	½	81.081505	−2.037800E+08	−2.127400	0.000	
3	^{6}Li	6	3/2	15.662901	3.936516E+07	0.821920	7.590	6.90E−04
3	^{7}Li	7	3/2	41.365263	1.039622E+08	3.256000	92.410	−3.00E−02
4	^{9}Be	9	3/2	14.958065	−3.759372E+07	−1.177400	100.000	5.20E−02
5	^{10}B	10	3	11.438334	2.874767E+07	1.800700	19.580	7.40E−02
5	^{11}B	11	3/2	34.149209	8.582632E+07	2.688000	13.660	3.55E−02
6	^{13}C	13	½	26.762880	6.726245E+07	0.702199	1.109	
7	^{14}N	14	1	7.688717	1.932385E+07	0.403470	99.630	−2.60E−02
7	^{15}N	15	½	10.785205	−2.710618E+07	−0.282980	0.370	
8	^{17}O	17	5/2	14.429565	−3.626545E+07	−1.893000	0.037	−0.26
9	^{19}F	19	½	100.133028	2.516617E+08	2.627270	100.000	
10	^{21}Ne	21	3/2	8.402637	−2.111813E+07	−0.661400	0.257	
11	^{23}Na	23	3/2	28.154041	7.075882E+07	2.216100	100.000	1.45E−01
12	^{25}Mg	25	5/2	6.513428	−1.637003E+07	−0.854490	10.130	
13	^{27}Al	27	5/2	27.734799	6.970515E+07	3.638500	100.000	1.49E−01
14	^{29}Si	29	½	21.143925	−5.314048E+07	−0.554770	4.700	
15	^{31}P	31	½	43.086698	1.082887E+08	1.130500	100.000	
16	^{33}S	33	3/2	8.163414	2.051690E+07	0.642570	0.760	−6.40E−02
17	^{35}Cl	35	3/2	10.429102	2.621119E+07	0.820910	75.530	−7.89E−02
17	^{37}Cl	37	3/2	8.680861	2.181738E+07	0.683300	24.470	−6.21E−02
19	^{39}K	39	3/2	4.967008	1.248345E+07	0.390970	93.100	1.10E−01
19	^{41}K	41	3/2	2.726220	6.851738E+06	0.214590	6.880	
20	^{43}Ca	43	7/2	7.161425	−1.799862E+07	−1.315300	0.145	
21	^{45}Sc	45	7/2	25.858008	6.498826E+07	4.749200	100.000	−2.20E−01
22	^{47}Ti	47	5/2	5.999742	−1.507900E+07	−0.787100	7.280	
22	^{49}Ti	49	7/2	6.001157	−1.508255E+07	−1.102200	5.510	
23	^{50}V	50	6	10.612235	2.667146E+07	3.341300	0.240	
23	^{51}V	51	7/2	27.980355	7.032230E+07	5.139000	99.760	−4.00E−02

Table A3.1 (cont.)

Z	Elem.	A	I	f_{NMR} at 2.5 T (MHz)	Gyromagnetic ratio γ (s^{-1}/T)	Dipole moment $\mu/\mu_N = gI$	Abundance (%)	Quadrupole moment (barn)
24	^{53}Cr	53	3/2	6.016003	−1.511987E+07	−0.473540	9.550	
25	^{55}Mn	55	5/2	26.252205	6.597899E+07	3.444000	100.000	5.50E−01
26	^{57}Fe	57	1/2	3.439313	8.643937E+06	0.090240	2.190	
27	^{59}Co	59	7/2	25.134406	6.316965E+07	4.616300	100.000	4.00E−01
28	^{61}Ni	61	3/2	9.511470	−2.390493E+07	−0.748680	1.190	
29	^{63}Cu	63	3/2	28.211211	7.090251E+07	2.220600	69.090	−1.60E−01
29	^{65}Cu	65	3/2	30.222304	7.595694E+07	2.378900	30.910	−1.50E−01
30	^{67}Zn	67	5/2	6.656809	1.673039E+07	0.873300	4.110	1.50E−01
32	^{73}Ge	73	9/2	3.713007	−9.331803E+06	−0.876790	7.760	−2.00E−01
45	^{103}Rh	103	1/2	3.350129	−8.419792E+06	−0.087900	100.000	
46	^{105}Pd	105	5/2	4.870836	−1.224175E+07	−0.639000	22.230	
47	^{107}Ag	107	1/2	4.307145	−1.082504E+07	−0.113010	51.820	
47	^{109}Ag	109	1/2	4.951635	−1.244482E+07	−0.12992	48.180	
54	^{129}Xe	129	1/2	29.441116	−7.399359E+07	−0.77247	26.401	
54	^{131}Xe	131	3/2	8.727486	2.193456E+07	0.68697	21.180	−1.20E−01
78	^{195}Pt	195	1/2	22.883019	5.751130E+07	0.6004	33.800	

A.4 Thermophysical Properties of Selected Target Materials

Table A4.1 lists low-temperature thermophysical properties that are important for polarized targets. Unfortunately there are very few direct measurements of target material characteristics below 3 K; there is data for pure materials but almost none for glasses that are mixtures of alcohol or diol with water or pinacol, for example.

The Debye temperature Θ_D describes the low-temperature phonon spectrum that is important for the electron spin-lattice relaxation as was discussed in Section 3.4.1. All mechanisms coupling the electron spin with the lattice at low temperatures depend on the acoustic phonon spectrum and therefore on the Debye temperature or the related Debye frequency, with a typical temperature and field dependence whose measurement yields the contributions of the various mechanisms. This method was used in Ref. [11] to determine the Debye temperatures of glassy matric materials, including glycerol:water and toluene:CHCl$_3$.

Data on Debye temperature exists for crystalline materials but for few glassy substances. We have included in the table LiD and also Xe because of the possible use of hyperpolarized ^{129}Xe obtained by rapid evaporation of irradiated solid polarized by DNP.

Table A4.1 Thermophysical characteristics of some polarized target materials.

Material	Phase	M (g/mol)	Θ_D (K)	T_{mp} (K)	T_g (K)	T_s (K)
LiD ([a])	fcc crystal	9	611	965		
Xenon	fcc crystal	136	64	161		
Water (ice) ([b])	Glass	18	192	273	136	
Ethanol ([c])	monocl. crystal	46	284	159	95	62
Ethanol-d6 ([c])	monocl. crystal	52	268	159	95	62
H$_2$O:glycerol ([d])	Glass/trityl ([e])		105			
H$_2$O:glycerol ([d])	Glass/TEMPO ([e])		112			

([a]) From Ref. [9]; ([b]) Θ_D calculated from specific heat; ([c]) for ethanol and ethanol-d6 the listed Θ_D is for the crystalline phase – in the glass phase the T^3-dependent part of the specific heat suggests $\Theta_D \approx 200$ K [10]; ([d]) 1:1 by mass; ([e]) low radical concentrations from 0.5 to 1.5 mM.

The glass properties for ethanol and water are also included in Table 7.3. The Debye temperatures for the ethanols were determined for the crystalline phases from the measured specific heats [10]. As the T^3-component of the specific heat is of the same order of magnitude for both glass and crystalline phases, it is tempting to view the acoustic phonons in the glass phase as representative of a material with similar Debye temperature. The glass state data of Ref. [10] extends down to 1.8 K and shows below 3 K a T^3-dependent component that suggests a spectrum with acoustic phonons typical of dielectric crystalline materials; the other components in the specific heat are due to local vibration modes and other possible degrees of freedom.

The Debye temperatures for glycerol-water glasses were determined by fitting of the temperature dependence of the spin-lattice relaxation time to a model that includes direct, Raman, local vibration mode, Orbach process and thermally activated models that were discussed in Chapter 3. The data covers temperatures down to 10 K, and the X-band relaxation times were determined by the pulse recovery method [11].

The thermophysical properties of organic glasses at low temperatures are not well known, and it would be interesting for polarized target material development to obtain experimental results down to 0.1 K on the heat capacity, thermal conductivity, Kapitza conductance and acoustic velocity.

A.5 Properties of ^{3}He, ^{4}He and Their Mixtures

Gaseous and liquid ^{3}He, ^{4}He and their mixtures at low temperatures are called 'quantum fluids' because their subtle behavior at low temperatures can be understood only by quantum mechanical treatment. The pure fluids do not solidify at saturated vapor pressure even at absolute zero temperature and therefore they have no triple point, but they have a two-phase region (liquid coexisting with saturated gas) and therefore a critical point. Their most basic properties are given in Table A5.1:

A.5 Properties of ^{3}He, ^{4}He and Their Mixtures

Table A5.1 Basic properties of liquid ^{3}He and ^{4}He.

Property	^{3}He	^{4}He
Molar mass (g/mol)	3.01603	4.002603
Molar volume of liquid at 0 K (cm^3/mol)	36.83	27.58
Latent heat of evaporation at 0 K (J/mol)	20.56	59.62

Liquid ^{4}He has a phase transition into superfluid state at 2.17 K temperature at saturated vapor pressure. As the spin of the ^{4}He atom is zero, the system of atoms obeys Bose-Einstein statistics and therefore the superfluid transition can be considered as a kind of Bose-Einstein condensation. However, because the density of the liquid is high and atomic interactions are strong, the characteristic features of the condensation cannot be seen as dramatically as can be expected for a weakly interacting system. The superfluid state of the liquid can be modelled by the phenomenological two-fluid model which invokes intermixed normal and superfluid liquids, the relative concentrations of which depend on the temperature. Microscopically many of the superfluid properties can be understood from the dispersion relation (energy vs. atomic momentum), which has a minimum at $E/k \approx 5$ K. Excitations in the linear part of the dispersion curve are called phonons because they behave much like acoustic waves, whereas around the minimum energy they are called rotons because they can be physically understood as elementary vortices of rotation, with a slow or zero velocity of propagation.

The superfluid phase is characterized by a very high heat conductivity (up to a critical flux) above 0.5 K, by second (thermal) sound waves, by zero viscosity up to a critical velocity, by fountain pressure and by film creep phenomena, among those which are most important for the design of ^{4}He or dilution refrigerators.

Liquid ^{3}He also becomes superfluid but only at 0.9 mK temperature at saturated vapor pressure and zero field. The superfluid characteristics of ^{3}He are of little practical concern for the design of cooling systems for polarized targets, but it is interesting to note that the Fermi-Dirac statistics, obeyed by the system of spin 1/2 atoms of ^{3}He, very strongly influences the superfluid properties, so that there is little in common with superfluid ^{4}He. The superfluid ^{3}He features, however, a high thermal conductivity and low viscosity. In a magnetic field several different superfluid phases can be distinguished.

A.5.1 Phase Diagram of ^{3}He/^{4}He Mixtures

Figure A5.1 shows the solubility lines and the lambda-transition line for the binary mixtures of ^{3}He and ^{4}He. A mixture with an average molar concentration $X_3 = n_3/(n_3 + n_4)$ and temperature such that their intersection falls between the solubility lines will spontaneously separate into two phases with concentrations given by the solubility lines. The lines defining the concentrations of the dilute and concentrated phases join at the junction with a third line below that the mixture has a superfluid component. Above this line there

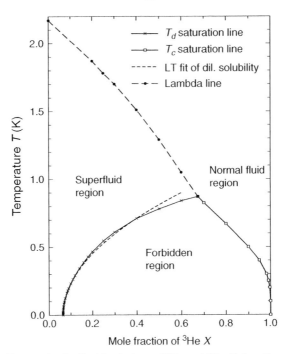

Figure A5.1 Phase diagram for the liquid solutions of ^{3}He and ^{4}He. Below the saturation lines there is the forbidden region in which no stable mixture can exist. The junction of these lines is called tricritical point. A mixture brought into the forbidden region separates spontaneously into two saturated phases, with the lighter ^{3}He-rich phase floating on top of the heavier dilute phase

is no superfluid component. The term lambda-transition refers to the shape of the specific heat anomaly, which resembles the Greek letter lambda at the transition temperature. The junction of phase diagram lines is located at $T = 0.871 \pm 0.002\,\text{K}$ and molar concentration $X_3 = 0.6735 \pm 0.0030$ [12].

The maximum solubility at $T = 0$ has been determined experimentally by several laboratories and the results vary from $X_3 = 0.064$ to 0.068. The tables compiled by Radebaugh [13] are based on the value 0.064 which deviates from the more recent value of 0.066 [14].

The dilute solubility line based on the more recent experimental data obeys rather closely the equation [15]

$$X_m(T) = 0.066\left(1 + 10\left(\frac{T}{\text{K}}\right)^2\right), \tag{A5.1}$$

below 150 mK. This equation is plotted also in Figure A5.1. A more precise polynomial fit of Kuerten [16]

$$X_m(T) = 0.066 + 0.506\left(\frac{T}{\text{K}}\right)^2 - 0.249\left(\frac{T}{\text{K}}\right)^3 + 18.2\left(\frac{T}{\text{K}}\right)^4 - 74.2\left(\frac{T}{\text{K}}\right)^5 \tag{A5.2}$$

is useful up to 0.5 K mixing chamber temperature.

A.5.2 Thermodynamic Properties

Enthalpy

The data on the enthalpy of pure liquid ^{3}He has been compiled by Radebaugh [13] in a wide range of temperatures. The compiled data, tabulated between 0 K and 1.6 K, fits accurately enough (for our design purposes) with the expression

$$H_3 = \frac{bT^2}{1 + T/T_3}, \quad (A5.3)$$

where $b = 13.03 \pm 0.03$ J/(Mol·K^2) and $T_3 = 323.4 \pm 1.6$ mK. Figure A5.2 shows this expression. The deviation from the data compiled by Radebaugh is less than 4% below 1.3 K and reaches 7% at 1.5 K. These data are also shown in the plot.

The specific heat of liquid ^{3}He is obtained by taking the temperature derivative of the above expression. It is admittedly clear that more accurate results could be expected if one would fit the specific heat function using available experimental data, and then obtain the enthalpy by integration.

The enthalpy per mole of ^{3}He of dilute solutions is also presented in Figure A5.2. The enthalpy depends on the concentration of the solution in addition to the temperature, and for the design purposes of dilution refrigerators two types of enthalpy functions are useful: Enthalpy along the dilute solubility line $H_{dm}(T)$, and enthalpy at constant osmotic pressure $H_\pi(T, \pi_m)$. The first is needed for the calculation of the cooling power available in the mixing chamber, and the second is used for the determination of the dilute stream temperature in the main heat exchanger. $H_{dm}(T)$ and $H_\pi(T, \pi_m)$ have been tabulated by Radebaugh [13] in a wide temperature range that very well covers the design needs of dilution refrigerators.

The enthalpy along the solubility line was fit to the expression

$$H_{dm}(T) = \frac{aT^2}{1 + (T/T_d)^{3/2}}, \quad (A5.4)$$

where $a = 97.56 \pm 0.24$ J/(Mol·K^2) and $T_d = 266.3 \pm 1.1$ mK. This deviates less than 3% from the data compiled by Radebaugh, also shown in the plot.

The enthalpy in the dilute stream of a dilution refrigerator deviates from that on the solubility line because the concentration drops with increased temperature so as to keep the osmotic pressure constant, and consequently the enthalpy changes. If the dilute stream channel is small, the concentration may also drop due to the finite diffusion constant, but we shall assume here that the channel is so large that the osmotic pressure is practically constant. The osmotic enthalpy is obtained by integrating the osmotic specific heat which has been tabulated by Radebaugh; the lines in Figure A5.2 show the resulting osmotic enthalpies for mixing chamber temperatures of 10 mK, 50 mK and 200 mK. We note that the osmotic enthalpies become larger than the enthalpy on the solubility curve; this fact helps in keeping the dilute stream temperature low before entering into the still and thus in avoiding losses in concentration due to reduced diffusion constant.

The specific heat of the concentrated solutions has not been measured and therefore its enthalpy is not known. In design calculations it is assumed that the enthalpy of the

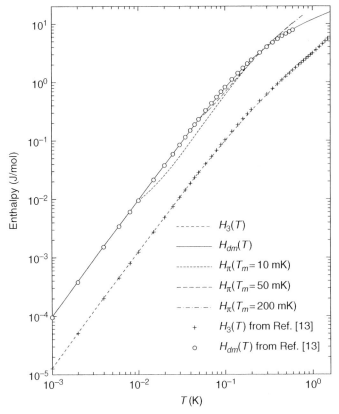

Figure A5.2 Enthalpy of pure ³He and of dilute solutions of ³He in ⁴He. See text for the explanation of the different lines

concentrated phase is equal to that of the pure ³He, which is a good approximation at temperatures below 200 mK.

In the circulated helium there is often more than 30% of ⁴He and therefore the enthalpy balance in a heat exchanger is determined from the total enthalpy of the mixture, which has two phases below 0.8 K. The phase separation increases the enthalpy significantly, and it is clear that this has an adverse effect in the cooling power. However, when inspecting the expression of cooling power at optimum flow, the enthalpy of the concentrated stream enters in the determination of the exchanger outlet temperature T_o, which determines the cooling power via the heat transfer parameter. At low temperatures the concentrated stream enthalpy per mole of circulated ³He is estimated from

$$H_c(T_c) \cong \left(1 - \frac{X_l}{1-X_l}\frac{X_4}{1-X_4}\right) H_3(T_c) + \frac{X_l}{1-X_l}\frac{X_4}{1-X_4} H_d(T_c), \tag{A5.5}$$

where X_l is the ^{3}He concentration on the dilute solubility curve, and X_4 is the molar fraction of ^{4}He in the circulated helium. The deterioration of the maximum cooling power in the asymptotic low-temperature region is proportional to

$$\frac{\dot{Q}(T_m, X_4)}{\dot{Q}(T_m, X_4 = 0)} = \left(\frac{H_3(T_m)}{H_c(T_m)}\right)^2 \cong \left(1 + 6.49 \frac{X_l}{1 - X_l} \frac{X_4}{1 - X_4}\right)^{-2}. \quad (A5.6)$$

which can be substantial when $X_4 > 0.3$. At higher temperatures the cooling power is slightly less deteriorated than that given by the above expression.

Osmotic Pressure

The osmotic pressure $\pi(T, X)$ has been determined and tabulated by Radebaugh [13] using thermodynamic data. These tables give the value of $\pi(0, X = 0.064)$ on the solubility line as 1,629 Pa, which deviates from the more recent experimental value of $\pi(0, X=0.064) = 2,240$ Pa [17–19]. The osmotic pressure is sensitive to the thermodynamics of the dilute solutions and it has been used for revising the thermodynamic data below 150 mK [16, 20], [21].

At constant concentration the osmotic pressure well below T_F has a quadratic temperature dependence, which changes towards linear dependence of Van't Hoff's law above T_F. As the concentration on the solubility line does not have a strong temperature dependence below 50 mK, the osmotic pressure in the mixing chamber of a dilution refrigerator behaves as [18]

$$\pi(X_l, T) = \left[21.74 + 978\left(\frac{T}{K}\right)^2\right] \text{mbar}; \quad (A5.7)$$

this extrapolates to a slightly low value at $T = 0$ probably because the experimental temperature range was limited to 25 mK. Bloyet et al. [18] suggested the use of osmotic pressure on the solubility line as a secondary thermometer below 300 mK [18, 22, 23].

The osmotic pressure is measured by using the heat flush effect where the fountain pressure p_f of superfluid ^{4}He becomes equal to the osmotic pressure. The dilute phase is connected via a long capillary to a heated cell; the capillary passes through a thermal 'grounding' at 0.6 K below which the thermal conductivity becomes very small so that the heat leak to the mixing chamber remains negligible. When the fountain pressure is equal to the osmotic pressure, there is no ^{3}He in the superfluid that then becomes isothermal due to its high thermal conductivity. The heat flush is based on the fact that when the superfluid component of a mixture is not accelerated, we have everywhere

$$\nabla\left(p_f + \pi + p_{\text{mech}}\right) = 0. \quad (A5.8)$$

Assuming that the gravitational pressure gradient is small or negligible, the expression integrates to

$$\pi = -p_f - \Delta p_{\text{mech}} \quad (A5.9)$$

in the heated cell; here Δp_{mech} is the gravitational pressure difference between the phase boundary in the mixing chamber and the liquid-vapor surface of the heated cell.

The design of dilution refrigerators would greatly profit from improved experimental data and theoretical models on the osmotic pressure, and from the thermodynamic data which can be derived from the osmotic pressure.

Vapor Pressure of ^{3}He

The absolute temperature scale between 0.65 K and 3.2 K is defined by the International Temperature Scale of 1990 (ITS-90) as the relation between the vapor pressure in Pa of ^{3}He and the temperature T_{90} in K [24]:

$$\frac{T_{90}}{K} = A_0 + \sum_{i=1}^{9} A_i \left[\frac{\ln \frac{p}{Pa} - B}{C} \right]^i , \text{ where}$$

$A_0 = 1.053447$
$A_1 = 0.980106$
$A_2 = 0.676380$
$A_3 = 0.372692$
$A_4 = 0.151656$ \hfill (A5.10)
$A_5 = -0.002263$
$A_6 = 0.006596$
$A_7 = 0.088966$
$A_8 = -0.004770$
$A_9 = -0.054943$
$B = 7.3$
$C = 4.3.$

A new provisional low-temperature scale PLTS-2000 is based on the melting curve of ^{3}He between 0.65 and 1 K. The reported deviations of the thermodynamic temperature from ITS-90 below 1 K are marked, but above 1 K the deviations are less than 1 mK from the new scale now named PTB-2006 [25].

The formally accepted ITS-90 based on the vapor pressure of the helium isotopes has been thus corrected and is going to be replaced by PTB-2006, although the formal SI scale is still based on ITS-90.

Gaseous Helium

Gaseous helium obeys the laws of ideal gas at elevated temperatures. This yields the specific heat

$$C_p = \frac{5}{2}R = 20.786 \frac{\text{J}}{\text{mol} \cdot \text{K}}, \tag{A5.11}$$

which is same for both isotopes and their mixtures. Deviations at low temperatures are small and have no consequences for heat exchanger design. The specific heat does not depend on the pressure, obviously.

A.5.3 Transport Properties

Thermal Conductivity of Liquid Helium Mixtures

The thermal conductivity of ^{3}He, ^{4}He and their two dilute solutions is plotted in Figure A5.3 from 1 mK to 1 K. For the purpose of comparison, the thermal conductivities of some solids are also plotted. The data is fit to functions with an accuracy of about 5% and extrapolated to the limits of the plot, but the validity of some of these extrapolations is questionable.

The experimental thermal conductivity of ^{3}He between 50 K and 0.6 K [30] features a broad minimum of 5.8×10^{-3} W/(K·m) at about 0.23 K, and a rise at lower temperatures by [31]

$$\kappa_3 = 3.6 \cdot 10^{-4} \frac{\text{K}}{T} \frac{\text{W}}{\text{Km}} \tag{A5.12}$$

below about 50 mK. This is understood in the terms of the Fermi fluid model where collisions between quasiparticles have a probability proportional to $1/T^2$ and specific heat behaves linearly with T.

The dilute solutions have a similar rise at low temperatures, with

$$\kappa_d(1.32\%) = 1.1 \cdot 10^{-4} \frac{\text{K}}{T} \frac{\text{W}}{\text{Km}}, \tag{A5.13}$$

$$\kappa_d(5.0\%) = 2.0 \cdot 10^{-4} \frac{\text{K}}{T} \frac{\text{W}}{\text{Km}}, \tag{A5.14}$$

but they have a much more complex behavior [32] at higher temperatures where the ^{4}He phonons begin to contribute to the transport of heat. The measured data extends from 30 mK to 0.5 K.

The saturated dilute solution has a fairly constant thermal conductivity of 25 mW/(K·m) above 30 mK, and the conductivity behaves as

$$\kappa_d(6.4\%) = 3.0 \cdot 10^{-4} \frac{\text{K}}{T} \frac{\text{W}}{\text{Km}} \tag{A5.15}$$

below 15 mK. At temperatures above 200 mK the thermal conductivity in dilute solutions scales roughly as $1/X$ with concentration.

^{4}He has a totally different behavior below 1 K where very high heat currents are possible because the phonon mean free path is limited only by boundary scattering. The conductivity in capillary pipes of diameter D behaves as

$$\kappa_4 = 3.0 \cdot 10^5 \frac{D}{m}\left(\frac{T}{K}\right)^3 \frac{W}{Km} \tag{A5.16}$$

below 0.6 K. At temperatures below 0.2 K, however, the effective diameter increases because the boundary scattering becomes specular [32]. The line in Figure A5.3 is the apparent conductivity of ^{4}He in a capillary of 2.5 mm diameter, in which the conductivities of the dilute solutions were also measured.

In pure ^{4}He the two-fluid model describes heat transport by counterflowing currents of superfluid and normal fluid. The thermal conduction is then due to the normal fluid that carries entropy and obeys the law of Poiseuille, giving an apparent thermal conductivity

$$\kappa_4 = \frac{\rho^2 S_4^2 T d^2}{8\eta_n} \tag{A5.17}$$

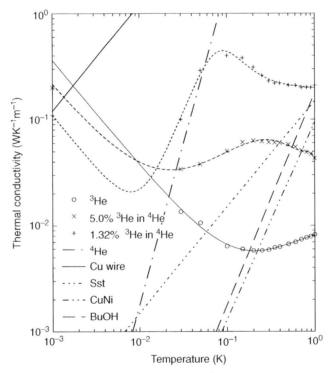

Figure A5.3 Experimental thermal conductivities of ^{3}He, ^{4}He and two dilute solutions, compared with selected solids: Cu wire = commercial copper wire [26]; Sst = stainless steel tube [27]; CuNi = 70% Cu / 30% Ni cupronickel alloy tube [28]; BuOH = glassy butanol / 5% water beads, extrapolated from 1.08 K [29] according to the T^2-law, which is obeyed by most glassy materials in a pipe of diameter d; here ρ is the density, S the entropy and η_n the viscosity of normal ^{4}He. This is valid for very low heat fluxes; at higher fluxes the thermal gradient is no longer linearly related with the heat flux, and one must use the empirical formula

$$\frac{\dot{Q}/\text{W}}{A/\text{cm}^2} = \left[\frac{X(T_{\text{cold}}) - X(T_{\text{hot}})}{\ell/\text{cm}}\right]^{0.29}, \qquad (A5.18)$$

where A is the cross section available for heat transport, ℓ is the length of the channel, and the dimensionless function $X(T)$ is

$$X(T) = 520\left\{1 - \exp\left[3(2.16 - T/\text{K})\right]^{2.5}\right\}. \qquad (A5.19)$$

Viscosity of Liquid ^{3}He and Dilute Solutions

The viscosity of pure ^{3}He has been measured by Black, Hall and Thompson [33] between 50 mK and 2 K and by Abel, Anderson and Wheatley [34] below 50 mK; a compilation of these data is plotted in Figure A5.4. The asymptotic behavior of the viscosity follows the theoretically predicted $1/T^2$-law. The data fits reasonably well with the function

$$\mu_3 = 1.88\left(\frac{\text{K}}{T}\right)^2\left[1 + 5.1\left(\frac{T}{\text{K}}\right) + 11.8\left(\frac{T}{\text{K}}\right)^2\right] \cdot 10^{-6} \frac{\text{g}}{\text{cm}\cdot\text{s}}, \qquad (A5.20)$$

which is accurate enough for the design of dilution refrigerators.

The viscosity of dilute solutions has been measured by different methods, which give widely diverging results. The vibrating crystal measurements fail to show the theoretically predicted [35] $1/T^2$-law and show significantly lower viscosities at all temperatures compared with other methods, whereas those made with a low-frequency vibrating wire [36] and flow methods [37] tend to agree in the temperature range 30 mK–0.9 K among themselves and with the theory within about 20%. The lines shown in Figure A5.4 are fits to the data of Ref. [37]. The 5% solution has a trend towards the $1/T^2$-law

$$\mu_3 = 0.143\left(\frac{\text{K}}{T}\right)^2\left[1 + 25.4\left(\frac{T}{\text{K}}\right) + 94.6\left(\frac{T}{\text{K}}\right)^2\right] \cdot 10^{-6} \frac{\text{g}}{\text{cm}\cdot\text{s}}, \qquad (A5.21)$$

but the data for 1.3% solution shows insignificant trend towards this.

The viscosity of the saturated dilute solution (on the phase separation line) has been measured by Zeegers [38] both by the flow method and by the vibrating wire viscometer. His results below 80 mK obey

$$\mu_3 = \left[(2.7 \pm 0.2)\left(\frac{\text{K}}{T}\right)^2 + (34 \pm 3)\left(\frac{\text{K}}{T}\right)\right] \cdot 10^{-8} \frac{\text{g}}{\text{cm}\cdot\text{s}}. \qquad (A5.22)$$

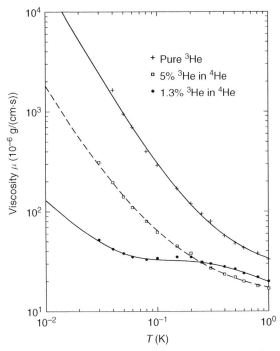

Figure A5.4 Viscosity of pure ^{3}He and dilute solutions with 5% and 1.3% of ^{3}He in ^{4}He. The lines are numeric fits to the experimental data discussed in the text

The viscosity controls the convectional heat transport by the dilute solution, and its low value above 50 mK contributes to the heat absorption capability of the mixing chamber, as was discussed in Chapter 8. The viscosity of dilute solutions, however, does not seem to control alone the flow of ^{3}He in the dilute stream of the dilution refrigerators, where the flow at high velocities appears rather to be limited by diffusion or mutual friction between ^{3}He and superfluid ^{4}He. In circular tubes the mutual friction seems to appear only above a critical velocity, to be discussed below.

Mass Diffusion Constant and Mutual Friction

The two-fluid model relates the thermal conductivity by internal convection to the mass diffusion constant of ^{3}He, as was discussed in Chapter 8. Table A5.2 shows the diffusion constant based on the total thermal conductivity coefficient, which is believed to be a good estimate down to 0.6 K and to give the right order of magnitude down to 0.4 K.

The diffusion of ^{3}He in isothermal dilute solution results in the concentration gradient

$$\frac{\partial X}{\partial z} \cong \frac{\dot{n}_3 V_4}{AD}, \tag{A5.23}$$

which adds on that due to the thermal gradient. The design of the dilute stream close to the still must be such that the gradient due to diffusion is not too large at the desired mass flow rate.

The diffusion constant is related to the thermal conductivity by

$$D = \kappa_d \frac{V_4^0 R X}{\left(S_4^0\right)^2}. \qquad (A5.24)$$

This relation, shown in Table A5.2, is rather accurate above 0.6 K and below a few % concentration; it is expected to be of the right order of magnitude down to 0.4 K temperature.

At temperatures below 0.25 K the transport of ^{3}He in the dilute solution obeys laws that deviate strongly from classical diffusion. Also, at low flow velocity the 'mechanical vacuum' model of Wheatley [39] may be only qualitatively correct. The resulting concentration gradient does not obey Eq. A5.23 but [40]

$$\frac{\partial X}{\partial z} = (3.2 \pm 0.3) \cdot 10^{-8} \left(\frac{\dot{n}_3}{A} \frac{m^2 s}{mol}\right)^{2.8 \pm 0.4} \frac{1}{m}, \qquad (A5.25)$$

as was discussed in Chapter 8. This unusual type of diffusion in not theoretically understood yet but is tentatively explained by the creation of vortices in ^{4}He, which pin to the wall and scatter the quasiparticles.

Gaseous Helium

The isobaric specific heat c_p is that of nearly ideal monoatomic gas:

$$c_p = \frac{5}{2} R = 20.7862 \frac{J}{mol\ K} \qquad (A5.26)$$

for both isotopes of helium down to the boiling point. This is used for the precooling heat exchangers, notably to get the dimensionless Prandtl number Pr:

$$\Pr = \frac{c_p \mu}{\kappa} = 0.67. \qquad (A5.27)$$

The design of precooling heat exchangers also depends on the viscosity and thermal conductivity of gaseous helium. Both of these parameters depend linearly on the square root

Table A5.2 Experimental thermal conductivity [32] and diffusion constant of two dilute solutions.

	$X_3 = 0.013$		$X_3 = 0.050$	
T(K)	κ_d(W/m/K)	D(cm²/s)	κ_d(W/m/K)	D(cm²/s)
0.4	0.21	2,060	0.060	2,260
0.5	0.21	540	0.058	573
0.6	0.20	172	0.055	182
0.7	0.20	63	0.050	61
0.8	0.19	19.0	0.048	18.5

of the temperature and of the atomic mass, for the case of an ideal gas of hard spheres. For helium the thermal conductivity is

$$\kappa = 1.499 \left[\frac{T}{300\,\text{K}} \frac{M}{4\,\text{g/mol}} \right]^{1/2} \frac{\text{mW}}{\text{cmK}}, \quad (A5.28)$$

which holds quite well down to 10 K for both isotopes. Below this temperature quantum effects become important [41] and the conductivity of ^{4}He drops faster with temperature, with a value of 0.0607 mW/(cmK) at 3 K. The thermal conductivity of ^{3}He obeys the law down to 3 K and then drops roughly linearly to zero at absolute zero. The thermal conductivity is independent of pressure down to the range where molecular effects become important, i.e. the mean free path becomes of the order of the size of the gap where the conductivity is determined.

The viscosity behaves similarly

$$\mu = 196.4 \cdot \left[\frac{T}{300\,\text{K}} \frac{M}{\text{g/mol}} \right]^{1/2} 10^{-6} \frac{\text{g}}{\text{cm} \cdot \text{s}} \quad (A5.29)$$

for an ideal gas in zeroth order. The experimental viscosity goes down slightly faster with temperature

$$\mu = 196.4 \left(1 + 0.3 \log_{10} \frac{T}{300\,\text{K}} \right) \cdot \left[\frac{T}{300\,\text{K}} \frac{M}{\text{g/mol}} \right]^{1/2} 10^{-6} \frac{\text{g}}{\text{cm} \cdot \text{s}} \quad (A5.30)$$

down to 10 K. The viscosity of ^{3}He has a maximum at 2 K, whereas that of ^{4}He goes down faster than the viscosity of ideal gas. These can be understood from molecular corrections treated quantum mechanically [41].

A.5.4 Heat Transfer between Sintered Powder and Liquid

The Kapitza surface boundary resistance gives a severe limitation for heat transfer at low temperatures and therefore heat exchangers need extended surfaces. The heat transfer between the phonon systems in the lattice of a solid and those in helium liquid, given by the Kapitza conductance law of Eq. 8.2', is rewritten here:

$$\dot{Q} = \sigma_c \alpha \left(T_L^4 - T_{He}^4 \right), \quad (A5.31)$$

where σ_c is the contact area. The same coefficient α applies in both directions of heat flow, but there is some dependence on the composition of the liquid and on the temperature range where the measurement has been made. Table A5.3 gives the coefficient under the conditions indicated, between the liquid and sintered powder surfaces.

As the thermal conductance at the boundary gets very low at low temperatures, the heat exchangers need to have extended surfaces. Sintered powders are a practical solution to

A.5 Properties of ³He, ⁴He and Their Mixtures

Table A5.3 Coefficient of the Kapitza conductance in the region where $n \approx 4$.

Solid	Liquid	α (W/K⁴ m²)	Temp. range (mK)	Ref.
Sintered Cu 325 mesh	Conc. ³He	20	10–500	
Sintered Cu 325 mesh	Dil. ³He	25	10–500	
Sintered Cu 325 mesh	⁴He	100	100–1,000	
Sintered Cu 1 µm, h=0.5 mm	Dil. ³He	500	<25 mK	[42]
Sintered Cu 1 µm, h=1 mm	Dil. ³He	2000	<20 mK	
Sintered Ag 1 µm	Pure ³He	6	<13 mK	[43]
CuNi	Conc. ³He	80		

[1] The hydrodynamic surface area is the area determined from the flow resistance of the powder for a gas such as N_2 or He. It has been found that for most sintered powders the hydrodynamic area is about three times smaller than the area determined by gas adsorption, and that the hydrodynamic area gives the same value for the Kapitza resistance as that measured for bulk samples [44].

this, but this technique also has limitations because of the finite thermal conductivities of both the helium solutions and the metallic materials used. We shall develop here a simplified model for estimating the influence of the thermal conductivity in the effective surface area that is available for heat exchange. We shall consider random isotropic powder with a filling factor η and surface-to-volume diameter d_e (see Eq. 8.1)

$$d_e = 6\eta \frac{V}{\sigma}, \qquad (A5.32)$$

where V is the total volume filled by the sinter, and σ is its hydrodynamic surface area.[1]

Let us first consider the effect of thermal conduction in the helium filling the pores of the powder. We may simplify the pores by replacing them with cylindrical holes with total cross-sectional area

$$A_{\text{eff}} = A(1-\eta), \qquad (A5.33)$$

where A is the geometric surface of the sinter, and surface area per unit depth of

$$\frac{\sigma}{t} = \frac{6\eta}{d} A, \qquad (A5.34)$$

where t is the thickness of the sinter. The heat flow into the walls of the holes is now obtained from

$$\dot{Q} = \frac{\sigma}{t} \alpha \int_0^t \left[T(z)^4 - T_0^4 \right] dz, \qquad (A5.35)$$

where the temperature $T(z)$ at depth z has a gradient which is determined by the thermal conductivity of the fluid. At the entrance into the pores, in particular, the gradient has the value determined by

$$\dot{Q} = \kappa(T_1) A_{\text{eff}} \left. \frac{dT}{dz} \right|_{z=0}, \quad (A5.36)$$

and at z it is given by

$$\dot{Q} - \frac{\sigma}{t} \int_0^z \left[T(z)^4 - T_0^4 \right] dz = \kappa(T(z)) A_{\text{eff}} \frac{dT(z)}{dz}. \quad (A5.37)$$

Solving $dT/dz|_{z=0}$ from these we may use it for defining the thermal penetration depth

$$\lambda_T = \frac{T_1 - T_0}{dT/dz|_{z=0}}. \quad (A5.38)$$

When the thermal conductivity does not vary strongly between T_1 and T_0, this can be expressed approximately as

$$\lambda_T^2 \cong \frac{d_e(1-\eta)}{6\eta} \frac{\kappa(T_1)}{\alpha T_1^3} \quad (A5.39)$$

for the case that T_1/T_0 deviates significantly from 1 and $t < \lambda T$; when the ratio is close to 1, the thermal penetration depth is

$$\lambda_T^2 \cong \frac{d_e(1-\eta)}{6\eta} \frac{\kappa(T)}{4\alpha T^3}. \quad (A5.40)$$

The approximations made above are admittedly rather imprecise, but for the present purpose it is sufficient to obtain an estimate which is accurate to ±50% in the effective exchange area.

The thermal penetration depth can now be used for estimating the effective thermal exchange surface area for infinitely thick layer of sintered metal with infinite thermal conductivity, by the simple approximation

$$\sigma_{\text{eff}} = \frac{\sigma}{V} \lambda_T A = N(d_e, \eta) \left(\frac{T_\sigma}{T} \right)^n A, \quad (A5.41)$$

where we have separated the power-law temperature dependence from the numeric coefficient N, which depends also on the thermal conduction coefficient and Kapitza conductance constant. We note that the area A covered by the sintered powder is enhanced by a temperature-dependent factor, which is very large for fine powders at low temperatures. The coefficient N, the power n and the reference temperature T_σ are given for three powders in Table A5.4.

We note that for ^{4}He the surface enhancement is very small because of the low thermal conductivity in the pores of the fine powder, and that this is independent of temperature because the thermal conductivity and Kapitza resistance have the same temperature dependence.

Table A5.4 Enhancement N of the contact surface area by overlaid sinter of depth much larger than the thermal penetration depth; see Eq. A5.41. The filling factor is 0.4 for the Cu powder and 0.5 for the Ag powders; all three grain sizes correspond to commercially available powders. At temperatures below 15 mK the thermal conductivity of the ^{3}He mixtures is taken to follow the $1/T$-law, whereas above 30 mK it is assumed to be constant.

T-range (mK)	Liquid	$N(d_e,\eta)$ Cu, d_e=18 μm	$N(d_e,\eta)$ Ag, d_e=1 μm	$N(d_e,\eta)$ Ag, d_e=70 nm	T_σ (mK)	n
< 15	^{3}He	6,000	26,000	98,200	10	2
< 15	^{3}He/^{4}He	4,900	21,200	80,200	10	2
> 30	^{3}He	155	671	2,536	100	3/2
> 30	^{3}He/^{4}He	283	1225	4,630	100	3/2
	^{4}He	51	47.4	47.4	–	0

The corresponding parameters for the thermal penetration depths are defined by rewriting

$$\lambda_T = \Lambda_T \left(\frac{T_\sigma}{T}\right)^n ; \tag{A5.42}$$

these parameters are given in Table A5.5.

We may define in a similar way the effective length of a metal fin projecting into the fluid, under the assumption that the fluid is isothermal and that the heat transfer is limited by Kapitza resistance and by the thermal conductivity of the fin material. Let us assume fins with cross-sectional area A_f and perimeter P_f which yield the equation

$$\lambda_T^2 \cong \frac{A_f}{P_f} \frac{\kappa(T_1)}{\alpha T_1^3} \tag{A5.43}$$

for fins with constant cross-sectional area and perimeter. For a powder we have approximately $A_f/P_f = \eta^2 d_e/6$, which yields

$$\lambda_T \cong 0.693 \frac{K}{T} \text{ mm } (18 \, \mu\text{m Cu}, \eta = 0.4);$$

$$\lambda_T \cong 0.163 \frac{K}{T} \text{ mm } (1 \, \mu\text{m Ag}, \eta = 0.5); \tag{A5.43'}$$

$$\lambda_T \cong 0.0432 \frac{K}{T} \text{ mm } (70 \text{ nm Ag}, \eta = 0.5).$$

This shows that the copper powders that are easy to sinter to a thickness of about 1 mm have a conductivity which will not limit the heat transfer at temperatures below 0.5 K, whereas in the silver powders the powder itself starts to limit the heat transfer if layers of 1 mm thickness are used at temperatures above 50 mK. Here the value 16 (T/K) W/(K·m) was used for the thermal conductivity of the silver in the powder.

By comparing the thermal penetration depths of the helium and the metal, we see that with 18 μm Cu powder the conductivity of the copper begins to limit the useful thickness of

Table A5.5 Thermal penetration depths in helium filling the pores of isothermal sintered powder. At temperatures below 15 mK the thermal conductivity of the ^{3}He mixtures is taken to follow the $1/T$-law, whereas above 30 mK it is assumed to be constant.

T-range (mK)	Liquid	Λ_T (mm) Cu, $d_e=18$ μm	Λ_T (mm) Ag, $d_e=1$ μm	Λ_T (mm) Ag, $d_e=70$ nm	T_σ (mK)	n
< 15	^{3}He	90	17	4.5	10	2
< 15	^{3}He/^{4}He	73.5	14.1	3.73	10	2
> 30	^{3}He	1.16	0.224	0.059	100	3/2
> 30	^{3}He/^{4}He	2.1	0.41	0.108	100	3/2
	^{4}He	0.38	0.0212	0.00148	–	0

the sinter only below about 10 mK to 13 mK temperature, whereas at higher temperatures it is always the fluid which puts a limitation to the heat transfer. For the silver powders this happens at 8 mK to 10 mK temperature.

Another way of using Table A5.4 and Eq. A5.42 is to determine the temperature at which the thermal penetration depth becomes equal to the thickness of the sintered layer, in the temperature domain where the conductivity of the fluid limits the thermal penetration. Choosing a practical thickness of $t = 1$ mm and requiring that $t = \lambda_T$, yields the temperatures given in Table A5.6.

Because the fabrication of sintered layers thinner than about 0.5 mm is impractical, the usefulness of 1 μm Ag powder is limited to temperatures below 50 mK and that of the Cu powders to below 300 mK. The high conductivity of copper, however, enables enhancement of the heat transfer even with 1 mm layers at temperatures close to 1 K.

If the operation of a dilution refrigerator does not require high power absorption capability below 20 mK, the sintered copper heat exchangers are the best choice. If operation below 10 mK is desired, the addition of sintered silver heat exchangers will give definite benefits. Sintered silver layer thicknesses above 1 mm, however, are only useful in the last exchanger units when operating below 10 mK mixing chamber temperatures.

If an external heat load needs to be absorbed by the mixing chamber, the sintered contact thickness is not the only parameter to be considered, because the heat transport from the contact to the phase boundary may limit the overall heat transfer coefficient more than the large surface of a thick sintered contact. It is therefore better to distribute the sintered contact in the mixing chamber volume so that convection in the dilute phase and conduction in both phases match the heat current which comes from the contact surfaces.

A.6 Lineshape Functions and Their Moments

In the following we shall review some integrable symmetric functions that are convenient for modelling and fitting symmetric and slightly asymmetric NMR signals, whose shapes are dominated by dipolar interactions. The dipolar interactions lead to lineshapes that cannot be presented in closed analytical form, and therefore trial functions must be used.

Table A5.6 Temperatures at which the thermal penetration depth is equal to 1 mm for sintered layers of three commercial powders. For the 70 nm Ag powder the parameters valid below 15 mK were used because the thermal conductivity of the fluid is close to the asymptotic behavior already at 20 mK.

Powder	Cu, d_e=18 μm	Ag, d_e=1 μm	Ag, d_e=70 nm
$T(^3\text{He})$, mK	110	37	21
$T(^3\text{He}/^4\text{He})$, mK	164	55	19

Another reason for trial functions is that only rarely the line broadening is due to only dipolar interactions; other causes are magnetic field inhomogeneity, non-uniform magnetization due to target or bead shape, non-uniform polarization, pseudo-scalar or pseudo-dipolar interactions with anisotropy and anisotropy of chemical shift. These have been discussed in Chapter 5.

The functions are presented with the variable x, which can be the frequency difference $f-f_0$ or angular frequency difference $\omega-\omega_0$ or any linear scaled version of these. When scaling, however, care must be taken to change the normalization constant accordingly. The normalization constants of the symmetric functions are obtained from

$$C = 2\int_0^\infty f(x)\,dx. \tag{A6.1}$$

A6.1 Gaussian Lineshape

The normalized Gaussian function

$$f(x) = \frac{1}{C}e^{-x^2/2\sigma^2}; \quad C = \sigma\sqrt{2\pi} \tag{A6.2}$$

gives the following results for the even moments:

$$\langle x^2 \rangle = \sigma^2; \quad \langle x^4 \rangle = 3\sigma^4; \quad \langle x^{2n} \rangle = 1\cdot 3\cdots(2n-1)\sigma^{2n}. \tag{A6.3}$$

The odd moments are all zero by definition that the line is symmetric. The second moment is related to the RMS width by

$$\Delta_{RMS} = 2\sqrt{\langle x^2 \rangle} = 2\sigma \tag{A6.4}$$

and the full width at half-maximum (FWHM) is

$$\Delta_{FWHM} = \sigma 2\sqrt{2\ln 2} = 2.35482\,\sigma. \tag{A6.5}$$

The ratio of the fourth moment and the square of the second moment, often called the lineshape parameter, is the simplest dimensionless measure of the shape. For the Gaussian it is

$$\frac{\langle x^4 \rangle}{(\langle x^2 \rangle)^2} = 3. \tag{A6.6}$$

This parameter can be easily determined for an experimental lineshape and it gives a rather sensitive test for the assumption that an experimental lineshape is close to a Gaussian. Unfortunately the ratio evaluated for an experimental signal is also sensitive to the possible contributions to the signal of the dispersion part of the susceptibility; this can be substantial in the wings and can therefore increase the ratio.

A6.2 Lorenzian Lineshape

The second and higher even moments of the normalized Lorentzian function

$$f(x) = \frac{\delta}{\pi} \frac{1}{\delta^2 + x^2} \tag{A6.7}$$

diverge. The Lorentzian is a special case of two classes of functions with the definite integrals [45]

$$\int_0^\infty \frac{x^{m-1}}{1+x^n} dx = \frac{\pi}{n \sin \frac{m\pi}{n}} \tag{A6.8}$$

and

$$\int_0^\infty \frac{x^a}{(m+x^b)^c} dx = m^{\frac{a+1-bc}{b}} \frac{\Gamma\left(\frac{a+1}{b}\right) \Gamma\left(c - \frac{a+1}{b}\right)}{b \, \Gamma(c)} \text{ with } \begin{array}{l} a > -1 \\ b > 0 \\ m > 0 \\ c > \frac{a+1}{b} \end{array}, \tag{A6.9}$$

where the gamma function is defined in the next section.

A convenient normalized symmetric lineshape function is obtained by setting $a = 0$, $b = 2$, $m = \delta^2$:

$$f(x) = \frac{\delta^{2c-1}}{C} \frac{1}{(\delta^2 + x^2)^c}; \tag{A6.10}$$

here C is the normalization constant that is obtained from

$$\int_{-\infty}^\infty f(x) dx = 2 \int_0^\infty f(x) dx = \frac{2\delta^{2c-1}}{C} \int_0^\infty \frac{dx}{(\delta^2 + x^2)^c} = 1, \tag{A6.11}$$

which yields

$$C = \Gamma\left(\frac{1}{2}\right) \frac{\Gamma\left(c - \frac{1}{2}\right)}{\Gamma(c)} = \sqrt{\pi} \frac{\Gamma\left(c - \frac{1}{2}\right)}{\Gamma(c)}. \tag{A6.12}$$

This function has the peak value of

$$f(0) = \frac{1}{\delta C} \tag{A6.13}$$

and FWHM width

$$\text{FWHM} = 2\delta\sqrt{2^{1/c} - 1}; \quad (A6.14)$$

The second moment is finite for all $c > 3/2$ and is obtained by using the recursion formulae given in the next section:

$$\langle x^2 \rangle = 2\int_0^\infty x^2 f(x)dx = \frac{2\delta^{2c-1}}{C}\int_0^\infty \frac{x^2 dx}{(\delta^2 + x^2)^c} = \frac{\delta^2}{C}\Gamma\left(\frac{3}{2}\right)\frac{\Gamma\left(c - \frac{3}{2}\right)}{\Gamma(c)} = \frac{\delta^2}{2c - 3}. \quad (A6.15)$$

The fourth moment is finite for all $c > 5/2$ and is calculated similarly

$$\langle x^4 \rangle = \frac{2\delta^{2c-1}}{C}\int_0^\infty \frac{x^4 dx}{(\delta^2 + x^2)^c} = \frac{\delta^4}{C}\Gamma\left(\frac{5}{2}\right)\frac{\Gamma\left(c - \frac{5}{2}\right)}{\Gamma(c)} = \frac{3\delta^4}{(2c-3)(2c-5)}. \quad (A6.16)$$

The lineshape parameter is now obtained easily

$$\frac{\langle x^4 \rangle}{\langle x^2 \rangle^2} = 3\frac{c - \frac{3}{2}}{c - \frac{5}{2}} \quad (A6.17)$$

and is seen to converge towards the same value 3 as the Gaussian lineshape, when c tends to infinity.

As the gamma function is numerically easily expressed for all half-integer values of c, we shall list the normalizing constant, FWHM, second moment and lineshape parameter for $1 \le c \le 5$ in Table A6.1.

Lineshape functions are needed for the simulation of NMR signals arising from circuit theoretical expressions of the susceptibility, on the one hand, and on the other hand for fitting of experimental signals with such theoretical expressions. In the fitting programs the parameter c can be chosen to be free within a range around the value obtained from the experimental lineshape parameter. The dispersion part of the susceptibility can be obtained using the Kramers-Krönig transform, which has to be integrated numerically.

Slightly asymmetric lineshapes can be generated by multiplying the lineshape function by a low-order antisymmetric polynomial, or by adding another function that is antisymmetric, such as

$$f(x) = \frac{\delta^{2d-1}}{C'}\frac{x^n}{(\delta^2 + x^2)^d}; \quad (A6.18)$$

where $n \le d-c$ is an odd positive integer. Such functions integrate to zero and their even moments are zero.

Table A6.1 Normalizing constants, linewidths and lineshape parameters for the generalized lineshape functions of Eq. A.6.10.

c	C	FWHM/δ	$\langle x^2 \rangle / \delta^2$	$\langle x^4 \rangle / \langle x^2 \rangle^2$
1	π	2		
$1\frac{1}{2}$	2	1.53284		
2	$\frac{1}{2}\pi$	1.28719	1	
$2\frac{1}{2}$	$\frac{4}{3}$	1.13050	$\frac{1}{2}$	
3	$\frac{3}{8}\pi$	1.01965	$\frac{1}{3}$	9
$3\frac{1}{2}$	$\frac{2^4}{15}$	0.93598	$\frac{1}{4}$	6
4	$\frac{5}{2^4}\pi$	0.86996	$\frac{1}{5}$	5
$4\frac{1}{2}$	$\frac{2^5}{35}$	0.81616	$\frac{1}{6}$	$4\frac{1}{2}$
5	$\frac{2^5}{35}\pi$	0.77123	$\frac{1}{7}$	$4\frac{1}{5}$

A.7 Gamma Function

The gamma function, an extension of the factorial function [45–47], is defined by

$$\Gamma(n) = \int_0^\infty t^{n-1} e^{-t} dt, \quad \text{if} \quad n = 1, 2, \ldots, \quad \Gamma(n+1) = n! \tag{A7.1}$$

which is not convenient for numerical evaluation because the integral converges slowly. The argument n can be any integer, real or complex number; we are here interested only in the positive real numbers. The gamma function needs to be known only between 1 and 2 because it can be calculated recursively from

$$\Gamma(n+1) = n\Gamma(n) \quad \text{or} \quad \Gamma(n) = \frac{1}{n}\Gamma(n+1). \tag{A7.2}$$

In the interval from 1 to 2 it can be numerically calculated to the precision of $< 5 \times 10^{-6}$ from a polynomial of sixth degree

$$P_k(n) = \sum_{k=0}^{6} A_k n^k, \tag{A7.3}$$

with coefficients
 $A_0 = 0.999998938$
 $A_1 = -0.57644001$
 $A_2 = 0.975135246$
 $A_3 = -0.81371653$

$A_4 = 0.657600132$
$A_5 = -0.31562574$
$A_6 = 0.07304891$.

The gamma function has the special values of

$$\Gamma(1) = \Gamma(2) = 1 \; ; \; \Gamma\left(\frac{1}{2}\right) = \sqrt{\pi} \; ; \; \Gamma\left(\frac{3}{2}\right) = \frac{1}{2}\sqrt{\pi} \; ; \; \Gamma'(1) = -\gamma = -0.577215664... \quad (A7.4)$$

and it has poles at 0 and at all negative integers.

Double-precision floating-point implementations of the gamma function and its logarithm are available in most scientific computing software and special functions libraries.

A.8 Transmission Lines

A8.1 Rectangular Waveguides

The common transverse modes of propagation in a rectangular guide are TE_{mn} and TM_{mn} where T refers to transverse and E or M refer to electric or magnetic field that has only transverse components with respect to the axis of the guide. The subscripts m and n refer to the number of half-wavelengths along the wide side and narrow side of the rectangular guide, respectively. In such a guide TE_{10} is the fundamental mode that has the lowest cut-off frequency, and the results below are given for this mode; in these formulas a is the length of the wide side of the guide. Low-frequency limit of propagation takes place at the cut-off angular frequency

$$\omega_c = \frac{\pi}{a\sqrt{\mu_0 \varepsilon}} = \frac{\pi c}{a}. \quad (A8.1)$$

Cut-off frequency is then $f_c = c/(2a)$ and the cut-off wavelength is

$$\lambda_c = 2a. \quad (A8.2)$$

The phase velocity $u_p = \omega/\beta$ for an approximately lossless line is

$$u_p = \frac{c}{\sqrt{1 - \left(\frac{\pi c}{a\omega}\right)^2}}. \quad (A8.3)$$

Group velocity of a pulse propagated along the guide is

$$u_g = \frac{c^2}{u_p} = c\sqrt{1 - \left(\frac{\pi c}{a\omega}\right)^2}. \quad (A8.4)$$

The impedance of the waveguide transmission line is

$$Z_g = \frac{\max(E_y)}{\max(H_x)} = \frac{\omega \mu_0}{\beta} = \mu_0 u_p = \frac{\mu_0 c}{\sqrt{1 - \left(\frac{\pi c}{a\omega}\right)^2}} = \frac{Z_{\text{free}}}{\sqrt{1 - \left(\frac{\pi c}{a\omega}\right)^2}}. \quad (A8.5)$$

The impedance of free space is

$$Z_{free} = \sqrt{\frac{\mu_0}{\varepsilon_0}} \cong 377\,\Omega$$

The attenuation constant α of the waveguide is

$$\alpha = \frac{2}{\omega\mu_0\beta ab}\sqrt{\frac{\omega\mu}{2\sigma}}\left[\frac{\pi^2 b}{a^2} + \frac{\pi^2}{2a} + \frac{\beta^2 a}{2}\right], \quad (A8.6)$$

where the propagation constant is, from Eq. 9.18a,

$$\beta^2 = \left(\frac{\omega}{c}\right)^2 - \left(\frac{\pi}{a}\right)^2. \quad (A8.7)$$

Table A8.1 lists the standard waveguides that are likely to be used in polarized targets, with the cut-off frequencies of the two lowest order modes and with the inner dimensions that can be used for the calculation of attenuation. The V band and E band standard guides can be fitted with UG-385 flanges.

Table A8.1 Characteristics of the rectangular waveguides covering the mm-wave range used commonly in polarized targets and EPR spectrometers. The X band and Ka band guides are used as oversized waveguides to transmit over long straight sections with losses lower than the single-mode guides. The RCSC waveguide names are those given by Radio Components Standardization Committee. Electronic Industries Alliance (EIA) uses WRxxx code, where xxx is the guide inner width in 1/100 of an inch; E band guide WG26 is therefore equivalent to WR12 and V band guide WG25 is equivalent to WR15.

RCSC waveguide name	Frequency band name	Recommended frequency band of operation (GHz)	Cut-off frequency of lowest order mode (GHz)	Cut-off frequency of next mode (GHz)	Inner dimensions of waveguide opening (mm × mm)
WG16	X band	8.20–12.40	6.557	13.114	22.9 × 10.2
WG17	–	10.00–15.00	7.869	15.737	19.1 × 9.53
WG18	Ku band	12.40–18.00	9.488	18.976	15.8 × 7.90
WG19	–	15.00–22.00	11.572	23.143	13.0 × 6.48
WG20	K band	18.00–26.50	14.051	28.102	10.7 × 4.32
WG21	–	22.00–33.00	17.357	34.715	8.64 × 4.32
WG22	Ka band	26.50–40.00	21.077	42.154	7.11 × 3.56
WG23	Q band	33.00–50.00	26.346	52.692	5.68 × 2.84
WG24	U band	40.00–60.00	31.391	62.782	4.78 × 2.39
WG25	V band	50.00–75.00	39.875	79.750	3.76 × 1.88
WG26	E band	60.00–90.00	48.373	96.746	3.10 × 1.55
WG27	W band	75.00–110.00	59.015	118.030	2.54 × 1.27
WG28	F band	90.00–140.00	73.768	147.536	2.03 × 1.02
WG29	D band	110.00–170.00	90.791	181.583	1.65 × 0.826

A8.2 Round Waveguides

For round waveguides the modes are designated with the same transversity codes TE or TM as the rectangular ones, but the meaning of the subscripts is different: in TE_{an} or TM_{an} the subscript a gives the order of the Bessel function that characterizes the solution of the wave equation, and the subscript n gives the number of the root of the particular Bessel function. The zeros of the Bessel functions are obtained from

$$J_a(p_n) = 0 \quad \text{for } TM_{an} \text{ modes;}$$
$$J'_a(p_n) = 0 \quad \text{for } TE_{an} \text{ modes,}$$

and the values for these roots can be found from the tables of Bessel functions [46]. Selected values are given in Table A8.2 for the lowest modes.

A8.3 Coaxial Lines

In the following we shall use the normal circuit bloc symbols R, L, G and C for the characteristics of the cable of length λ and consider frequencies above which the R and G are small in comparison with the reactances. The approximate formulas refer to the cases where the attenuation in the line can be neglected, and for frequencies above about 10 kHz where the TEM_{00} is the only mode of propagation, up to the low-frequency cut-off of the next mode TEM_{11}.

Let us assume a coaxial line of length λ, diameter of the inner conductor d, inner diameter of the outer conductor D, dielectric constant of the insulator $\varepsilon = \varepsilon_0 \varepsilon_r$ and magnetic permeability of the insulator $\mu = \mu_0 \mu_r$. Then we have the following results: Shunt capacitance per unit length

$$\frac{C}{\ell} = \frac{2\pi\varepsilon_0\varepsilon_r}{\ln(D/d)} \tag{A8.8}$$

Table A8.2 Roots of some Bessel functions.

Mode	a	n	p_n
TE_{11}	1	1	1.84118
TM_{01}	0	1	2.40482
TE_{21}	2	1	3.05424
TM_{11}	1	1	3.83171
TE_{01}	0	1	3.83171
TE_{31}	3	1	4.20119
TM_{21}	2	1	5.13562
TE_{12}	1	2	5.33144

Series inductance per unit length

$$\frac{L}{\ell} = \frac{\mu_0 \mu_r}{2\pi} \ln(D/d) \tag{A8.9}$$

Propagation constant

$$\gamma = \alpha + i\beta = \sqrt{(R+i\omega L)(G+i\omega C)}/\ell \tag{A8.10}$$

Phase constant

$$\beta = \mathrm{Im}\{\gamma\} \cong \omega\sqrt{LC}/\ell = \frac{2\pi}{\lambda} \tag{A8.11}$$

Attenuation constant α

$$\alpha = \mathrm{Re}\{\gamma\} \cong \frac{R}{Z_0 \ell} \tag{A8.12}$$

Characteristic impedance

$$Z_0 = \sqrt{\frac{R+i\omega L}{G+i\omega C}} \cong \sqrt{\frac{L}{C}} \tag{A8.13}$$

Impedance Z through a series cable of length ℓ terminated by an impedance Z_r

$$Z = Z_0 \frac{Z_r + Z_0 \tanh \gamma \ell}{Z_0 + Z_r \tanh \gamma \ell} \tag{A8.14}$$

Impedance of line terminated with a short circuit ($Z_r = 0$)

$$Z = Z_0 \tanh \gamma \ell \cong iZ_0 \tan \beta \ell \tag{A8.15}$$

Impedance of open-circuited line ($Z_r = \infty$)

$$Z = Z_0 \coth \gamma \ell \cong iZ_0 \cot \beta \ell \tag{A8.16}$$

Impedance of a line an odd number of quarter wavelengths long

$$Z = Z_0 \frac{Z_r + Z_0 \coth \alpha \ell}{Z_0 + Z_r \coth \alpha \ell} \cong \frac{Z_0^2}{Z_r} \tag{A8.17}$$

Impedance of a line an integral number of half wavelengths long

$$Z = Z_0 \frac{Z_r + Z_0 \tanh \alpha \ell}{Z_0 + Z_r \tanh \alpha \ell} \cong Z_r \tag{A8.18}$$

Low-frequency cut-off of the next lowest mode TE_{11} is approximately

$$f_c = \frac{1}{\pi \frac{D+d}{2}\sqrt{\mu\varepsilon}} = \frac{c}{\pi \bar{D}\sqrt{\mu_r \varepsilon_r}}, \tag{A8.19}$$

Table A8.3 Characteristics of standard 50 Ω semi-rigid coaxial lines with solid Cu jacket, PTFE insulator and silver-plated center conductor.

Cable type (= outer diameter in inches)	Maximum practical frequency (GHz)	Speed of propagation (% of speed of light)	Attenuation at 5 GHz (dB/km)
0.141	35	69.5	29
0.085	60	69.5	46
0.047	65	69.5	78

where D is the inner diameter of the outer conductor and d is the outer diameter of the inner conductor, and speed of light is that in the insulator that separates the two conductors.

The resistance R in the above equations is that of the skin layer of Eq. 9.19. The attenuation therefore increases proportional with the square root of the frequency and with the inverse of the cable diameter. Table A8.3 gives the key characteristics of the most popular 50 Ω semi-rigid coaxial cable sizes with solid copper jacket, PTFE insulator and silver-plated center conductor.

The series Q-meter NMR circuit uses most often the 0.141″ semi-rigid coaxial line in the tuned cable connecting the RT circuitry to the part of line inside the refrigerator. In the case of very small NMR signals, in particular the TE signal in a deuterated target, the $\lambda/2$ cable length needs to be very stable over the period of the thermal equilibrium calibration signal measurement, which may be several days if not weeks. The tuned cable then needs to be thermally stabilized, because the electrical characteristics of the PTFE insulation suffer from a sensitivity to phase transitions that occur in the crystallized components of the material [48]. The thermal change of the electrical length of the cable is about –80 ppm/°C between 12 °C and 18 °C, which is the most sensitive region [48], and about –10 ppm/°C around 27 °C, which has been the set point of thermally stabilized cable set of the SMC-polarized target [49].

Inside the refrigerator the coaxial lines need to be fabricated of materials with low thermal conductivity. As these materials need to be non-magnetic, cables are available in the 0.085″ size with BeCu outer and inner conductors, with the inner conductor silver plated. Isothermal sections of the coaxial line inside the cryostat are often made of the 0.047″ size cable with a Cu jacket. All these cables can be equipped with SMA connectors; hermetic feedthroughs are available for the passage of the coaxial lines through vacuum tight walls. For the two larger cables the maximum practical frequency is limited by the cut-off of the next higher mode TE_{11}. The SMA connectors limit the frequency ranges to below about 30 GHz depending on the type and precision of the connector.

References

[1] B. N. Taylor, *Guide for the Use of the International System of Units (SI)*, National Institute of Standards and Technology (NIST) NIST Special Publication 811, 1995.

[2] B. N. Taylor, *The International System of Units (SI)*, National Institute of Standards and Technology (NIST) NIST Special Publication 330, 2001.

[3] W. K. H. Panofsky, M. Phillips, *Classical Electricity and Magnetism*, Addison-Wesley, Reading, MA, 1962.

[4] Particle Data Group, M. Tanabashi, K. Hagiwara, et al., Review of particle physics, *Phys. Rev. D* **98** (2018).

[5] Wikipedia contributors, 2019 redefinition of the SI base units, *Wikipedia, The Free Encyclopedia*, https://en.wikipedia.org/w/index.php?title=2019_redefinition_of_the_SI_base_units&oldid=915062971.

[6] M. Borghini, *Choice of substances for polarized proton targets*, CERN Yellow Report CERN 66–3, 1966.

[7] D. J. E. Ingram, *Free Radicals as Studied by E.S.R.*, Butterworth, London, 1958.

[8] A. S. Lilly Thankamony, J. J. Wittmann, M. Kaushik, B. Corzilius, Dynamic nuclear polarization for sensitivity enhancement in modern solid-state NMR, *Progress in Nuclear Magnetic Resonance Spectroscopy* **102–103** (2017) 120–195.

[9] R. L. Smith, J. Miser, *Compilation of the Properties of Lithium Hydride*, NASA Technical Memorandum NASA TM X-483, 1963.

[10] C. Talón, M. A. Ramos, G. J. Cuello, et al., Low-temperature specific heat and glassy dynamics of a polymorphic molecular solid, *Phys. Rev. B* **58** (1998) 745–755.

[11] Y. Zhou, B. E. Bowler, G. R. Eaton, S. S. Eaton, Electron spin lattice relaxation rates for $S = 1/2$ molecular species in glassy matrices or magnetically dilute solids at temperatures between 10 and 300 K, *J. Magnetic Resonance* **139** (1999) 165–174.

[12] T. A. Alvesalo, P. M. Berglund, S. T. Islander, G. R. Pickett, W. Zimmermann, Specific heat of liquid He^3/He^4 mixtures near the junction of the λ and phase-separation curves. I, *Phys. Rev. A* **4** (1971) 2354–2368.

[13] R. Radebaugh, *Thermodynamic properties of ^3He-^4He solutions with applications to the ^3He-^4He dilution refrigerator*, National Bureau of Standards Technical Note 362, 1967.

[14] G. E. Watson, J. D. Reppy, R. C. Richardson, Low-Temperature density and solubility of He3 in liquid He4 under pressure, *Phys. Rev.* **188** (1969) 384–394.

[15] C. A. M. Castelijns, Flow Properties of ^3He in Dilute ^3He-^4He Mixtures at Temperatures between 10 and 150 mK, Dr. Techn. Thesis, 1986, Department of Physics, Eindhoven University of Technology.

[16] J. G. M. Kuerten, C. A. M. Castelijns, A. T. A. M. de Waele, H. M. Gijsman, Thermodynamic properties of liquid ^3He-^4He mixtures at zero pressure for temperatures below 250 mK and ^3He concentrations below 8%, *Cryogenics* **25** (1985) 419.

[17] J. Landau, J. T. Tough, N. R. Brubacker, D. O. Edwards, Osmotic pressure of degenerate He^3-He^4 mixtures, *Phys. Rev.* **A2** (1970) 2472.

[18] D. Bloyet, A. C. Ghozlan, E. J.-A. Varoquaux, Osmotic pressure secondary thermometer for dilution refrigerators, in: K.D. Timmerhaus, et al. (Eds.) *Low Temperature Physics – LT13*, Plenum Press, New York and London, 1974, pp. 503–509.

[19] A. Ghozlan, E. Varoquaux, Propriétés osmotiques et magnétiques des solutions d'hélium-3 dans l'hélium-4 superfluide, *Ann. de Phys.* **4** (1979) 239–327.

[20] J. G. M. Kuerten, ^3He-^4He II Mixtures: Thermodynamic and Hydrodynamic Properties, Dr. Techn. Thesis, 1987, Department of Physics, Eindhoven University of Technology

[21] A. T. A. M. de Waele, J. G. M. Kuerten, Thermodynamics of liquid ^3He-^4He mixtures, *Physica B* **160** (1989) 143–153.

[22] R. L. Rosenbaum, Y. Eckstein, J. Landau, Thermometry using the osmotic pressure of mixtures of He^3 in superfluid He^4, *Cryogenics* **14** (1974) 21–24.

[23] R. P. Hudson, H. Marshak, R. J. Soulen, D. B. Utton, Review paper: recent advances in thermometry below 300 mK, *Journal of Low Temperature Physics* **20** (1975) 1–102.

[24] H. Preston-Thomas, The international temperature scale of 1990 (ITS-90), *Metrologia* **27** (1990) 3–10.

[25] J. Engert, B. Fellmuth, K. Jousten, A new ^{3}He vapour-pressure based temperature scale from 0.65 K to 3.2 K consistent with the PLTS-2000, *Metrologia* **44** (2007) 40–53.

[26] M. Suomi, A. C. Anderson, B. Holmström, Heat transfer below 0.2°K, *Physica* **38** (1968) 67–80.

[27] O. V. Lounasmaa, *Experimental Principles and Methods below 1 K*, Academic Press, New York, 1974.

[28] H.A. Fairbank, D.M. Lee, Thermal conductivity of 70–30 cupro-nickel alloy from 0.3° to 4.0°K, *Rev. Sci Instrum.* **31** (1960) 660.

[29] E. Boyes, G. R. Court, B. Craven, R. Gamet, P. J. Hayman, Measurements of the effect of absorbed power on the polarization attainable in a butanol target, in: G. Shapiro (Ed.) *Proc. 2nd Int. Conf. on Polarized Targets*, LBL, University of California, Berkeley, Berkeley, 1971, pp. 403–406.

[30] A. C. Anderson, J. I. Connolly, O. E. Vilches, J. C. Wheatley, Experimental thermal conductivity of helium-3, *Phys. Rev.* **147** (1966) 86–93.

[31] W. R. Abel, R. T. Johnson, J. C. Wheatley, W. Zimmermann, Thermal conductivity of pure He^3 and of dilute solutions of He^3 in He^4 at low temperatures, *Phys. Rev. Lett.* **18** (1967) 737–740.

[32] W. R. Abel, J. C. Wheatley, Experimental thermal conductivity of two dilute solutions of He^3 in superfluid He^4, *Phys. Rev. Lett.* **21** (1968) 1231–1234.

[33] M. A. Black, H. E. Hall, K. Thompson, The viscosity of liquid helium-3, *Journal of Physics C: Solid State Physics* **4** (1971) 129–142.

[34] W. R. Abel, A. C. Anderson, J. C. Wheatley, Propagation of zero sound in liquid ^{3}He at low temperatures, *Phys. Rev. Lett.* **17** (1966) 74–78.

[35] G. Baym, W. F. Saam, Phonon-quasiparticle interactions in dilute solutions of He^3 in superfluid He^4. II phonon Boltzmann equation and first viscosity, *Phys. Rev.* **171** (1968) 172–178.

[36] D. J. Fisk, H. E. Hall, The viscosity of ^{3}He-^{4}He solutions, *13th Int. Conf. of Low Temp. Phys. (LT13)*, 1972, pp. 568–570.

[37] K. A. Kuenhold, D. B. Crum, R. E. Sarwinski, The viscosity of dilute solutions of ^{3}He in ^{4}He at low temperatures, *13th Int. Conf. of Low Temp. Phys. (LT13)*, 1972, pp. 563–567.

[38] J. Zeegers, Critical Velocities and Mutual Friction in ^{3}He-^{4}He Mixtures at Temperatures Below 100 mK, Dr. Techn. Thesis, 1991, Physics Department, Eindhoven University of Technology

[39] J. C. Wheatley, O. E. Vilches, W. R. Abel, Principles and methods of dilution refrigeration, *Physics* **4** (1968) 1–64.

[40] A. T. A. M. de Waele, J. C. M. Keltjens, C. A. M. Castelijns, H. M. Gijsma, Flow properties of ^{3}He moving through ^{4}He-II at temperatures below 150 mK, *Phys. Rev.* **B28** (1983) 5350–5353.

[41] L. Monchick, E. A. Mason, R. J. Munn, F. J. Smith, Transport properties of gaseous He^3 and He^4, *Phys. Rev.* **139** (1965) A1076–A1082.

[42] G. Burghart, Baseline Design of the Cryogenic System for EURECA, Dr. Techn. Thesis, 2010, Atominstitut (E141), Technical University of Vienna
[43] K. Andres, W. O. Sprenger, Kapitza resistance measurements between ^{3}He and silver at very low temperatures, in: M. Krusius, M. Vuorio (Eds.) *14th International Conference on Low Temperature Physics*, North-Holland, Otaniemi, 1975, pp. 123–126.
[44] T. O. Niinikoski, Dilution refrigeration: new concepts, in: K. Mendelssohn (Ed.) *6th Int. Cryogenic Engineering Conf.*, IPC Science and Technology Press, Guilford, 1976, pp. 102–111.
[45] F. W. J. Olver, D. W. Lozier, R. F. Boisvert, C. W. Clark, *NIST Handbook of Mathematical Functions*, Cambridge University Press, Cambridge, 2010.
[46] E. Jahnke, F. Emde, *Tables of Functions*, 4 ed., Dover Publications, Inc., New York, 1945.
[47] M. Abramowitz, I. A. Stegun, *Handbook of Mathematical Functions with Formulas, Graphs and Mathematical Tables*, NIST 1972.
[48] S. K. Dhawan, Understanding effect of teflon room temperature phase transition on coax cable delay in order to improve the measurement of TE signals of deuterated polarized targets, *IEEE Trans. Nucl. Sci.* **39** (1992) 1331–1335.
[49] Spin Muon Collaboration (SMC), D. Adams, B. Adeva, et al., The polarized double-cell target of the SMC, *Nucl. Instr. and Meth. in Phys. Res. A* **437** (1999) 23–67.

Index

absorption
  RF power, 118
    due to Provotorov, 119
acceptance drift
  geometric and dead time, 439
acceptance of detector, 284
  in each kinematic bin, 436
acceptor, 100
acoustic phonon
  emitted energy flux, 340
  velocity of, 231
acoustic velocity, 83
adiabatic
  transition, 29
adiabatic condition
  ignoring relaxation, 442
adiabatic demagnetization, 33
  cooling by, 35
  in rotating frame, 35, 38, 95
adiabatic passage
  in rotating frame, 39, 96
ADRF, 38, *See* adiabatic demagnetization
alcohol, 288
alignment, 17, 19, 20
alkanes, 288
amine, 288
aminoxyl
  4-hydroxy-TEMPO, 148
  OH-TEMPO, 148
  oxo-TEMPO, 148
  TEMPO, 148
  TEMPOL, 148
ammonia
  second-order spin couplings, 225
ammonia beads
  irradiation, 312
  preparation by freezing liquid droplets, 311
  preparation by slow freezing and crushing, 311
ammonium borohydride, 288
analyte
  molecule under study, 309

angular momentum
  classical definition of, 2
  intrinsic, 3
  law of conservation, 2
  orbital, 3
  quantum mechanical definition, 3
  vector, 2
anhydrous magnesium sulphate
  water removal, 303
anisotropic hyperfine tensor, 227
anomalous magnetic moment, 4, 98
APRF, 40, *See* adiabatic passage in rotating frame
  at constant entropy, 443
  density matrix with two temperatures, 443
  high-temperature models, 444
  magnetic ordering
    loss of reversibility, 444
  mixing field, 443
  of polarized proton targets, 460
  polarization loss mechanisms, 444
APRF reversal of deuteron spins, 445
APRF reversal of proton polarization
  superradiance, 445
Arrhenius law, 293
attenuation
  in oversized waveguide, 400
  rectangular guide, 400
azephenylenyls, 148

beam normalization, 439
beam polarization, 437
beam transmission, 290
Bessel function, 401
  root of, 401
BHHA-Cr(V), 149
  proton hyperfine spectrum, 228
binomial coefficients, 108
biradical, 147, 309
  in cross effect, 183
  TOTAPOL, 147
bisdiphenylallyl (BDPA), 148

bisdiphenylene-b-phenylallyl, 148
Bloch equations, 87
Bloch function
　of conduction electron, 235
Bloch state
　of conduction electron, 235
Bohr magneton, 4
boiling heat transfer
　at low pressure, 344
　correlation for ^{4}He, 345
　Cu-dilute solution, 345
　dryout heat flux limit, 346
　forced convection, 346
　nucleate, 345
　pool boiling, 346
　propanediol-^{3}He, 345
bolometer
　microwave detector, 140
　used for study of FM, 193
Boltzmann constant, 18, 279
Boltzmann distribution, 18, 24, 59, 66, 246
Boltzmann factor, 18
Boltzmann ratios, 189
Born approximation, 452
boron compounds, 288
Bose-Einstein condensation
　^{4}He, 481
Bose-Einstein statistics, 80
Breit-Rabi polarimeter, 431
bremsstrahlung
　energy loss by, 286
Brillouin function, 19, 21, 29, 30
brute force polarization
　^{27}Al, ^{93}Nb, ^{165}Ho, 422
　metal hydrides, 422
　of nuclei in metals, 422
　solid ^{3}He, 423
　solid HD, 423
BWO, 463, *See* backward wave oscillator

canonical
　density of protons, 115
　distribution, 24
CGS Gaussian system of units, 471
charged glassy beads
　discharging, 307
chemical DNP, 425–427
chemical potential
　of ^{4}He, 367
chemical shift, 55, 477
　anisotropic, 56
　magic angle spinning (MAS), 229
chemical spin polarization
　atomic hydrogen, 425
　CIDNP, 427
　parahydrogen induced polarization, 427

Clausius-Clapeyron equation, 381
coaxial line
　characteristics, 503
　electrical length, 250
　phase transition in PTFE, 267
　propagation constant, 250
　semi-rigid, 267
　thermal drift of attenuation, 267
　thermal drift of length, 267
cobalt-iron alloy
　magnet pole pieces, 409
coherent scattering
　polarized slow neutrons, 290
coherent slow neutron scattering, 452
cold neutron beam
　polarization by magnetic supermirrors, 452
commutation relations, 5
commutation rules, 32
　exponential operators, 12
COMPASS collaboration
　triple target, 450
concentrated stream
　energy balance, 358
conduction electrons
　cooling of, 36
　paramagnetic, 100
configurational degrees of freedom
　loss of, 292
contamination
　by water frost, 331
convectional heat transfer
　hydrodynamic surface area, 347
　in dilute solution, 346
cooling cycle
　dilution refrigerator, 355
　evaporation refrigerator
　　^{3}He, 350
　　^{4}He, 350
cooling of the rotating sample
　in MAS NMR, 463
cooling power
　evaporation refrigerator, 351
counterflow heat exchanger
　effectiveness, 375
coupling constant
　spin-orbit, 51
coupling resistance, 249
Cr(V) complexes
　in stable pure form, 304
　preparation with diols, 302
Cr(V)-diol reaction
　sensitivity to light and water, 302
critical temperature
　for magnetic ordering, 460
critical velocity
　for zero viscosity, 481

cross effect
  for MAS-DNP, 183
  simulation using TEMPO, 184
cross effect DNP, 463
cross relaxation, 46, 79
  flip-flop terms, 133
  inhomogeneous broadening, 167
  nuclear spins, 115
  rate parameter, 184
  role of nuclear spins, 135
  spin pair flip-flop transitions, 133
cross relaxation parameter, 183
cross section
  difference for opposite orientations, 284
  unpolarized, 284
cross-relaxation
  between different nuclear spin species, 239
crystallization
  hindered nucleation, 292
Curie law, 21, 78
  correction to, 22
Curie temperature, 22, 23, 32
Curie-Weiss law, 22
cut-off wavelength
  rectangular guide, 399

damping resistance
  resonant circuit, 252
Debye
  cut-off, 126
  frequency, 231
  model, 80
  temperature, 80, 132, 341
Debye frequency, 479
Debye model
  phonons in glassy materials, 231
Debye temperature, 479, 480
density matrix, 23, 24, 73, 155
  high-temperature approximation, 25
  linearized, 25
  off-diagonal elements, 25
deuterated ammonia
  post-irradiation, 314
deuteron NMR signal
  errors in asymmetry, 269
devitrification
  microcalorimetry, 296
  stability against, 299
differential solid effect, 463
diffusion
  nuclear spin, 230
  of nuclear spin polarization, 242
diffusion barrier
  for nuclear spins, 242
diffusion constant
  in dilute stream, 367

dilute solubility line
  of phase diagram, 482
dilute solution
  degenerate Fermi fluid, 366
  isenthalpic expansion, 366
dilute stream flow
  limited by diffusion, 366
dilution factor
  for pure chemical substance, 286
  of PT material, 284
  of target
    mean over kinematic bins, 437
dilution refrigerator
  cooling power, 357
  counterflow heat exchanger, 357
  design of flow channels, 362
  enthalpy in the dilute stream, 483
  evaporator and condenser, 355
  expansion device, 355
  main heat exchanger, 356
  maximum cooling power, 357, 361
  mixing chamber, 356
  optimum flow, 360
  optimum flow of ^{3}He, 357
  phase boundary, 356
  principle of operation, 355
  still, 355
diol, 288
diphenyl picryl hydrazyl (DPPH), 148
dipolar broadening
  fourth moment, 207
  of EPR line, 111
  of NMR line, 206
  second moment, 206
dipolar energy
  quantum mechanical, 45
dipolar field, 32
  local, 112
dipolar frequency, 69, 72, 113
dipolar interaction
  energy, 24
dipolar lineshape
  at high polarization, 207
  moments at high polarization:, 208
dipolar spin-spin interactions
  secular part, 164
dipolar temperature, 61, 133
dipolar width, 113, 114
  EPR line, 115
Dirac term
  scalar interaction, 184
direct process
  relaxation rate due to, 84
dispersion relation
  for ^{4}He superfluid, 481
dissolution-DNP (dDNP), 183, 465

dissolution NMR
    DNP before rapid melting, 309
Dittus and Bölter correlation, 376
divacancy, 100
DNP, 155
    by cross effect, 182
    by Overhauser effect, 187
    cross effect at low temperatures, 183
    diffusion barrier, 180
    dynamic cooling of the spin-spin interactions, 155
    enhancement of MAS NMR signal, 309
    in fluorene
        using photoexcited phenanthrene, 321
    in irradiated Ca(OH)$_2$, 319
    in irradiated CaF$_2$, 319
    in irradiated glassy butanol, 320
    in scintillating plastics, 308
    irradiated deuterated butanol, 321
    leakage factor, 179
    of hyperfine nuclei, 184
    Overhauser effect, 155
        in liquid state, 190
        in solids, 190
    solid effect, 155
    time evolution, 174
    using cross effect, 309
    via electron spin locking, 429
    with frequency modulation, 191
    with inhomogeneously broadened EPR line, 167
DNP in crystalline materials
    using substitutional paramagnetic ions, 322
DNP MAS NMR spectroscopy, 462
donor, 100
double balanced mixer, 249
doublet
    non-Kramers, 125
D-state probability
    deuteron, 201
dynamic cooling
    inhomogeneous broadening, 163
    of electron dipolar interactions
        at high temperature, 161
dynamic cooling of nuclear spins
    phenomenology, 156
dynamic nuclear polarization. See DNP

effective damping
    frequency sensitivity, 252
effective field, 38
effectiveness method
    design of heat exchangers, 374
EHBA-Cr(V), 149
    preparation, 304
    proton hyperfine spectrum, 228
eigenfunction, 25
    of spin operator, 6
eigenvalues
    of spin 1 Hamiltonian, 216

EIO, 463, See extended interaction oscillator
elastic coherent scattering
    forward peaked, 455
elastic peak
    for free protons, 289
elastic scattering
    elastic peak, 439
electric field
    asymmetry parameter, 49
    gradient, 49
electric quadrupole moments
    of nuclei, 200, 203
electromagnetic compatibility (EMC), 279
electromagnetic interference (EMI)
    control of, 279, 382
electron dipolar interactions
    neglect of, 172
electron-nucleus spin interaction
    dipolar, 227
    hyperfine, 227
Ellis-Jaffe sum rule, 450
EMC solenoid
    main and trim coils, 418
EMI control
    design of signal paths, 280
    hardening by circuit design, 280
    RF interferences, 279
energy conservation
    in dynamic cooling, 165
energy reservoirs
    electron dipolar, 156
    electron Zeeman, 156
    nuclear Zeeman, 156
enthalpy
    dilute solution, 357
    helium isotopes and their mixtures, 483–485
    of helium mixtures, 483
    of the concentrated stream, 484
    osmotic, 483
enthalpy-pressure diagram
    Joule-Thomson cycle, 351
entropy, 28, 31, 35
    additive quantity, 28
    equilibrium, 27
    excess between liquid and crystal, 293
    in high magnetic field, 29, 30
    loss of, 39
    maximum for a spin system, 28
    measurement of, 29
    of ^{4}He, 367
    of a composite system, 28
    of a spin system, 24
    quantum statistical definition, 28
    relation with polarization, 460
    with dipolar interactions, 32
entropy theory
    glass transition, 293

EPR saturation, 172
EPR spectrometer
　high field, 138
　V-band, 139
　X-band, 135
EPR spectroscopy, 135
　in situ, 140
EPR spectrum
　from NEDOR signal, 146
equilibrium thermodynamics
　in a glass, 292
equivalent surface-to-volume
　heat transfer surface, 339
equivalent surface-to-volume diameter
　flow resistance parameter, 340
European Muon Collaboration (EMC)
　CERN NA2 twin target, 448
evaporation refrigerator
　with ^{3}He-^{4}He mixture, 353
exchange interaction, 40, 55, 58
exchange narrowing, 118

Fabry-Perot interferometer
　semiconfocal, 139
false asymmetry
　estimation and mitigation, 439–441
fast neutron beam
　polarization using polarized proton target, 290
F-center, 100
F-centers
　in glassy organic materials, 149
Fermi contact interaction, 54
Fermi distribution function, 236
Fermi pseudopotential, 453, 458
Fermi surface, 57
　of conduction electrons, 236
Fermi-Dirac statistics, 481
fibre-optic links
　replacing galvanic lines, 280
FID. *See* free induction decay
　time scale, 88
FID envelope, 67
figure of merit
　for a scattering experiment, 436
　material dependent part, 285
　maximization in intense beam, 285
　of PT, 284
　of the target, 437
filling factor
　of probe coil, 261
film boiling
　critical heat flux, 380
finite element (FE)
　numeric simulation, 413
flash evaporation, 353
flip-flop term, 46
fountain pressure
　in ^{4}He superfluid, 481

free electron
　trapped by solute molecule, 150
　trapped in a cavity, 150
free induction decay, 62
free precession signal
　non-exponential, 88
free radical, 147
　radiolytic, 99
　stable, 99
free radicals
　table, 475
free-volume model
　glass transition, 293
Fremy's salt, 148
frequency modulation
　during DNP, 463
　effect on EPR spectrum, 143
　effect on NEDOR signal, 146
　repopulation of hyperfine lines, 195
frozen spin operation
　coolant density measurement, 440
　during field rotation, 379
　transverse field mode, 379
fusion with polarized fuel, 290

gamma function, 498, 500
gaseous helium
　specific heat, 486
Gaussian function
　normalized, 71, 497
gel matrix
　immobilizing radicals, 465
general relativity, 2
g-factor
　anisotropic, 101, 114
　coupling with the lattice, 101
　Dirac value of, 4, 98
　free nucleon, 199
　quark
　　in naïve quark model, 200
　structural, 4
Gibbs free energy, 28
glass
　organic, 132
glass former
　characteristics, 292
glass transition
　1,2-propanediol, 299
　microcalorimetry, 296
glass transition temperature, 292
glassy material, 70, 115
　delayed heat release, 299
　mould casting, 308
glassy materials
　chemically doped, 291
　properties of the matrix, 292

glassy target beads
  preparation of, 306
Glebsch-Gordan coefficients, 49, 201
gluon, 2
  contribution to nucleon spin, 200
Gomberg's triphenylmethyl radical, 148
Gorter's formula
  for relaxation time, 236
gradient tensor
  electric field, 48
group velocity
  rectangular guide, 399
g-shift
  paramagnetic electron, 100
  unpaired electron, 101
gyromagnetic factor, 45
gyromagnetic ratio, 3, 21, 199
  sign of, 5
gyrotron, 463

hadron, 2
Hamiltonian, 47
  dipolar, 31, 45
    secular part, 33
  electron exchange, 58, 118
  Hermitian, 69
  phonon, 82
  secular part, 47
  spin, 13, 17
  spin orbit, 51
  Zeeman, 17, 33, 45
heat capacity
  of spin system, 131
heat exchanger
  continuous-flow, 358
heat flush effect, 355
  for measurement of osmotic pressure, 485
heat transfer
  in sintered sponge, 359
  sintered powder and liquid, 492
heat transport
  by evaporating superfluid film, 349
  subcooled ^{4}He superfluid, 349
Heaviside-Lorentz system of units, 473
helium gas purifiers, 386–388
  operation, 388
helium pump
  leak testing, 386
  Root's blower, 384
  rotary blade pump, 385
helium recirculation systems
  active purifiers, 385
  adsorbent purifiers, 386
  pump oil purge, 385
  shaft seal, 386
high-intensity beam
  direct heating of target, 438
  radiation damage, 438

high-resolution NMR technique
  pulsed NMR, 462
high-temperature approximation
  interacting spin system, 31
horn antenna
  coupling structure, 463
hydraulic surface area, 370
hydrodynamic surface area, 493
hydrogen
  free molecule wavefunction, 223
  solid spin conversion, 224
hyperfine constant
  isotropic, 107
hyperfine interaction, 54
  dipolar term, 184
  neglect in DNP, 173
hyperfine splitting, 107
hyperfine tensor, 107
  anisotropic, 54, 107, 109, 114
  isotropic, 54
hyperpolarization, 421, 460
  deuterated glucose, 466

impedance
  rectangular guide, 399
incomplete inner shells, 98
indirect coupling between nuclear spins, 55
indirect nuclear spin interaction
  pseudo-dipolar, 226
  pseudo-exchange, 226
indirect spin interaction
  NMR lineshape, 222
integral theorem of Cauchy, 64
integrated beam fluxes
  through target, 436
integrated NMR signal
  linearity, 270
International System of units (SI), 471
inverse temperature, 34
irradiated ammonia
  preparation, 311
irradiated crystalline materials
  F-centers, 151
  H-centers, 151
irradiated lithium hydride
  structure and properties, 314
irradiated materials
  radical yield, 310
irradiated PT materials, 309
irradiation
  safety aspects with liquid N_2, 322
irradiation damage
  bleaching by light, 326
  in irradiated ammonia, 327
  repair by annealing, 325
irradiation effects
  in situ, 324

isospin mirror nuclei, 201
isotropic hyperfine tensor, 227
isotropic space, 2, 6
ITS-90 scale
  possible improvement using NMR, 278
  vapor pressure of helium, 276
ITS-90 temperature scale, 381

Kapitza conductance, 171, 492
  Cu-^{3}He, 341
  Cu-^{4}He, 341
  deuterated butanol-dilute solution, 344
  epoxy-^{3}He, 343
  FEP-dilute solution, 343
  glassy butanol-^{4}He, 343
  paramagnetic salts-^{4}He, 342
Kapitza conductances, 341
Kapitza resistance
  acoustic phonon mismatch, 340
  heat transfer, 306
  solid ortho-hydrogen, 224
  thermal boundary resistance, 340
kinematical bin, 436
Knight shift, 55, 56, 229, 477
Korringa constant, 36, 238
  effect of impurities, 238
  with weak coupling, 239
Korringa law, 422, 428
Korringa relation
  between relaxation and Knight shift, 238
Kozeny-Carman equation, 370
Kramers doublet
  in YES, 424
Kramers-Krönig equations, 247
Kramers-Krönig relations, 64, 65, 76
Kramers-Krönig transform, 499

Lagrange multipliers, 34
lambda-transition, 481
Landé g-factor, 51, 101
large superconducting magnets
  examples, 417
Larmor frequency, 17
  chemical shift, 229
Larmor precession, 11
  classical treatment, 12
  frequency of, 21, 246, 477
lattice interactions, 76
lattice stress
  interstitial radiolytic centers, 314
leakage factor, 173
leakage mechanism
  for nuclear spin polarization, 94
level gauge
  for liquid helium, 384
Ley's radical, 182

linear distortion
  of wide NMR signal, 269
linear response
  approximation, 119
  theory, 60, 162, 204
lineshape
  Gaussian, 134
  Lorentzian with cut-offs, 114
  quadrupolar, 49, 50, 210
    in high field, 210
lineshape functions, 496
  moments of, 67, 68, 496
linewidth
  EPR contribution of nuclear spins, 115
Liouville formalism, 78
liquid nitrogen
  explosion following irradiation, 322
lithium deuteride ^{6}LiD
  irradiated, 288
lithium hydrides
  DNP results, 315–319
  high radiation length, 288
LMN
  crystal growth, 323
local field, 33, 44, 72
Lorentzian function, 71, 498

Mach number
  at freezing radius of atomic beam, 430
magnet pole pieces
  shimming, 412
magnet system
  CEN-Saclay Nucleon-Nucleon experiments, 419
magnetic cooling, 29
magnetic dipole interaction, 43–48
  energy of, 17
magnetic dipole moment
  of nuclei, 199, 200
  of nucleon, 199
magnetic dipole operator
  for a nucleus, 200
magnetic energy levels, 9
magnetic field
  transverse, 48
magnetic induction, 43, 113
magnetic moment
  density, 20
magnetic moments
  table, 476
magnetic ordering
  due to exchange interaction, 58
  loss of entropy, 460
magnetic resonance, 58
magnetic scalar potential, 43
magnetization, 21
  evolution of, 16
  static, 20, 246

main heat exchanger
  axial conduction, 365
  viscous heating, 365
MAS-DNP
  sensitivity enhancement, 229
mass diffusion
  of dilute solutions, 490
mass flowmeters
  calorimetric, 383
matrix representation
  of spin operator, 7–9
  spin 1 vector, 216
Maxwell-Boltzmann distribution, 157, 159
mechanical vacuum model, 491
memory functions, 78
metallo-organic complexes, 477
metallo-organic compounds, 100, 149
methane, 288
  spin conversion in solid, 224
microwave counter
  with harmonic downconversion, 408
microwave power
  measurement by calorimeter, 408
microwave source
  backward wave oscillator (BWO)
    carcinotron, 393
  extended interaction oscillator (EIO), 394
  IMPATT diode source, 395
  klystron, 394
  measurement of frequency and power, 408
miniature gas turbines
  for MAS NMR, 462
mixing field, 34
modified Lorenzian function
  moments of, 499
modified Lorenzian functions, 498
molecular dioxygen (O_2), 147
molecular field effect, 23
molecular spin effects, 57
molecular spin isomers, 222
MRI contrast enhancement, 464
MRI sensitivity enhancement, 464
multimode cavity, 463
multiple scattering, 285, 437, 438
multiple targets in same beam
  mitigation of drift sources, 448
mutual friction
  in dilute solution below 0.25 K, 491
  in dilute stream, 366

ND_3
  irradiated, 288
NEDOR spectroscopy, 145
nitronyl nitroxides, 148
NMR circuit
  crossed-coil, 272

NMR coil
  Q-factor, 252
NMR line broadening
  Van Vleck formulas, 227
NMR measurement of polarization
  principles, 245
NMR saturation
  by polarization measurement
    neglect of, 172
NMR signal
  ab initio size, 253
  calibration for polarization measurement, 275
  continuous-wave (CW), 260
  saturation, 256
  signal-to-noise ratio, 254
non-equilibrium phenomenon
  liquid-glass transition, 292
normalized Gaussian function, 497
nuclear magnetic cooling, 35
nuclear magnetic ordering, 36, 39, 40, 460
  in Ag, 40
  in $AuIn_2$, 40
  in $Ca(OH)_2$, 41
  in CaF_2, 41
  in Cu, 40
  in LiH, 41
  in Rh, 40
nuclear magnetic phases
  by neutron diffraction experiments, 461
nuclear magneton, 477
nuclear Overhauser effect
  NOE sequence
    I. Solomon, 429
nuclear spin polarization
  by brute force, 421
nuclear spin relaxation
  in dielectric solids, 235
  in metals, 235
  polarization dependence, 232
nuclear spin-lattice relaxation
  shell-of-influence model, 230
nucleation of crystal growth
  effect of impurities, 297
  in slow cooling, 296
nucleon, 2
Nusselt number, 374, 375

operation in high beam flux
  frozen spin mode, 373
operation in high beam fluxes
  choice of refrigeration method, 372
operator
  exponential, 12
  ladder, 6
optical pumping
  by collisional spin-exchange (SEOP), 432

by metastability-exchange (MEOP), 431, 432
  ground-state and metastable atoms and
      ions, 431
  in dense gas, 430
optically pumped ^{3}He
  MRI contrast medium, 291
optimum filter
  pre-filtering and cross-correlation, 446
Orbach expansion
  lattice potential, 124
orbital moment
  nuclear, 53
  quenched, 53
orbital motion
  contribution to nucleon spin, 200
oriented single crystal
  ellipsoidal, 461
ortho-hydrogen, 223
osmotic pressure
  loss in dilute stream, 366
  of dilute solutions, 485
Overhauser effect, 427, 464
  DNP, 463
  in liquid phase using TEMPO, 464
  in metals and liquids, 427

para-hydrogen, 223
paramagnetic compounds, 147
paramagnetic ion
  Nd^{3+} in $LaAlO_3$, 149
  Nd^{3+} in LaF_3, 149
  Nd^{3+} in LMN, 149
  Tm^{2+} in CaF_2, 149
paramagnetic oxygen
  magnetic liquid, 147
paramagnetic substance
  diluted, 98
paramagnetic susceptibility
  static, 21, 99
paramagnetism, 98
partition function, 24, 28, 34
parton, 2
Pascal's triangle, 108
Penning trap, 4, 477
  single-electron, 98
perdeuterated propanediol
  reduction of potassium dichromate, 303
permeability
  of free space, 20
perturbation theory
  time-dependent, 16
phase diagram of ^{3}He/^{4}He mixtures, 481
phase sensitive detection
  real part of Q-meter signal, 249
phase separator
  ^{4}He liquid-gas, 378

phase velocity
  rectangular guide, 399
phonon, 80
  ballistic, 130
phonon bottleneck, 130
  DNP speed limiting, 165
  Kochelaev model, 132
  with frequency modulation, 132
phonon energy, 82
phonon spectrum, 479
phonons
  in ^{4}He superfluid, 481
Planck constant, 2
PLTS-2000
  provisional low temperature scale, 486
  provisional temperature scale, 277
point defect
  interstitial, 100
  substitutional, 100
Poiseuille flow
  laminar region, 369
Poisson's equation
  vector form, 412
polarization, 17, 18
  adiabatic reversal of, 34
  measurement, 22, 60
  measurement of, 21
  tensor. *See* alignment
  vector, 17
polarization asymmetry
  in scattering experiments, 283
polarization by spin selection
  in atomic beam, 430
polarization of nuclei in large biomolecules, 454
polarization reversal
  by adiabatic passage in rotating frame
      (APRF), 442
  by DNP, 441
  by field rotation, 450
  by spin rotation, 441
  diabatic, 37
  efficiency of, 39
  optimum filter theory, 446
polarization reversal frequency
  optimization, 445
polarized target refrigerator
  SMC double cell target, 379
  Virginia-Basel-SLAC target, 377
polarizing agent (PA)
  in MAS-DNP, 309
  generic name for a free radical, 464
porphyrexide (PX), 148
porphyrindene (PB), 148
powder
  of small crystals, 70

power absorption
 due to spin flips, 59
power supplies
 separated analog and digital parts, 280
Poynting vector, 400
Prandtl number, 374
 for gaseous helium, 491
precooling heat exchangers
 pressure drop, 376
pressure measurement
 capacitive gauge, 382
 McLeod gauge, 383
 Pirani gauge, 383
 vacuum gauges, 383
probe coil
 filling factor, 258
 reduced current, 265
 sampling function, 257
 size compared with wavelength, 258
proton NMR spectrum
 hyperfine protons, 228
Provotorov equations, 38, 60, 63, 72, 74, 75, 88, 160, 161
 generalized, 78
pseudo-dipolar coupling, 57
pseudo-exchange coupling, 57
pseudomagnetic field
 due to strong interaction, 459
pseudomagnetism, 457
PT refrigerator
 target holder insert, 379
 target insert, 378
PTB-2006 scale
 forthcoming new temperature scale, 486
 provisional scale, 382
pulse response
 Fourier transform, 205
pycril-N-aminocarbazyl (PAC), 148

Q-curve
 drift of, 255
Q-meter
 circuit optimization, 265
 complex circuit theory, 257
 Liverpool circuit, 249
 optimum tuning, 266
 parallel tuned, 249
 probe coil
  series tuned, 256
 Q-curve, 251
 series tuned, 249
 signal expansion, 251
quadrupole energy
 sign of, 214
quadrupole interaction, 48–50
 asymmetry parameter, 214
 coupling parameters, 221
 in low field, 215

quadrupole moments
 table, 476
quadrupole resonance
 due to magnetic interaction, 215
Quantum Chromodynamics (QCD)
 testing based on nucleon spin, 200
quantum fluids, 480
quantum statistical treatment, 16, 120
quantum statistics, 27
 of a system of spins, 17
quark, 2
quark model, 199
quenching of the angular momentum, 52
quenching rate
 glass formation, 292

radiation damage
 chemically doped hydrocarbons, 325
 in lithium hydrides, 329
radiation length, 286
radiation resistance
 of PT materials, 324
radicals
 alcoxy, 150
 in irradiated alcohols, 150
 in irradiated alkanes, 150
 in irradiated ammonia, 151
 in irradiated aryls, 150
radiolytic centers, 147
radiolytic free radical, 466
radiolytic radicals
 in glassy organic materials, 149
Raman process
 relaxation rate due to, 84
rare earth group, 477
rare spin species
 polarization measurement, 276
rectangular waveguide
 characteristics, 501
refrigerator for polarized target
 precooling heat exchangers, 373
relative permeability
 magnet material, 410
relaxation
 Blume and Orbach process, 128
 cross-relaxation, 90
 dipolar, 87
 direct leakage, 165
 direct process, 124, 128
 effective rate with phonon bottleneck, 132
 electron spin-lattice, 3, 101, 122
 electron spin-spin, 122
 nuclear spin, 230
 nuclear spin-lattice, 94
 of transverse magnetization, 87
 Orbach process, 127, 128
 Orbach-Aminov process, 127

Raman process, 125
  spin-lattice, 18, 76, 78
    experimental, 128
    Waugh and Slichter process, 128
relaxation measurement
  NEDOR method, 129
  pulse recovery method, 128
relaxation mechanisms, 16
relaxation of the dipolar energy
  loss of entropy, 460
relaxation time
  determined from pulse recovery, 480
  dipolar
    from saturated lineshape, 138
    from pulse recovery, 137
    from saturation spectrum, 137
    from signal height, 137
  in situ NEDOR measurement, 146
reluctance
  of magnetic circuit, 410
Remotely Enhanced Liquids for Image Contrast
      (RELIC), 464
reservoir
  dipolar, 38
  Zeeman, 38
residual gas analyzer, 384
resonator impedance, 249
reversal of nuclear spin polarization, 94
Reynolds number, 374
RF field
  transverse, 25
RF filters
  in power feeds, 280
rotating frame, 15, 23, 161
  spin resonance in, 12
  transformation, 13
rotation operators, 5
rotons
  in ^{4}He superfluid, 481
round waveguide, 404
  circular polarization, 401
  dominant mode, 401
  guide wavelength, 404
  phase velocity, 404
  propagation constant, 404
  propagation impedance
    TE_{11} mode, 404
    TM_{01} mode, 405
round waveguides, 503
Rudermann-Kittel interaction, 40

saturation, 16, 47, 72, 155
  BPP phenomenological model, 119
  effect of frequency modulation on, 193
  effect on DNP, 174
  electron spin resonance, 118
  error in polarization measurement, 261
  function, 119
  NMR lineshape, 76
  Provotorov equations, 119
  strong, 133
  time constant, 260
saturation decay
  experimental
    deuteron NMR signal, 264
saturation error
  deuteron NMR signal measurement, 262
saturation function, 119
saturation of NMR signal
  error due to, 256
saturation parameter
  in cross effect, 182
scattering amplitude
  neutron beam traversing a polarized target, 458
Schrödinger equation, 13
  time-dependent, 9
scintillating detector
  polarized by DNP, 308
second (thermal) sound waves, 481
second quantization, 83
self-inductance
  definition, 257
semi-rigid coaxial cables
  replacing flexible cables, 280
series Q-meter
  capacitively coupled, 272
  design for deuteron NMR, 268
  design for proton NMR, 268
  improvements, 271
  probe coil design, 270
  using a hybrid bridge, 274
  with quadrature mixer, 274
shock wave
  at atomic beam skimmer, 430
sintered heat exchanger
  design and evaluation, 369
  thermal penetration, 371
sintered heat exchangers
  continuous approximation, 371
sintered powder
  extended surface, 359
skin depth, 257, 400
  anomalous, 400
slow drift
  beam and acceptance, 448
SMA connectors, 280
small-angle neutron scattering
  time resolved, 456
SMC magnet system
  solenoid and dipole coils, 419
SMC polarized target
  microwave isolator, 380
  mixing chamber, 380
  sintered copper heat exchanger, 380

SMC solenoid
  main coil and trim coils, 418
solenoid winding
  gap filling technique, 417
solid effect, 47, 174
  cavity losses, 179
  differential, 181
  diffusion barrier, 180
  leakage factor, 179
  phonon bottleneck, 181
  rate equations, 175
  resolved, 175
  semi-phenomenological model, 174
  transition rates, 175
solid effect DNP, 463
solutions of ^{3}He in ^{4}He
  spontaneous separation in two phases, 355
sources of noise
  RF oscillator, 279
  thermal, 279
spallation neutron source, 457
specific heat, 33, 36
  of a spin system, 29
specific heat anomaly
  at glass transition, 292
specific heat of liquid ^{3}He, 483
spectral diffusion, 135
spectral redistribution
  cross relaxation, 159
spherical polar coordinate system, 43
spin
  components of, 2
  effective, 3
  intrinsic, 2
  maximum projection of, 2
  quantum number, 7
  temperature, 17
  vector, 2
spin contrast variation, 290
  small-angle scattering of neutrons, 454
spin filter
  transmission of neutrons, 451
  transmitted neutron polarization, 451
spin filtering, 290
spin filtering of neutron beams, 450
spin filtering of neutrons
  by Sm, ^{3}He, TiH$_2$, 452
spin flip coil
  neutron beam polarization reversal, 455
spin packet, 135
  dipolar width, 135
spin packets
  quadrupolar lineshape, 212
  with individual Zeeman temperatures, 183

spin pairs
  like, 46
  unlike, 46, 47
spin polarization
  matrix, 25
spin refrigerator, 463
  by rotating YES crystals, 424
  polarized target, 323
spin temperature, 25
  in low effective field, 23
  in rotating frame, 25
  inverse, 21
  measurement, 28
spin-orbit coupling, 106
spin-orbit interaction, 50
split-coil solenoid magnet
  SLAC E143 experiment, 419
stable atomic hydrogen
  as polarized electron target, 433
stable isotopes, 476
standard waveguides, 502
state vector
  spin 1 exact solution, 220
static field
  internal, 21
statistical accuracy
  dominated by systematic errors, 438
  of polarization asymmetry, 436
statistical error
  of target asymmetry, 437
statistical fluctuations
  multiplicative and additive components, 446
statistical uncertainty
  minimization, 438
storage ring
  muon, 4
strong saturation
  transverse field, 159
substructure, 98
superconducting magnet, 413
  rectangular conductor, 413
  round conductor, 413
  with split coils, 413
superconducting magnets
  control and protection, 417
superconducting solenoid
  end compensation, 415
  field on axis, 414
  winding accuracy, 415
supercooled liquid, 292
superfluid ^{3}He, 481
superfluid film creep, 349, 353, 481
superfluid state
  ^{4}He, 481
superisolation, 330

superradiance
  of NMR circuit, 271
surface-to-volume diameter, 370, 493
susceptibility
  complex, 66
  static, 20, 67, 119
systematic error
  due to slow drift
    beam and acceptance, 438

Taler and Taler correlation, 376
target asymmetry, 436
target cavity
  bolometer, 407
  coupling slots, 407
  microwave field strength, 407
  multimode structure, 407
target dilution factor, 437
target length, 437
target materials
  storage and handling, 330
target thickness, 437
target thickness variation
  in dilution refrigerator, 440
  in evaporative cooling, 440
Taylor expansion, 25
Taylor series
  of density matrix, 31
temperature
  dipolar, 159
  Zeeman, 159
temperature measurement
  during NMR signal calibration, 381
  of helium bath, 276
  using resistance thermometers, 382
tetramethylsilane marker, 229
thermal conductivity
  in gaseous helium, 492
  of helium isotope mixtures, 487
thermal equilibrium, 18
  between spins and lattice, 275
  calibration, 22
thermal exchange
  effective surface area, 494
thermal mixing, 36, 39, 93
thermal penetration depth, 494
  dilute solution, 371
thermodynamic functions, 27–28
thermodynamics
  at high polarization, 34
  second law of, 28
thermomolecular effect, 353
toluene
  glass former, 301

transition metal ions, 100
transition metals, 98, 477
transition probability, 16, 119
transport properties
  liquid-glass transition, 293
transverse field, 38
  strong limit, 94
transverse susceptibility, 246
  absorption part, 59, 76
  complex, 246
  dispersion part, 59
  general features
    complex, 205
  in NMR, 204
triplet state
  optically induced, 100
triradical, 309
tris(2,4,6-trichlorophenyl)methyl radical
  TTM, 148
trityl radicals
  Finland-D36, 148
  OX063, 148
  triphenylmethyl, 148
tune shift
  attenuation in coaxial line, 267
  coaxial line, 252
twin target
  acceptance drift elimination, 450
two-fluid model
  for ^{4}He superfluid, 481
two-phase boiling
  heat transfer, 306
two-phase region
  liquid and gas, 480

units
  CGS Gaussian, 1
  MKSA, 1
  Système International (SI), 1
unpolarized cross section, 437
unpolarized dummy
  target tests, 441

V_1-center, 100
vacuum-insulated dewar
  for biological samples
    low loss, 330
Van Vleck formulas
  electron-nucleus interaction, 227
Van Vleck's
  second moment, 113
Van Vleck's method of moments, 68
vapor pressure
  of ^{3}He, 276
  ITS-90 temperature scale, 486